Sustainable Squirrel Conservation

Moriz Steiner • Falk Huettmann

Sustainable Squirrel Conservation

A Modern Reassessment of Family *Sciuridae*

Springer

Moriz Steiner
Department of Animal Science
Wageningen University and Research
Wageningen, The Netherlands

Falk Huettmann
-EWHALE lab-
Biology and Wildlife Department
Institute of Arctic Biology
University of Alaska Fairbanks (UAF)
Fairbanks, AK, USA

ISBN 978-3-031-23546-7 ISBN 978-3-031-23547-4 (eBook)
https://doi.org/10.1007/978-3-031-23547-4

This Springer imprint is published by the registered company Springer Nature Switzerland AG
The registered company address is: Gewerbestrasse 11, 6330 Cham, Switzerland

Foreword

Emiliano Mori is a Researcher at the National Research Council (Institute of Research on Terrestrial Ecosystems) in Italy. He is a globally working, interdisciplinary, widely published scientist based in Florence, Italy. His research is mostly conducted in Italy, but it also includes several works from Asia and Antarctica.

Rodents constitute a key component of ecosystems, being important seed dispersers and representing important prey for carnivore species, as well as bioindicators of sustainable environmental management. Despite this, these small mammals are often neglected in conservation programs, deserving little attention in worldwide research with respect to other taxa. This might be due to the fact that most rodents are crop pests or unwelcome to most humans; this attitude may have brought rodents in a shaded area for both conservation and basic ecology studies. Thus, preserving rodents is far from being an easy issue, also considering that they are rarely taken into account by legislations.

Amongst rodents, squirrels include over 280 species worldwide showing a wide range of anatomical sizes and functions, from 15 grams of the pigmy tree squirrels to over 8 kg in the largest marmots. Squirrels have coevolved with humans for millennia, and their aspect is so cute that everyone is attracted to them, making this group the best one (or at least one of the best) to start discussing on how to conserve rodents!

Squirrels are specialised to thrive in a number of different environments ranging from deciduous forests to harsh environments, mountain tops, and urban green areas, by means of natural selection. Therefore, they occupy several ecological niches throughout the world, from the northernmost latitudes all the way to the tropics. Many species are also well-adapted to human-modified environments and may be fed by humans in urban areas, as grey squirrels introduced to Europe.

A grey squirrel *Sciurus carolinensis* in an urban park in Turin (NW Italy). This species is well-adapted to human-modified environments. (Photo Emiliano Mori, CNR-IRET Sesto Fiorentino (Italy))

Squirrels provide humans and habitats with a number of ecosystem services improving environmental functioning, such as being keystone herbivores, seed predators, and seed dispersers.

The interaction between squirrels and their habitat is so deep that they are widely considered as ecosystem engineers, as being able to change the physical status of the area where they live, e.g. through borrowing activity and germination through scatter/ladder hoarding. Therefore, their conservation is pivotal to preserve habitats and local biotic diversity.

The book by Moriz Steiner and Falk Huettmann is therefore a fundamental milestone to summarise current conservation status and gaps of worldwide squirrel species. Both authors are highly qualified for this topic.

Moriz Steiner is a really promising graduating student, who has worked for a number of institutions dealing with animal science, and he recently authored a work on squirrel taxonomy. He is a great thinker, active across several disciplines and an aspiring young squirrelogist. Falk Huettmann is a naturalist professor who has already authored a number of books and book chapters: furthermore, he is an associate editor and reviews for over 70 open access journals. He is currently involved in wildlife ecology in Alaska, but he has carried out research work in over 60 countries throughout the world.

Together with the authors' undeniable expertise, the main strength point of this work is that it highlighted knowledge and gaps by deftly intermixing data taken on field work with modern digital technologies and new-generation approaches. Digital approach, online data, the creation of open access models, SDMs, and the Cloud are fundamental for all squirrel species. This book is definitely important, as modern tools (e.g. open access data and open source ML) have been used to assess species conservation and life fate in a new, holistic fashion.

This book started from a deep literature review on squirrel taxonomy and passed through evolutionary data of this group, spatial behaviour and conservation status. Taxonomy is partly a subjective matter, with over 30 available species concepts, but it is pivotal for ecological and especially conservation works. Boundaries at and around species concept are fuzzy; therefore, a complete non-arbitrary species delimitation is almost impossible. This evident lack of objectivity in taxonomy has produced a splitting frenzy trend, which is potentially threatening species conservation. On the other hand, "species lumpers" can be accused of taxonomic inertia which may prevent important management and conservation actions towards potentially endangered taxa. Researchers who are prone to splitting and those who are prone to lumping are neither wrong nor right, but each case is specific and deserves specific attention.

This problem is particularly evident with squirrels, as a global research effort and coordination towards the squirrel world is still lacking.

Interactions with the environment and particularly with humans have been assessed, with interesting issues on the near future of these rodents. In particular, four habitat types, i.e. cities, old-growth forests, tropics, and islands are under high risk for the conservation of squirrels. All of these habitats are used by a series of squirrel species. This book provided a reality opinion on the squirrels in the Anthropocene and the failed institutions and sciences, together with a brief but factual opinion on the future fate of squirrels.

IUCN estimates of threat categories of small mammals are often over-optimistic, as well as those provided by national red lists. Data Deficient species are often ignored or marginalised as they receive less conservative attention than Critically Endangered and Endangered species, even though population numbers are likely lower than assumed and the population trend is unknown.

I have been really pleased to prepare the foreword and I accepted their invitation with no hesitation, as squirrels and small mammals in general are the main topic of my current research line. My hope is that reading this book would bring small mammals out of the marginalisation they undergo within research projects and national/international legislations, to fill the discrepancy between being important ecosystem components and being overlooked.

Emiliano Mori,

Consiglio Nazionale delle Ricerche,
Via Madonna del Piano 10, 50019 Sesto Fiorentino
Florence, Italy

Preface

While cute and ubiquitous, squirrels are globally marginalized; it is as if they would not exist in our busy lives and appear to be taken as granted by everyone. They are not appreciated much by the bird feeder community and used as target practice by teenagers and whoever wants to "have some fun" (according to several Facebook groups, e.g., "Extreme Squirrel Hunting"). People hardly stop their cars for them, and squirrel roadkill, fur sales, and bush meat dominate. An effective legal protection, policy, and law enforcement or relevant squirrel-related budgets and education are usually next to nil. A large portion of the squirrel species are Data Deficient and endangered, yet the western "developed" and rich society seems to overlook the need for consideration and protection.

Squirrels have no curriculum and are hardly on the agenda in any wildlife management, conservation, or forestry programs, with climate change or without.

Having been in nature for most of our lives, and having worked fully engaged on wildlife conservation aspects, we found that very bothersome. Are we the only ones?

So how to change it?

Here we tried to start that discussion and effort, with a more meaningful and progressive science agenda and a large amount of underlying digital data and models about the global squirrels, for squirrels. We start with acknowledging the global *status quo*, and then can form a policy agenda to move forward in good and sustainable terms. We believe, it can help mankind to position itself better, and more sustainably, for a better governance and conservation regime. Arguably the world is in a conservation crisis. Our outlook on using squirrels is more than gloomy, but let's wish for the best. Here we attempt a sound analysis for betterment.

Wageningen, the Netherlands — Moriz Steiner
Fairbanks, AK, USA — Falk Huettmann

Acknowledgments

We acknowledge the public data delivery and open source code for carrying out this study, e.g. GBIF.org. FH appreciates the help of H. Berrios and E. Huettmann as well as the Chrome team. This is EWHALE lab publication #300.

MS appreciates the consitent support of J. Cise, as well as S. Gruber and A. Steiner.

Contents

About the Authors

Moriz Steiner graduated from Wageningen University and Research (Netherlands) in the field of Animal Science, where he was also a former teaching assistant for behavioral ecology. He has conducted several research internships in Institutions such as the Max-Planck-Institute (MPI) for plant breeding research (Germany), MPI for Dynamics of Complex Technical Systems (Germany), Alfred-Wegener Institute (AWI) for Polar and Marine Research (Germany), Eurac – Institute for Alpine Environment (Italy), Fondazione Edmund Mach (Forest Ecology and biogeochemical cycles – Italy), and EWHALE lab (Institute of Alaska Fairbanks UAF – Institute of Arctic Biology). Additional 9 months of remote work on an alpine farm have passioned Moriz to work toward a more sustainable future living and to conserve nature. His most recent published squirrel taxonomy papers have stimulated him to conduct further research in this field. The author shows a strong personal interest in Ecology and Conservation Management in the Arctic Regions as well as the use of modern software technology for an increased conservation success of nature to preserve it for future generations. Throughout the process of writing this book, Moriz has developed and significantly improved his expertise in ecological modeling. In addition, he also contributes to a couple of ecological modeling papers in Central Asia (including a project on Lynx, and small mammal diseases). Moriz is also a member of the IUCN Small Mammal Specialist Group (SMSG) and the IUCN Species Survival Commission (SSC).

Falk Huettmann is a 'digital naturalist' concerned with global sustainability. Being a well-rounded and published Professor in wildlife ecologist employed with the University of Alaska Fairbanks in his -EWHALE lab- he has worked in over 60 multi-year project locations worldwide during a time period of approximately 20 years on over 2,000 biodiversity species. Falk worked on pelagic seabirds in the North Atlantic, marbled murrelets in coastal old-growth forest, grizzly bear habitat in the Canadian Rocky Mountains, snow leopard data issues for the Hindu Kush-Himalaya, migratory birds in Russia and China, Matschie's Tree Kangaroo in Papua New Guinea, and over 30 species in the ocean surrounding Antarctica, including the largest and remote Marine Protected Area (MPA) of the Ross Sea. He teaches nationally certified online classes and international workshops and his expertise is in rapid assessment methods such as open access digital data, geographic information systems (GIS), data mining, and disease perspectives all footed in ecological economics, predictive risk models, and associated governance schemes. He has published over 300 publications, including 9 books about the tropics, the three poles (Arctic, Antarctic, and Hindu Kush-Himalayas), oceans, landscape ecology, machine learning/AI, the International Polar Year (IPY) as well as open access data and open source code with ISO-compliant metadata.

Part I
Introduction to Squirrels of the World and Their Conservation Trends

Chapter 1
Taxonomy for the Squirrels of the World: Hairy Classifications and Conservation Category Games

Abstract This chapter represents an overview list of all the extant squirrel species with the illustrated discrepancies between the number of accepted species among different authoritative institutions and entities such as ITIS, GBIF, Encyclopedia of Life (EOL), MANIS (VertNet), GenBank (NCBI), IUCN Red List, IDigBIO, iNaturalist, Mammal Diversity Database (MDD), "Squirrels of the world" (book – Thorington et al. 2012), "Squirrels – The animal answer guide" (book – Thorington and Ferrell 2006), Illustrated Checklist of the mammals of the world" (book – Burgin et al. 2020), and "The handbook of the mammals of the world" (book – Wilson and Mittermeier 2011). Also, here we present generally obvious taxonomic discrepancies in the order of Rodentia, and specifically, the Family of the squirrels (Sciuridae) using a digital "Big Data" approach. The squirrels of this world are owned by nobody and are a public trust resource. They are managed by governmental entities, usually done in a democratic fashion. But when around 10 to 20% of all squirrel species are highly endangered, or under high risk of extinction, or worse, it indicates a failure of their management. One would think it urgently calls for an increase in conservation efforts and public awareness to be able to preserve these species for future generations and the integrity as part of the global ecosystem, yet no such efforts can really be observed anywhere. Those were never done even, nor are they on the horizon. Here, some modern solutions are presented to strengthen recent science-based proposed changes with the greater aim to contribute to a uniformly and mutually accepted and defendable taxonomic species list and finally for more successful conservation management. This is done by addressing widely outdated taxonomic misalignments (e.g. taxonomic classifications mostly disagreed species and subspecies taxonomies among different institutions and their taxonomic lists. Therefore, here we summarize virtually all of the existing publicly available data at hand, make the compiled data and findings openly available, and present them in a clean form. Additionally, we are linking every species with its conservation status and population trend (assigned by IUCN Red List and Burgin et al. 2020) and depict the result in a crisp table to maximize the understanding of our findings. Finally, we discuss the wide lack of appropriate conservation classification and the over-positive classification policies. The taxonomic species overview of the different institutions and their species lists are provided as an insight into the relevance of this subject.

Keywords Squirrels · Sciuridae · Taxonomy · Institutional discrepancy · Big Data · Synthesis

1.1 Introduction

Squirrels are arguably cute, almost every child around the world is aware of these little mammals. In the Western world, squirrels are often associated with acorns, hoarding behavior, trees, and discovering nature. However, even though there is a broad awareness of the existence of some species of these mammals, little interest is invested in their species-specific differences and description, their subsequent conservation, global well-being, and even basic research (excluding 3–5 out of approximately 307 species, ~98% remain understudied). A global overview of all known squirrel species can be seen as illustrated in Table 1.1.

In this book, a modern assessment of this global species group is laid out. This is done in order to gain better insights into their basic distribution, conservation, extinction risk, economic importance, position in modern society, and their future trend for 2100. It's more based on a meaningful sustainability. To define for the global public what generally is considered a squirrel, it reads according to the Cambridge dictionary (November 2021), that a squirrel is "*a small animal covered in fur with a long tail. Squirrels climb trees and feed on nuts and seeds.*" However, already that is not so correct in multiple aspects, e.g.

Supplementary Information The online version contains supplementary material available at https://doi.org/10.1007/978-3-031-23547-4_1.

M. Steiner, F. Huettmann, *Sustainable Squirrel Conservation*, https://doi.org/10.1007/978-3-031-23547-4_1

Table 1.1 Institutional taxonomic species list

Squirrel species list			Presence of the according squirrel species in the species list of the corresponding agency															
Scientific name	Common name/ Alternative names	TSN (Taxonomic Serial Number)	ITIS	GBIF	Encyclopedia of Life (EOL)	MA-NIS (VertNet)	GenBank (NCBI)	IUCN Red List	IUCN Red List (old)	IDigBIO	iNaturalist	Mammal Diversity Database (MDD)	Squirrels of the world (book)	The handbook of the mammals of the world (book)	Squirrels-The Animal Answer Guide	Illustrated Checklist of the Mammals of the World	"Yes" count	"No" count
Aeretes melanopterus	Northern Chinese Flying Squirrel	632487	Yes	Yes	Yes	Yes	Yes	Yes	Yes	Yes	Yes	Yes	Yes	Yes	Yes	Yes	13	0
Aeromys tephromelas	Black Flying Squirrel	632488	Yes	Yes	Yes	Yes	Yes	Yes	Yes	Yes	Yes	Yes	Yes	Yes	Yes	Yes	13	0
Aeromys thomasi	Thomas's Flying Squirrel	632489	Yes	Yes	Yes	Yes	No	Yes	Yes	Yes	Yes	Yes	Yes	Yes	Yes	Yes	12	1
Ammospermophilus harrisii	Harris' Antelope Squirrel	180179	Yes	Yes	Yes	Yes	Yes	Yes	Yes	Yes	Yes	Yes	Yes	Yes	Yes	Yes	13	0
Ammospermophilus insularis	Espiritu Santo Island Antelope Squirrel	632323	Yes	Yes	Yes	Yes	Yes	No	Yes	Yes	No	No	Yes	No	Yes	No	8	5
Ammospermophilus interpres	Texas Antelope Squirrel	180180	Yes	Yes	Yes	Yes	Yes	Yes	Yes	Yes	Yes	Yes	Yes	Yes	Yes	Yes	13	0
Ammospermophilus leucurus	White-tailed Antelope Squirrel	180181	Yes	Yes	Yes	No	Yes	Yes	Yes	Yes	Yes	Yes	Yes	Yes	Yes	Yes	12	1
Ammospermophilus nelsoni	Nelson's Antelope Squirrel	180182	Yes	Yes	Yes	Yes	Yes	Yes	Yes	Yes	Yes	Yes	Yes	Yes	Yes	Yes	13	0
Atlantoxerus getulus	Barbary Ground Squirrel	632324	Yes	Yes	Yes	Yes	Yes	Yes	Yes	Yes	Yes	Yes	Yes	Yes	Yes	Yes	13	0
Belomys pearsonii	Hairy-footed Flying Squirrel	632490	Yes	Yes	Yes	Yes	Yes	Yes	Yes	Yes	Yes	Yes	Yes	Yes	Yes	Yes	13	0
Biswamoyopterus biswasi	Nam-dapha Flying Squirrel	632491	Yes	Yes	Yes	No	Yes	Yes	Yes	Yes	Yes	Yes	Yes	Yes	Yes	Yes	12	1

Biswamoyopterus gaoligongensis	Mount Gaoligong Flying Squirrel	NA	No	No	No	No	No	No	No	No	Yes	Yes	No	No	No	Yes	3	10
Biswamoyopterus laoensis	Laotian Giant Flying Squirrel	930360	Yes	Yes	Yes	No	No	Yes	Yes	Yes	Yes	Yes	No	Yes	No	Yes	9	4
Callosciurus adamsi	Ear-spot Squirrel	632325	Yes	Yes	Yes	Yes	Yes	Yes	Yes	Yes	Yes	Yes	Yes	Yes	Yes	Yes	13	0
Callosciurus albescens	Kloss's Squirrel	NA	No	No	No	No	No	Yes	No	No	No	No	No	No	Yes	No	2	11
Callosciurus baluensis	Kinabalu Squirrel	632327	Yes	Yes	Yes	No	Yes	Yes	Yes	Yes	Yes	Yes	Yes	Yes	Yes	Yes	12	1
Callosciurus caniceps	Gray-bellied Squirrel	632328	Yes	No	Yes	Yes	Yes	Yes	Yes	No	Yes	Yes	Yes	Yes	Yes	Yes	11	2
Callosciurus erythraeus	Pallas's Squirrel	632329	Yes	Yes	Yes	Yes	Yes	Yes	Yes	Yes	Yes	Yes	Yes	Yes	Yes	Yes	13	0
Callosciurus finlaysonii	Finlay-son's Squirrel	632330	Yes	Yes	Yes	Yes	Yes	Yes	Yes	Yes	Yes	Yes	Yes	Yes	Yes	Yes	13	0
Callosciurus honkhoaiensis	Hon Khoai Squirrel	NA	No	No	No	No	Yes	No	No	No	No	Yes	No	No	No	Yes	3	10
Callosciurus inornatus	Inornate Squirrel	632331	Yes	Yes	Yes	Yes	Yes	Yes	Yes	Yes	Yes	Yes	Yes	Yes	Yes	Yes	13	0
Callosciurus melanogaster	Mentawai Squirrel	632332	Yes	Yes	Yes	Yes	No	Yes	Yes	Yes	Yes	Yes	Yes	Yes	Yes	Yes	12	1
Callosciurus nigrovittatus	Black-striped Squirrel	632333	Yes	Yes	Yes	Yes	Yes	Yes	Yes	Yes	Yes	Yes	Yes	Yes	Yes	Yes	13	0
Callosciurus notatus	Plantain Squirrel	632334	Yes	Yes	Yes	Yes	Yes	Yes	Yes	Yes	Yes	Yes	Yes	Yes	Yes	Yes	13	0
Callosciurus orestes	Borneo Black-banded Squirrel	632335	Yes	Yes	Yes	Yes	Yes	Yes	Yes	Yes	Yes	Yes	Yes	Yes	Yes	Yes	13	0
Callosciurus phayrei	Phayre's Squirrel	632336	Yes	Yes	Yes	No	Yes	Yes	Yes	Yes	Yes	Yes	Yes	Yes	Yes	Yes	12	1
Callosciurus prevostii	Prevost's Squirrel	632337	Yes	Yes	Yes	Yes	Yes	Yes	Yes	Yes	Yes	Yes	Yes	Yes	Yes	Yes	13	0
Callosciurus pygerythrus	Irrawaddy Squirrel	632338	Yes	Yes	Yes	Yes	Yes	Yes	Yes	Yes	Yes	Yes	Yes	Yes	Yes	Yes	13	0

(continued)

Table 1.1 (continued)

Squirrel species list			Presence of the according squirrel species in the species list of the corresponding agency															
Scientific name	Common name/ Alternative names	TSN (Taxonomic Serial Number)	ITIS	GBIF	Encyclopedia of Life (EOL)	MA-NIS (VertNet)	GenBank (NCBI)	IUCN Red List	IUCN Red List (old)	IDigBIO	iNaturalist	Mammal Diversity Database (MDD)	Squirrels of the world (book)	The handbook of the mammals of the world (book)	Squirrels-The Animal Answer Guide	Illustrated Checklist of the Mammals of the World	"Yes" count	"No" count
Callosciurus quinquestriatus	Anderson's Squirrel	632339	Yes	Yes	Yes	Yes	No	Yes	Yes	Yes	Yes	Yes	Yes	Yes	Yes	Yes	12	1
Callospermophilus lateralis	Golden-mantled Ground Squirrel	930305	Yes	Yes	Yes	No	Yes	Yes	Yes	Yes	Yes	Yes	Yes	Yes	No	Yes	11	2
Callospermophilus madrensis	Sierra Madre Ground Squirrel	930303	Yes	Yes	Yes	Yes	Yes	Yes	Yes	Yes	Yes	Yes	Yes	Yes	No	Yes	12	1
Callospermophilus saturatus	Cascade Golden-mantled Ground Squirrel	930304	Yes	Yes	Yes	No	Yes	Yes	Yes	Yes	Yes	Yes	Yes	Yes	No	Yes	11	2
Cynomys gunnisoni	Gunnison's Prairie Dog	180184	Yes	Yes	Yes	Yes	Yes	Yes	Yes	No	Yes	Yes	Yes	Yes	Yes	Yes	12	1
Cynomys leucurus	White-tailed Prairie Dog	180185	Yes	Yes	Yes	Yes	Yes	Yes	Yes	Yes	Yes	Yes	Yes	Yes	Yes	Yes	13	0
Cynomys ludovicianus	Arizona black-tailed prairie dog	180186	Yes	Yes	Yes	Yes	Yes	Yes	Yes	Yes	Yes	Yes	Yes	Yes	Yes	Yes	13	0
Cynomys mexicanus	Mexican Prairie Dog	632340	Yes	Yes	Yes	Yes	Yes	Yes	Yes	Yes	Yes	Yes	Yes	Yes	Yes	Yes	13	0
Cynomys parvidens	Utah Prairie Dog	180187	Yes	Yes	Yes	Yes	Yes	Yes	Yes	Yes	Yes	Yes	Yes	Yes	Yes	Yes	13	0
Dremomys everetti	Bornean Mountain Ground Squirrel	632341	Yes	Yes	Yes	Yes	Yes	Yes	Yes	Yes	Yes	No	Yes	No	Yes	No	10	3
Dremomys gularis	Red-throated Squirrel	930262	Yes	Yes	Yes	No	Yes	Yes	Yes	Yes	Yes	Yes	Yes	Yes	Yes	Yes	12	1

Dremomys lokriah	Orange-bellied Himala-yan Squirrel	632342	Yes	Yes	Yes	Yes	Yes	Yes	Yes	Yes	Yes	Yes	Yes	Yes	Yes	Yes	13	0
Dremomys pernyi	Perny's Long-nosed Squirrel	632343	Yes	Yes	Yes	Yes	Yes	Yes	Yes	Yes	Yes	Yes	Yes	Yes	Yes	Yes	13	0
Dremomys pyrrhomerus	Red-hipped Squirrel	632344	Yes	Yes	Yes	Yes	Yes	Yes	Yes	Yes	Yes	Yes	Yes	Yes	Yes	Yes	13	0
Dremomys rufigenis	Asian Red-cheeked Squirrel	632345	Yes	Yes	Yes	Yes	Yes	Yes	Yes	Yes	Yes	Yes	Yes	Yes	Yes	Yes	13	0
Eoglaucomys fimbriatus	Kashmir Flying Squirrel	930259	Yes	Yes	Yes	Yes	Yes	Yes	Yes	Yes	Yes	Yes	Yes	Yes	Yes	Yes	13	0
Epixerus ebii	African Palm Squirrel	632346	Yes	Yes	Yes	Yes	No	Yes	Yes	Yes	Yes	Yes	Yes	Yes	Yes	Yes	12	1
Epixerus wilsoni	Biafran palm squirrel	NA	No	No	No	No	Yes	No	No	No	No	No	No	No	No	No	1	12
Eupetaurus cinereus	Woolly Flying Squirrel	632492	Yes	Yes	Yes	No	Yes	Yes	Yes	Yes	Yes	Yes	Yes	Yes	Yes	Yes	12	1
Eutamias sibiricus	Siberian Chipmunk	NA	No	No	No	No	No	Yes	No	No	Yes	Yes	No	No	No	Yes	4	9
Euxerus erythropus	Striped Ground Squirrel	NA	No	No	No	No	No	No	No	No	Yes	Yes	No	Yes	No	Yes	4	9
Exilisciurus concinnus	Philippine Pygmy Squirrel	632348	Yes	Yes	Yes	Yes	Yes	Yes	Yes	Yes	Yes	Yes	Yes	Yes	Yes	Yes	13	0
Exilisciurus exilis	Least Pygmy Squirrel	632349	Yes	Yes	Yes	Yes	Yes	Yes	Yes	Yes	Yes	Yes	Yes	Yes	Yes	Yes	13	0
Exilisciurus whiteheadi	Tufted Pygmy Squirrel	632350	Yes	Yes	Yes	Yes	Yes	Yes	Yes	Yes	Yes	Yes	Yes	Yes	Yes	Yes	13	0
Funambulus insignis	NA	234289	No	No	No	No	No	No	No	Yes	No	No	No	No	No	No	1	12
Funambulus layardi	Layard's Palm Squirrel	632351	Yes	Yes	Yes	Yes	Yes	Yes	Yes	Yes	Yes	Yes	Yes	Yes	Yes	Yes	13	0

(continued)

Table 1.1 (continued)

Squirrel species list			Presence of the according squirrel species in the species list of the corresponding agency															
Scientific name	Common name/ Alternative names	TSN (Taxonomic Serial Number)	ITIS	GBIF	Encyclopedia of Life (EOL)	MA-NIS (VertNet)	GenBank (NCBI)	IUCN Red List	IUCN Red List (old)	IDigBIO	iNaturalist	Mammal Diversity Database (MDD)	Squirrels of the world (book)	The handbook of the mammals of the world (book)	Squirrels-The Animal Answer Guide	Illustrated Checklist of the Mammals of the World	"Yes" count	"No" count
Funambulus obscurus	Dusky Striped Squirrel	NA	No	No	No	No	No	Yes	No	No	No	Yes	No	No	No	Yes	3	10
Funambulus palmarum	Indian Palm Squirrel	632352	Yes	Yes	Yes	Yes	Yes	Yes	Yes	Yes	Yes	Yes	Yes	Yes	Yes	Yes	13	0
Funambulus pennantii	Northern Palm Squirrel	632353	Yes	Yes	Yes	Yes	Yes	Yes	Yes	Yes	Yes	Yes	Yes	Yes	Yes	Yes	13	0
Funambulus sublineatus	Dusky Palm Squirrel	632354	Yes	Yes	Yes	Yes	Yes	Yes	Yes	Yes	Yes	Yes	Yes	Yes	Yes	Yes	13	0
Funambulus tristriatus	Jungle Palm Squirrel	632355	Yes	Yes	Yes	Yes	No	Yes	Yes	Yes	Yes	Yes	Yes	Yes	Yes	Yes	12	1
Funisciurus anerythrus	Thomas's Rope Squirrel	632356	Yes	Yes	Yes	Yes	Yes	Yes	Yes	Yes	Yes	Yes	Yes	Yes	Yes	Yes	13	0
Funisciurus bayonii	Lunda Rope Squirrel	632357	Yes	Yes	Yes	Yes	No	Yes	Yes	Yes	Yes	Yes	Yes	Yes	Yes	Yes	12	1
Funisciurus carruthersi	Car-ruther's Mountain Squirrel	632358	Yes	Yes	Yes	Yes	Yes	Yes	Yes	Yes	Yes	Yes	Yes	Yes	Yes	Yes	13	0
Funisciurus congicus	Congo Rope Squirrel	632359	Yes	Yes	Yes	Yes	No	Yes	Yes	Yes	Yes	Yes	Yes	Yes	Yes	Yes	12	1
Funisciurus duchaillui	Du Chaillu's Rope Squirrel	930333	Yes	Yes	Yes	No	No	Yes	Yes	No	Yes	Yes	Yes	Yes	No	Yes	9	4
Funisciurus isabella	Lady Burton's Rope Squirrel	632360	Yes	Yes	Yes	Yes	No	Yes	Yes	Yes	Yes	Yes	Yes	Yes	Yes	Yes	12	1

Funisciurus lemniscatus	Ribboned Rope Squirrel	632361	Yes	Yes	Yes	Yes	No	Yes	Yes	Yes	Yes	Yes	Yes	Yes	Yes	Yes	12	1
Funisciurus leucogenys	Red-cheeked Rope Squirrel	632362	Yes	Yes	Yes	Yes	No	Yes	Yes	Yes	Yes	Yes	Yes	Yes	Yes	Yes	12	1
Funisciurus mystax	NA	NA	No	No	No	Yes	No	No	No	No	No	No	No	No	No	No	1	12
Funisciurus pyrropus	Fire-footed Rope Squirrel	632363	Yes	Yes	Yes	Yes	Yes	Yes	Yes	Yes	Yes	Yes	Yes	Yes	Yes	Yes	13	0
Funisciurus substriatus	Kintampo Rope Squirrel	632364	Yes	Yes	Yes	No	No	Yes	Yes	Yes	Yes	Yes	Yes	Yes	Yes	Yes	11	2
Geomys bursarius	NA	NA	No	No	No	Yes	No	No	No	No	No	No	No	No	No	No	1	12
Geosciurus inauris	South African Ground Squirrel	NA	No	No	No	No	No	No	No	No	Yes	Yes	No	Yes	No	Yes	4	9
Geosciurus princeps	Damara Ground Squirrel	NA	No	No	No	No	No	No	No	No	Yes	Yes	No	Yes	No	Yes	4	9
Glaucomys oregonensis	Hum-boldt's Flying Squirrel	NA	No	No	No	No	Yes	No	No	No	Yes	Yes	No	No	No	Yes	4	9
Glaucomys sabrinus	Northern Flying Squirrel	180169	Yes	Yes	Yes	Yes	Yes	Yes	Yes	Yes	Yes	Yes	Yes	Yes	Yes	Yes	13	0
Glaucomys volans	Southern Flying Squirrel	180170	Yes	Yes	Yes	Yes	Yes	Yes	Yes	Yes	Yes	Yes	Yes	Yes	Yes	Yes	13	0
Glyphotes simus	Sculptor Squirrel	632365	Yes	Yes	Yes	Yes	Yes	Yes	Yes	Yes	Yes	Yes	Yes	Yes	Yes	Yes	13	0
Graphiurus murinus	NA	NA	No	No	No	Yes	No	No	No	No	No	No	No	No	No	No	1	12
Guerlinguetus aestuans	Guianan squirrel	NA	No	No	No	No	Yes	No	No	No	No	No	No	No	No	No	1	12
Guerlinguetus brasiliensis	NA	NA	No	No	No	No	Yes	No	No	No	No	No	No	No	No	No	1	12
Hadrosciurus igniventris	North Amazo-nan red squirrel	NA	No	No	No	No	Yes	No	No	No	No	No	No	No	No	No	1	12
Hadrosciurus pyrrhinus	Junin red squirrel	NA	No	No	No	No	Yes	No	No	No	No	No	No	No	No	No	1	12

(continued)

Table 1.1 (continued)

Squirrel species list			Presence of the according squirrel species in the species list of the corresponding agency															
Scientific name	Common name/ Alternative names	TSN (Taxonomic Serial Number)	ITIS	GBIF	Encyclopedia of Life (EOL)	MA-NIS (VertNet)	GenBank (NCBI)	IUCN Red List	IUCN Red List (old)	IDigBIO	iNaturalist	Mammal Diversity Database (MDD)	Squirrels of the world (book)	The handbook of the mammals of the world (book)	Squirrels-The Animal Answer Guide	Illustrated Checklist of the Mammals of the World	"Yes" count	"No" count
Hadrosciurus spadiceus	Southern Amazon red squirrel	NA	No	No	No	No	Yes	No	No	No	No	No	No	No	No	No	1	12
Heliosciurus gambianus	Gambian Sun Squirrel	632366	Yes	Yes	Yes	Yes	Yes	Yes	Yes	Yes	Yes	Yes	Yes	Yes	Yes	Yes	13	0
Heliosciurus multicolor	NA	930896	No	No	No	Yes	No	No	No	Yes	No	No	No	No	No	No	2	11
Heliosciurus mutabilis	Mutable Sun Squirrel	632367	Yes	Yes	Yes	Yes	No	Yes	Yes	Yes	Yes	Yes	Yes	Yes	Yes	Yes	12	1
Heliosciurus punctatus	Small Sun Squirrel	632368	Yes	Yes	Yes	Yes	No	Yes	Yes	Yes	Yes	Yes	Yes	Yes	Yes	Yes	12	1
Heliosciurus rufobrachium	Red-legged Sun Squirrel	632369	Yes	Yes	Yes	Yes	Yes	Yes	Yes	Yes	Yes	Yes	Yes	Yes	Yes	Yes	13	0
Heliosciurus ruwenzorii	Ruwen-zori Sun Squirrel	632370	Yes	Yes	Yes	Yes	Yes	Yes	Yes	Yes	Yes	Yes	Yes	Yes	Yes	Yes	13	0
Heliosciurus sanguines	NA	NA	No	No	No	Yes	No	No	No	No	No	No	No	No	No	No	1	12
Heliosciurus undulatus	Zanj Sun Squirrel	632371	Yes	Yes	Yes	No	Yes	Yes	Yes	Yes	Yes	Yes	Yes	Yes	Yes	Yes	12	1
Hylopetes alboniger	Particolored Flying Squirrel	632493	Yes	Yes	Yes	Yes	Yes	Yes	Yes	Yes	Yes	Yes	Yes	Yes	Yes	Yes	13	0
Hylopetes bartelsi	Bartel's Flying Squirrel	632495	Yes	Yes	Yes	No	No	Yes	Yes	Yes	Yes	Yes	Yes	Yes	Yes	Yes	11	2
Hylopetes electilis	Hainan Flying Squirrel	930616	No	No	No	Yes	No	No	No	Yes	No	Yes	No	No	No	Yes	4	9
Hylopetes lepidus	Gray-cheeked flying squirrel	NA	No	No	No	Yes	Yes	No	No	No	No	No	Yes	No	Yes	No	4	9

Hylopetes nigripes	Palawan Flying Squirrel	632498	Yes	Yes	Yes	Yes	Yes	Yes	Yes	Yes	Yes	Yes	Yes	Yes	Yes	Yes	13	0
Hylopetes phayrei	Indochi-nese Flying Squirrel	632499	Yes	Yes	Yes	Yes	Yes	Yes	Yes	Yes	Yes	Yes	Yes	Yes	Yes	Yes	13	0
Hylopetes platyurus	Gray-cheeked flying squirrel	930260	Yes	Yes	Yes	Yes	No	Yes	Yes	Yes	Yes	Yes	Yes	Yes	Yes	Yes	12	1
Hylopetes sagitta	Arrow Flying Squirrel	930343	Yes	Yes	Yes	No	No	Yes	Yes	Yes	Yes	Yes	No	Yes	No	Yes	9	4
Hylopetes sipora	Sipora Flying Squirrel	632500	Yes	Yes	Yes	No	No	Yes	Yes	Yes	Yes	Yes	Yes	Yes	Yes	Yes	11	2
Hylopetes spadiceus	Red-cheeked Flying Squirrel	632501	Yes	Yes	Yes	No	Yes	Yes	Yes	Yes	Yes	Yes	Yes	Yes	Yes	Yes	12	1
Hylopetes winstoni	Suma-tran Flying Squirrel	632502	Yes	Yes	Yes	No	No	Yes	Yes	Yes	Yes	Yes	Yes	Yes	Yes	Yes	11	2
Hyosciurus heinrichi	Montane Long-nosed Squirrel	632372	Yes	Yes	Yes	Yes	Yes	Yes	Yes	Yes	Yes	Yes	Yes	Yes	Yes	Yes	13	0
Hyosciurus ileile	Lowland Long-nosed Squirrel	632373	Yes	Yes	Yes	Yes	Yes	Yes	Yes	Yes	Yes	Yes	Yes	Yes	Yes	Yes	13	0
Ictidomys mexicanus	Mexican Ground Squirrel	930307	Yes	Yes	Yes	No	Yes	Yes	Yes	Yes	Yes	Yes	Yes	Yes	No	Yes	11	2
Ictidomys parvidens	Rio Grande Ground Squirrel	930267	Yes	Yes	Yes	No	Yes	No	Yes	Yes	Yes	Yes	Yes	Yes	No	Yes	10	3
Ictidomys tridecemlineatus	Thirteen-lined Ground Squirrel	930308	Yes	Yes	Yes	No	Yes	Yes	Yes	Yes	Yes	Yes	Yes	Yes	No	Yes	11	2
Iomys horsfieldii	Javanese Flying Squirrel	632503	Yes	Yes	Yes	Yes	Yes	Yes	Yes	Yes	Yes	Yes	Yes	Yes	Yes	Yes	13	0

(continued)

Table 1.1 (continued)

Squirrel species list			Presence of the according squirrel species in the species list of the corresponding agency															
Scientific name	Common name/ Alternative names	TSN (Taxonomic Serial Number)	ITIS	GBIF	Encyclopedia of Life (EOL)	MA-NIS (VertNet)	GenBank (NCBI)	IUCN Red List	IUCN Red List (old)	IDigBIO	iNaturalist	Mammal Diversity Database (MDD)	Squirrels of the world (book)	The handbook of the mammals of the world (book)	Squirrels-The Animal Answer Guide	Illustrated Checklist of the Mammals of the World	"Yes" count	"No" count
Iomys sipora	Menta-wai Flying Squirrel	632504	Yes	Yes	Yes	No	No	Yes	Yes	Yes	Yes	Yes	Yes	Yes	Yes	Yes	11	2
Lariscus hosei	Four-striped Ground Squirrel	632374	Yes	Yes	Yes	Yes	No	Yes	Yes	Yes	Yes	Yes	Yes	Yes	Yes	Yes	12	1
Lariscus insignis	Three-striped Ground Squirrel	632375	Yes	Yes	Yes	Yes	Yes	Yes	Yes	Yes	Yes	Yes	Yes	Yes	Yes	Yes	13	0
Lariscus niobe	Niobe Ground Squirrel	632376	Yes	Yes	Yes	No	Yes	Yes	Yes	Yes	Yes	Yes	Yes	Yes	Yes	Yes	12	1
Lariscus obscurus	Menta-wai Three-striped Squirrel	632377	Yes	Yes	Yes	No	Yes	Yes	Yes	Yes	Yes	Yes	Yes	Yes	Yes	Yes	12	1
Marmota baibacina	Gray Marmot	632378	Yes	Yes	Yes	Yes	Yes	Yes	Yes	Yes	Yes	Yes	Yes	Yes	Yes	Yes	13	0
Marmota bobak	Bobak Marmot	632379	Yes	Yes	Yes	Yes	Yes	Yes	Yes	Yes	Yes	Yes	Yes	Yes	Yes	Yes	13	0
Marmota broweri	Alaska Marmot	180138	Yes	Yes	Yes	Yes	Yes	Yes	Yes	Yes	Yes	Yes	Yes	Yes	Yes	Yes	13	0
Marmota caligata	Hoary Marmot	180139	Yes	Yes	Yes	Yes	Yes	Yes	Yes	Yes	Yes	Yes	Yes	Yes	Yes	Yes	13	0
Marmota camtschatica	Black-capped Marmot	632380	Yes	Yes	Yes	No	Yes	Yes	Yes	Yes	Yes	Yes	Yes	Yes	Yes	Yes	12	1
Marmota caudata	Long-tailed Marmot	632381	Yes	Yes	Yes	Yes	Yes	Yes	Yes	Yes	Yes	Yes	Yes	Yes	Yes	Yes	13	0
Marmota flaviventris	Yellow-bellied Marmot	180140	Yes	Yes	Yes	Yes	Yes	Yes	Yes	Yes	Yes	Yes	Yes	Yes	Yes	Yes	13	0
Marmota himalayana	Hima-layan Marmot	632382	Yes	Yes	Yes	Yes	Yes	Yes	Yes	Yes	Yes	Yes	Yes	Yes	Yes	Yes	13	0

Marmota kastschenkoi	Forest Steppe Marmot	930334	Yes	Yes	Yes	No	No	No	Yes	No	No	Yes	Yes	Yes	No	Yes	7	6
Marmota marmota	Alpine Marmot	632383	Yes	Yes	Yes	Yes	Yes	Yes	Yes	Yes	Yes	Yes	Yes	Yes	Yes	Yes	13	0
Marmota menzbieri	Menz-bier's Marmot	632384	Yes	Yes	Yes	Yes	Yes	Yes	Yes	Yes	Yes	Yes	Yes	Yes	Yes	Yes	13	0
Marmota monax	Wood-chuck	180137	Yes	Yes	Yes	Yes	Yes	Yes	Yes	Yes	Yes	Yes	Yes	Yes	Yes	Yes	13	0
Marmota olympus	Olympic Marmot	180141	Yes	Yes	Yes	Yes	Yes	Yes	Yes	Yes	Yes	Yes	Yes	Yes	Yes	Yes	13	0
Marmota sibirica	Tarbagan Marmot	632385	Yes	Yes	Yes	Yes	Yes	Yes	Yes	Yes	Yes	Yes	Yes	Yes	Yes	Yes	13	0
Marmota vancouverensis	Vancou-ver Island Marmot	180142	Yes	Yes	Yes	Yes	Yes	Yes	Yes	Yes	Yes	Yes	Yes	Yes	Yes	Yes	13	0
Menetes berdmorei	Indochi-nese Ground Squirrel	632386	Yes	Yes	Yes	Yes	Yes	Yes	Yes	Yes	Yes	Yes	Yes	Yes	Yes	Yes	13	0
Microsciurus alfari	Central Ameri-can Dwarf Squirrel	632387	Yes	Yes	Yes	Yes	Yes	Yes	Yes	Yes	Yes	Yes	Yes	Yes	Yes	Yes	13	0
Microsciurus flaviventer	Amazon Dwarf Squirrel	632388	Yes	Yes	Yes	Yes	Yes	Yes	Yes	Yes	Yes	Yes	Yes	Yes	Yes	Yes	13	0
Microsciurus isthmuis	NA	NA	No	No	No	Yes	No	No	No	No	No	No	No	No	No	No	1	12
Microsciurus mimulus	Western Dwarf Squirrel	632389	Yes	Yes	Yes	Yes	Yes	Yes	Yes	Yes	Yes	Yes	Yes	Yes	Yes	Yes	13	0
Microsciurus napi	NA	NA	No	No	No	Yes	No	No	No	No	No	No	No	No	No	No	1	12
Microsciurus otinus	NA	NA	No	No	No	No	Yes	No	No	No	No	No	No	No	No	No	1	12
Microsciurus sabanillae	Sabanil-lae Dwarf Squirrel	NA	No	No	No	No	Yes	No	No	No	No	No	No	No	No	No	1	12
Microsciurus santanderensis	Santan-der Dwarf Squirrel	632390	Yes	Yes	Yes	No	No	Yes	Yes	No	Yes	Yes	Yes	Yes	Yes	Yes	10	3
Microsciurus similis	NA	NA	No	No	No	Yes	Yes	No	No	No	No	No	No	No	No	No	2	11
Myosciurus pumilio	African Pygmy Squirrel	632391	Yes	Yes	Yes	Yes	Yes	Yes	Yes	Yes	Yes	Yes	Yes	Yes	Yes	Yes	13	0

(continued)

Table 1.1 (continued)

Squirrel species list			Presence of the according squirrel species in the species list of the corresponding agency															
Scientific name	Common name/ Alternative names	TSN (Taxonomic Serial Number)	ITIS	GBIF	Encyclopedia of Life (EOL)	MA-NIS (VertNet)	GenBank (NCBI)	IUCN Red List	IUCN Red List (old)	IDigBIO	iNaturalist	Mammal Diversity Database (MDD)	Squirrels of the world (book)	The handbook of the mammals of the world (book)	Squirrels-The Animal Answer Guide	Illustrated Checklist of the Mammals of the World	"Yes" count	"No" count
Nannosciurus melanotis	Black-eared Squirrel	632392	Yes	Yes	Yes	Yes	Yes	Yes	Yes	Yes	Yes	Yes	Yes	Yes	Yes	Yes	13	0
Nannosciurus surrutilus	NA	NA	No	No	No	Yes	No	No	No	No	No	No	No	No	No	No	1	12
Neotamias alpinus	Alpine Chipmunk	NA	No	No	No	No	No	Yes	No	No	Yes	Yes	No	No	No	Yes	4	9
Neotamias amoenus	Yellow-pine Chip-munk	NA	No	No	No	No	No	Yes	No	No	Yes	Yes	No	No	No	Yes	4	9
Neotamias bulleri	Buller's Chipmunk	NA	No	No	No	No	No	Yes	No	No	Yes	Yes	No	No	No	Yes	4	9
Neotamias canipes	Gray-footed Chipmunk	NA	No	No	No	No	No	Yes	No	No	Yes	Yes	No	No	No	Yes	4	9
Neotamias cinereicollis	Gray-collared Chip-munk	NA	No	No	No	No	No	Yes	No	No	Yes	Yes	No	No	No	Yes	4	9
Neotamias dorsalis	Cliff Chip-munk	NA	No	No	No	No	No	Yes	No	No	Yes	Yes	No	No	No	Yes	4	9
Neotamias durangae	Durango Chip-munk	NA	No	No	No	No	No	Yes	No	No	Yes	Yes	No	No	No	Yes	4	9
Neotamias merriami	Merriam's Chipmunk	NA	No	No	No	No	No	Yes	No	No	Yes	Yes	No	No	No	Yes	4	9
Neotamias minimus	Least Chipmunk	NA	No	No	No	No	No	Yes	No	No	Yes	Yes	No	No	No	Yes	4	9
Neotamias obscurus	California Chipmunk	NA	No	No	No	No	No	Yes	No	No	Yes	Yes	No	No	No	Yes	4	9
Neotamias ochrogenys	Yellow-cheeked Chipmunk	NA	No	No	No	No	No	Yes	No	No	Yes	Yes	No	No	No	Yes	4	9
Neotamias palmeri	Palmer's Chipmunk	NA	No	No	No	No	No	Yes	No	No	Yes	Yes	No	No	No	Yes	4	9
Neotamias panamintinus	Panamint Chipmunk	NA	No	No	No	No	No	Yes	No	No	Yes	Yes	No	No	No	Yes	4	9
Neotamias quadrimaculatus	Long-eared Chipmunk	NA	No	No	No	No	No	Yes	No	No	Yes	Yes	No	No	No	Yes	4	9

Neotamias quadrivittatus	Colorado Chipmunk	NA	No	No	No	No	No	Yes	No	No	Yes	Yes	No	No	No	Yes	4	9
Neotamias ruficaudus	Red-tailed Chipmunk	NA	No	No	No	No	No	Yes	No	No	Yes	Yes	No	No	No	Yes	4	9
Neotamias rufus	Hopi Chipmunk	NA	No	No	No	No	No	Yes	No	No	Yes	Yes	No	No	No	Yes	4	9
Neotamias senex	Allen's Chipmunk	NA	No	No	No	No	No	Yes	No	No	Yes	Yes	No	No	No	Yes	4	9
Neotamias siskiyou	Siskiyou Chipmunk	NA	No	No	No	No	No	Yes	No	No	Yes	Yes	No	No	No	Yes	4	9
Neotamias solivagus	Sierra del Carmen Chipmunk	NA	No	No	No	No	No	No	No	No	No	Yes	No	No	No	Yes	2	11
Neotamias sonomae	Sonoma Chipmunk	NA	No	No	No	No	No	Yes	No	No	Yes	Yes	No	No	No	Yes	4	9
Neotamias speciosus	Lodgepole Chipmunk	NA	No	No	No	No	No	Yes	No	No	Yes	Yes	No	No	No	Yes	4	9
Neotamias townsendii	Town-send's Chipmunk	NA	No	No	No	No	No	Yes	No	No	Yes	Yes	No	No	No	Yes	4	9
Neotamias umbrinus	Uinta Chipmunk	NA	No	No	No	No	No	Yes	No	No	Yes	Yes	No	No	No	Yes	4	9
Notocitellus adocetus	Tropical Ground Squirrel	930298	Yes	Yes	Yes	No	Yes	Yes	Yes	Yes	Yes	Yes	Yes	Yes	No	Yes	11	2
Notocitellus annulatus	Ring-tailed Ground Squirrel	930299	Yes	Yes	Yes	No	Yes	Yes	Yes	Yes	Yes	Yes	Yes	Yes	No	Yes	11	2
Otospermophilus atricapillus	Baja California Rock Squirrel	930300	Yes	Yes	Yes	Yes	Yes	No	Yes	Yes	Yes	Yes	Yes	Yes	No	Yes	11	2
Otospermophilus beecheyi	California Ground Squirel	930301	Yes	Yes	Yes	No	Yes	Yes	Yes	No	Yes	Yes	Yes	Yes	No	Yes	10	3
Otospermophlius variegatus	Rock Squirrel	930302	Yes	Yes	Yes	No	Yes	Yes	Yes	Yes	Yes	Yes	Yes	Yes	No	Yes	11	2
Paraxerus alexandri	Alexan-der's Bush Squirrel	632393	Yes	Yes	Yes	Yes	Yes	Yes	Yes	Yes	Yes	Yes	Yes	Yes	Yes	Yes	13	0
Paraxerus boehmi	Boehm's Bush Squirrel	632394	Yes	Yes	Yes	Yes	No	Yes	Yes	Yes	Yes	Yes	Yes	Yes	Yes	Yes	12	1

(continued)

Table 1.1 (continued)

Squirrel species list			Presence of the according squirrel species in the species list of the corresponding agency															
Scientific name	Common name/ Alternative names	TSN (Taxonomic Serial Number)	ITIS	GBIF	Encyclopedia of Life (EOL)	MA-NIS (VertNet)	GenBank (NCBI)	IUCN Red List	IUCN Red List (old)	IDigBIO	iNaturalist	Mammal Diversity Database (MDD)	Squirrels of the world (book)	The handbook of the mammals of the world (book)	Squirrels-The Animal Answer Guide	Illustrated Checklist of the Mammals of the World	"Yes" count	"No" count
Paraxerus cepapi	Smith's Bush Squirrel	632395	Yes	Yes	Yes	Yes	Yes	Yes	Yes	Yes	Yes	Yes	Yes	Yes	Yes	Yes	13	0
Paraxerus cooperi	Cooper's Mountain Squirrel	632396	Yes	Yes	Yes	No	No	Yes	Yes	Yes	Yes	Yes	Yes	Yes	Yes	Yes	11	2
Paraxerus flavovittis	Striped Bush Squirrel	632397	Yes	Yes	Yes	Yes	No	Yes	Yes	Yes	Yes	Yes	Yes	Yes	Yes	Yes	12	1
Paraxerus lucifer	Black and Red Bush Squirrel	632398	Yes	Yes	Yes	Yes	No	Yes	Yes	Yes	Yes	Yes	Yes	Yes	Yes	Yes	12	1
Paraxerus ochraceus	Ochre Bush Squirrel	632399	Yes	Yes	Yes	Yes	Yes	Yes	Yes	Yes	Yes	Yes	Yes	Yes	Yes	Yes	13	0
Paraxerus palliatus	Red Bush Squirrel	632400	Yes	Yes	Yes	Yes	No	Yes	Yes	Yes	Yes	Yes	Yes	Yes	Yes	Yes	12	1
Paraxerus poensis	Green Bush Squirrel	632401	Yes	Yes	Yes	Yes	No	Yes	Yes	Yes	Yes	Yes	Yes	Yes	Yes	Yes	12	1
Paraxerus vexillarius	Svynnerton's Bush Squirrel	632402	Yes	Yes	Yes	Yes	Yes	Yes	Yes	Yes	Yes	Yes	Yes	Yes	Yes	Yes	13	0
Paraxerus vincenti	Vincent's Bush Squirrel	632403	Yes	Yes	Yes	No	No	Yes	Yes	Yes	Yes	Yes	Yes	Yes	Yes	Yes	11	2
Petaurillus emiliae	Lesser Pygmy Flying Squirrel	632505	Yes	Yes	Yes	No	No	Yes	Yes	Yes	Yes	Yes	Yes	Yes	Yes	Yes	11	2
Petaurillus hosei	Hose's Pygmy Flying Squirrel	632506	Yes	Yes	Yes	No	No	Yes	Yes	Yes	Yes	Yes	Yes	Yes	Yes	Yes	11	2
Petaurillus kinlochii	Selangor Pygmy Flying Squirrel	632507	Yes	Yes	Yes	Yes	Yes	Yes	Yes	Yes	Yes	Yes	Yes	Yes	Yes	Yes	13	0

Petaurista albiventer	White-bellied Giant Flying Squirrel	NA	No	No	No	No	Yes	No	No	No	No	Yes	No	Yes	No	Yes	4	9
Petaurista alborufus	Red and White Giant Flying Squirrel	632508	Yes	Yes	Yes	Yes	Yes	Yes	Yes	Yes	Yes	Yes	Yes	Yes	Yes	Yes	13	0
Petaurista caniceps	Gray-headed Flying Squirrel	930328	Yes	Yes	Yes	No	No	No	Yes	Yes	No	Yes	Yes	Yes	No	Yes	8	5
Petaurista elegans	Spotted Giant Flying Squirrel	632509	Yes	Yes	Yes	No	Yes	Yes	Yes	Yes	Yes	Yes	Yes	Yes	Yes	Yes	12	1
Petaurista grandis	Formosan Giant Flying Squirrel	NA	No	No	No	No	No	No	No	No	No	Yes	No	No	No	Yes	2	11
Petaurista hainana	Hainan Giant Flying Squirrel	NA	No	No	No	No	Yes	No	No	No	No	Yes	No	Yes	No	Yes	4	9
Petaurista lena	Taiwan Giant Flying Squirrel	NA	No	No	No	No	No	No	No	No	No	Yes	No	Yes	No	Yes	3	10
Petaurista leucogenys	Japanese Giant Flying Squirrel	632510	Yes	Yes	Yes	No	Yes	Yes	Yes	Yes	Yes	Yes	Yes	Yes	Yes	Yes	12	1
Petaurista magnificus	Hodgson's Giant Flying Squirrel	632511	Yes	Yes	Yes	Yes	No	Yes	Yes	Yes	Yes	Yes	Yes	Yes	Yes	Yes	12	1
Petaurista marica	Indochinese Giant Flying Squirrel	NA	No	No	No	No	No	No	No	No	No	Yes	No	No	No	Yes	2	11
Petaurista mechukaensis	Mechuka Giant Flying Squirrel	NA	No	No	No	No	No	Yes	No	No	Yes	Yes	No	Yes	No	Yes	5	8

(continued)

Table 1.1 (continued)

Squirrel species list			Presence of the according squirrel species in the species list of the corresponding agency															
Scientific name	Common name/ Alternative names	TSN (Taxonomic Serial Number)	ITIS	GBIF	Encyclopedia of Life (EOL)	MA-NIS (VertNet)	GenBank (NCBI)	IUCN Red List	IUCN Red List (old)	IDigBIO	iNaturalist	Mammal Diversity Database (MDD)	Squirrels of the world (book)	The handbook of the mammals of the world (book)	Squirrels-The Animal Answer Guide	Illustrated Checklist of the Mammals of the World	"Yes" count	"No" count
Petaurista mishmiensis	Mishmi Hills Giant Flying Squirrel	NA	No	No	No	No	No	Yes	No	No	Yes	Yes	No	Yes	No	Yes	5	8
Petaurista nobilis	Bhutan Giant Flying Squirrel	632512	Yes	Yes	Yes	No	No	Yes	Yes	Yes	Yes	Yes	Yes	Yes	Yes	Yes	11	2
Petaurista petaurista	Red Giant Flying Squirrel	632513	Yes	Yes	Yes	Yes	Yes	Yes	Yes	Yes	Yes	Yes	Yes	Yes	Yes	Yes	13	0
Petaurista philippensis	Indian Giant Flying Squirrel	632514	Yes	Yes	Yes	Yes	Yes	Yes	Yes	Yes	Yes	Yes	Yes	Yes	Yes	Yes	13	0
Petaurista siangensis	Mebo Giant Flying Squirrel	NA	No	No	No	No	No	No	No	No	No	Yes	No	Yes	No	Yes	3	10
Petaurista sybilla	Chindwin Giant Flying Squirrel	NA	No	No	No	No	No	No	No	No	No	Yes	No	No	No	Yes	2	11
Petaurista xanthotis	Chinese Giant Flying Squirrel	632515	Yes	Yes	Yes	No	Yes	Yes	Yes	Yes	Yes	Yes	Yes	Yes	Yes	Yes	12	1
Petaurista yunanensis	Yunnan Giant Flying Squirrel	NA	No	No	No	No	Yes	No	No	No	No	Yes	No	Yes	No	Yes	4	9
Petinomys crinitus	Mindanao Flying Squirrel	632516	Yes	Yes	Yes	Yes	No	Yes	Yes	Yes	Yes	Yes	Yes	Yes	Yes	Yes	12	1
Petinomys fuscocapillus	Travan-core Flying Squirrel	632517	Yes	Yes	Yes	Yes	Yes	Yes	Yes	Yes	Yes	Yes	Yes	Yes	Yes	Yes	13	0

Petinomys genibarbis	Whis-kered Flying Squirrel	632518	Yes	Yes	Yes	No	No	Yes	Yes	Yes	Yes	Yes	Yes	Yes	Yes	Yes	11	2
Petinomys hageni	Hagen's Flying Squirrel	632519	Yes	Yes	Yes	Yes	No	Yes	Yes	Yes	Yes	Yes	Yes	Yes	Yes	Yes	12	1
Petinomys lugens	Sipora Flying Squirrel	632520	Yes	Yes	Yes	Yes	No	Yes	Yes	Yes	Yes	Yes	Yes	Yes	Yes	Yes	12	1
Petinomys mindanensis	Mindanao Flying Squirrel	930261	Yes	Yes	Yes	No	No	Yes	Yes	Yes	Yes	Yes	Yes	Yes	Yes	Yes	11	2
Petinomys momonga	NA	NA	No	No	No	No	No	No	No	Yes	No	No	No	No	No	No	1	12
Petinomys morris	NA	NA	No	No	No	No	No	No	No	Yes	No	No	No	No	No	No	1	12
Petinomys sagitta	Arrow Flying Squirrel	NA	No	No	No	No	No	No	No	No	No	No	No	No	Yes	No	1	12
Petinomys setosus	Tem-minck's Flying Squirrel	632522	Yes	Yes	Yes	Yes	Yes	Yes	Yes	Yes	Yes	Yes	Yes	Yes	Yes	Yes	13	0
Petinomys vordermanni	Vorder-mann's Flying Squirrel	632523	Yes	Yes	Yes	Yes	No	Yes	Yes	Yes	Yes	Yes	Yes	Yes	Yes	Yes	12	1
Poliocitellus franklinii	Franklin's Ground Squirrel	930309	Yes	Yes	Yes	No	Yes	Yes	Yes	Yes	Yes	Yes	Yes	Yes	No	Yes	11	2
Prosciurillus abstrusus	Secretive Dwarf Squirrel	632404	Yes	Yes	Yes	Yes	Yes	Yes	Yes	Yes	Yes	Yes	Yes	Yes	Yes	Yes	13	0
Prosciurillus alstoni	Alston's Squirrel	930329	Yes	Yes	Yes	No	No	Yes	Yes	No	Yes	Yes	Yes	Yes	No	Yes	9	4
Prosciurillus leucomus	Whitish Dwarf Squirrel	632405	Yes	Yes	Yes	Yes	Yes	Yes	Yes	Yes	Yes	Yes	Yes	Yes	Yes	Yes	13	0
Prosciurillus murinus	Celebes Dwarf Squirrel	632406	Yes	Yes	Yes	No	Yes	Yes	Yes	Yes	Yes	Yes	Yes	Yes	Yes	Yes	12	1
Prosciurillus rosenbergii	Sanghir Squirrel	930263	Yes	Yes	Yes	No	No	Yes	Yes	Yes	Yes	Yes	Yes	Yes	Yes	Yes	11	2
Prosciurillus topapuensis	Mount Topapu Squirrel	930330	Yes	Yes	Yes	No	No	Yes	Yes	No	Yes	Yes	Yes	Yes	No	Yes	9	4

(continued)

Table 1.1 (continued)

Squirrel species list			Presence of the according squirrel species in the species list of the corresponding agency															
Scientific name	Common name/ Alternative names	TSN (Taxonomic Serial Number)	ITIS	GBIF	Encyclopedia of Life (EOL)	MA-NIS (VertNet)	GenBank (NCBI)	IUCN Red List	IUCN Red List (old)	IDigBIO	iNaturalist	Mammal Diversity Database (MDD)	Squirrels of the world (book)	The handbook of the mammals of the world (book)	Squirrels-The Animal Answer Guide	Illustrated Checklist of the Mammals of the World	"Yes" count	"No" count
Prosciurillus weberi	Weber's Dwarf Squirrel	632407	Yes	Yes	Yes	No	Yes	Yes	Yes	Yes	Yes	Yes	Yes	Yes	Yes	Yes	12	1
Protoxerus aubinnii	Slender-tailed Squirrel	632408	Yes	Yes	Yes	Yes	No	Yes	Yes	Yes	Yes	Yes	Yes	Yes	Yes	Yes	12	1
Protoxerus stangeri	Forest Giant Squirrel	632409	Yes	Yes	Yes	Yes	Yes	Yes	Yes	Yes	Yes	Yes	Yes	Yes	Yes	Yes	13	0
Pteromys momonga	Japanese Flying Squirrel	632524	Yes	Yes	Yes	Yes	Yes	Yes	Yes	Yes	Yes	Yes	Yes	Yes	Yes	Yes	13	0
Pteromys volans	Siberian Flying Squirrel	632525	Yes	Yes	Yes	No	Yes	Yes	Yes	Yes	Yes	Yes	Yes	Yes	Yes	Yes	12	1
Pteromyscus pulverulentus	Smoky Flying Squirrel	632526	Yes	Yes	Yes	Yes	Yes	Yes	Yes	Yes	Yes	Yes	Yes	Yes	Yes	Yes	13	0
Ratufa affinis	Pale Giant Squirrel	632410	Yes	Yes	Yes	Yes	Yes	Yes	Yes	Yes	Yes	Yes	Yes	Yes	Yes	Yes	13	0
Ratufa bicolor	Black Giant Squirrel	632411	Yes	Yes	Yes	Yes	Yes	Yes	Yes	Yes	Yes	Yes	Yes	Yes	Yes	Yes	13	0
Ratufa indica	Indian Giant Squirrel	632412	Yes	Yes	Yes	Yes	Yes	Yes	Yes	Yes	Yes	Yes	Yes	Yes	Yes	Yes	13	0
Ratufa macroura	Sri Lankan Giant Squirrel	632413	Yes	Yes	Yes	Yes	Yes	Yes	Yes	Yes	Yes	Yes	Yes	Yes	Yes	Yes	13	0
Ratufa malabarica	NA	NA	No	No	No	Yes	No	No	No	No	No	No	No	No	No	No	1	12
Rheithrosciurus macrotis	Tufted Ground Squirrel	632414	Yes	Yes	Yes	Yes	Yes	Yes	Yes	Yes	Yes	Yes	Yes	Yes	Yes	Yes	13	0
Rhinosciurus laticaudatus	Shrew-faced Squirrel	632415	Yes	Yes	Yes	Yes	Yes	Yes	Yes	Yes	Yes	Yes	Yes	Yes	Yes	Yes	13	0

Rubrisciurus rubriventer	Sulawesi Giant Squirrel	632416	Yes	Yes	Yes	Yes	Yes	Yes	Yes	Yes	Yes	Yes	Yes	Yes	Yes	Yes	13	0
Sciurillus pusillus	Neotropical Pygmy Squirrel	632417	Yes	Yes	Yes	Yes	Yes	Yes	Yes	Yes	Yes	Yes	Yes	Yes	Yes	Yes	13	0
Sciurotamias davidianus	Père David's Rock Squirrel	632418	Yes	Yes	Yes	Yes	Yes	Yes	Yes	Yes	Yes	Yes	Yes	Yes	Yes	Yes	13	0
Sciurotamias forresti	Forrest's Rock Squirrel	632419	Yes	Yes	Yes	No	Yes	Yes	Yes	Yes	Yes	Yes	Yes	Yes	Yes	Yes	12	1
Sciurus aberti	Abert's Squirrel	180173	Yes	Yes	Yes	Yes	Yes	Yes	Yes	Yes	Yes	Yes	Yes	Yes	Yes	Yes	13	0
Sciurus aestuans	Guianan Squirrel	632420	Yes	Yes	Yes	Yes	No	Yes	Yes	Yes	Yes	Yes	Yes	Yes	Yes	Yes	12	1
Sciurus alleni	Allen's Squirrel	632421	Yes	Yes	Yes	Yes	Yes	Yes	Yes	Yes	Yes	Yes	Yes	Yes	Yes	Yes	13	0
Sciurus alphonsei	NA	NA	No	No	No	Yes	No	No	No	No	No	No	No	No	No	No	1	12
Sciurus anomalus	Caucasian Squirrel	632422	Yes	Yes	Yes	Yes	Yes	Yes	Yes	Yes	Yes	Yes	Yes	Yes	Yes	Yes	13	0
Sciurus apache	NA	NA	No	No	No	Yes	No	No	No	No	No	No	No	No	No	No	1	12
Sciurus arizonensis	Arizona Gray Squirrel	180174	Yes	Yes	Yes	Yes	Yes	Yes	Yes	Yes	Yes	Yes	Yes	Yes	Yes	Yes	13	0
Sciurus aureogaster	Mexican gray squirrel	552498	Yes	Yes	Yes	Yes	Yes	Yes	Yes	Yes	Yes	Yes	Yes	Yes	Yes	Yes	13	0
Sciurus carolinensis	Eastern Gray Squirrel	180175	Yes	Yes	Yes	Yes	Yes	Yes	Yes	Yes	Yes	Yes	Yes	Yes	Yes	Yes	13	0
Sciurus cinerea	NA	NA	No	No	No	Yes	No	No	No	No	No	No	No	No	No	No	1	12
Sciurus colliaei	Collie's Squirrel	632423	Yes	Yes	Yes	Yes	Yes	Yes	Yes	Yes	Yes	Yes	Yes	Yes	Yes	Yes	13	0
Sciurus deppei	Deppe's Squirrel	632424	Yes	Yes	Yes	Yes	Yes	Yes	Yes	Yes	Yes	Yes	Yes	Yes	Yes	Yes	13	0
Sciurus erythropus	NA	NA	No	No	No	No	No	No	No	Yes	No	No	No	No	No	No	1	12
Sciurus flammifer	Fiery Squirrel	632425	Yes	Yes	Yes	Yes	Yes	Yes	Yes	Yes	Yes	Yes	Yes	Yes	Yes	Yes	13	0
Sciurus fossor	NA	NA	No	No	No	Yes	No	No	No	No	No	No	No	No	No	No	1	12
Sciurus giganteus	NA	NA	No	No	No	No	No	No	No	Yes	No	No	No	No	No	No	1	12

(continued)

Table 1.1 (continued)

Squirrel species list			Presence of the according squirrel species in the species list of the corresponding agency															
Scientific name	Common name/ Alternative names	TSN (Taxonomic Serial Number)	ITIS	GBIF	Encyclopedia of Life (EOL)	MA-NIS (VertNet)	GenBank (NCBI)	IUCN Red List	IUCN Red List (old)	IDigBIO	iNaturalist	Mammal Diversity Database (MDD)	Squirrels of the world (book)	The handbook of the mammals of the world (book)	Squirrels-The Animal Answer Guide	Illustrated Checklist of the Mammals of the World	"Yes" count	"No" count
Sciurus gilvigularis	Yellow-throated Squirrel	632426	Yes	Yes	Yes	Yes	Yes	Yes	Yes	Yes	Yes	Yes	Yes	Yes	Yes	Yes	13	0
Sciurus granatensis	Red-tailed Squirrel	632427	Yes	Yes	Yes	Yes	Yes	Yes	Yes	Yes	Yes	Yes	Yes	Yes	Yes	Yes	13	0
Sciurus griseus	Western Gray Squirrel	180176	Yes	Yes	Yes	Yes	Yes	Yes	Yes	Yes	Yes	Yes	Yes	Yes	Yes	Yes	13	0
Sciurus hudsonicus	NA	NA	No	No	No	Yes	No	No	No	No	No	No	No	No	No	No	1	12
Sciurus ignitus	Bolivian Squirrel	632428	Yes	Yes	Yes	Yes	Yes	Yes	Yes	Yes	Yes	Yes	Yes	Yes	Yes	Yes	13	0
Sciurus igniventris	Northern Amazon Red Squirrel	632429	Yes	Yes	Yes	Yes	No	Yes	Yes	Yes	Yes	Yes	Yes	Yes	Yes	Yes	12	1
Sciurus ingrami	NA	NA	No	No	No	Yes	No	No	No	No	No	No	No	No	No	No	1	12
Sciurus kaibabensis	NA	NA	No	No	No	Yes	No	No	No	No	No	No	No	No	No	No	1	12
Sciurus lis	Japanese Squirrel	632430	Yes	Yes	Yes	Yes	Yes	Yes	Yes	Yes	Yes	Yes	Yes	Yes	Yes	Yes	13	0
Sciurus meridionalis	Calabrian Black Squirrel	NA	No	No	No	No	No	No	No	No	Yes	Yes	No	No	No	Yes	3	10
Sciurus nayaritensis	Mexican Fox Squirrel	180177	Yes	Yes	Yes	Yes	Yes	Yes	Yes	Yes	Yes	Yes	Yes	Yes	Yes	Yes	13	0
Sciurus niger	Eastern Fox Squirrel	180172	Yes	Yes	Yes	Yes	Yes	Yes	Yes	Yes	Yes	Yes	Yes	Yes	Yes	Yes	13	0
Sciurus oculatus	Peters's Squirrel	632431	Yes	Yes	Yes	Yes	Yes	Yes	Yes	Yes	Yes	Yes	Yes	Yes	Yes	Yes	13	0
Sciurus olsoni	NA	NA	No	No	No	No	No	No	No	Yes	No	No	No	No	No	No	1	12
Sciurus poliopus	NA	NA	No	No	No	Yes	No	No	No	No	No	No	No	No	No	No	1	12
Sciurus pucheranii	Andean Squirrel	632432	Yes	Yes	Yes	Yes	Yes	Yes	Yes	Yes	Yes	Yes	Yes	Yes	Yes	Yes	13	0
Sciurus pyrenaicus	NA	NA	No	No	No	Yes	No	No	No	No	No	No	No	No	No	No	1	12
Sciurus pyrrhinus	Junin Red Squirrel	632433	Yes	Yes	Yes	Yes	No	Yes	Yes	Yes	Yes	Yes	Yes	Yes	Yes	Yes	12	1

Sciurus richmondi	Rich-mond's Squirrel	632434	Yes	Yes	Yes	Yes	No	Yes	Yes	Yes	Yes	Yes	Yes	Yes	Yes	Yes	12	1
Sciurus sanborni	Sanborn's Squirrel	632435	Yes	Yes	Yes	Yes	No	Yes	Yes	Yes	Yes	Yes	Yes	Yes	Yes	Yes	12	1
Sciurus sinaloensis	NA	NA	No	No	No	Yes	No	No	No	No	No	No	No	No	No	No	1	12
Sciurus socialis	NA	NA	No	No	No	Yes	No	No	No	No	No	No	No	No	No	No	1	12
Sciurus spadiceus	Southern Amazon Red Squirrel	632436	Yes	Yes	Yes	Yes	No	Yes	Yes	Yes	Yes	Yes	Yes	Yes	Yes	Yes	12	1
Sciurus stramineus	Guayaquil Squirrel	632437	Yes	Yes	Yes	Yes	No	Yes	Yes	Yes	Yes	Yes	Yes	Yes	Yes	Yes	12	1
Sciurus variabilis	NA	NA	No	No	No	No	No	No	No	Yes	No	No	No	No	No	No	1	12
Sciurus variegatoides	Variega-ted Squirrel	632438	Yes	Yes	Yes	Yes	Yes	Yes	Yes	Yes	Yes	Yes	Yes	Yes	Yes	Yes	13	0
Sciurus vulgaris	Eurasian Red Squirrel	632439	Yes	Yes	Yes	Yes	Yes	Yes	Yes	Yes	Yes	Yes	Yes	Yes	Yes	Yes	13	0
Sciurus yucatanensis	Yucatan Squirrel	632440	Yes	Yes	Yes	Yes	Yes	Yes	Yes	Yes	Yes	Yes	Yes	Yes	Yes	Yes	13	0
Simosciurus nebouxii	NA	NA	No	No	No	No	Yes	No	No	No	No	No	No	No	No	No	1	12
Simosciurus stramineus	Guayaquil squirrel	NA	No	No	No	No	Yes	No	No	No	No	No	No	No	No	No	1	12
Spermophilopsis leptodactylus	Long-clawed Ground Squirrel	632441	Yes	Yes	Yes	Yes	Yes	Yes	Yes	Yes	Yes	Yes	Yes	Yes	Yes	Yes	13	0
Spermophilus adocetus	Tropical Ground Squirrel	NA	No	No	No	Yes	No	No	No	No	No	No	No	No	Yes	No	2	11
Spermophilus alashanicus	Alashan Ground Squirrel	632443	Yes	Yes	Yes	Yes	Yes	Yes	Yes	Yes	Yes	Yes	Yes	Yes	Yes	Yes	13	0
Spermophilus annulatus	Ring-tailed Ground Squirrel	NA	No	No	No	Yes	No	No	No	No	No	No	No	No	Yes	No	2	11
Spermophilus armatus	Uinta Ground Squirrel	NA	No	No	No	Yes	No	No	No	No	No	No	No	No	Yes	No	2	11
Spermophilus atricapillus	Baja California Rock Squirrel	NA	No	No	No	No	No	No	No	No	No	No	No	No	Yes	No	1	12

(continued)

Table 1.1 (continued)

Squirrel species list			Presence of the according squirrel species in the species list of the corresponding agency															
Scientific name	Common name/ Alternative names	TSN (Taxonomic Serial Number)	ITIS	GBIF	Encyclopedia of Life (EOL)	MA-NIS (VertNet)	GenBank (NCBI)	IUCN Red List	IUCN Red List (old)	IDigBIO	iNaturalist	Mammal Diversity Database (MDD)	Squirrels of the world (book)	The handbook of the mammals of the world (book)	Squirrels-The Animal Answer Guide	Illustrated Checklist of the Mammals of the World	"Yes" count	"No" count
Spermophilus beecheyi	California Ground Squirel	NA	No	No	No	Yes	No	No	No	No	No	No	No	No	Yes	No	2	11
Spermophilus beldingi	Belding's Ground Squirrel	NA	No	No	No	Yes	No	No	No	No	No	No	No	No	Yes	No	2	11
Spermophilus brevicauda	Brandt's Ground Squirrel	930266	Yes	Yes	Yes	No	Yes	Yes	Yes	Yes	Yes	Yes	Yes	Yes	Yes	Yes	12	1
Spermophilus brunneus	Idaho Ground Squirrel	NA	No	No	No	No	No	No	No	No	No	No	No	No	Yes	No	1	12
Spermophilus canus	Merriam's Ground Squirrel	NA	No	No	No	No	No	No	No	No	No	No	No	No	Yes	No	1	12
Spermophilus citellus	European Ground Squirrel	632446	Yes	Yes	Yes	Yes	Yes	Yes	Yes	Yes	Yes	Yes	Yes	Yes	Yes	Yes	13	0
Spermophilus columbianus	Colum-bian Ground Squirrel	NA	No	No	No	Yes	No	No	No	No	No	No	No	No	Yes	No	2	11
Spermophilus dauricus	Daurian Ground Squirrel	632447	Yes	Yes	Yes	Yes	Yes	Yes	Yes	Yes	Yes	Yes	Yes	Yes	Yes	Yes	13	0
Spermophilus elegans	Wyoming Ground Squirrel	NA	No	No	No	No	No	No	No	No	No	No	No	No	Yes	No	1	12
Spermophilus erythrogenys	Red-cheeked Ground Squirrel	632448	Yes	Yes	Yes	Yes	Yes	Yes	Yes	Yes	Yes	Yes	Yes	Yes	Yes	Yes	13	0
Spermophilus franklinii	Franklin's Ground Squirrel	NA	No	No	No	Yes	No	No	No	No	No	No	No	No	Yes	No	2	11
Spermophilus fulvus	Yellow Ground Squirrel	632449	Yes	Yes	Yes	Yes	Yes	Yes	Yes	Yes	Yes	Yes	Yes	Yes	Yes	Yes	13	0

Spermophilus lateralis	Golden-mantled Ground Squirrel	NA	No	No	No	Yes	No	No	No	No	No	No	No	No	Yes	No	2	11
Spermophilus madrensis	Sierra Madre Ground Squirrel	NA	No	No	No	No	No	No	No	No	No	No	No	No	Yes	No	1	12
Spermophilus major	Russet Ground Squirrel	632451	Yes	Yes	Yes	Yes	Yes	Yes	Yes	Yes	Yes	Yes	Yes	Yes	Yes	Yes	13	0
Spermophilus mexicanus	Mexican Ground Squirrel	NA	No	No	No	No	No	No	No	No	No	No	No	No	Yes	No	1	12
Spermophilus mohavensis	Mohave Ground Squirrel	NA	No	No	No	Yes	No	No	No	No	No	No	No	No	Yes	No	2	11
Spermophilus mollis	Piute Ground Squirrel	NA	No	No	No	Yes	No	No	No	No	No	No	No	No	Yes	No	2	11
Spermophilus musicus	Caucasian Mountain Ground Squirrel	632452	Yes	Yes	Yes	Yes	Yes	Yes	Yes	Yes	Yes	Yes	Yes	Yes	Yes	Yes	13	0
Spermophilus navigator	NA	NA	No	No	No	Yes	No	No	No	No	No	No	No	No	No	No	1	12
Spermophilus nilkaensis	Tien Shan Ground Squirrel	NA	No	No	No	No	No	No	No	No	Yes	Yes	No	Yes	No	Yes	4	9
Spermophilus odessanus	NA	NA	No	No	No	No	Yes	No	No	No	No	No	No	No	No	No	1	12
Spermophilus pallidicauda	Pallid Ground Squirrel	930268	Yes	Yes	Yes	Yes	Yes	Yes	Yes	Yes	Yes	Yes	Yes	Yes	Yes	Yes	13	0
Spermophilus parryii	Arctic Ground Squirrel	NA	No	No	No	Yes	No	No	No	No	No	No	No	No	Yes	No	2	11
Spermophilus plesius	NA	NA	No	No	No	Yes	No	No	No	No	No	No	No	No	No	No	1	12
Spermophilus perotensis	Perote Ground Squirrel	NA	No	No	No	No	No	No	No	No	No	No	No	No	Yes	No	1	12
Spermophilus pygmaeus	Little Ground Squirrel	632454	Yes	Yes	Yes	Yes	Yes	Yes	Yes	Yes	Yes	Yes	Yes	Yes	Yes	Yes	13	0

(continued)

Table 1.1 (continued)

Squirrel species list			Presence of the according squirrel species in the species list of the corresponding agency															
Scientific name	Common name/ Alternative names	TSN (Taxonomic Serial Number)	ITIS	GBIF	Encyclopedia of Life (EOL)	MA-NIS (VertNet)	GenBank (NCBI)	IUCN Red List	IUCN Red List (old)	IDigBIO	iNaturalist	Mammal Diversity Database (MDD)	Squirrels of the world (book)	The handbook of the mammals of the world (book)	Squirrels-The Animal Answer Guide	Illustrated Checklist of the Mammals of the World	"Yes" count	"No" count
Spermophilus ralli	Tien Shan Ground Squirrel	930269	Yes	Yes	Yes	No	No	Yes	Yes	No	No	No	Yes	No	Yes	No	6	7
Spermophilus relictus	Relict Ground Squirrel	632455	Yes	Yes	Yes	Yes	Yes	Yes	Yes	Yes	Yes	Yes	Yes	Yes	Yes	Yes	13	0
Spermophilus richardsonii	Richardson's Ground Squirrel	NA	No	No	No	Yes	No	No	No	No	No	No	No	No	Yes	No	2	11
Spermophilus saturatus	Cascade Golden-mantled Ground Squirrel	NA	No	No	No	Yes	No	No	No	No	No	No	No	No	Yes	No	2	11
Spermophilus spilosoma	Spotted Ground Squirrel	NA	No	No	No	Yes	No	No	No	No	No	No	No	No	Yes	No	2	11
Spermophilus suslicus	Speckled Ground Squirrel	632456	Yes	Yes	Yes	Yes	Yes	Yes	Yes	Yes	Yes	Yes	Yes	Yes	Yes	Yes	13	0
Spermophilus taurensis	Taurus Ground Squirrel	930306	Yes	Yes	Yes	No	Yes	Yes	Yes	No	Yes	Yes	Yes	Yes	No	Yes	10	3
Spermophilus tereticaudus	Round-tailed Ground Squirrel	NA	No	No	No	Yes	No	No	No	No	No	No	No	No	Yes	No	2	11
Spermophilus townsendii	Townsend's Ground Squirrel	NA	No	No	No	No	No	No	No	No	No	No	No	No	Yes	No	1	12
Spermophilus tridecemlineatus	Thirteen-lined Ground Squirrel	NA	No	No	No	Yes	No	No	No	No	No	No	No	No	Yes	No	2	11

Spermophilus undulatus	Long-tailed Ground Squirrel	NA	No	No	No	Yes	No	No	No	No	No	No	No	No	Yes	No	2	11
Spermophilus variegatus	Rock Squirrel	NA	No	No	No	Yes	No	No	No	No	No	No	No	No	Yes	No	2	11
Spermophilus washingtoni	Washing-ton Ground Squirrel	NA	No	No	No	Yes	No	No	No	No	No	No	No	No	Yes	No	2	11
Spermophilus xanthoprymnus	Asia Minor Ground Squirrel	632458	Yes	Yes	Yes	Yes	Yes	Yes	Yes	Yes	Yes	Yes	Yes	Yes	Yes	Yes	13	0
Sundasciurus altitudinis	Sumatran Mountain Squirrel	930331	Yes	Yes	Yes	No	Yes	No	Yes	No	No	Yes	Yes	Yes	No	Yes	8	5
Sundasciurus brookei	Brooke's Squirrel	632459	Yes	Yes	Yes	Yes	Yes	Yes	Yes	Yes	Yes	Yes	Yes	Yes	Yes	Yes	13	0
Sundasciurus davensis	Davao Squirrel	632460	Yes	Yes	Yes	No	No	Yes	Yes	No	Yes	Yes	Yes	Yes	Yes	Yes	10	3
Sundasciurus everetti	Bornean Mountain Ground Squirrel	NA	No	No	No	No	No	No	No	No	Yes	Yes	No	Yes	No	Yes	4	9
Sundasciurus fraterculus	Fraternal Squirrel	632473	Yes	Yes	Yes	Yes	Yes	Yes	Yes	Yes	Yes	Yes	Yes	Yes	Yes	Yes	13	0
Sundasciurus hippurus	Horse-tailed Squirrel	632461	Yes	Yes	Yes	Yes	Yes	Yes	Yes	Yes	Yes	Yes	Yes	Yes	Yes	Yes	13	0
Sundasciurus hoogstraali	Busuanga Squirrel	632462	Yes	Yes	Yes	Yes	Yes	Yes	Yes	Yes	Yes	Yes	Yes	Yes	Yes	Yes	13	0
Sundasciurus jentinki	Jentink's Squirrel	632463	Yes	Yes	Yes	Yes	Yes	Yes	Yes	Yes	Yes	Yes	Yes	Yes	Yes	Yes	13	0
Sundasciurus juvencus	Northern Palawan Tree Squirrel	632464	Yes	Yes	Yes	Yes	Yes	Yes	Yes	Yes	Yes	Yes	Yes	Yes	Yes	Yes	13	0
Sundasciurus lowii	Low's Squirrel	632465	Yes	Yes	Yes	Yes	Yes	Yes	Yes	Yes	Yes	Yes	Yes	Yes	Yes	Yes	13	0
Sundasciurus mindanensis	Mindanao Squirrel	632466	Yes	Yes	Yes	Yes	Yes	Yes	Yes	Yes	Yes	Yes	Yes	Yes	Yes	Yes	13	0
Sundasciurus moellendorffi	Culion Tree Squirrel	632467	Yes	Yes	Yes	No	Yes	Yes	Yes	Yes	Yes	Yes	Yes	Yes	Yes	Yes	12	1
Sundasciurus natunensis	Natuna Squirrel	NA	No	No	No	No	No	No	No	No	No	Yes	No	No	No	No	1	12

(continued)

Table 1.1 (continued)

Squirrel species list			Presence of the according squirrel species in the species list of the corresponding agency															
Scientific name	Common name/ Alternative names	TSN (Taxonomic Serial Number)	ITIS	GBIF	Encyclopedia of Life (EOL)	MA-NIS (VertNet)	GenBank (NCBI)	IUCN Red List	IUCN Red List (old)	IDigBIO	iNaturalist	Mammal Diversity Database (MDD)	Squirrels of the world (book)	The handbook of the mammals of the world (book)	Squirrels-The Animal Answer Guide	Illustrated Checklist of the Mammals of the World	"Yes" count	"No" count
Sundasciurus philippinensis	Philippine Tree Squirrel	632468	Yes	Yes	Yes	Yes	Yes	Yes	Yes	Yes	Yes	Yes	Yes	Yes	Yes	Yes	13	0
Sundasciurus rabori	Palawan Montane Squirrel	632469	Yes	Yes	Yes	Yes	Yes	Yes	Yes	Yes	Yes	Yes	Yes	Yes	Yes	Yes	13	0
Sundasciurus robinsoni	Robinson's Squirrel	NA	No	No	No	No	Yes	No	No	No	No	Yes	No	No	No	No	2	11
Sundasciurus samarensis	Samar Squirrel	632470	Yes	Yes	Yes	Yes	Yes	Yes	Yes	Yes	Yes	Yes	Yes	Yes	Yes	Yes	13	0
Sundasciurus steerii	Southern Palawan Tree Squirrel	632471	Yes	Yes	Yes	Yes	Yes	Yes	Yes	Yes	Yes	Yes	Yes	Yes	Yes	Yes	13	0
Sundasciurus tahan	Upland Squirrel	930332	Yes	Yes	Yes	No	Yes	No	Yes	Yes	No	Yes	Yes	Yes	No	Yes	9	4
Sundasciurus tenuis	Slender Squirrel	632472	Yes	Yes	Yes	Yes	Yes	Yes	Yes	Yes	Yes	Yes	Yes	Yes	Yes	Yes	13	0
Syntheosciurus brochus	Bangs's Mountain Squirrel	632474	Yes	Yes	Yes	No	Yes	Yes	Yes	Yes	Yes	Yes	Yes	Yes	Yes	Yes	12	1
Syntheosciurus poasensis	NA	NA	No	No	No	No	No	No	No	Yes	No	No	No	No	No	No	1	12
Tamias alpinus	Alpine Chipmunk	180189	Yes	Yes	Yes	Yes	Yes	No	Yes	Yes	No	No	Yes	Yes	Yes	No	9	4
Tamias amoenus	Yellow-pine Chipmunk	180190	Yes	Yes	Yes	Yes	Yes	No	Yes	Yes	No	No	Yes	Yes	Yes	No	9	4
Tamias ater	NA	NA	No	No	No	Yes	No	No	No	No	No	No	No	No	No	No	1	12
Tamias bulleri	Buller's Chipmunk	632475	Yes	Yes	Yes	Yes	Yes	No	Yes	Yes	No	No	Yes	Yes	Yes	No	9	4
Tamias canipes	Gray-footed Chipmunk	180191	Yes	Yes	Yes	Yes	Yes	No	Yes	Yes	No	No	Yes	Yes	Yes	No	9	4

Tamias cinereicollis	Gray-collared Chipmunk	180192	Yes	Yes	Yes	Yes	Yes	No	Yes	Yes	No	No	Yes	Yes	Yes	No	9	4
Tamias dorsalis	Cliff Chipmunk	180193	Yes	Yes	Yes	Yes	Yes	No	Yes	Yes	No	No	Yes	Yes	Yes	No	9	4
Tamias durangae	Durango Chipmunk	632476	Yes	Yes	Yes	Yes	Yes	No	Yes	Yes	No	No	Yes	Yes	Yes	No	9	4
Tamias merriami	Merriam's Chipmunk	180194	Yes	Yes	Yes	Yes	Yes	No	Yes	Yes	No	No	Yes	Yes	Yes	No	9	4
Tamias minimus	Least Chipmunk	180195	Yes	Yes	Yes	Yes	Yes	No	Yes	Yes	No	No	Yes	Yes	Yes	No	9	4
Tamias obscurus	California Chipmunk	180196	Yes	Yes	Yes	Yes	Yes	No	Yes	Yes	No	No	Yes	Yes	Yes	No	9	4
Tamias ochrogenys	Yellow-cheeked Chipmunk	180197	Yes	Yes	Yes	Yes	Yes	No	Yes	Yes	No	No	Yes	Yes	Yes	No	9	4
Tamias palmeri	Palmer's Chipmunk	180198	Yes	Yes	Yes	Yes	Yes	No	Yes	Yes	No	No	Yes	Yes	Yes	No	9	4
Tamias panamintinus	Panamint Chipmunk	180199	Yes	Yes	Yes	Yes	Yes	No	Yes	Yes	No	No	Yes	Yes	Yes	No	9	4
Tamias quadrimaculatus	Long-eared Chipmunk	180200	Yes	Yes	Yes	Yes	Yes	No	Yes	Yes	No	No	Yes	Yes	Yes	No	9	4
Tamias quadrivittatus	Colorado Chipmunk	180201	Yes	Yes	Yes	Yes	Yes	No	Yes	Yes	No	No	Yes	Yes	Yes	No	9	4
Tamias ruficaudus	Red-tailed Chipmunk	180202	Yes	Yes	Yes	Yes	Yes	No	Yes	Yes	No	No	Yes	Yes	Yes	No	9	4
Tamias rufus	Hopi Chipmunk	552503	Yes	Yes	Yes	Yes	Yes	No	Yes	Yes	No	No	Yes	Yes	Yes	No	9	4
Tamias senex	Shadow Chipmunk	180203	Yes	Yes	Yes	Yes	Yes	No	Yes	Yes	No	No	Yes	Yes	Yes	No	9	4
Tamias sibiricus	Siberian Chipmunk	632477	Yes	Yes	Yes	Yes	Yes	No	Yes	Yes	No	No	Yes	Yes	Yes	No	9	4
Tamias siskiyou	Siskiyou Chipmunk	180204	Yes	Yes	Yes	Yes	Yes	No	Yes	Yes	No	No	Yes	Yes	Yes	No	9	4
Tamias sonomae	Sonoma Chipmunk	180205	Yes	Yes	Yes	Yes	Yes	No	Yes	Yes	No	No	Yes	Yes	Yes	No	9	4
Tamias speciosus	Lodgepole Chipmunk	180206	Yes	Yes	Yes	Yes	Yes	No	Yes	Yes	No	No	Yes	Yes	Yes	No	9	4
Tamias striatus	Eastern Chipmunk	180207	Yes	Yes	Yes	Yes	Yes	Yes	Yes	Yes	Yes	Yes	Yes	Yes	Yes	Yes	13	0
Tamias townsendii	Town-send's Chipmunk	180208	Yes	Yes	Yes	No	Yes	No	Yes	Yes	No	No	Yes	Yes	Yes	No	8	5

(continued)

Table 1.1 (continued)

Squirrel species list			Presence of the according squirrel species in the species list of the corresponding agency															
Scientific name	Common name/ Alternative names	TSN (Taxonomic Serial Number)	ITIS	GBIF	Encyclopedia of Life (EOL)	MA-NIS (VertNet)	GenBank (NCBI)	IUCN Red List	IUCN Red List (old)	IDigBIO	iNaturalist	Mammal Diversity Database (MDD)	Squirrels of the world (book)	The handbook of the mammals of the world (book)	Squirrels-The Animal Answer Guide	Illustrated Checklist of the Mammals of the World	"Yes" count	"No" count
Tamias umbrinus	Uinta Chipmunk	180209	Yes	Yes	Yes	Yes	Yes	No	Yes	Yes	No	No	Yes	Yes	Yes	No	9	4
Tamiasciurus douglasii	Douglas's Squirrel	180167	Yes	Yes	Yes	Yes	Yes	Yes	Yes	Yes	Yes	Yes	Yes	Yes	Yes	Yes	13	0
Tamiasciurus fremonti	Fremont's Squirrel	NA	No	No	No	Yes	No	No	No	No	No	Yes	No	Yes	No	Yes	4	9
Tamiasciurus hudsonicus	Red Squirrel	180166	Yes	Yes	Yes	Yes	Yes	Yes	Yes	Yes	Yes	Yes	Yes	Yes	Yes	Yes	13	0
Tamiasciurus mearnsi	Mearns's Squirrel	632478	Yes	Yes	Yes	Yes	Yes	Yes	Yes	Yes	No	No	Yes	No	Yes	No	9	4
Tamiops maritimus	Maritime Striped Squirrel	632479	Yes	Yes	Yes	No	Yes	Yes	Yes	Yes	Yes	Yes	Yes	Yes	Yes	Yes	12	1
Tamiops mcclellandii	Himala-yan Striped Squirrel	930264	Yes	Yes	Yes	Yes	Yes	Yes	Yes	Yes	Yes	Yes	Yes	Yes	Yes	Yes	13	0
Tamiops rodolphii	Cambodian Striped Squirrel	930265	Yes	Yes	Yes	Yes	Yes	Yes	Yes	Yes	Yes	Yes	Yes	Yes	Yes	Yes	13	0
Tamiops swinhoei	Swinhoe's Striped Squirrel	632482	Yes	Yes	Yes	Yes	Yes	Yes	Yes	Yes	Yes	Yes	Yes	Yes	Yes	Yes	13	0
Thomomys talpoides	NA	NA	No	No	No	Yes	No	No	No	No	No	No	No	No	No	No	1	12
Trogopterus xanthipes	Complex-Toothed Flying Squirrel	632527	Yes	Yes	Yes	Yes	Yes	Yes	Yes	Yes	Yes	Yes	Yes	Yes	Yes	Yes	13	0
Urocitellus armatus	Uinta Ground Squirrel	930314	Yes	Yes	Yes	No	Yes	Yes	Yes	Yes	Yes	Yes	Yes	Yes	No	Yes	11	2
Urocitellus beldingi	Belding's Ground Squirrel	930315	Yes	Yes	Yes	No	Yes	Yes	Yes	Yes	Yes	Yes	Yes	Yes	No	Yes	11	2
Urocitellus brunneus	Idaho Ground Squirrel	930316	Yes	Yes	Yes	Yes	Yes	Yes	Yes	Yes	Yes	Yes	Yes	Yes	No	Yes	12	1

Urocitellus canus	Merriam's Ground Squirrel	930317	Yes	Yes	Yes	Yes	Yes	Yes	Yes	Yes	Yes	Yes	Yes	Yes	No	Yes	12	1
Urocitellus columbianus	Colum-bian Ground Squirrel	930318	Yes	Yes	Yes	Yes	Yes	Yes	Yes	Yes	Yes	Yes	Yes	Yes	No	Yes	12	1
Urocitellus elegans	Wyoming Ground Squirrel	930319	Yes	Yes	Yes	Yes	Yes	Yes	Yes	Yes	Yes	Yes	Yes	Yes	No	Yes	12	1
Urocitellus endemicus	Southern Idaho Ground Squirrel	NA	No	No	No	No	No	Yes	No	No	No	No	No	No	No	No	1	12
Urocitellus mollis	Piute Ground Squirrel	930320	Yes	Yes	Yes	No	Yes	Yes	Yes	Yes	Yes	Yes	Yes	Yes	No	Yes	11	2
Urocitellus nancyae	Town-send's Ground Squirrel	NA	No	No	No	No	No	Yes	No	No	No	No	No	No	No	No	1	12
Urocitellus parryii	Arctic Ground Squirrel	180135	Yes	Yes	Yes	No	Yes	Yes	Yes	Yes	Yes	Yes	Yes	Yes	No	Yes	11	2
Urocitellus richardsonii	Rich-ardson's Ground Squirrel	930322	Yes	Yes	Yes	No	Yes	Yes	Yes	Yes	Yes	Yes	Yes	Yes	No	Yes	11	2
Urocitellus townsendii	Town-send's Ground Squirrel	930323	Yes	Yes	Yes	Yes	Yes	Yes	Yes	Yes	Yes	Yes	Yes	Yes	No	Yes	12	1
Urocitellus undulatus	Long-tailed Ground Squirrel	930324	Yes	Yes	Yes	No	Yes	Yes	Yes	Yes	Yes	Yes	Yes	Yes	No	Yes	11	2
Urocitellus washingtoni	Washing-ton Ground Squirrel	930325	Yes	Yes	Yes	No	Yes	Yes	Yes	Yes	Yes	Yes	Yes	Yes	No	Yes	11	2
Xerospermophilus mohavensis	Mohave Ground Squirrel	930310	Yes	Yes	Yes	No	Yes	Yes	Yes	Yes	Yes	Yes	Yes	Yes	No	Yes	11	2
Xerospermophilus perotensis	Perote Ground Squirrel	930311	Yes	Yes	Yes	Yes	Yes	Yes	Yes	Yes	Yes	Yes	Yes	Yes	No	Yes	12	1

(continued)

Table 1.1 (continued)

Squirrel species list			Presence of the according squirrel species in the species list of the corresponding agency															
Scientific name	Common name/ Alternative names	TSN (Taxonomic Serial Number)	ITIS	GBIF	Encyclopedia of Life (EOL)	MA-NIS (VertNet)	GenBank (NCBI)	IUCN Red List	IUCN Red List (old)	IDigBIO	iNaturalist	Mammal Diversity Database (MDD)	Squirrels of the world (book)	The handbook of the mammals of the world (book)	Squirrels-The Animal Answer Guide	Illustrated Checklist of the Mammals of the World	"Yes" count	"No" count
Xerospermophilus spilosoma	Spotted Ground Squirrel	930312	Yes	Yes	Yes	No	Yes	Yes	Yes	Yes	Yes	Yes	Yes	Yes	No	Yes	11	2
Xerospermophilus tereticaudus	Round-tailed Ground Squirrel	930313	Yes	Yes	Yes	No	Yes	Yes	Yes	Yes	Yes	Yes	Yes	Yes	No	Yes	11	2
Xerus erythropus	Striped Ground Squirrel	632483	Yes	Yes	Yes	Yes	Yes	Yes	Yes	Yes	No	No	Yes	No	Yes	No	9	4
Xerus inauris	South African Ground Squirrel	632484	Yes	Yes	Yes	Yes	Yes	Yes	Yes	Yes	No	No	Yes	No	Yes	No	9	4
Xerus princeps	Damara Ground Squirrel	632485	Yes	Yes	Yes	No	Yes	Yes	Yes	Yes	No	No	Yes	No	Yes	No	8	5
Xerus rutilus	Unstriped Ground Squirrel	632486	Yes	Yes	Yes	Yes	Yes	Yes	Yes	Yes	Yes	Yes	Yes	Yes	Yes	Yes	13	0
Sum of "Yes"			286	285	286	263	250	285	286	284	286	304	285	292	278	302		

Blue text indicates hyperlinks to the institution-specific website for the taxonomic records the latter part of the definition is not true for all squirrels (e.g. ground squirrels). Not only the general definition shows flaws, but there are also major discrepancies among institutions (see Tables 1.1 and 1.2 for suggested adjustments). Here in this study, and all other chapters of this book, we utilize ITIS.gov as the taxonomic reference, as it is believed to be the global leader in species taxonomy.

Scientifically, squirrels are species belonging to the family Sciuridae, these species can be roughly classified into tree squirrels, ground squirrels, chipmunks, marmot & groundhogs, flying squirrels, and prairie dogs (Thorington et al. 2005). These species' weight ranges from 12–26 g (see African pygmy squirrel (*Myosciurus pumilio*) and Least pygmy squirrel (*Exilisciurus exilis*)) to more than 8 kg (several marmot species (*Marmota spp.*)) (Kryštufek and Vohralík 2013), with body sizes between 10 cm and 14 cm (African pygmy squirrel and Least pygmy squirrel) and 1.3 m (see Bhutan giant flying squirrel (*Petaurista nobilis*)) (Choudhury 2002; Jackson 2012; Payne et al. 1985; Thorington et al. 2012). These small mammal species occur naturally on all continents except Antarctica and Australia & Oceania and inhabit almost all global ecosystems (Bertolino and Lurz 2013; Thorington et al. 2012), but with some notable exceptions such as Iceland, Greenland, and Papua New Guinea for instance.

1.1.1 Why Study and Manage Squirrels for Conservation and beyond?

Here we try to present a state-of-the-art conservation assessment, its tools, and data with applications on squirrels (Family Sciuridae) as a generic species model example. We hope that with the methods presented here, we can eventually reach beyond 'just' squirrels. But one might wonder, why squirrels to start out with? To answer this question thoroughly, it has several societal, ecological, and economic reasons, which we discuss next.

Firstly, and with crucial importance, squirrels naturally occur on every continent except Antarctica and Australia & Oceania and are therefore one of the most commonly known small mammals around the globe (Mendes et al. 2019). They are part of a public image and narrative (typical example found with e.g. Walt Disney's Flora and Ulysees – Wikipedia 2021). The fact that the majority of the global society is aware of the existence of this animal family facilitates the understanding of research findings from the public, and the general "caring" of the society is assumingly greater. It is commonly known that conservation activities are better supported for broadly known and often charismatic animal species (Ducarme et al. 2013; Skibins et al. 2013). Additionally, many people feel easily connected with squirrels as they might have had their own special encounter with them, up close and personal, outdoors in nature, or near the front door. Squirrels can easily be observed from your armchair at home! However, despite all the public awareness of this diverse family, the conservation for them is still widely lagging behind, hardly any legal text is specifically devoted to them, anywhere (Steiner and Huettmann 2021). Convictions or fines paid, related to squirrels are virtually absent, law enforcement is minuscule, and associated budgets (if any available) are almost negligible (see Chap. 11 in Steiner and Huettmann 2023). These situations and statements will be strongly supported by modern research findings shown throughout this book, where we also present the latest conservation assessment tools. These assessment tools aim to identify and address this widely missing conservation attention for animal species and families such as squirrels, with the ultimate goal to improve their conservation approach and preserve these species and their habitats for future generations.

Secondly, the reason why we chose squirrels as the model species group for this series of studies is that they have a highly undervalued ecological impact. Squirrels are meso-predators and eco-engineers after all; they operate on a global scale (Dalton 2017; Ewacha et al. 2016). Every child knows the story of the squirrel gathering nuts in autumn and burying them for the winter reserve (see books like Potter 2002); it is a classic but widely underestimated example of seed dispersal crucial for any forest landscape! And as the squirrel might forget some nuts/seeds in the 'cache' or does not consume all its stored treasures, the seeds and nuts turn into seedlings and finally into new trees. This is not only a children's story but also an important event observed in nature of course (Smith 1970). And this is not the only ecological function of squirrels. Other important characteristics of squirrels include their predation on Songbirds (*Passeriformes*), Hares (*Lepus*), Chipmunks (*Tamias*), Lizards (*Lacertilia*), Mice (*Mus*), Rats (*Rattus*), Turkeys (*Meleagris*), Snakes (*Serpentes*), Crabs (*Brachyura*), Frogs (*Anura*), Salamanders (*Urodela*) and other conspecifics (Squirrels (*Sciuridae*)) (Callahan 1993). Squirrels are known to participate in infanticide making them a hot topic of research on social mammals (Dobson 1990; Ebensperger and Blumstein 2007; Trulio 1996). Additionally, it has widely been shown that squirrels specifically learn to prey on avian nests (Pelech et al. 2010), and Lemmings (*Dicrostonyx kilangmiutak*) (Boonstra et al. 1990), which makes them an important but widely overlooked mesopredator included in many ecosystems worldwide. Squirrels are however also the prey of many other species such as the Northern goshawk (*Accipiter gentilis*), Cooper's Hawk (*A. cooperii*), Sharp-shinned Hawk (*A. striatus*), Red-tailed Hawk (*Buteo jamaicensis*), Mexican spotted Owl (*Strix occidentalis lucida*), Great-horned Owl (*Bubo virginianus*). Further, several mammal predators such as Coyote (*Canis latrans*), Wolf (*C. lupus*), Lynx (*Lynx canadensis*), Eurasian lynx (*Lynx lynx*) (Bizhanova et al. 2022), Bobcat (*L. rufus*), Gray Fox (*Urocyon cinereoargen*) (Schauffert et al. 2002), as well as Wild Boar (*Sus scrofa*) (Keuling et al. 2018), stray cats (Bertram and Moltu 1986), and dogs (Benson 2013) are known to prey on squirrels. All these prey-predator interactions between squirrels and many different species and families seem to show the holistic importance of squirrels for global ecosystems and food systems. It is noteworthy that the upper-mentioned predator and prey species of squirrels are mostly valid for North America. To include global predator and prey species of squirrels, many more species must be considered. An additional and even more relevant aspect of this food web is that squirrels also serve as a reservoir for zoonotic diseases (Lane et al. 2005; Salkeld et al. 2008, Steiner and Huettmann 2021, 2023 – Chap. 5).

The third major reason for using squirrels as the study family is their global economic importance and their contribution to human consumption. As mentioned above and presented in Steiner and Huettmann 2023 (Chap. 3), squirrels are ubiquitous. This means that many countries and nations benefit financially from squirrels, as squirrel kills are often sold to/by hunters, furbearers, trappers, or similar, if the local authority even considers them as valuable to put a pricetag on them, and not just give free killing rights with no limits to "everyone" (see Chap. 11 – Steiner and Huettmann 2023). Additionally, squirrels are often consumed by us humans (Ahmadi et al. 2018; Beard 2009; Fa et al. 2002 & see a first-person story from a bush hunter of Sierra Leone (West Africa) in Appendix 11.3 in Steiner and Huettmann 2023). Squirrels are in many regions of the world considered "bush meat" and are often also sold as such on markets (Ahmadi et al. 2018; Fa et al. 2002; Sanamxay et al. 2013). It is a vast grey economy hardly fathomed yet for its importance, but it certainly puts much food on the table for rural people. In addition to that, their fur is also frequently sold commercially (Garofalo et al. 2018; Schlesinger 2017); typical examples as shown in Alaska. In order for squirrel populations to sustain all this human pressure, which is not always legal (Vanitharani 2018), and certainly not managed really, or accounted for, the situation certainly requires strict control of conservation managing agencies and similar. If the conservation for these hunted squirrels is not fully intact, it will not be possible to hunt these species in the future as they can easily get overhunted and extinct in the wild, all while 39% of the population trends are unknown and 25% are decreasing – see Results Section below. It is a *laissez-faire* approach and thus not the best professional practice in the year 2023. These conservation actions are especially important for squirrel species that are hunted intensively (e.g. Jordan 1971; Kline 1964; Koli et al. 2013), even though their population trend is unknown (e.g. Hairy-footed Gliding Squirrel (*Belomys pearsonii*), Mishmi Hill Giant Gliding Squirrel (*Petaurista mishmiensis*)) or even confirmed to be declining (e.g. Namdapha Flying Squirrel (*Biswamoyopterus biswasi*), Hodgson's Giant Gliding Squirrel (*Petaurista magnificus*), Bhutan Giant Gliding Squirrel (*Petaurista nobilis*), Indian Giant Gliding Squirrel (*Petaurista philippensis*), and Red Giant Gliding Squirrel (*Petaurista petaurista*)) (Krishna et al. 2016; Sanamxay et al. 2013).

The upper-mentioned economic importance is also originating from the global funds invested in conservation activities as these are for some countries also quite significant (see an example for GDP and AID funds distribution on a global scale – Wikipedia 2022), however, even though the GDP is somewhat rising, squirrels are hardly ever considered, and hardly benefitting (see Steiner and Huettmann 2023 – Chaps. 10 and 14 for details; Rich 2014).

1.2 A Brief History of the Squirrel's Taxonomy.

During the early days of western species classification in the 1700s, several squirrel species have already been 'officially' described back then. Common examples of this include the species of Southern Flying Squirrel (*Glaucomys volans*), Mentawai Flying Squirrel (*Iomys sipora*), Woodchuck (*Marmota monax*), Sri Lankan Giant Squirrel (*Ratufa macroura*), Siberian Chipmunk (*Tamias sibiricus*), and Eastern Chipmunk (*Tamias striatus*). The reason for mentioning this in a conservation book is because most of these species' classifications from almost 280–330 years ago have not been revised thoroughly ever since, while science, genetic research, and ecological niche information have vastly progressed e.g. Steiner and Huettmann (2021) for overview. This underlines the fact that many of these classifications are by now outdated and obsolete as many modern taxonomic investigation tools have not been used centuries ago (genetic analysis, machine learning (ML), AI, internet, etc.). The fact that those taxonomies are based on Latin and ancient Greek, while the official language of science is now 'English' adds to the injury.

Additionally, in recent years, many new taxonomic changes have been proposed (e.g. for Espiritu Santo Island Antelope Squirrel (*Ammospermophilus insularis*), Mount Gaoligong Flying Squirrel (*Biswamoyopterus gaoligongensis*), Laotian Giant Flying Squirrel (*Biswamoyopterus laoensis*), Hon Khoai Squirrel (*Callosciurus honkhoaiensis*), Dusky Striped Squirrel (*Funambulus obscurus*), Yunnan Giant Flying Squirrel (*Petaurista yunanensis*), Taurus Ground Squirrel (*Spermophilus taurensis*), Sumatran Mountain Squirrel (*Sundasciurus altitudinis*), Bornean Mountain Ground Squirrel (*Sundasciurus everetti*), and Upland Squirrel (*Sundasciurus tahan*) – see Table 1.3). These proposed changes can easily lead to taxonomic chaos if they are not accepted mutually among all major institutions or updated by a global squirrel authority. Thus, this *'laissez-faire'* attitude can easily lead to major taxonomic discrepancies and conservation management problems (Steiner and Huettmann 2021). Therefore, it is of crucial importance for global institutions to keep up with modern research and to keep their taxonomic species lists as accurate and up-to-date as possible. These latter aspects are necessary for all global species, especially small mammals like squirrels, which is the reason why we think squirrels are a good role model for such questions and studies for moving forward in an alternative but more appropriate directions.

The book "*Illustrated Checklist of the mammals of the world*" (Burgin et al. 2020), introduces itself with a great overview of how taxonomy developed in the last 250+ years. It presents the initial boom of the continuously increased number of described species and genera which greatly contributed to the understanding and the taxonomic classification of our world's mammals. It seems that the number of described species increased drastically and disproportionally over the last years, which however did not happen steadily. The initial boom eventually flattened off quite a bit until the discovery of genetic studies for uses in the identification of the origin of species (early 1900s) (Griffiths et al. 2005). This new era where the species' kinship among the species of interest is studied by identifying the genetic distance/difference between the species at stake has had an immense impact on the global species taxonomy, while the conservation arguably got worse. With these new methods which allow scientists to identify how related species and subspecies are to each other, many studied species, and especially subspecies are proposed to be reclassified. Examples from the Sciuridae family are Humboldt's Flying Squirrel (*Glaucomys oregonensis*), Hainan Flying Squirrel (*Hylopetes electilis*), Rio Grande Ground Squirrel (*Ictidomys parvidens*), Forest Steppe Marmot (*Marmota kastschenkoi*), Sierra del Carmen Chipmunk (*Neotamias solivagus*), Gray-headed Flying Squirrel (*Petaurista caniceps*), Formosan Giant Flying Squirrel (*Petaurista grandis*), Hainan Giant Flying Squirrel (*Petaurista hainana*), Taiwan Giant Flying Squirrel (*Petaurista lena*), Indochinese Giant Flying Squirrel (*Petaurista marica*), Mechuka Giant Flying Squirrel (*Petaurista mechukaensis*), Mishmi Hills Giant Flying Squirrel (*Petaurista mishmiensis*), Mebo Giant Flying Squirrel (*Petaurista siangensis*), Chindwin Giant Flying Squirrel (*Petaurista sybilla*), Calabrian Black Squirrel (*Sciurus meridionalis*), Tien Shan Ground Squirrel (*Spermophilus nilkaensis*), Natuna Squirrel (*Sundasciurus natunensis*), Robinson's Squirrel (*Sundasciurus robinsoni*), Fremont's Squirrel (*Tamiasciurus fremonti*), and Mearns's Squirrel (*Tamiasciurus mearnsi*) (see Table 1.3 for details). Subspecies have been underlined because in recent years many subspecies have been genetically studied and proposed to be elevated to the species level. Examples from the Sciuridae family are mentioned above.

This mentioned new genetic approach of the industrial society to taxonomy has however also its downsides. A common pitfall of such genetic analyses is that their results are rarely verified and a large number of even highly-cited studies are found to be incorrect or erroneous (Kwon 2023). Updates and retractions are very common in this discipline. That's due to the fact when genetic evidence has been found so that the species/subspecies are distinct 'enough', it is often immediately considered to be a new/different species (e.g. Espiritu Santo Island Antelope Squirrel (*Ammospermophilus insularis*)) (Álvarez-Castañeda 2007). However, to ensure the most correct and valid taxonomic classifications, several other aspects, apart from genetics, should likely be included and considered as well, e.g. morphology, history, ecology, and behavior (Steiner and Huettmann 2021). This has been beautifully described as:

> Current taxonomy focuses on multicharacter integrative approaches, considering all potentially useful sources of information provided by the various fields of biology – by Ohl (2007).

An additional aspect that seems important to mention in this context is the correlation between the taxonomy of species occurring in a certain region of the world and the taxonomic approach of this region, or how it has been influenced by it. To be more concrete, different regions around the world have a different approach to taxonomy than others and this can still be recognized in modern taxonomy. It is broadly acknowledged that Europe has an approach based on the early European taxonomic pioneers (e.g. Carl Linnaeus, Michael Rogers, and Thomas Oldfield) (Brockway 1979; Burgin et al. 2020). This is the same for many so-called colonial nations but is just slightly delayed (Brockway 1979; Figueiredo and Smith 2010). U.S.'s and subsequent Canada's taxonomic classifications are often DNA-driven (Hebert and Gregory 2005; see Pyle et al. 2021 for 'issues'), and China has been observed to look for independent solutions, and thereby an independent taxonomic classification for species gets applied (Jansen 2009). Like China being on the receiving end of western taxonomies, Latin America, Australia, and New Zealand have a different approach to taxonomy and even revised the EU-centric concept (Ebach and Michaux 2017). Overall one ends up with quite a chaotic mosaic of taxonomy, even for the same species. Whoever has the power sets the taxonomy, using their science and their concensus. And this does not appear to be changing any time soon!

1.3 Current Taxonomic Status and the Inherent Institutional Discrepancy and Funding Dilemma

As one of the major contributors to this chapter, this section discusses Table 1.1, which depicts a detailed assessment overview of all globally significant taxonomic institutions including Integrated Taxonomic Information System (ITIS), Global Biodiversity Information Facility (GBIF), Encyclopedia of Life (EOL), VertNet (Manis), GenBank (NCBI), The International Union for Conservation of Nature Red List of Threatened Species (IUCN Red List), Integrated Digitized Biocollections (IDigBIO), iNaturalist, and Mammal Diversity Database (MDD). Additionally, some broadly known mammal- and squirrel books are included Squirrels of the world (Thorington et al. 2012), The handbook of the mammals of the world (Wilson and Mittermeier 2011), Squirrels: the animal answer guide (Thorington and Ferrell 2006), and Illustrated Checklist of the Mammals of the World (Burgin et al. 2020), which include the squirrel's taxonomic classification as a reference.

This taxonomic species overview in Table 1.1 clearly depicts that there are still major discrepancies among the most acknowledged taxonomic institutions, differing by up to 54 species between the institution/book with the least accounted number of species (NCBI), and the highest (MDD) with 250 and 304 species respectively.

However, a closer look at Table 1.1 additionally reveals the completeness of each institutional-/books taxonomic list and it allows us to easily compare the different lists with each other. Table 1.1 also contains a species count at the very bottom to depict how many species are included in each institution's/book's taxonomic list. In addition, it also contains a species count on the far right of the table to facilitate the comparison of a species and its presence in the different institutions/books. These species counts illustrate how many institutes, and books agree on the taxonomy of one species and include it in their taxonomic species list. The species count on the far right of each row provides a first impression of the taxonomic discrepancies per species among institutes. Whereas the species count at the bottom of the table provides a holistic overview of all species included, and not, by the included institutions and books to underline the major differences among the number of accepted species.

1.3.1 Data Mining of Globally Distributed Species Including the Handling of Big Data

Table 1.1 has been created with the purpose to assess and illustrate the taxonomic discrepancies among the world's leading institutions and organizations with the addition of some taxonomically relevant books for squirrels. This table has been created by conducting extensive literature research and online data compilation. By using the open access online sources of taxonomic databases we sought to obtain the most modern and updated version of the databases from each included institution. Online databases can be continuously updated, enhancing possibilities to keep up with the continuous changes happening in taxonomy. This has a great advantage over books that commonly use a cut-off date for their included taxonomic classifications. An additional upside of an online taxonomic database is that it is commonly openly accessible and thus dynamic, in contrast to books.

However, this mentioned aspect of the continuous possibility to be up to date carries also a possible disadvantage with it, the online data is never final which can make it difficult for researchers and other people to take the final version from these websites. A great example can be seen in Table 1.1 for IUCN Red List, where one dataset has been obtained in mid-2020 (August 2020) and another dataset has been obtained in early 2021 (25 March 2021). As can be observed in the two lists, the taxonomy of 31 species changed already just in a period of a few months (see Table 1.1). Possible solutions for this are presented in the final section of this chapter which focuses on presenting solutions to overcome these kinds of issues.

The individual online taxonomic databases can be accessed via the following URLs (www.itis.gov, www.gbif.org, www.eol.org, www.vertnet.org, www.ncbi.nlm.nih.gov, www.iucnredlist.org, www.idigbio.org, www.inaturalist.org, and www.mammaldiversity.org), with the specific direction to the squirrel databases via the hyperlinks in Table 1.1 (online version).

As taxonomic data has been collected for around 285 species for each institution, this included many times the handling of big data sets. A great example of large data sets for taxonomic databases is VertNet with its Mammalia dataset (https://doi.org/10.7946/P2TG68) (with 120+ million excel cells containing taxonomic information). As it is not possible to obtain only the taxonomic information from only squirrels in some cases, the data must first be filtered out of the large data sets. The most important aspect of data filtering is that it is done correctly and that no valuable data is lost. Therefore, it must be assured that also misspellings or alternative forms of writing are considered and not lost during the filtering process.

Consequently, there are several advantages and disadvantages of big datasets. The greatest advantage of big datasets is that they contain more data which in many cases provides more information. Besides that, a large dataset with multiple variables has a greater audience, and subsequent impact, than very small and specific datasets (Junqué de Fortuny et al. 2013).

The major disadvantages of large datasets are that they require a great amount of storage, and they are slower handled by computers and computer models/maps. Also, once data sets get very big it is hardly possible to check every single record, which can result in a dataset with hidden incorrect data, which might affect the outcome for which the data has been used. The pro side is, that when approaches such as data mining are used, a small amounts of incorrect data are not a big deal, as with machine learning these incorrect data are often correctly handled and discarded anyway.

An alternative to big downloadable datasets is using online-only databases which allow an interactive overview of the species/family of interest. These online databases are also known as web-based electronic catalogs (ELECATs) (Chavan et al. 2005). A great example of this is www.ITIS.gov. This type of database usually provides a pretty good overview of the species/genus/family of interest, with the only downside of its impractical use in case the dataset wants to be used for future and outside applications (e.g. for Species Distribution Models – SDMs), as it cannot just be downloaded from the web portal.

Combining the downloadable databases with the online interactive ones can provide a good all-around experience and application if the interface is easy to use and the data is presented transparently. A typical example of these hybrid databases is the IUCN Red List, which however lacks a great overview possibility for multiple species/whole families. It has no ISO-compliant metadata either. The IUCN Red List additionally lacks public access to all of the underlying mammalian distribution point data (which includes squirrels), hindering the assessment and mapping process of these species and making it impossible to actually verify the correctness and precision of the provided range data (polygons). (IUCN regulations cut-off date 10 April 2021). Transparent and repeatable science, and such decisions, are thereby not guaranteed whatsoever, unfortunately.

1.3.2 Completeness of Table 1.1 and the Continuous Change of a Highly Dynamic Taxonomy

With Table 1.1, we attempt to create the most complete taxonomic species overview for all global squirrel species to date. It has never been done before in such a comparative manner. By presenting the world's most significant taxonomic squirrel databases and their taxonomic lists in a very clean and comparative way, it aims to support the comparison between some of the major databases available for squirrels. With this approach, the authors support the development of an improved taxonomy, being a discipline where problems cannot suddenly be completely erased but the continuous endorsement of crucial changes may have a significant impact long-term. Thus far, a true oversight there is missing, nor mutually agreed upon (e.g. see D'Elía et al. 2019).

The institutions and books used to compile Table 1.1 are not all solely focused on taxonomy. Some of them also focus mainly on other aspects, where taxonomy is included in order to assign the data to a certain species (see details in Sect. 1.4). Because these institutions/books do not all solely focus on taxonomy, it might also not be the focus of the institution to be as up-to-date as possible. However, it then becomes the foundation of their underlying actions. Some institutions make this very clear by placing a disclaimer on their data, stating that this data should not be taken as a taxonomic reference (e.g. NCBI – GenBank). However, we consider it still crucial to remain current with the latest updates and when data are readily available. This latter aspect is discussed in more detail in Sect. 1.4.

1.4 The Importance of Taxonomic Correctness for Successful Species Conservation to Convince in the Public Eye

Taxonomy is widely promoted as the critical basis of any scientific approach to studying the biotic world (e.g. Vane-Wright et al. 1991). What this means is that for every scientific study it must be defined to which biotic species, genus, or family the study is applicable. However, if these taxonomic classifications are obsolete, outdated, incorrect, or even non-existing it might seem obvious that this is a frequent source of occurring errors or mistakes. These mistakes are subsequently extrapolated into all different fields dealing with biotic species and successful management appears impossible, nor serious. It is hard to defend when public money is spent. A great example where these mistakes can be observed is in the discipline of ecology and specifically, conservation ecology and conservation policy (Dubois 2003). To better illustrate this, we provide two examples below:

(i) Example of the importance of a correct taxonomy for general biotic research

If highly valued research has been conducted with extensive funding behind, for one or multiple species (e.g. whole-genome sequencing (WGS) in GenBank -NCBI), and the species classification attributed to the studied species is incorrect or outdated, the research findings are to a great extend useless. This is because the new research findings cannot be correctly assigned to a valid classified species, making the research findings pretty useless – or wrong – for future studies. This issue becomes even more noticeable and problematic when taxonomic information is taken from one institution (e.g. ITIS) and modern research findings from another (e.g. GenBank) and the taxonomic classification then does not match and overline (Brehm et al. 2008).

(ii) Example of the importance of correct taxonomy for conservation ecology

No doubt, taxonomic discrepancies among institutions can have severe influences on biotic research. But these discrepancies might have an even larger impact on the species and their conservation itself. Imagine the following scenario:

An institution (A) classifies a species in a certain way, this same species is however classified differently by another institution (B). It might be called differently (scientific- and/or common name), belong to a different genus, have different allocated subspecies, split into two species, etc. (as we see all the time, here presented for squirrels). For this certain species, it has been identified that there is an extinction threat.

Conclusively, the government of the country in which this threatened species is occurring decides to initiate highly valuable conservation activities to avoid the extinction of this species. However, in case the governmental agency has based the conservation initiative on the taxonomy of a certain institution (A), that might differ from other institutions (B). This institution (B) names and classifies the species differently which can easily result in problems, these problems can subsequently even influence conservation activities as these conservation activities might even be meant for another species/subspecies. Additionally, conservation ecologists, citizens, and other conservation management respondents might use again different taxonomic institutions as a reference, leading to additional confusion as none really agree with each other. Besides the initial taxonomic confusion, the executing force often acts in different geographical regions (as different institutions often include different distribution ranges for species), and thus, might even be act on a different/ wrong species (as common names are often obsolete and confusing among different institutions and similar between species, see Table 1.1). It is a typical case for species occurring across borders and jurisdictions, with different languages for instance. On top of this, the species' subspecies might not be included in one institution's classification and therefore might lack action in the field where conservation activities have been initiated. Several more scenarios are possible which all can be caused by taxonomic discrepancies among different institutions, especially with outdated classifications. The following papers address some widely known conservation failures caused by taxonomic misalignments and discrepancies, in support of the statements above Dubois (2003), May (1990), and Raup (1995).

Possible solutions to diminish or even resolve these discrepancies among institutions are presented in the final section of this study (Sect. 1.7.1).

1.5 Genera and Species of Squirrels in Immediate Need of a Well-Thought-Out Taxonomic Update

This section aims to present all squirrel genera and species in immediate need of a taxonomic revision and update. The evaluation of which species/genera are included in these upcoming tables is based on Table 1.1. This was never been done before for squirrels and thus, it is important to present these species in this section as it calls for important changes and directly addresses the taxonomic issues.

1.5.1 Generic Genera and Species of Squirrels in Immediate Need of Taxonomic Updates/ Revisions

To introduce this issue, Table 1.2 addresses all species and genera of the family Sciuridae which are in immediate need of taxonomic revision/update.

Table 1.2 presents all global squirrel species with major discrepancies among the included authorities and institutions, calling for urgent taxonomic revision and updates. This table has been created by compiling the 44 most discrepant species from Table 1.1. Every species/genus which differs by ≥25% among the institutions/books is included in this table. This table aims to highlight the most discrepant species and to present them as crisp and clean as possible. This table should additionally act as a prioritization tool and subsequent call to action for squirrel taxonomy and its responsible entities. Thus far, it is

Table 1.2 Generic genera and species of squirrels in immediate need of taxonomic updates/revisions

Scientific name	Common name/ Alternative names	TSN (Taxonomic Serial Number – according to ITIS)	Percentage of acceptance	"Yes" count	"No" count	Reason for Discrepancy	Comments
Ammospermophilus insularis	Espiritu Santo Island Antelope Squirrel	632323	61.5	8	5	Species – subspecies level	This species is mentioned as subspecies of "*Ammospermophilus leucurus*" in Illustrated Checklist (book). Has been recently (2007-2008) studied genetically and apparently too small differences have been found for a distinct species. Thus, it was proposed to be a subspecies of its closest representative (*Ammospermophilus leucurus insularis*).
Biswamoyopterus gaoligongensis	Mount Gaoligong Flying Squirrel	NA	23.1	3	10	Species description too recent	Has been firstly described in 2019 and all *Biswamoyopterus* may be conspecific (Li et al. 2019)
Biswamoyopterus laoensis	Laotian Giant Flying Squirrel	930360	69.2	9	4	Species description too recent	Has been firstly described in September 2013 after discovering it on the bushmeat market in Laos (Li et al. 2019; Sanamxay et al. 2013).
Callosciurus honkhoaiensis	Hon Khoai Squirrel	NA	23.1	3	10	Species description too recent	Has been firstly described in 2018 and is not yet accepted mutually (Nguyen et al. 2018).
Dremomys everetti	Bornean Mountain Ground Squirrel	632341	76.9	10	3	Tautology species	Not accepted by the institutions that accept *Sundasciurus everetti.*
Eutamias sibiricus	Siberian Chipmunk	NA	30.8	4	9	Distribution interference	Scientific name from EOL is mentioned in brackets as old/alternative name for the genus "*Tamias*". Is the only member of the genus "*Tamias*" living in Europe/ Siberia and has been reclassified as "*Eutamias*".
Euxerus erythropus	Striped Ground Squirrel	NA	30.8	4	9	Genus – subgenus level	Genus is considered as a subgenus of "*Xerus"* from 1940 onwards. According to MDD, it moved from *Xerus* to *Euxerus* (Kryštufek et al. 2016).
Funambulus obscurus	Dusky Striped Squirrel	NA	23.1	3	10	Species – subspecies level	Species has been split from *F. sublineatus* according to MDD and Dissanayake and Oshida (2012).
Funisciurus duchaillui	Du Chaillu's Rope Squirrel	930333	69.2	9	4	Species – subspecies level	Species has been split from *F. isabella* according to MDD and Brugière et al. (2005).
Geosciurus inauris	South African Ground Squirrel	NA	30.8	4	9	Genus – subgenus level	The genus "*Geosciurus*" is by most institutions recognized as subgenus of "*Xerus*". According to MDD the species moved from *Xerus* to *Geosciurus* (Kryštufek et al. 2016).
Geosciurus princeps	Damara Ground Squirrel	NA	30.8	4	9	Genus – subgenus level	The genus "*Geosciurus*" is by most institutions recognized as subgenus of "*Xerus*". According to MDD the species moved from *Xerus* to *Geosciurus* (Kryštufek et al. 2016).

(continued)

Table 1.2 (continued)

Scientific name	Common name/ Alternative names	TSN (Taxonomic Serial Number – according to ITIS)	Percentage of acceptance	"Yes" count	"No" count	Reason for Discrepancy	Comments
Glaucomys oregonensis	Humboldt's Flying Squirrel	NA	30.8	4	9	Species – subspecies level	This is mentioned as subspecies of "*Glaucomys sabrinus*" in The mammals of the world (book). According to MDD and Arbogast et al. (2017) it has been split from *G. sabrinus.*
Hylopetes electilis	Hainan Flying Squirrel	930616	30.8	4	9	Species – subspecies level	According to MDD and Song & Fa-Hong (2013), the species has been split from *H. phayrei*. It is mentioned as species in IDigBIO, in other institutions (e.g. ITIS, GBIF, The mammals of the world (book)) it is mentioned as subspecies "*Hylopetes phayrei electilis*". TSN has been obtained from ITIS subspecies level.
Hylopetes lepidus	Gray-cheeked flying squirrel	NA	30.8	4	9	Common name	Same common name as "*Hylopetes platyurus*".
Hylopetes sagitta	Arrow Flying Squirrel	930343	69.2	9	4	Genus update, Common name	According to MDD and Jackson and Thorington Jr. (2012), it includes *lepidus* and moved from *Petinomys* to *Hylopetes*. Common name from The mammals of the world (book), Illustrated Checklist (book) is "Grey-cheeked Flying Squirrel".
Ictidomys parvidens	Rio Grande Ground Squirrel	930267	76.9	10	3	Genus update, Species – subspecies level	According to MDD and Helgen et al. (2009), it has been split from *I. mexicanus* and moved to *Ictidomys* from *Spermophilus*. Synonym of "*Spermophilus mexicanus parvidens*", accordingly to ITIS.gov. It is not listed on IUCN.org.
Marmota kastschenkoi	Forest Steppe Marmot	930334	53.8	7	6	Species – subspecies level	Is traditionally (until 1956) considered as subspecies of Gray marmot (*M. baibacina*) (Kryštufek and Vohralík 2013).
Microsciurus santanderensis	Santander Dwarf Squirrel	632390	76.9	10	3	Non-taxonomic institution discrepancy	This species is mostly not accepted by the institutions which do not focus on taxonomy (VertNet, GenBank (NCBI), IDigBIO). Additionally, its precise taxonomic allocation is not certain as recent revisions were not able to assign this species to the most suitable genus (MDD).
Neotamias	Chipmunks	NA	30.8	4	9	Genus – subgenus level	Scientific name from EOL is mentioned in brackets as old/alternative name for the genus "*Tamias*". Generally considered as subgenus of "*Tamias*". Moved from *Tamias* to *Neotamias* according to MDD and Patterson and Norris (2016).

(continued)

Table 1.2 (continued)

Scientific name	Common name/ Alternative names	TSN (Taxonomic Serial Number – according to ITIS)	Percentage of acceptance	"Yes" count	"No" count	Reason for Discrepancy	Comments
Otospermophilus beecheyi	California Ground Squirel	930301	76.9	10	3	Species description too recent	Synonym of "*Spermophilus beecheyi*", accordingly to ITIS. Formerly placed in *Spermophilus*, as *Spermophilus beecheyi*, it was reclassified as part of the genus *Otospermophilus* in 2009 as it became clear that *Spermophilus,* as previously defined, was not a natural (monophyletic) group (Helgen et al. 2009).
Petaurista albiventer	White-bellied Giant Flying Squirrel	930642	30.8	4	9	Species – subspecies level	Commonly seen as *Petaurista petaurista albiventer.* According to MDD, this genus and species are experiencing discrepancies and the final taxonomic classification has not yet been found (Li and Feng 2017).
Petaurista caniceps	Gray-headed Flying Squirrel	930328	61.5	8	5	Species – subspecies level, Common name	Common name from The mammals of the world (book), Illustrated Checklist (book) is "Grey-headed Giant Flying Squirrel". Synonym of "*Petaurista elegans caniceps*", accordingly to ITIS. In the past it was recognized as subspecies, in recent studies it has been reinforced to elevate it to the species level and thus, split it off from *P. elegans* (Li et al. 2013).
Petaurista hainana	Hainan Giant Flying Squirrel	931274	30.8	4	9	Species – subspecies level	Commonly seen as *Petaurista philippensis hainana*, and recently split off (Yu et al. 2006).
Petaurista lena	Taiwan Giant Flying Squirrel	NA	23.1	3	10	Species – subspecies level	Commonly perceived as *Petaurista alborufus lena*, and recently split off (Li et al. 2013).
Petaurista mechukaensis	Mechuka Giant Flying Squirrel	NA	38.5	5	8	Species description too recent	Has been redescribed in 2007 and 2009 as it showed major discrepancies, but still not updated everywhere (Choudhury 2007; Choudhury 2009b).
Petaurista mishmiensis	Mishmi Hills Giant Flying Squirrel	NA	38.5	5	8	Species description too recent, Common name	Common name from NCBI is "Mishmi Giant Flying Squirrel". Has been firstly described in 2009 and has not been updated everywhere yet (Choudhury 2009a).
Petaurista siangensis	Mebo Giant Flying Squirrel	NA	23.1	3	10	Species description too recent	Has been firstly described in 2013 and has not yet been accepted as valid species mutually amongst all institutions (Choudhury 2013).
Petaurista yunanensis	Yunnan Giant Flying Squirrel	NA	30.8	4	9	Species – subspecies level, Species description too recent	Has lately (2006 and onwards) been identified as individual species after century-long discrepancies (Yu et al. 2006). Thus, it is still not yet accepted/ updated in a few institutions and their taxonomic lists.
Prosciurillus alstoni	Alston's Squirrel	930329	69.2	9	4	Species – subspecies level, Common name	Common name from IUCN, iNaturalist is "Alston's Sulawesi Dwarf Squirrel". According to MDD and Musser et al. (2010), it has been split off from *P. leucomus.*

(continued)

Table 1.2 (continued)

Scientific name	Common name/ Alternative names	TSN (Taxonomic Serial Number – according to ITIS)	Percentage of acceptance	"Yes" count	"No" count	Reason for Discrepancy	Comments
Prosciurillus topapuensis	Mount Topapu Squirrel	930330	69.2	9	4	Species – subspecies level, Common name	Common name from NCBI is "Roux's Sulawesi Dwarf Squirrel". According to MDD and Musser et al. (2010), it has been split off from *P. leucomus*.
Sciurus meridionalis	Calabrian Black Squirrel	NA	23.1	3	10	Species – subspecies level	Has been recently proposed to be split off from *S. vulgaris* (MDD; Wauters et al. 2017).
Spermophilus nilkaensis	Tien Shan Ground Squirrel	NA	30.8	4	9	Species – subspecies level, Tautology species	Commonly seen as *Spermophilus relictus nilkaensis*. According to MDD, the species was formerly known as *S. ralli* (now a synonym of *relictus*) and has been included under *S. relictus*, but is considered a distinct species following recent publications (Kryštufek and Vohralík 2012).
Spermophilus ralli	Tien Shan Ground Squirrel	930269	46.2	6	7	Tautology species, Common name	Often mistaken as *Spermophilus relictus* and shares the common name Tien Shan ground squirrel.
Spermophilus taurensis	Taurus Ground Squirrel	930306	76.9	10	3	Species description too recent, Common name	Common name from The mammals of the world (book), Illustrated Checklist (book) is "Tauren Ground Squirrel". Has been firstly identified as distinct species in 2007, and has not yet been mutually accepted amongst all institutions (Gündüz et al. 2007).
Sundasciurus altitudinis	Sumatran Mountain Squirrel	930331	61.5	8	5	Species description too recent	Has only been identified as distinct species and cut off from *S. tenuis* in 2010 and is not yet accepted everywhere (den Tex et al. 2010; Hinckley et al. 2020).
Sundasciurus davensis	Davao Squirrel	632460	76.9	10	3	Non-taxonomic institution discrepancy	This species is mostly not accepted by institutions that do not focus on taxonomy (VertNet, GenBank (NCBI), IDigBIO).
Sundasciurus everetti	Bornean Mountain Ground Squirrel	NA	30.8	4	9	Species description too recent	Has been recently proposed for re-description from *Dremomys everetti* to the genus *Sundasciurus* (Hawkins et al. 2016).
Sundasciurus tahan	Upland Squirrel	930332	69.2	9	4	Species – subspecies level, Species description too recent	Formerly considered as a subspecies of the Slender squirrel (*Sundasciurus tenuis*) until 2010, then it has been described as separate species (den Tex et al. 2010).
Tamias	Chipmunks	180188	69.2	9	4	Genus – subgenus level	The species of this genus are listed with a different scientific name on IUCN.org (*Neotamias*) – Accordingly to ITIS.gov the genus *Neotamias* is "invalid". All species from this genus are considered invalid according to ITIS.
Tamiasciurus fremonti	Fremont's Squirrel	NA	30.8	4	9	Species – subspecies level	"*Tamiasciurus fremonti*" is by most institutions considered as a subspecies of "*Tamiasciuris hudsonicus*", but has been proposed to be split off (Hope et al. 2016).

(continued)

Table 1.2 (continued)

Scientific name	Common name/ Alternative names	TSN (Taxonomic Serial Number – according to ITIS)	Percentage of acceptance	"Yes" count	"No" count	Reason for Discrepancy	Comments
Tamiasciurus mearnsi	Mearns's Squirrel	632478	69.2	9	4	Species – subspecies level	This species is sometimes classified as a subspecies of "*Tamiasciurus douglasii*".
Xerus erythropus	Striped Ground Squirrel	632483	69.2	9	4	Genus – subgenus level	Common name from EOL is "Cape Ground Squirrel". Taxonomy dates back to 1803, seemingly quite outdated. For institutions/books considering the subgenus "*Geosciurus*" as valid, this genus is not mentioned in their lists.
Xerus inauris	South African Ground Squirrel	632484	69.2	9	4	Genus – subgenus level	Common name from EOL is "Geoffroy's Ground Squirrel". Synonym of "*Sciurus inauris*", accordingly to ITIS. For institutions/ books considering the subgenus "*Geosciurus*" as valid, this genus is not mentioned in their lists.
Xerus princeps	Damara Ground Squirrel	632485	61.5	8	5	Genus – subgenus level	For institutions/books considering the subgenus "*Geosciurus*" as valid, this genus is not mentioned in their lists.

Blue text indicates hyperlinks to the institution-specific website for the taxonomic records

easy to show that squirrel taxonomy is indeed chaotic, not well agreed upon, and widely approached in a '*laissez-faire*' way (as stated in Steiner and Huettmann 2021; see also Mori et al. 2019).

The descriptive categories used to reason the discrepancy among institutions in Table 1.2 are "Common name", "Distribution interference", "Genus – subgenus level", "Non-taxonomic institution discrepancy", "Species – subspecies level", "Species description too recent", and "Tautology species". The category "Common name" is used for species where the common name creates discrepancies. The category "Distribution interference" is used for species where the taxonomy has been updated to include the species' distribution in the taxonomy which is not accepted among all institutions. The category "Genus – subgenus level" is used for species where the genus is either commonly known as the species' subgenus or vice versa. The category "Non-taxonomic institution discrepancy" is used for species that are mostly not accepted by institutions/books that generally do not specifically focus on taxonomy. The category "Species – subspecies level" is used for species with a discrepancy on whether it is classified as species or subspecies. The category "Species description too recent" is used for species that have been (re)described only recently and have not yet been accepted/updated by all institutions/books. The category "Tautology species" is used for species that share the same name (scientific or common).

1.5.2 Suggested Individual Taxonomic Revisions for Each Institution

To further address this issue, Table 1.3 connects all species and genera of the family Sciuridae. We find that these are in immediate need of a taxonomic revision/update, with the proposed changes specified for each institution/book.

Table 1.3 presents for each institution/book included in Table 1.1 the suggestion revisions/updates based on the discrepancies found in Table 1.1 (listings not mutually acknowledged by all institutions/books). Thus, it includes all global squirrel species except the ones that are mutually accepted by all institutions/books. These discrepancies include the scientific and common names of the species. By presenting these species individually for each institution/book, this table aims to be as constructive and solution-oriented as possible in order to drastically diminish the discrepancies among institutions. With this, we seek to present the most recent taxonomic status and strive for a mutual agreement on a complete taxonomic list of all global squirrels. Included institutions can utilize this table as a tool to overview and understand all other major taxonomic lists for squirrels and ideally improve their own accordingly.

Table 1.3 Taxonomic revision suggestions individually for each institution

Squirrel species list			Institution-specific suggested revision													Institution count	
Scientific name	Common name/ Alternative names	TSN (Taxonomic Serial Number)	ITIS	GBIF	Encyclopedia of Life (EOL)	MANIS (VertNet)	GenBank (NCBI)	IUCN Red List	IDigBIO	iNaturalist	Mammal Diversity Database (MDD)	Squirrels of the world (book)	The handbook of the mammals of the world (book)	Squirrels – The Animal Answer Guide (book)	Illustrated Checklist of the Mammals of the World	Revision needed	Revision NOT needed
Aeromys thomasi	Thomas's Flying Squirrel	632489	–	–	–	–	Non-taxonomic institution discrepancy	–	–	–	–	–	–	–	–	0	13
Ammospermophilus insularis	Espiritu Santo Island Antelope Squirrel	632323	Species description too recent	Species description too recent	Species description too recent	Species description too recent	Species description too recent	–	Species description too recent	–	Species description too recent	Species description too recent	–	Species description too recent	–	8	5
Ammospermophilus leucurus	White-tailed Antelope Squirrel	180181	–	–	–	Non-taxonomic institution discrepancy	–	–	–	–	–	–	–	–	–	1	12
Biswamoyopterus biswasi	Namdapha Flying Squirrel	632491	–	–	–	Non-taxonomic institution discrepancy	–	–	–	–	–	–	–	–	–	1	12
Biswamoyopterus gaoligongensis	Mount Gaoligong Flying Squirrel	NA	Species description too recent	Species description too recent	Species description too recent	Species description too recent	Species description too recent	Species description too recent	Species description too recent	–	–	Species description too recent	Species description too recent	Species description too recent	–	10	3
Biswamoyopterus laoensis	Laotian Giant Flying Squirrel	930360	–	–	–	Species description too recent	Species description too recent	–	–	–	–	Species description too recent	–	Species description too recent	–	4	9
Callosciurus albescens	Kloss's Squirrel	NA	–	–	–	–	–	Species – subspecies level	–	–	–	–	–	Species – subspecies level	–	2	11
Callosciurus baluensis	Kinabalu Squirrel	632327	–	–	–	Non-taxonomic institution discrepancy	–	–	–	–	–	–	–	–	–	1	12
Callosciurus caniceps	Gray-bellied Squirrel	632328	–	Non-taxonomic institution discrepancy	–	–	–	–	Non-taxonomic institution discrepancy	–	–	–	–	–	–	2	11
Callosciurus honkhoaiensis	Hon Khoai Squirrel	NA	Species description too recent	Species description too recent	Species description too recent	Species description too recent	–	Species description too recent	Species description too recent	Species description too recent	–	Species description too recent	Species description too recent	Species description too recent	–	10	3
Callosciurus melanogaster	Mentawai Squirrel	632332	–	–	–	–	Non-taxonomic institution discrepancy	–	–	–	–	–	–	–	–	1	12
Callosciurus phayrei	Phayre's Squirrel	632336	–	–	–	Non-taxonomic institution discrepancy	–	–	–	–	–	–	–	–	–	1	12

Callosciurus quinquestriatus	Anderson's Squirrel	632339	–	–	–	–	Non-taxonomic institution discrepancy	–	–	–	–	–	–	–	–	1	12
Callospermophilus lateralis	Golden-mantled Ground Squirrel	930305	–	–	–	Non-taxonomic institution discrepancy	–	–	–	–	–	–	–	Non-taxonomic institution discrepancy	–	2	11
Callospermophilus madrensis	Sierra Madre Ground Squirrel	930303	–	–	–	–	–	–	–	–	–	–	–	Non-taxonomic institution discrepancy	–	1	12
Callospermophilus saturatus	Cascade Golden-mantled Ground Squirrel	930304	–	–	–	Non-taxonomic institution discrepancy	–	–	–	–	–	–	–	Non-taxonomic institution discrepancy	–	2	11
Cynomys gunnisoni	Gunnison's Prairie Dog	180184	–	–	–	–	–	–	Non-taxonomic institution discrepancy	–	–	–	–	–	–	1	12
Dremomys everetti	Bornean Mountain Ground Squirrel	632341	–	–	–	–	–	–	–	–	Genus name invalid	–	Genus name invalid	–	Genus name invalid	3	10
Dremomys gularis	Red-throated Squirrel	930262	–	–	–	Non-taxonomic institution discrepancy	–	–	–	–	–	–	–	–	–	1	12
Epixerus ebii	African Palm Squirrel	632346	–	–	–	–	Non-taxonomic institution discrepancy	–	–	–	–	–	–	–	–	1	12
Epixerus wilsoni	Biafran palm squirrel	NA	–	–	–	–	Species – subspecies level	–	–	–	–	–	–	–	–	1	12
Eupetaurus cinereus	Woolly Flying Squirrel	632492	–	–	–	Non-taxonomic institution discrepancy	–	–	–	–	–	–	–	–	–	1	12
Eutamias sibiricus	Siberian Chipmunk	NA	Distribution range and history	Distribution range and history	Distribution range and history	Distribution range and history	Distribution range and history	–	Distribution range and history	–	–	Distribution range and history	Distribution range and history	Distribution range and history	–	9	4
Euxerus erythropus	Striped Ground Squirrel	NA	Genus update	Genus update	Genus update	Genus update	Genus update	Genus update	Genus update	–	–	Genus update	–	Genus update	–	9	4
Funambulus insignis	NA	NA	–	–	–	–	–	–	Species – subspecies level	–	–	–	–	–	–	1	12
Funambulus obscurus	Dusky Striped Squirrel	NA	Species description too recent	Species description too recent	Species description too recent	Species description too recent	Species description too recent	–	Species description too recent	Species description too recent	–	Species description too recent	Species description too recent	Species description too recent	–	10	3

(continued)

Table 1.3 (continued)

Squirrel species list			Institution-specific suggested revision													Institution count	
Scientific name	Common name/ Alternative names	TSN (Taxonomic Serial Number)	ITIS	GBIF	Encyclopedia of Life (EOL)	MANIS (VertNet)	GenBank (NCBI)	IUCN Red List	IDigBIO	iNaturalist	Mammal Diversity Database (MDD)	Squirrels of the world (book)	The handbook of the mammals of the world (book)	Squirrels – The Animal Answer Guide (book)	Illustrated Checklist of the Mammals of the World	Revision needed	Revision NOT needed
Funambulus tristriatus	Jungle Palm Squirrel	632355	–	–	–	–	Non-taxonomic institution discrepancy	–	–	–	–	–	–	–	–	1	12
Funisciurus bayonii	Lunda Rope Squirrel	632357	–	–	–	–	Non-taxonomic institution discrepancy	–	–	–	–	–	–	–	–	1	12
Funisciurus congicus	Congo Rope Squirrel	632359	–	–	–	–	Non-taxonomic institution discrepancy	–	–	–	–	–	–	–	–	1	12
Funisciurus duchaillui	Du Chaillu's Rope Squirrel	930333	–	–	–	Split species	Split species	–	Split species	–	–	–	–	Split species	–	4	9
Funisciurus isabella	Lady Burton's Rope Squirrel	632360	–	–	–	–	Non-taxonomic institution discrepancy	–	–	–	–	–	–	–	–	1	12
Funisciurus lemniscatus	Ribboned Rope Squirrel	632361	–	–	–	–	Non-taxonomic institution discrepancy	–	–	–	–	–	–	–	–	1	12
Funisciurus leucogenys	Red-cheeked Rope Squirrel	632362	–	–	–	–	Non-taxonomic institution discrepancy	–	–	–	–	–	–	–	–	1	12
Funisciurus mystax	NA	NA	–	–	–	Species – subspecies level	–	–	–	–	–	–	–	–	–	1	12
Funisciurus substriatus	Kintampo Rope Squirrel	632364	–	–	–	Non-taxonomic institution discrepancy	Non-taxonomic institution discrepancy	–	–	–	–	–	–	–	–	2	11
Geomys bursarius	NA	NA	–	–	–	Family name invalid	–	–	–	–	–	–	–	–	–	1	12
Geosciurus inauris	South African Ground Squirrel	NA	Genus update	Genus update	Genus update	Genus update	Genus update	Genus update	Genus update	–	–	Genus update	–	Genus update	–	9	4
Geosciurus princeps	Damara Ground Squirrel	NA	Genus update	Genus update	Genus update	Genus update	Genus update	Genus update	Genus update	–	–	Genus update	–	Genus update	–	9	4

Glaucomys oregonensis	Humboldt's Flying Squirrel	NA	Species – subspecies level	Species – subspecies level	Species – subspecies level	Species – subspecies level	–	Species – subspecies level	Species – subspecies level	–	–	Species – subspecies level	Species – subspecies level	Species – subspecies level	–	9	4
Graphiurus murinus	NA	NA	–	–	–	Family name invalid	–	–	–	–	–	–	–	–	–	1	12
Guerlinguetus aestuans	Guianan squirrel	NA	–	–	–	–	Non-taxonomic institution discrepancy	–	–	–	–	–	–	–	–	1	12
Guerlinguetus brasiliensis	NA	NA	–	–	–	–	Non-taxonomic institution discrepancy	–	–	–	–	–	–	–	–	1	12
Hadrosciurus igniventris	North Amazonan red squirrel	NA	–	–	–	–	Non-taxonomic institution discrepancy	–	–	–	–	–	–	–	–	1	12
Hadrosciurus pyrrhinus	Junin red squirrel	NA	–	–	–	–	Non-taxonomic institution discrepancy	–	–	–	–	–	–	–	–	1	12
Hadrosciurus spadiceus	Southern Amazon red squirrel	NA	–	–	–	–	Non-taxonomic institution discrepancy	–	–	–	–	–	–	–	–	1	12
Heliosciurus multicolor	NA	930896	–	–	–	Species – subspecies level	–	–	Species – subspecies level	–	–	–	–	–	–	2	11
Heliosciurus mutabilis	Mutable Sun Squirrel	632367	–	–	–	–	Non-taxonomic institution discrepancy	–	–	–	–	–	–	–	–	1	12
Heliosciurus punctatus	Small Sun Squirrel	632368	–	–	–	–	Non-taxonomic institution discrepancy	–	–	–	–	–	–	–	–	1	12
Heliosciurus sanguines	NA	NA	–	–	–	Non-taxonomic institution discrepancy	–	–	–	–	–	–	–	–	–	1	12
Heliosciurus undulatus	Zanj Sun Squirrel	632371	–	–	–	Non-taxonomic institution discrepancy	–	–	–	–	–	–	–	–	–	1	12
Hylopetes bartelsi	Bartel's Flying Squirrel	632495	–	–	–	Non-taxonomic institution discrepancy	Non-taxonomic institution discrepancy	–	–	–	–	–	–	–	–	2	11
Hylopetes electilis	Hainan Flying Squirrel	930616	Species – subspecies level	Species – subspecies level	Species – subspecies level	–	Species – subspecies level	Species – subspecies level	–	Species – subspecies level	–	Species – subspecies level	Species – subspecies level	Species – subspecies level	–	9	4

(continued)

Table 1.3 (continued)

Squirrel species list			Institution-specific suggested revision													Institution count	
Scientific name	Common name/ Alternative names	TSN (Taxonomic Serial Number)	ITIS	GBIF	Encyclopedia of Life (EOL)	MANIS (VertNet)	GenBank (NCBI)	IUCN Red List	IDigBIO	iNaturalist	Mammal Diversity Database (MDD)	Squirrels of the world (book)	The handbook of the mammals of the world (book)	Squirrels – The Animal Answer Guide (book)	Illustrated Checklist of the Mammals of the World	Revision needed	Revision NOT needed
Hylopetes lepidus	Gray-cheeked flying squirrel	NA	–	–	–	Species name invalid	Species name invalid	–	–	–	–	Species name invalid	–	Species name invalid	–	4	9
Hylopetes platyurus	Jentink's Flying Squirrel	930260	–	–	–	–	Non-taxonomic institution discrepancy	–	–	–	–	–	–	–	–	1	12
Hylopetes sagitta	Arrow Flying Squirrel	930343	–	–	–	Genus update	Genus update	–	–	–	–	Genus update	–	Genus update	–	4	9
Hylopetes sipora	Sipora Flying Squirrel	632500	–	–	–	Genus update	Genus update	–	–	–	–	–	–	–	–	2	11
Hylopetes spadiceus	Red-cheeked Flying Squirrel	632501	–	–	–	Genus update	–	–	–	–	–	–	–	–	–	1	12
Hylopetes winstoni	Sumatran Flying Squirrel	632502	–	–	–	Genus update	Genus update	–	–	–	–	–	–	–	–	2	11
Ictidomys mexicanus	Mexican Ground Squirrel	930307	–	–	–	Non-taxonomic institution discrepancy	–	–	–	–	–	–	–	Non-taxonomic institution discrepancy	–	2	11
Ictidomys parvidens	Rio Grande Ground Squirrel	930267	Species – subspecies level	Species – subspecies level	Species – subspecies level	–	Species – subspecies level	–	Species – subspecies level	Species – subspecies level	–	Species – subspecies level	Species – subspecies level	–	Species – subspecies level	9	4
Ictidomys tridecemlineatus	Thirteen-lined Ground Squirrel	930308	–	–	–	Non-taxonomic institution discrepancy	–	–	–	–	–	–	–	Non-taxonomic institution discrepancy	–	2	11
Iomys sipora	Mentawai Flying Squirrel	632504	–	–	–	Non-taxonomic institution discrepancy	Non-taxonomic institution discrepancy	–	–	–	–	–	–	–	–	2	11
Lariscus hosei	Four-striped Ground Squirrel	632374	–	–	–	–	Non-taxonomic institution discrepancy	–	–	–	–	–	–	–	–	1	12
Lariscus niobe	Niobe Ground Squirrel	632376	–	–	–	Non-taxonomic institution discrepancy	–	–	–	–	–	–	–	–	–	1	12

Lariscus obscurus	Mentawai Three-striped Squirrel	632377	–	–	–	Non-taxonomic institution discrepancy	–	–	–	–	–	–	–	–	–	1	12
Marmota camtschatica	Black-capped Marmot	632380	–	–	–	Non-taxonomic institution discrepancy	–	–	–	–	–	–	–	–	–	1	12
Marmota kastschenkoi	Forest Steppe Marmot	930334	–	–	–	Species – subspecies level	Species – subspecies level	Species – subspecies level	Species – subspecies level	Species – subspecies level	–	–	–	Species – subspecies level	–	6	7
Microsciurus isthmuis	NA	NA	–	–	–	Species – subspecies level	–	–	–	–	–	–	–	–	–	1	12
Microsciurus napi	NA	NA	–	–	–	Species – subspecies level	–	–	–	–	–	–	–	–	–	1	12
Microsciurus otinus	NA	NA	–	–	–	–	Species – subspecies level	–	–	–	–	–	–	–	–	1	12
Microsciurus sabanillae	Sabanillae Dwarf Squirrel	NA	–	–	–	–	Species – subspecies level	–	–	–	–	–	–	–	–	1	12
Microsciurus santanderensis	Santander Dwarf Squirrel	632390	–	–	–	Non-taxonomic institution discrepancy	Non-taxonomic institution discrepancy	–	Non-taxonomic institution discrepancy	–	–	–	–	–	–	3	10
Microsciurus similis	NA	NA	–	–	–	Species – subspecies level	Species – subspecies level	–	–	–	–	–	–	–	–	2	11
Nannosciurus surrutilus	NA	NA	–	–	–	Genus name invalid	–	–	–	–	–	–	–	–	–	1	12
Neotamias – all species included	Chipmunks	NA	Genus – subgenus level	Genus – subgenus level	Genus – subgenus level	Genus – subgenus level	Genus – subgenus level	–	Genus – subgenus level	–	–	Genus – subgenus level	Genus – subgenus level	Genus – subgenus level	–	9	4
Neotamias solivagus	Sierra del Carmen Chipmunk	NA	Species – subspecies level	Species – subspecies level	Species – subspecies level	Species – subspecies level	Species – subspecies level	Species – subspecies level	Species – subspecies level	Species – subspecies level	–	Species – subspecies level	Species – subspecies level	Species – subspecies level	–	11	2
Notocitellus adocetus	Tropical Ground Squirrel	930298	–	–	–	Genus update	–	–	–	–	–	–	–	Genus update	–	2	11
Notocitellus annulatus	Ring-tailed Ground Squirrel	930299	–	–	–	Genus update	–	–	–	–	–	–	–	Genus update	–	2	11
Otospermophilus atricapillus	Baja California Rock Squirrel	930300	–	–	–	–	–	Genus update	–	–	–	–	–	Genus update	–	2	11
Otospermophilus beecheyi	California Ground Squirrel	930301	–	–	–	Non-taxonomic institution discrepancy	–	–	Non-taxonomic institution discrepancy	–	–	–	–	Non-taxonomic institution discrepancy	–	3	10

(continued)

Table 1.3 (continued)

Squirrel species list			Institution-specific suggested revision													Institution count	
Scientific name	Common name/ Alternative names	TSN (Taxonomic Serial Number)	ITIS	GBIF	Encyclopedia of Life (EOL)	MANIS (VertNet)	GenBank (NCBI)	IUCN Red List	IDigBIO	iNaturalist	Mammal Diversity Database (MDD)	Squirrels of the world (book)	The handbook of the mammals of the world (book)	Squirrels – The Animal Answer Guide (book)	Illustrated Checklist of the Mammals of the World	Revision needed	Revision NOT needed
Otospermophilus variegatus	Rock Squirrel	930302	–	–	–	Non-taxonomic institution discrepancy	–	–	–	–	–	–	–	Non-taxonomic institution discrepancy	–	2	11
Paraxerus boehmi	Boehm's Bush Squirrel	632394	–	–	–	–	Non-taxonomic institution discrepancy	–	–	–	–	–	–	–	–	1	12
Paraxerus cooperi	Cooper's Mountain Squirrel	632396	–	–	–	Non-taxonomic institution discrepancy	Non-taxonomic institution discrepancy	–	–	–	–	–	–	–	–	2	11
Paraxerus flavovittis	Striped Bush Squirrel	632397	–	–	–	–	Non-taxonomic institution discrepancy	–	–	–	–	–	–	–	–	1	12
Paraxerus lucifer	Black and Red Bush Squirrel	632398	–	–	–	–	Non-taxonomic institution discrepancy	–	–	–	–	–	–	–	–	1	12
Paraxerus palliatus	Red Bush Squirrel	632400	–	–	–	–	Non-taxonomic institution discrepancy	–	–	–	–	–	–	–	–	1	12
Paraxerus poensis	Green Bush Squirrel	632401	–	–	–	–	Non-taxonomic institution discrepancy	–	–	–	–	–	–	–	–	1	12
Paraxerus vincenti	Vincent's Bush Squirrel	632403	–	–	–	Non-taxonomic institution discrepancy	Non-taxonomic institution discrepancy	–	–	–	–	–	–	–	–	2	11
Petaurillus emiliae	Lesser Pygmy Flying Squirrel	632505	–	–	–	Non-taxonomic institution discrepancy	Non-taxonomic institution discrepancy	–	–	–	–	–	–	–	–	2	11
Petaurillus hosei	Hose's Pygmy Flying Squirrel	632506	–	–	–	Non-taxonomic institution discrepancy	Non-taxonomic institution discrepancy	–	–	–	–	–	–	–	–	2	11
Petaurista albiventer	White-bellied Giant Flying Squirrel	930642	–	–	–	–	Species – subspecies level	–	–	–	–	–	Species – subspecies level	–	Species – subspecies level	3	10

Petaurista caniceps	Gray-headed Flying Squirrel	930328	–	–	–	Species – subspecies level	Species – subspecies level	Species – subspecies level	–	Species – subspecies level	–	–	–	Species – subspecies level	–	5	8
Petaurista elegans	Spotted Giant Flying Squirrel	632509	–	–	–	Non-taxonomic institution discrepancy	–	–	–	–	–	–	–	–	–	1	12
Petaurista grandis	Formosan Giant Flying Squirrel	NA	Species – subspecies level	Species – subspecies level	Species – subspecies level	Species – subspecies level	Species – subspecies level	Species – subspecies level	Species – subspecies level	Species – subspecies level	–	Species – subspecies level	Species – subspecies level	Species – subspecies level	–	11	2
Petaurista hainana	Hainan Giant Flying Squirrel	931274	Species – subspecies level	Species – subspecies level	Species – subspecies level	Species – subspecies level	–	Species – subspecies level	Species – subspecies level	Species – subspecies level	–	Species – subspecies level	–	Species – subspecies level	–	9	4
Petaurista lena	Taiwan Giant Flying Squirrel	NA	Species – subspecies level	Species – subspecies level	Species – subspecies level	Species – subspecies level	Species – subspecies level	Species – subspecies level	Species – subspecies level	Species – subspecies level	–	Species – subspecies level	–	Species – subspecies level	–	10	3
Petaurista leucogenys	Japanese Giant Flying Squirrel	632510	–	–	–	Non-taxonomic institution discrepancy	–	–	–	–	–	–	–	–	–	1	12
Petaurista magnificus	Hodgson's Giant Flying Squirrel	632511	–	–	–	–	Non-taxonomic institution discrepancy	–	–	–	–	–	–	–	–	1	12
Petaurista marica	Indochinese Giant Flying Squirrel	NA	Species – subspecies level	Species – subspecies level	Species – subspecies level	Species – subspecies level	Species – subspecies level	Species – subspecies level	Species – subspecies level	Species – subspecies level	–	Species – subspecies level	Species – subspecies level	Species – subspecies level	–	11	2
Petaurista mechukaensis	Mechuka Giant Flying Squirrel	NA	Species – subspecies level	Species – subspecies level	Species – subspecies level	Species – subspecies level	Species – subspecies level	–	Species – subspecies level	–	–	Species – subspecies level	–	Species – subspecies level	–	8	5
Petaurista mishmiensis	Mishmi Hills Giant Flying Squirrel	NA	Species – subspecies level	Species – subspecies level	Species – subspecies level	Species – subspecies level	Species – subspecies level	–	Species – subspecies level	–	–	Species – subspecies level	–	Species – subspecies level	–	8	5
Petaurista nobilis	Bhutan Giant Flying Squirrel	632512	–	–	–	Non-taxonomic institution discrepancy	Non-taxonomic institution discrepancy	–	–	–	–	–	–	–	–	2	11
Petaurista siangensis	Mebo Giant Flying Squirrel	NA	Species – subspecies level	Species – subspecies level	Species – subspecies level	Species – subspecies level	Species – subspecies level	Species – subspecies level	Species – subspecies level	Species – subspecies level	–	Species – subspecies level	–	Species – subspecies level	–	10	3
Petaurista sybilla	Chindwin Giant Flying Squirrel	NA	Species – subspecies level	Species – subspecies level	Species – subspecies level	Species – subspecies level	Species – subspecies level	Species – subspecies level	Species – subspecies level	Species – subspecies level	–	Species – subspecies level	Species – subspecies level	Species – subspecies level	–	11	2
Petaurista xanthotis	Chinese Giant Flying Squirrel	632515	–	–	–	Non-taxonomic institution discrepancy	–	–	–	–	–	–	–	–	–	1	12
Petaurista yunanensis	Yunnan Giant Flying Squirrel	NA	Species description too recent	Species description too recent	Species description too recent	Species description too recent	–	Species description too recent	Species description too recent	Species description too recent	–	Species description too recent	–	Species description too recent	–	9	4

(continued)

Table 1.3 (continued)

Squirrel species list			Institution-specific suggested revision													Institution count	
Scientific name	Common name/ Alternative names	TSN (Taxonomic Serial Number)	ITIS	GBIF	Encyclopedia of Life (EOL)	MANIS (VertNet)	GenBank (NCBI)	IUCN Red List	IDigBIO	iNaturalist	Mammal Diversity Database (MDD)	Squirrels of the world (book)	The handbook of the mammals of the world (book)	Squirrels – The Animal Answer Guide (book)	Illustrated Checklist of the Mammals of the World	Revision needed	Revision NOT needed
Petinomys crinitus	Mindanao Flying Squirrel	632516	–	–	–	–	Non-taxonomic institution discrepancy	–	–	–	–	–	–	–	–	1	12
Petinomys genibarbis	Whiskered Flying Squirrel	632518	–	–	–	Non-taxonomic institution discrepancy	Non-taxonomic institution discrepancy	–	–	–	–	–	–	–	–	2	11
Petinomys hageni	Hagen's Flying Squirrel	632519	–	–	–	–	Non-taxonomic institution discrepancy	–	–	–	–	–	–	–	–	1	12
Petinomys lugens	Sipora Flying Squirrel	632520	–	–	–	–	Non-taxonomic institution discrepancy	–	–	–	–	–	–	–	–	1	12
Petinomys mindanensis	Mindanao Flying Squirrel	930261	–	–	–	Non-taxonomic institution discrepancy	Non-taxonomic institution discrepancy	–	–	–	–	–	–	–	–	2	11
Petinomys momonga	NA	NA	–	–	–	–	–	–	Non-taxonomic institution discrepancy	–	–	–	–	–	–	1	12
Petinomys morris	NA	NA	–	–	–	–	–	–	Non-taxonomic institution discrepancy	–	–	–	–	–	–	1	12
Petinomys sagitta	Arrow Flying Squirrel	NA	–	–	–	–	–	–	–	–	–	–	–	Genus name invalid	–	1	12
Petinomys vordermanni	Vorder-mann's Flying Squirrel	632523	–	–	–	–	Non-taxonomic institution discrepancy	–	–	–	–	–	–	–	–	1	12
Poliocitellus franklinii	Franklin's Ground Squirrel	930309	–	–	–	Non-taxonomic institution discrepancy	–	–	–	–	–	–	–	Non-taxonomic institution discrepancy	–	2	11
Prosciurillus alstoni	Alston's Squirrel	930329	–	–	–	Non-taxonomic institution discrepancy	Non-taxonomic institution discrepancy	–	Non-taxonomic institution discrepancy	–	–	–	–	Non-taxonomic institution discrepancy	–	4	9

Prosciurillus murinus	Celebes Dwarf Squirrel	632406	–	–	–	Non-taxonomic institution discrepancy	–	–	–	–	–	–	–	–	–	1	12
Prosciurillus rosenbergii	Sanghir Squirrel	930263	–	–	–	Non-taxonomic institution discrepancy	Non-taxonomic institution discrepancy	–	–	–	–	–	–	–	–	2	11
Prosciurillus topapuensis	Mount Topapu Squirrel	930330	–	–	–	Non-taxonomic institution discrepancy	Non-taxonomic institution discrepancy	–	Non-taxonomic institution discrepancy	–	–	–	–	Non-taxonomic institution discrepancy	–	4	9
Prosciurillus weberi	Weber's Dwarf Squirrel	632407	–	–	–	Non-taxonomic institution discrepancy	–	–	–	–	–	–	–	–	–	1	12
Protoxerus aubinnii	Slender-tailed Squirrel	632408	–	–	–	–	Non-taxonomic institution discrepancy	–	–	–	–	–	–	–	–	1	12
Pteromys volans	Siberian Flying Squirrel	632525	–	–	–	Non-taxonomic institution discrepancy	–	–	–	–	–	–	–	–	–	1	12
Ratufa malabarica	NA	NA	–	–	–	Species – subspecies level	–	–	–	–	–	–	–	–	–	1	12
Sciurotamias forresti	Forrest's Rock Squirrel	632419	–	–	–	Non-taxonomic institution discrepancy	–	–	–	–	–	–	–	–	–	1	12
Sciurus aestuans	Guianan Squirrel	632420	–	–	–	–	Non-taxonomic institution discrepancy	–	–	–	–	–	–	–	–	1	12
Sciurus alphonsei	NA	NA	–	–	–	Species – subspecies level	–	–	–	–	–	–	–	–	–	1	12
Sciurus apache	NA	NA	–	–	–	Species – subspecies level	–	–	–	–	–	–	–	–	–	1	12
Sciurus cinerea	NA	NA	–	–	–	Species – subspecies level	–	–	–	–	–	–	–	–	–	1	12
Sciurus erythropus	NA	NA	–	–	–	–	–	–	Species name invalid	–	–	–	–	–	–	1	12
Sciurus fossor	NA	NA	–	–	–	Species name invalid	–	–	–	–	–	–	–	–	–	1	12
Sciurus giganteus	NA	NA	–	–	–	–	–	–	Extinct	–	–	–	–	–	–	1	12

(continued)

Table 1.3 (continued)

Squirrel species list			Institution-specific suggested revision													Institution count	
Scientific name	Common name/ Alternative names	TSN (Taxonomic Serial Number)	ITIS	GBIF	Encyclopedia of Life (EOL)	MANIS (VertNet)	GenBank (NCBI)	IUCN Red List	IDigBIO	iNaturalist	Mammal Diversity Database (MDD)	Squirrels of the world (book)	The handbook of the mammals of the world (book)	Squirrels – The Animal Answer Guide (book)	Illustrated Checklist of the Mammals of the World	Revision needed	Revision NOT needed
Sciurus hudsonius	NA	NA	–	–	–	Species name invalid	–	–	–	–	–	–	–	–	–	1	12
Sciurus igniventris	Northern Amazon Red Squirrel	632429	–	–	–	–	Non-taxonomic institution discrepancy	–	–	–	–	–	–	–	–	1	12
Sciurus ingrami	NA	NA	–	–	–	Species – subspecies level	–	–	–	–	–	–	–	–	–	1	12
Sciurus kaibabensis	NA	NA	–	–	–	Species – subspecies level	–	–	–	–	–	–	–	–	–	1	12
Sciurus meridionalis	Calabrian Black Squirrel	NA	Species – subspecies level	Species – subspecies level	Species – subspecies level	Species – subspecies level	Species – subspecies level	Species – subspecies level	Species – subspecies level	–	–	Species – subspecies level	Species – subspecies level	Species – subspecies level	–	10	3
Sciurus olsoni	NA	NA	–	–	–	–	–	–	Extinct	–	–	–	–	–	–	1	12
Sciurus poliopus	NA	NA	–	–	–	Species – subspecies level	–	–	–	–	–	–	–	–	–	1	12
Sciurus pyrenaicus	NA	NA	–	–	–	Species name invalid	–	–	–	–	–	–	–	–	–	1	12
Sciurus pyrrhinus	Junin Red Squirrel	632433	–	–	–	–	Non-taxonomic institution discrepancy	–	–	–	–	–	–	–	–	1	12
Sciurus richmondi	Richmond's Squirrel	632434	–	–	–	–	Non-taxonomic institution discrepancy	–	–	–	–	–	–	–	–	1	12
Sciurus sanborni	Sanborn's Squirrel	632435	–	–	–	–	Non-taxonomic institution discrepancy	–	–	–	–	–	–	–	–	1	12
Sciurus sinaloensis	NA	NA	–	–	–	Species – subspecies level	–	–	–	–	–	–	–	–	–	1	12
Sciurus socialis	NA	NA	–	–	–	Species – subspecies level	–	–	–	–	–	–	–	–	–	1	12
Sciurus spadiceus	Southern Amazon Red Squirrel	632436	–	–	–	–	Non-taxonomic institution discrepancy	–	–	–	–	–	–	–	–	1	12

Sciurus stramineus	Guayaquil Squirrel	632437	–	–	–	–	Non-taxonomic institution discrepancy	–	–	–	–	–	–	–	–	1	12
Sciurus variabilis	NA	NA	–	–	–	–	–	–	Species – subspecies level	–	–	–	–	–	–	1	12
Simosciurus nebouxii	NA	NA	–	–	–	–	Genus name invalid	–	–	–	–	–	–	–	–	1	12
Simosciurus stramineus	Guayaquil squirrel	NA	–	–	–	–	Genus name invalid	–	–	–	–	–	–	–	–	1	12
Spermophilus adocetus	Tropical Ground Squirrel	NA	–	–	–	Genus update	–	–	–	–	–	–	–	Genus update	–	2	11
Spermophilus annulatus	Ring-tailed Ground Squirrel	NA	–	–	–	Genus update	–	–	–	–	–	–	–	Genus update	–	2	11
Spermophilus armatus	Uinta Ground Squirrel	NA	–	–	–	Genus update	–	–	–	–	–	–	–	Genus update	–	2	11
Spermophilus atricapillus	Baja California Rock Squirrel	NA	–	–	–	–	–	–	–	–	–	–	–	Genus update	–	1	12
Spermophilus beecheyi	California Ground Squirrel	NA	–	–	–	Genus update	–	–	–	–	–	–	–	Genus update	–	2	11
Spermophilus beldingi	Belding's Ground Squirrel	NA	–	–	–	Genus update	–	–	–	–	–	–	–	Genus update	–	2	11
Spermophilus brevicauda	Brandt's Ground Squirrel	930266	–	–	–	Non-taxonomic institution discrepancy	–	–	–	–	–	–	–	–	–	1	12
Spermophilus brunneus	Idaho Ground Squirrel	NA	–	–	–	–	–	–	–	–	–	–	–	Genus update	–	1	12
Spermophilus canus	Merriam's Ground Squirrel	NA	–	–	–	–	–	–	–	–	–	–	–	Genus update	–	1	12
Spermophilus columbianus	Columbian Ground Squirrel	NA	–	–	–	Genus update	–	–	–	–	–	–	–	Genus update	–	2	11
Spermophilus elegans	Wyoming Ground Squirrel	NA	–	–	–	–	–	–	–	–	–	–	–	Genus update	–	1	12
Spermophilus franklinii	Franklin's Ground Squirrel	NA	–	–	–	Genus update	–	–	–	–	–	–	–	Genus update	–	2	11
Spermophilus lateralis	Golden-mantled Ground Squirrel	NA	–	–	–	Genus update	–	–	–	–	–	–	–	Genus update	–	2	11

(continued)

Table 1.3 (continued)

Squirrel species list			Institution-specific suggested revision													Institution count	
Scientific name	Common name/ Alternative names	TSN (Taxonomic Serial Number)	ITIS	GBIF	Encyclopedia of Life (EOL)	MANIS (VertNet)	GenBank (NCBI)	IUCN Red List	IDigBIO	iNaturalist	Mammal Diversity Database (MDD)	Squirrels of the world (book)	The handbook of the mammals of the world (book)	Squirrels – The Animal Answer Guide (book)	Illustrated Checklist of the Mammals of the World	Revision needed	Revision NOT needed
Spermophilus madrensis	Sierra Madre Ground Squirrel	NA	–	–	–	–	–	–	–	–	–	–	–	Genus update	–	1	12
Spermophilus mexicanus	Mexican Ground Squirrel	NA	–	–	–	–	–	–	–	–	–	–	–	Genus update	–	1	12
Spermophilus mohavensis	Mohave Ground Squirrel	NA	–	–	–	Genus update	–	–	–	–	–	–	–	Genus update	–	2	11
Spermophilus mollis	Piute Ground Squirrel	NA	–	–	–	Genus update	–	–	–	–	–	–	–	Genus update	–	2	11
Spermophilus navigator	NA	NA	–	–	–	Species name invalid	–	–	–	–	–	–	–	–	–	1	12
Spermophilus nilkaensis	Tien Shan Ground Squirrel	NA	Species – subspecies level	Species – subspecies level	Species – subspecies level	Species – subspecies level	Species – subspecies level	Species – subspecies level	Species – subspecies level	–	–	Species – subspecies level	–	Species – subspecies level	–	9	4
Spermophilus odessanus	NA	NA	–	–	–	–	Species – subspecies level	–	–	–	–	–	–	–	–	1	12
Spermophilus parryii	Arctic Ground Squirrel	NA	–	–	–	Genus update	–	–	–	–	–	–	–	Genus update	–	2	11
Spermophilus plesius	NA	NA	–	–	–	Species – subspecies level	–	–	–	–	–	–	–	–	–	1	12
Spermophilus perotensis	Perote Ground Squirrel	NA	–	–	–	–	–	–	–	–	–	–	–	Genus update	–	1	12
Spermophilus ralli	Tien Shan Ground Squirrel	930269	Tautology species	Tautology species	Tautology species	–	–	Tautology species	–	–	–	Tautology species	–	Tautology species	–	6	7
Spermophilus richardsonii	Richardson's Ground Squirrel	NA	–	–	–	Genus update	–	–	–	–	–	–	–	Genus update	–	2	11
Spermophilus saturatus	Cascade Golden-mantled Ground Squirrel	NA	–	–	–	Genus update	–	–	–	–	–	–	–	Genus update	–	2	11
Spermophilus spilosoma	Spotted Ground Squirrel	NA	–	–	–	Genus update	–	–	–	–	–	–	–	Genus update	–	2	11

Spermophilus taurensis	Taurus Ground Squirrel	930306	–	–	–	Species description too recent	–	–	Species description too recent	–	–	–	–	Species description too recent	–	3	10
Spermophilus tereticaudus	Round-tailed Ground Squirrel	NA	–	–	–	Genus update	–	–	–	–	–	–	–	Genus update	–	2	11
Spermophilus townsendii	Townsend's Ground Squirrel	NA	–	–	–	–	–	–	–	–	–	–	–	Genus update	–	1	12
Spermophilus tridecemlineatus	Thirteen-lined Ground Squirrel	NA	–	–	–	Genus update	–	–	–	–	–	–	–	Genus update	–	2	11
Spermophilus undulatus	Long-tailed Ground Squirrel	NA	–	–	–	Genus update	–	–	–	–	–	–	–	Genus update	–	2	11
Spermophilus variegatus	Rock Squirrel	NA	–	–	–	Genus update	–	–	–	–	–	–	–	Genus update	–	2	11
Spermophilus washingtoni	Washington Ground Squirrel	NA	–	–	–	Genus update	–	–	–	–	–	–	–	Genus update	–	2	11
Sundasciurus altitudinis	Sumatran Mountain Squirrel	930331	–	–	–	Species description too recent	–	Species description too recent	Species description too recent	Species description too recent	–	–	–	Species description too recent	–	5	8
Sundasciurus davensis	Davao Squirrel	632460	–	–	–	Non-taxonomic institution discrepancy	Non-taxonomic institution discrepancy	–	Non-taxonomic institution discrepancy	–	–	–	–	–	–	3	10
Sundasciurus everetti	Bornean Mountain Ground Squirrel	NA	Species description too recent	Species description too recent	Species description too recent	Species description too recent	Species description too recent	Species description too recent	Species description too recent	–	–	Species description too recent	–	Species description too recent	–	9	4
Sundasciurus moellendorffi	Culion Tree Squirrel	632467	–	–	–	Non-taxonomic institution discrepancy	–	–	–	–	–	–	–	–	–	1	12
Sundasciurus natunensis	Natuna Squirrel	NA	Species – subspecies level	Species – subspecies level	Species – subspecies level	Species – subspecies level	Species – subspecies level	Species – subspecies level	Species – subspecies level	Species – subspecies level	–	Species – subspecies level	Species – subspecies level	Species – subspecies level	Species – subspecies level	12	1
Sundasciurus robinsoni	Robinson's Squirrel	NA	Species – subspecies level	Species – subspecies level	Species – subspecies level	Species – subspecies level	–	Species – subspecies level	Species – subspecies level	Species – subspecies level	–	Species – subspecies level	Species – subspecies level	Species – subspecies level	Species – subspecies level	11	2
Sundasciurus tahan	Upland Squirrel	930332	–	–	–	Species description too recent	–	Species description too recent	–	Species description too recent	–	–	–	Species description too recent	–	4	9
Syntheosciurus brochus	Bangs's Mountain Squirrel	632474	–	–	–	Non-taxonomic institution discrepancy	–	–	–	–	–	–	–	–	–	1	12
Syntheosciurus poasensis	NA	NA	–	–	–	–	–	–	Species name invalid	–	–	–	–	–	–	1	12

(continued)

Table 1.3 (continued)

Squirrel species list			Institution-specific suggested revision													Institution count	
Scientific name	Common name/ Alternative names	TSN (Taxonomic Serial Number)	ITIS	GBIF	Encyclopedia of Life (EOL)	MANIS (VertNet)	GenBank (NCBI)	IUCN Red List	IDigBIO	iNaturalist	Mammal Diversity Database (MDD)	Squirrels of the world (book)	The handbook of the mammals of the world (book)	Squirrels – The Animal Answer Guide (book)	Illustrated Checklist of the Mammals of the World	Revision needed	Revision NOT needed
	Tamias – all species except and lower-mentioned ones	180188	Genus update	Genus update	Genus update	Genus update	Genus update	–	Genus update	–	–	Genus update	Genus update	Genus update	–	9	4
Tamias ater	NA	NA	–	–	–	Species name invalid	–	–	–	–	–	–	–	–	–	1	12
Tamias townsendii	Townsend's Chipmunk	180208	Genus update	Genus update	Genus update	–	Genus update	–	Genus update	–	–	Genus update	Genus update	Genus update	–	8	5
Tamiasciurus fremonti	Fremont's Squirrel	NA	Species – subspecies level	Species – subspecies level	Species – subspecies level	–	Species – subspecies level	Species – subspecies level	Species – subspecies level	Species – subspecies level	–	Species – subspecies level	–	Species – subspecies level	–	9	4
Tamiasciurus mearnsi	Mearns's Squirrel	632478	–	–	–	–	–	–	–	Species – subspecies level	Species – subspecies level	–	Species – subspecies level	–	Species – subspecies level	4	9
Tamiops maritimus	Maritime Striped Squirrel	632479	–	–	–	Non-taxonomic institution discrepancy	–	–	–	–	–	–	–	–	–	1	12
Thomomys talpoides	NA	NA	–	–	–	Genus name invalid	–	–	–	–	–	–	–	–	–	1	12
Urocitellus armatus	Uinta Ground Squirrel	930314	–	–	–	Genus update	–	–	–	–	–	–	–	Genus update	–	2	11
Urocitellus beldingi	Belding's Ground Squirrel	930315	–	–	–	Genus update	–	–	–	–	–	–	–	Genus update	–	2	11
Urocitellus brunneus	Idaho Ground Squirrel	930316	–	–	–	–	–	–	–	–	–	–	–	Genus update	–	1	12
Urocitellus canus	Merriam's Ground Squirrel	930317	–	–	–	–	–	–	–	–	–	–	–	Genus update	–	1	12
Urocitellus columbianus	Columbian Ground Squirrel	930318	–	–	–	–	–	–	–	–	–	–	–	Genus update	–	1	12
Urocitellus elegans	Wyoming Ground Squirrel	930319	–	–	–	–	–	–	–	–	–	–	–	Genus update	–	1	12
Urocitellus endemicus	Southern Idaho Ground Squirrel	NA	–	–	–	–	–	Species – subspecies level	–	–	–	–	–	–	–	1	12

Urocitellus mollis	Piute Ground Squirrel	930320	–	–	–	Genus update	–	–	–	–	–	–	–	Genus update	–	2	11
Urocitellus nancyae	Townsend's Ground Squirrel	NA	–	–	–	–	–	Species – subspecies level	–	–	–	–	–	–	–	1	12
Urocitellus parryii	Arctic Ground Squirrel	180135	–	–	–	Genus update	–	–	–	–	–	–	–	Genus update	–	2	11
Urocitellus richardsonii	Richardson's Ground Squirrel	930322	–	–	–	Genus update	–	–	–	–	–	–	–	Genus update	–	2	11
Urocitellus townsendii	Townsend's Ground Squirrel	930323	–	–	–	–	–	–	–	–	–	–	–	Genus update	–	1	12
Urocitellus undulatus	Long-tailed Ground Squirrel	930324	–	–	–	Genus update	–	–	–	–	–	–	–	Genus update	–	2	11
Urocitellus washingtoni	Washington Ground Squirrel	930325	–	–	–	Genus update	–	–	–	–	–	–	–	Genus update	–	2	11
Xerospermophilus mohavensis	Mohave Ground Squirrel	930310	–	–	–	Genus update	–	–	–	–	–	–	–	Genus update	–	2	11
Xerospermophilus perotensis	Perote Ground Squirrel	930311	–	–	–	–	–	–	–	–	–	–	–	Genus update	–	1	12
Xerospermophilus spilosoma	Spotted Ground Squirrel	930312	–	–	–	Genus update	–	–	–	–	–	–	–	Genus update	–	2	11
Xerospermophilus tereticaudus	Round-tailed Ground Squirrel	930313	–	–	–	Genus update	–	–	–	–	–	–	–	Genus update	–	2	11
Xerus erythropus	Striped Ground Squirrel	632483	Genus update	Genus update	Genus update	Genus update	Genus update	Genus update	Genus update	–	–	Genus update	–	Genus update	–	9	4
Xerus inauris	South African Ground Squirrel	632484	Genus update	Genus update	Genus update	Genus update	Genus update	Genus update	Genus update	–	–	Genus update	–	Genus update	–	9	4
Xerus princeps	Damara Ground Squirrel	632485	Genus update	Genus update	Genus update	–	Genus update	Genus update	Genus update	–	–	Genus update	–	Genus update	–	8	5
Species count	**Revision needed**		**34**	**35**	**34**	**139**	**96**	**33**	**52**	**20**	**3**	**37**	**19**	**100**	**6**		
	Revision to add species		**28**	**29**	**28**	**92**	**77**	**26**	**38**	**20**	**3**	**30**	**16**	**64**	**5**		
	Revisions to remove species		**6**	**6**	**6**	**47**	**19**	**7**	**14**	**0**	**0**	**7**	**3**	**36**	**1**		
	Revision NOT needed		**187**	**186**	**187**	**82**	**125**	**188**	**169**	**201**	**218**	**184**	**201**	**121**	**215**		

Blue text indicates hyperlinks to the institution-specific website for the taxonomic records, green text indicates species records that are to be maintained, and red text indicates species records that are to be adjusted

The two colors in Table 1.3 (green and red) aim to indicate whether it is suggested to add a species to the institutional-specific taxonomic list (green) or to remove it (red). The descriptive categories within these colored cells indicate the reason for the suggested revision of the species for each institution/book.

The descriptive categories used to suggest institution-specific taxonomic revision in Table 1.3 are "Distribution interference", "Extinct", "Family name invalid", "Genus name invalid", "Genus update", "Genus – subgenus level", "Non-taxonomic institution discrepancy", "Species – subspecies level", "Species description too recent", "Species name invalid", and "Tautology species". The descriptive categories are also used in Table 1.2 (described in Sect. 1.5.1) and thus are not repeated here. Only the new categories are shortly explained below. The category "Extinct" is used for species that are commonly acknowledged as extinct but are still included in some taxonomic lists. The category "Family name invalid" is used for species that commonly do not belong to the squirrel family (Sciuridae). The category "Genus name invalid" is used for species that are assigned to a genus that does either not belong to the squirrel family, or that is commonly agreed not to exist. The category "Genus update" is used for species that require a revision of the assigned genus as this one has been changed recently, or in the past and has not been accepted mutually. The comments on the right-hand side of Table 1.3 reinforce and explain the descriptive categories used in the table. The institution count on the right-hand side indicates for how many institutions/books the taxonomic classification of a certain species is suggested to be reviewed. The species count at the bottom of the table indicates how many species are suggested to be reviewed for each institution. That summary aims to represent most cleanly and crisply where the taxonomic discrepancies lay and where and how these can be resolved/improved. The fact that everything is additionally institution-specific, provides the responsible institutions a great overview of where and what they can include in an upcoming taxonomic revision.

By analyzing Tables 1.1 and 1.3 it can be concluded that the most modern squirrel taxonomy list consists of approx. 306/307 species. This number can be obtained by taking the total number of accepted species from each institution in Table 1.1 and adding the suggested revisions from Table 1.3. This number is significantly bigger than any institution acknowledged thus far. The reason for this is that in recent years a few new species have been discovered in Asia, such as the Laotian Giant Flying Squirrel (*Biswamoyopterus laoensis*) and other members of this genus. Additionally, several subspecies have been proposed to be split and elevated to species level (e.g. see Sect. 1.2 and Table 1.3 for examples). These latest proposed species have often been acknowledged as subspecies in the past, but due to modern research techniques, including genetic and ecologic efforts, it was able to identify the most suitable classification for these species/subspecies.

Table 1.3 greatly depicts the institutions/books which align, more or less evidently, with the 306/307 species which seem to be the most complete taxonomic species set. Additionally, it greatly depicts how the specific institutions/books can improve to come closest to these 306/307 suggested species for a mutual agreement. Table 1.3 also depicts that the institutions/books which focus on taxonomy score the greatest to come close to the 306/307 species. The most recent book (Illustrated checklist of the mammals of the world by Burgin et al. (2020)) actually scores the second highest (302 species), supporting the approach that taxonomic books must be as up-to-date as possible with regular revisions/editions if possible.

1.6 The Conservation Situation of All Global Squirrel Species

Nature conservation is defined by the Cambridge dictionary (May 2021) as "*the protection of plants and animals, natural areas, and interesting and important structures and buildings, especially from the damaging effects of human activity*". This seems to have crucial importance in the modern world where nature is often disregarded over capitalistic economics and its resource extraction and space use (e.g. Czech 2000). Therefore, here we aim to present the global squirrel conservation situation by addressing encountered issues and presenting possible solutions to erase these.

1.6.1 The Various 'Conservation Status' Classes of the World's Squirrels

The concept of conservation status classes has been introduced by many authorities in the late 1900s where the Swiss-based IUCN Red List has mostly been the world-leading authority. These conservation status classes aim to categorize the global natural species into a ranked system. It implies law and order, and good management; based on the mindset from the nineteenth century. This system consists according to the IUCN Red List of 10 classes. These classes are 'Extinct (EX) – No known living individuals, Extinct in the wild (EW) – Known only to survive in captivity, or as a naturalized population outside its historic range, Critically Endangered (CR) – Extremely high risk of extinction in the wild, Endangered (EN) – High risk of extinction in the wild, Vulnerable (VU) – High risk of endangerment in the wild, Near Threatened (NT) – Likely to become endangered in the near future, Conservation Dependent (CD) – Low risk; is conserved to prevent being near threatened, certain events may lead it to be a higher risk level, and Least concern (LC) – Lowest risk; does not qualify for a higher risk category. Widespread and abundant taxa are included in this category.' Additional two classes are known which describe the absence of data 'Data Deficient (DD) – Not enough data to assess its risk of extinction, and Not evaluated (NE) – Has not yet been evaluated against the criteria.'

These classifications are broadly used to classify the extinction risk of species and are also applied to squirrels. An in-depth description of these categories can be found on the official website of the IUCN Red List (https://www.iucnredlist.org/assessment/process). It shows how species are perceived and managed, in the absence of a valid and modern management theory (Morden 2017). These are the conservation categories that are accepted almost globally. However, the country-specific conservation status categories and the species within might be different among countries (e.g. Australia, New Zealand, Russia, China, India, Europe, etc.). Table 1.4 presents the conservation status and population trend for all global squirrel species.

Table 1.4 depicts the conservation status assigned by one institution and one book (IUCN Red List, and "Illustrated Checklist of the mammals of the world" by Burgin et al. (2020)) for all global squirrel species. The reason why only two sources are mentioned in this table is that most other institutions/books either do not present a conservation status or reuse a classification from the IUCN Red list from an older cut-off date than if it is directly downloaded from the IUCN Red List website. By directly downloading the latest version of the species' taxonomic list from the IUCN Red List website it is assumed to obtain the most up-to-date version available. The other conservation status is provided by the "Illustrated Checklist of the mammals of the world" (Burgin et al. 2020).

At the bottom of Table 1.4, an overview is provided of how many species are assigned to each classification described above. To summarize, for the IUCN Red List, 5 species are considered NE, 33 species DD, 195 species LC, 22 species NT, 14 species VU, 15 species EN, 2 species CR, 0 species EW, and 0 species EX. For the book "Illustrated Checklist of the mammals of the world", 20 species are considered NE, 35 species DD, 191 species LC, 24 species NT, 15 species VU, 14 species EN, 3 species CR, 0 species EW, 0 species EX.

For a better understanding, this conservation status breakdown has been depicted for both the IUCN Red List and Illustrated Checklist of the Mammals of the World individually in Figs. 1.1 and 1.2 respectively.

1.6.2 Compilation of Conservation Status Classes for a Global and Large Taxonomic Family (Squirrels)

In this section, it will be explained how Table 1.4 has been compiled. This section has the purpose to support the presented table with a transparent compilation procedure to allow the readers to gain insight into how the table has been created and how it can be replicated for other species families if desired. Table 1.4 has been created by compiling the conservation class data from two different sources, the IUCN Red List website (www.iucnredlist.org) and the "Illustrated checklist of the mammals of the world" (Burgin et al. 2020). For the former one, it is possible to filter the database of the website by looking for the desired family name in the search bar (in this case, "Sciuridae"). Consecutively, the obtained data set, containing all the species from this family, can be either downloaded or displayed on the website. The obtained data can then be filled in a table such as Table 1.4 and presented likewise. For the latter one, the book, each species has simply been manually added from a physical version of the book into the table. The latter source can be difficult to access for the broad public as these books might not be available for everyone and exist sometimes only in hardcopy, which hinders the public to access this data in a useful and modern format. As emphasized in Sect. 1.3.1, an online database instead has a great advantage, it is digital and usually freely accessible to the broad public, thus allowing open and free data access. It is modern and more effective in that way reaching any citizen of the world with access to the internet.

1.6.3 Observed Discrepancies Among Conservation Status Classes Across Institutions and Authoritative Sources

This section aims to address the discrepancies observed between the two provided conservation status datasets in Table 1.4. As observed in Table 1.4 and also in Sect. 1.6.1, the number of species assigned to the same conservation classes differs, also in relation to the whole species number. Generally, it can be said that both lists should be up to date, as the book has only

Table 1.4 Conservation status and population trend of all global squirrel species

Scientific name	Squirrel Species name	Taxonomic serial number	Conservation status (IUCN)	Conservation status (Illustrated Checklist of the mammals of the world)	Population trend (IUCN Red List)	Comments
Aeretes melanopterus	Northern Chinese Flying Squirrel	632487	Near threatened	Near threatened	Decreasing	Population size and trend are unknown according to IUCN.org
Aeromys tephromelas	Black Flying Squirrel	632488	Data deficient	Data deficient	Unknown	Population size and trend are unknown
Aeromys thomasi	Thomas's Flying Squirrel	632489	Least concern	Least concern	Unknown	Population size and trend are unknown, still listed as "least concern" on IUCN.org
Ammospermophilus harrisii	Harris' Antelope Squirrel	180179	Least concern	Least concern	Unknown	Population size and trend are unknown, still listed as "least concern" according to IUCN.org
Ammospermophilus insularis	Espiritu Santo Island Antelope Squirrel	632323	Least concern	NA	Stable	
Ammospermophilus interpres	Texas Antelope Squirrel	180180	Least concern	Least concern	Unknown	Population size and trend are unknown, still listed as "least concern" on IUCN.org
Ammospermophilus leucurus	White-tailed Antelope Squirrel	180181	Least concern	Least concern	Stable	
Ammospermophilus nelsoni	Nelson's Antelope Squirrel	180182	Endangered	Endangered	Decreasing	Population size is unknown and population trend is listed as "decreasing". Habitat is highly restricted to its small ecological niche
Atlantoxerus getulus	Barbary Ground Squirrel	632324	Least concern	Least concern	Stable	
Belomys pearsonii	Hairy-footed Flying Squirrel	632490	Data deficient	Data deficient	Unknown	Population size and trend are unknown. Their depicted habitat on IUCN is highly scattered which does not support the species' conservation
Biswamoyopterus biswasi	Namdapha Flying Squirrel	632491	Critically Endangered	Critically Endangered	Decreasing	Population size is unknown and the population trend is listed as "decreasing" according to IUCN.org
Biswamoyopterus gaoligongensis	Mount Gaoligong Flying Squirrel	NA	NA	Not evaluated	NA	
Biswamoyopterus laoensis	Laotian Giant Flying Squirrel	930360	Data deficient	Data deficient	Unknown	Population size and trend are unknown. Additionally, the depicted distribution map indicates a very small restricted habitat
Callosciurus adamsi	Ear-spot Squirrel	632325	Near threatened	Near threatened	Decreasing	Population size is unknown and population trend is listed as "decreasing"
Callosciurus albescens	Kloss's Squirrel	NA	Data deficient	NA	Unknown	
Callosciurus baluensis	Kinabalu Squirrel	632327	Least concern	Least concern	Decreasing	Population size and trend are unknown, still listed as "least concern", even though the depicted distribution map indicates a small restricted habitat
Callosciurus caniceps	Gray-bellied Squirrel	632328	Least concern	Least concern	Stable	
Callosciurus erythraeus	Pallas's Squirrel	632329	Least concern	Least concern	Stable	

(continued)

Table 1.4 (continued)

Scientific name	Squirrel Species name	Taxonomic serial number	Conservation status (IUCN)	Conservation status (Illustrated Checklist of the mammals of the world)	Population trend (IUCN Red List)	Comments
Callosciurus finlaysonii	Finlayson's Squirrel	632330	Least concern	Least concern	Stable	
Callosciurus honkhoaiensis	Hon Khoai Squirrel	NA	NA	Not evaluated	NA	According to Illustrated Checklist, described too recently to evaluate the conservation status
Callosciurus inornatus	Inornate Squirrel	632331	Least concern	Least concern	Stable	
Callosciurus melanogaster	Mentawai Squirrel	632332	Vulnerable	Vulnerable	Decreasing	Population size is unknown and population trend is listed as "decreasing". Habitat is highly restricted to its small ecological niche
Callosciurus nigrovittatus	Black-striped Squirrel	632333	Least concern	Least concern	Stable	
Callosciurus notatus	Plantain Squirrel	632334	Least concern	Least concern	Increasing	One of the very few species with the population trend "increasing", according to IUCN.org
Callosciurus orestes	Borneo Black-banded Squirrel	632335	Least concern	Least concern	Stable	
Callosciurus phayrei	Phayre's Squirrel	632336	Least concern	Least concern	Unknown	Population size and trend are unknown, still listed as "least concern" according to IUCN.org
Callosciurus prevostii	Prevost's Squirrel	632337	Least concern	Least concern	Decreasing	Population size is unknown and the population trend is listed as "decreasing" according to IUCN.org and still listed as "least concern"
Callosciurus pygerythrus	Irrawaddy Squirrel	632338	Least concern	Least concern	Stable	
Callosciurus quinquestriatus	Anderson's Squirrel	632339	Least concern	Near threatened	Unknown	Population size and trend are unknown, still listed as "least concern"
Callospermophilus lateralis	Golden-mantled Ground Squirrel	930305	Least concern	Least concern	Stable	Synonym of "*Spermophilus lateralis*", according to ITIS.gov
Callospermophilus madrensis	Sierra Madre Ground Squirrel	930303	Near threatened	Near threatened	Unknown	Synonym of "*Spermophilus madrensis*", according to ITIS.gov. Population size and trend are unknown according to IUCN.org
Callospermophilus saturatus	Cascade Golden-mantled Ground Squirrel	930304	Least concern	Least concern	Stable	
Cynomys gunnisoni	Gunnison's Prairie Dog	180184	Least concern	Least concern	Decreasing	Population size is unknown and the population trend is listed as "decreasing" according to IUCN.org, even though it is considered as "least concern". Entire genus population is decreasing!
Cynomys leucurus	White-tailed Prairie Dog	180185	Least concern	Least concern	Decreasing	Population size is unknown and the population trend is listed as "decreasing" according to IUCN.org, even though it is considered as "least concern". Entire genus population is decreasing!

(continued)

Table 1.4 (continued)

Scientific name	Squirrel Species name	Taxonomic serial number	Conservation status (IUCN)	Conservation status (Illustrated Checklist of the mammals of the world)	Population trend (IUCN Red List)	Comments
Cynomys ludovicianus	Arizona black-tailed prairie dog	180186	Least concern	Least concern	Decreasing	Population size is unknown and the population trend is listed as "decreasing" according to IUCN.org, even though it is considered as "least concern". Entire genus population is decreasing!
Cynomys mexicanus	Mexican Prairie Dog	632340	Endangered	Endangered	Decreasing	Population size is unknown and the population trend is listed as "decreasing" according to IUCN.org. Entire genus population is decreasing!
Cynomys parvidens	Utah Prairie Dog	180187	Endangered	Endangered	Decreasing	Population size is unknown and the population trend is listed as "decreasing" according to IUCN.org. Entire genus population is decreasing!
Dremomys everetti	Bornean Mountain Ground Squirrel	632341	Least concern	NA	Stable	
Dremomys gularis	Red-throated Squirrel	930262	Data deficient	Data deficient	Unknown	Population size and trend are unknown with restricted habitat, according to IUCN.org
Dremomys lokriah	Orange-bellied Himalayan Squirrel	632342	Least concern	Least concern	Decreasing	Population size is unknown and the population trend is listed as "decreasing" according to IUCN.org and still listed as "least concern"
Dremomys pernyi	Perny's Long-nosed Squirrel	632343	Least concern	Least concern	Stable	
Dremomys pyrrhomerus	Red-hipped Squirrel	632344	Least concern	Least concern	Unknown	Population size and trend are unknown, still listed as "least concern" according to IUCN.org
Dremomys rufigenis	Asian Red-cheeked Squirrel	632345	Least concern	Least concern	Stable	
Eoglaucomys fimbriatus	Kashmir Flying Squirrel	930259	Least concern	Least concern	Unknown	Population size and trend are unknown, still listed as "least concern" by IUCN.org
Epixerus ebii	African Palm Squirrel	632346	Least concern	Least concern	Unknown	Population size and trend are unknown, still listed as "least concern" by IUCN.org
Eupetaurus cinereus	Woolly Flying Squirrel	632492	Endangered	Endangered	Unknown	Population size and trend are unknown according to IUCN.org
Eutamias sibiricus	Siberian Chipmunk	NA	Least concern	Least concern	Stable	
Euxerus erythropus	Striped Ground Squirrel	NA	NA	Least concern	NA	
Exilisciurus concinnus	Philippine Pygmy Squirrel	632348	Least concern	Least concern	Stable	
Exilisciurus exilis	Least Pygmy Squirrel	632349	Data deficient	Data deficient	Unknown	Population size and trend are unknown according to IUCN.org
Exilisciurus whiteheadi	Tufted Pygmy Squirrel	632350	Least concern	Least concern	Stable	
Funambulus layardi	Layard's Palm Squirrel	632351	Vulnerable	Vulnerable	Decreasing	Population trend is decreasing, which is reinforced by its small restricted and scattered habitat

(continued)

Table 1.4 (continued)

Scientific name	Squirrel Species name	Taxonomic serial number	Conservation status (IUCN)	Conservation status (Illustrated Checklist of the mammals of the world)	Population trend (IUCN Red List)	Comments
Funambulus obscurus	Dusky Striped Squirrel	NA	Vulnerable	Vulnerable	Decreasing	
Funambulus palmarum	Indian Palm Squirrel	632352	Least concern	Least concern	Increasing	One of the very few species with an increasing population trend according to IUCN.org
Funambulus pennantii	Northern Palm Squirrel	632353	Least concern	Least concern	Unknown	Population size and trend are unknown, still listed as "least concern" according to IUCN.org
Funambulus sublineatus	Dusky Palm Squirrel	632354	Vulnerable	Vulnerable	Decreasing	Is listed twice as "Dusky Striped Squirrel" and "Dusky-Striped-Squirrel". These two listings show a different distribution and both are listed with a declining population trend
Funambulus tristriatus	Jungle Palm Squirrel	632355	Least concern	Least concern	Decreasing	Population size is unknown and population trend is listed as "decreasing", however, it is still listed as "least concern" by IUCN.org
Funisciurus anerythrus	Thomas's Rope Squirrel	632356	Least concern	Least concern	Unknown	Population size and trend are unknown, still listed as "least concern" on IUCN.org
Funisciurus bayonii	Lunda Rope Squirrel	632357	Data deficient	Data deficient	Unknown	Population size and trend are unknown according to IUCN.org
Funisciurus carruthersi	Carruther's Mountain Squirrel	632358	Least concern	Least concern	Unknown	Population size and trend are unknown, still listed as "least concern" on IUCN.org
Funisciurus congicus	Congo Rope Squirrel	632359	Least concern	Least concern	Stable	
Funisciurus duchaillui	Du Chaillu's Rope Squirrel	930333	Data deficient	Data deficient	Unknown	Population size and trend are unknown according to IUCN.org
Funisciurus isabella	Lady Burton's Rope Squirrel	632360	Least concern	Least concern	Unknown	Population size and trend are unknown, still listed as "least concern" according to IUCN.org
Funisciurus lemniscatus	Ribboned Rope Squirrel	632361	Least concern	Least concern	Unknown	Population size and trend are unknown, still listed as "least concern" according to IUCN.org
Funisciurus leucogenys	Red-cheeked Rope Squirrel	632362	Least concern	Least concern	Unknown	Population size and trend are unknown, still listed as "least concern" according to IUCN.org
Funisciurus pyrropus	Fire-footed Rope Squirrel	632363	Least concern	Least concern	Stable	
Funisciurus substriatus	Kintampo Rope Squirrel	632364	Data deficient	Data deficient	Unknown	Population size and trend are unknown
Geosciurus inauris	South African Ground Squirrel	NA	NA	Least concern	NA	
Geosciurus princeps	Damara Ground Squirrel	NA	NA	Least concern	NA	
Glaucomys oregonensis	Humboldt's Flying Squirrel	NA	NA	Not evaluated	NA	
Glaucomys sabrinus	Northern Flying Squirrel	180169	Least concern	Least concern	Stable	
Glaucomys volans	Southern Flying Squirrel	180170	Least concern	Least concern	Stable	

(continued)

Table 1.4 (continued)

Scientific name	Squirrel Species name	Taxonomic serial number	Conservation status (IUCN)	Conservation status (Illustrated Checklist of the mammals of the world)	Population trend (IUCN Red List)	Comments
Glyphotes simus	Sculptor Squirrel	632365	Data deficient	Data deficient	Unknown	Population size and trend are unknown according to IUCN.org with a small restricted ecological distribution
Heliosciurus gambianus	Gambian Sun Squirrel	632366	Least concern	Least concern	Unknown	Population size and trend are unknown, still listed as "least concern"
Heliosciurus mutabilis	Mutable Sun Squirrel	632367	Least concern	Least concern	Unknown	Population size and trend are unknown, still listed as "least concern" according to IUCN.org
Heliosciurus punctatus	Small Sun Squirrel	632368	Data deficient	Data deficient	Unknown	Population size and trend are unknown according to IUCN.org
Heliosciurus rufobrachium	Red-legged Sun Squirrel	632369	Least concern	Least concern	Unknown	Population size and trend are unknown, still listed as "least concern" according to IUCN.org
Heliosciurus ruwenzorii	Ruwenzori Sun Squirrel	632370	Least concern	Least concern	Unknown	Population size and trend are unknown, still listed as "least concern" according to IUCN.org
Heliosciurus undulatus	Zanj Sun Squirrel	632371	Data deficient	Data deficient	Unknown	Population size and trend are unknown according to IUCN.org
Hylopetes alboniger	Particolored Flying Squirrel	632493	Least concern	Least concern	Decreasing	Population size is unknown and the population trend is listed as "decreasing" according to IUCN.org and still listed as "least concern"
Hylopetes bartelsi	Bartel's Flying Squirrel	632495	Data deficient	Data deficient	Unknown	Population size and trend are unknown, their distribution seems to be limited to a single volcano
Hylopetes electilis	Hainan Flying Squirrel	930616	NA	Not evaluated	NA	
Hylopetes nigripes	Palawan Flying Squirrel	632498	Near threatened	Near threatened	Decreasing	Population size and trend are unknown according to IUCN.org, their habitat is highly restricted to one island
Hylopetes phayrei	Indochinese Flying Squirrel	632499	Least concern	Least concern	Stable	
Hylopetes platyurus	Gray-cheeked flying squirrel	930260	Data deficient	Data deficient	Unknown	Population size and trend are unknown according to IUCN.org
Hylopetes sagitta	Arrow Flying Squirrel	930343	Data deficient	Data deficient	Unknown	Population size and trend are unknown according to IUCN.org
Hylopetes sipora	Sipora Flying Squirrel	632500	Endangered	Endangered	Decreasing	Population size and trend are unknown according to IUCN.org and depict an alarming distribution map where the occurring habitat is only one single small island!
Hylopetes spadiceus	Red-cheeked Flying Squirrel	632501	Least concern	Least concern	Unknown	Population size and trend are unknown, still listed as "least concern" according to IUCN.org
Hylopetes winstoni	Sumatran Flying Squirrel	632502	Data deficient	Data deficient	Unknown	Population size and trend are unknown according to IUCN.org and depict an alarming distribution map where the occurring habitat includes only a few square km
Hyosciurus heinrichi	Montane Long-nosed Squirrel	632372	Least concern	Least concern	Unknown	Population size and trend are unknown, still listed as "least concern" according to IUCN.org
Hyosciurus ileile	Lowland Long-nosed Squirrel	632373	Vulnerable	Vulnerable	Decreasing	Population trend is listed as "decreasing" (IUCN.org)

(continued)

Table 1.4 (continued)

Scientific name	Squirrel Species name	Taxonomic serial number	Conservation status (IUCN)	Conservation status (Illustrated Checklist of the mammals of the world)	Population trend (IUCN Red List)	Comments
Ictidomys mexicanus	Mexican Ground Squirrel	930307	Least concern	Least concern	Stable	
Ictidomys parvidens	Rio Grande Ground Squirrel	930267	NA	Not evaluated	NA	
Ictidomys tridecemlineatus	Thirteen-lined Ground Squirrel	930308	Least concern	Least concern	Stable	
Iomys horsfieldii	Javanese Flying Squirrel	632503	Least concern	Least concern	Stable	
Iomys sipora	Mentawai Flying Squirrel	632504	Endangered	Endangered	Decreasing	
Lariscus hosei	Four-striped Ground Squirrel	632374	Least concern	Least concern	Decreasing	Population size is unknown and population trend is listed as "decreasing", however, it is still listed as "least concern"
Lariscus insignis	Three-striped Ground Squirrel	632375	Least concern	Least concern	Decreasing	Population size is unknown and the population trend is listed as "decreasing"according to IUCN.org, even though it is considered as "least concern"
Lariscus niobe	Niobe Ground Squirrel	632376	Data deficient	Data deficient	Unknown	Population size and trend are unknown with restricted habitat, according to IUCN.org
Lariscus obscurus	Mentawai Three-striped Squirrel	632377	Near threatened	Near threatened	Decreasing	Population size is unknown and population trend is listed as "decreasing". Habitat is highly restricted to its small ecological niche
Marmota baibacina	Gray Marmot	632378	Least concern	Least concern	Unknown	Population size and trend are unknown, still listed as "least concern" according to IUCN.org
Marmota bobak	Bobak Marmot	632379	Least concern	Least concern	Stable	
Marmota broweri	Alaska Marmot	180138	Least concern	Least concern	Stable	
Marmota caligata	Hoary Marmot	180139	Least concern	Least concern	Stable	
Marmota camtschatica	Black-capped Marmot	632380	Least concern	Least concern	Unknown	Population size and trend are unknown, still listed as "least concern" according to IUCN.org with pretty scattered habitat distribution
Marmota caudata	Long-tailed Marmot	632381	Least concern	Least concern	Unknown	Population size and trend are unknown, still listed as "least concern" according to IUCN.org
Marmota flaviventris	Yellow-bellied Marmot	180140	Least concern	Least concern	Stable	
Marmota himalayana	Himalayan Marmot	632382	Least concern	Least concern	Unknown	Population size and trend are unknown, still listed as "least concern" according to IUCN.org
Marmota kastschenkoi	Forest Steppe Marmot	930334	Not evaluated	Not evaluated	NA	No data available on IUCN.org
Marmota marmota	Alpine Marmot	632383	Least concern	Least concern	Stable	

(continued)

Table 1.4 (continued)

Scientific name	Squirrel Species name	Taxonomic serial number	Conservation status (IUCN)	Conservation status (Illustrated Checklist of the mammals of the world)	Population trend (IUCN Red List)	Comments
Marmota menzbieri	Menzbier's Marmot	632384	Vulnerable	Vulnerable	Decreasing	Population size is unknown and population trend is listed as "decreasing", according to IUCN.org, with its small restricted habitat distribution
Marmota monax	Woodchuck	180137	Least concern	Least concern	Stable	
Marmota olympus	Olympic Marmot	180141	Least concern	Least concern	Decreasing	Population trend is decreasing with an estimated population size of 2,000-4,000 individuals and is still considered a "least concern" species according to IUCN.org!
Marmota sibirica	Tarbagan Marmot	632385	Endangered	Endangered	Decreasing	Population trend is decreasing with unknown population size. IUCN.org lists that no action recovery plan and systematic monitoring system exists even though their conservation trend is decreasing
Marmota vancouverensis	Vancouver Island Marmot	180142	Critically Endangered	Critically Endangered	Decreasing	Population trend is listed as "decreasing" with an estimated population size of 90 individuals. This species calls for immediate conservation efforts to avoid species extinction!
Menetes berdmorei	Indochinese Ground Squirrel	632386	Least concern	Least concern	Stable	
Microsciurus alfari	Central American Dwarf Squirrel	632387	Least concern	Least concern	Stable	
Microsciurus flaviventer	Amazon Dwarf Squirrel	632388	Least concern	Least concern	Unknown	Population size and trend are unknown, still listed as "least concern"
Microsciurus mimulus	Western Dwarf Squirrel	632389	Least concern	Least concern	Stable	
Microsciurus santanderensis	Santander Dwarf Squirrel	632390	Data deficient	Data deficient	Unknown	Population size and trend are unknown according to IUCN.org with restricted small ecological distribution
Myosciurus pumilio	African Pygmy Squirrel	632391	Least concern	Least concern	Unknown	Population size and trend are unknown, still listed as "least concern"
Nannosciurus melanotis	Black-eared Squirrel	632392	Least concern	Least concern	Decreasing	Population size is unknown and population trend is listed as "decreasing", however, it is still listed as "least concern"
Neotamias alpinus	Alpine Chipmunk	NA	Least concern	Least concern	Stable	
Neotamias amoenus	Yellow-pine Chipmunk	NA	Least concern	Least concern	Stable	
Neotamias bulleri	Buller's Chipmunk	NA	Vulnerable	Vulnerable	Decreasing	
Neotamias canipes	Gray-footed Chipmunk	NA	Least concern	Least concern	Stable	
Neotamias cinereicollis	Gray-collared Chipmunk	NA	Least concern	Least concern	Stable	
Neotamias dorsalis	Cliff Chipmunk	NA	Least concern	Least concern	Stable	

(continued)

Table 1.4 (continued)

Scientific name	Squirrel Species name	Taxonomic serial number	Conservation status (IUCN)	Conservation status (Illustrated Checklist of the mammals of the world)	Population trend (IUCN Red List)	Comments
Neotamias durangae	Durango Chipmunk	NA	Least concern	Least concern	Unknown	
Neotamias merriami	Merriam's Chipmunk	NA	Least concern	Least concern	Stable	
Neotamias minimus	Least Chipmunk	NA	Least concern	Least concern	Stable	
Neotamias obscurus	California Chipmunk	NA	Least concern	Least concern	Stable	
Neotamias ochrogenys	Yellow-cheeked Chipmunk	NA	Least concern	Least concern	Stable	
Neotamias palmeri	Palmer's Chipmunk	NA	Endangered	Endangered	Decreasing	
Neotamias panamintinus	Panamint Chipmunk	NA	Least concern	Least concern	Stable	
Neotamias quadrimaculatus	Long-eared Chipmunk	NA	Least concern	Least concern	Stable	
Neotamias quadrivittatus	Colorado Chipmunk	NA	Least concern	Least concern	Stable	
Neotamias ruficaudus	Red-tailed Chipmunk	NA	Least concern	Least concern	Stable	
Neotamias rufus	Hopi Chipmunk	NA	Least concern	Least concern	Stable	
Neotamias senex	Allen's Chipmunk	NA	Least concern	Least concern	Stable	
Neotamias siskiyou	Siskiyou Chipmunk	NA	Least concern	Least concern	Stable	
Neotamias solivagus	Sierra del Carmen Chipmunk	NA	NA	Not evaluated	NA	
Neotamias sonomae	Sonoma Chipmunk	NA	Least concern	Least concern	Stable	
Neotamias speciosus	Lodgepole Chipmunk	NA	Least concern	Least concern	Stable	
Neotamias townsendii	Townsend's Chipmunk	NA	Least concern	Least concern	Stable	
Neotamias umbrinus	Uinta Chipmunk	NA	Least concern	Least concern	Stable	
Notocitellus adocetus	Tropical Ground Squirrel	930298	Least concern	Least concern	Stable	
Notocitellus annulatus	Ring-tailed Ground Squirrel	930299	Least concern	Least concern	Unknown	Population size and trend are unknown, still listed as "least concern" according to IUCN.org
Otospermophilus atricapillus	Baja California Rock Squirrel	930300	Not evaluated	Endangered	NA	Not listed on IUCN.org
Otospermophilus beecheyi	California Ground Squirel	930301	Least concern	Least concern	Stable	
Otospermophilus variegatus	Rock Squirrel	930302	Least concern	Least concern	Stable	
Paraxerus alexandri	Alexander's Bush Squirrel	632393	Least concern	Least concern	Unknown	Population size and trend are unknown, still listed as "least concern"
Paraxerus boehmi	Boehm's Bush Squirrel	632394	Least concern	Least concern	Unknown	Population size and trend are unknown, still listed as "least concern"

(continued)

Table 1.4 (continued)

Scientific name	Squirrel Species name	Taxonomic serial number	Conservation status (IUCN)	Conservation status (Illustrated Checklist of the mammals of the world)	Population trend (IUCN Red List)	Comments
Paraxerus cepapi	Smith's Bush Squirrel	632395	Least concern	Least concern	Stable	
Paraxerus cooperi	Cooper's Mountain Squirrel	632396	Data deficient	Data deficient	Unknown	Population size and trend are unknown according to IUCN.org
Paraxerus flavovittis	Striped Bush Squirrel	632397	Least concern	Least concern	Unknown	Population size and trend are unknown according to IUCN.org
Paraxerus lucifer	Black and Red Bush Squirrel	632398	Data deficient	Data deficient	Unknown	Population size and trend are unknown according to IUCN.org
Paraxerus ochraceus	Ochre Bush Squirrel	632399	Least concern	Least concern	Stable	
Paraxerus palliatus	Red Bush Squirrel	632400	Least concern	Least concern	Unknown	Population size and trend are unknown, still listed as "least concern" according to IUCN.org
Paraxerus poensis	Green Bush Squirrel	632401	Least concern	Least concern	Unknown	Population size and trend are unknown, still listed as "least concern"
Paraxerus vexillarius	Svynnerton's Bush Squirrel	632402	Least concern	Near threatened	Unknown	Population size and trend are unknown according to IUCN.org, their distribution is depicted as highly scattered
Paraxerus vincenti	Vincent's Bush Squirrel	632403	Endangered	Endangered	Decreasing	Population size is unknown and population trend is listed as "decreasing". Habitat is highly restricted to its small ecological niche
Petaurillus emiliae	Lesser Pygmy Flying Squirrel	632505	Data deficient	Data deficient	Unknown	Population size and trend are unknown. The species' habitat is according to IUCN.org, constituted by two dangerously small distribution areas
Petaurillus hosei	Hose's Pygmy Flying Squirrel	632506	Data deficient	Data deficient	Unknown	Population size and trend are unknown
Petaurillus kinlochii	Selangor Pygmy Flying Squirrel	632507	Data deficient	Data deficient	Unknown	Population size and trend are unknown according to IUCN.org with restricted small ecological distribution
Petaurista albiventer	White-bellied Giant Flying Squirrel	NA	NA	Not evaluated	NA	
Petaurista alborufus	Red and White Giant Flying Squirrel	632508	Least concern	Least concern	Unknown	Population size and trend are unknown, still listed as "least concern" according to IUCN.org
Petaurista caniceps	Gray-headed Flying Squirrel	930328	Least concern	Least concern	Stable	
Petaurista elegans	Spotted Giant Flying Squirrel	632509	Least concern	Least concern	Stable	
Petaurista grandis	Formosan Giant Flying Squirrel	NA	NA	Not evaluated	NA	
Petaurista hainana	Hainan Giant Flying Squirrel	NA	NA	Not evaluated	NA	
Petaurista lena	Taiwan Giant Flying Squirrel	NA	NA	Not evaluated	NA	
Petaurista leucogenys	Japanese Giant Flying Squirrel	632510	Least concern	Least concern	Unknown	Population size and trend are unknown, still listed as "least concern"

(continued)

Table 1.4 (continued)

Scientific name	Squirrel Species name	Taxonomic serial number	Conservation status (IUCN)	Conservation status (Illustrated Checklist of the mammals of the world)	Population trend (IUCN Red List)	Comments
Petaurista magnificus	Hodgson's Giant Flying Squirrel	632511	Least concern	Least concern	Decreasing	Population size is unknown and population trend is listed as "decreasing", however, it is still listed as "least concern"
Petaurista marica	Indochinese Giant Flying Squirrel	NA	NA	Not evaluated	NA	
Petaurista mechukaensis	Mechuka Giant Flying Squirrel	NA	Data deficient	Data deficient	Unknown	
Petaurista mishmiensis	Mishmi Hills Giant Flying Squirrel	NA	Near threatened	Near threatened	Unknown	
Petaurista nobilis	Bhutan Giant Flying Squirrel	632512	Near threatened	Near threatened	Decreasing	Population size is unknown and population trend is listed as "decreasing"
Petaurista petaurista	Red Giant Flying Squirrel	632513	Least concern	Least concern	Decreasing	Population size is unknown and the population trend is listed as "decreasing" according to IUCN.org and still listed as "least concern"
Petaurista philippensis	Indian Giant Flying Squirrel	632514	Least concern	Least concern	Decreasing	Population size is unknown and population trend is listed as "decreasing", however, it is still listed as "least concern"
Petaurista siangensis	Mebo Giant Flying Squirrel	NA	NA	Not evaluated	NA	
Petaurista sybilla	Chindwin Giant Flying Squirrel	NA	NA	Not evaluated	NA	
Petaurista xanthotis	Chinese Giant Flying Squirrel	632515	Least concern	Least concern	Unknown	Population size and trend are unknown, still listed as "least concern"
Petaurista yunanensis	Yunnan Giant Flying Squirrel	NA	NA	Not evaluated	NA	
Petinomys crinitus	Mindanao Flying Squirrel	632516	Least concern	Least concern	Stable	
Petinomys fuscocapillus	Travancore Flying Squirrel	632517	Least concern	Least concern	Decreasing	Population size is unknown and the population trend is listed as "decreasing"according to IUCN.org, even though the species is considered "least concern"
Petinomys genibarbis	Whiskered Flying Squirrel	632518	Vulnerable	Vulnerable	Decreasing	Population size is unknown and the population trend is listed as "decreasing"according to IUCN.org with scattered depicted distribution
Petinomys hageni	Hagen's Flying Squirrel	632519	Data deficient	Data deficient	Decreasing	Population size is unknown and the population trend is decreasing. The habitat indicated by IUCN.org is very restricted to one small town and region
Petinomys lugens	Sipora Flying Squirrel	632520	Vulnerable	Vulnerable	Decreasing	Population size is unknown and the population trend is listed as "decreasing"according to IUCN. org facilitated by a population decline with extremely small restricted habitat on a single island
Petinomys mindanensis	Mindanao Flying Squirrel	930261	Least concern	Least concern	Unknown	Population size and trend are unknown, still listed as "least concern" according to IUCN.org

(continued)

Table 1.4 (continued)

Scientific name	Squirrel Species name	Taxonomic serial number	Conservation status (IUCN)	Conservation status (Illustrated Checklist of the mammals of the world)	Population trend (IUCN Red List)	Comments
Petinomys setosus	Temminck's Flying Squirrel	632522	Vulnerable	Vulnerable	Decreasing	Population size is unknown and the population trend is listed as "decreasing"according to IUCN.org
Petinomys vordermanni	Vordermann's Flying Squirrel	632523	Vulnerable	Vulnerable	Decreasing	Population size is unknown and the population trend is listed as "decreasing"according to IUCN.org with strongly scattered depicted distribution
Poliocitellus franklinii	Franklin's Ground Squirrel	930309	Least concern	Least concern	Decreasing	Population size is unknown and population trend is listed as "decreasing", however, it is still listed as "least concern"
Prosciurillus abstrusus	Secretive Dwarf Squirrel	632404	Data deficient	Data deficient	Unknown	Population size and trend are unknown according to IUCN.org with restricted small ecological distribution
Prosciurillus alstoni	Alston's Squirrel	930329	Near threatened	Near threatened	Unknown	Population size and trend are unknown
Prosciurillus leucomus	Whitish Dwarf Squirrel	632405	Least concern	Data deficient	Unknown	Population size and trend are unknown, still listed as "least concern" on IUCN.org. Illustrated Checklist lists this species as DD whereas IUCN lists it as LC
Prosciurillus murinus	Celebes Dwarf Squirrel	632406	Least concern	Least concern	Unknown	Population size and trend are unknown, still listed as "least concern"
Prosciurillus rosenbergii	Sanghir Squirrel	930263	Least concern	Least concern	Stable	Population size and trend are unknown, still listed as "least concern" according to IUCN.org with its highly restricted habitat to a single small island
Prosciurillus topapuensis	Mount Topapu Squirrel	930330	Near threatened	Not evaluated	Unknown	Population size and trend are unknown according to IUCN.org
Prosciurillus weberi	Weber's Dwarf Squirrel	632407	Endangered	Endangered	Unknown	Population size and the population trend are listed as "unknown". Habitat is highly restricted to its small ecological niche
Protoxerus aubinnii	Slender-tailed Squirrel	632408	Near threatened	Near threatened	Unknown	Population size and trend are unknown according to IUCN.org
Protoxerus stangeri	Forest Giant Squirrel	632409	Least concern	Least concern	Unknown	Population size and trend are unknown, still listed as "least concern"
Pteromys momonga	Japanese Flying Squirrel	632524	Least concern	Least concern	Unknown	Population size and trend are unknown, still listed as "least concern" according to IUCN.org
Pteromys volans	Siberian Flying Squirrel	632525	Least concern	Least concern	Decreasing	Population size is unknown and the population trend is listed as "decreasing" according to IUCN.org, even though the species is considered "least concern"
Pteromyscus pulverulentus	Smoky Flying Squirrel	632526	Endangered	Endangered	Decreasing	Population size is unknown and the population trend is listed as "decreasing"
Ratufa affinis	Pale Giant Squirrel	632410	Near threatened	Near threatened	Decreasing	Population size is unknown and the population trend is listed as "decreasing" according to IUCN.org
Ratufa bicolor	Black Giant Squirrel	632411	Near threatened	Near threatened	Decreasing	Population size is unknown and the population trend is listed as "decreasing"

(continued)

Table 1.4 (continued)

Scientific name	Squirrel Species name	Taxonomic serial number	Conservation status (IUCN)	Conservation status (Illustrated Checklist of the mammals of the world)	Population trend (IUCN Red List)	Comments
Ratufa indica	Indian Giant Squirrel	632412	Least concern	Least concern	Decreasing	Population size is unknown and the population trend is listed as "decreasing", however, it is still listed as "least concern"
Ratufa macroura	Sri Lankan Giant Squirrel	632413	Near threatened	Near threatened	Decreasing	Population size is unknown and the population trend is listed as "decreasing" according to IUCN.org
Rheithrosciurus macrotis	Tufted Ground Squirrel	632414	Vulnerable	Vulnerable	Decreasing	Population size is unknown and the population trend is listed as "decreasing" according to IUCN.org
Rhinosciurus laticaudatus	Shrew-faced Squirrel	632415	Near threatened	Near threatened	Decreasing	Population size is unknown and the population trend is listed as "decreasing" according to IUCN.org
Rubrisciurus rubriventer	Sulawesi Giant Squirrel	632416	Vulnerable	Vulnerable	Decreasing	Population size is unknown and the population trend is listed as "decreasing" according to IUCN.org
Sciurillus pusillus	Neotropical Pygmy Squirrel	632417	Least concern	Least concern	Unknown	Population size and trend are unknown, still listed as "least concern" according to IUCN.org
Sciurotamias davidianus	Père David's Rock Squirrel	632418	Least concern	Least concern	Unknown	Population size and trend are unknown, still listed as "least concern" according to IUCN.org
Sciurotamias forresti	Forrest's Rock Squirrel	632419	Least concern	Least concern	Unknown	Population size and trend are unknown, still listed as "least concern"
Sciurus aberti	Abert's Squirrel	180173	Least concern	Least concern	Stable	
Sciurus aestuans	Guianan Squirrel	632420	Least concern	Least concern	Unknown	Population size and trend are unknown, still listed as "least concern"
Sciurus alleni	Allen's Squirrel	632421	Least concern	Least concern	Decreasing	Population size is unknown and the population trend is listed as "decreasing", however, it is still listed as "least concern"
Sciurus anomalus	Caucasian Squirrel	632422	Least concern	Least concern	Decreasing	Population size is unknown and the population trend is listed as "decreasing", however, it is still listed as "least concern"
Sciurus arizonensis	Arizona Gray Squirrel	180174	Data deficient	Data deficient	Unknown	Population size and trend are unknown according to IUCN
Sciurus aureogaster	Mexican gray squirrel	552498	Least concern	Least concern	Stable	
Sciurus carolinensis	Eastern Gray Squirrel	180175	Least concern	Least concern	Increasing	One of the very few species with the population trend "increasing", according to IUCN.org
Sciurus colliaei	Collie's Squirrel	632423	Least concern	Least concern	Unknown	Population size and trend are unknown, still listed as "least concern" according to IUCN
Sciurus deppei	Deppe's Squirrel	632424	Least concern	Least concern	Stable	
Sciurus flammifer	Fiery Squirrel	632425	Data deficient	Data deficient	Unknown	Population size and trend are unknown according to IUCN
Sciurus gilvigularis	Yellow-throated Squirrel	632426	Data deficient	Data deficient	Unknown	Population size and trend are unknown according to IUCN
Sciurus granatensis	Red-tailed Squirrel	632427	Least concern	Least concern	Stable	

(continued)

Table 1.4 (continued)

Scientific name	Squirrel Species name	Taxonomic serial number	Conservation status (IUCN)	Conservation status (Illustrated Checklist of the mammals of the world)	Population trend (IUCN Red List)	Comments
Sciurus griseus	Western Gray Squirrel	180176	Least concern	Least concern	Unknown	Population size and trend are unknown, still listed as "least concern" on IUCN.org
Sciurus ignitus	Bolivian Squirrel	632428	Least concern	Least concern	Unknown	Population size and trend are unknown, still listed as "least concern"
Sciurus igniventris	Northern Amazon Red Squirrel	632429	Least concern	Least concern	Unknown	Population size and trend are unknown, still listed as "least concern" according to IUCN.org
Sciurus lis	Japanese Squirrel	632430	Least concern	Least concern	Stable	
Sciurus meridionalis	Calabrian Black Squirrel	NA	NA	Not evaluated	NA	
Sciurus nayaritensis	Mexican Fox Squirrel	180177	Least concern	Least concern	Unknown	Population size and trend are unknown, still listed as "least concern" according to IUCN.org.
Sciurus niger	Eastern Fox Squirrel	180172	Least concern	Least concern	Stable	
Sciurus oculatus	Peters's Squirrel	632431	Least concern	Least concern	Unknown	Population size and trend are unknown, still listed as "least concern" according to IUCN.org
Sciurus pucheranii	Andean Squirrel	632432	Data deficient	Data deficient	Unknown	Population size and trend are unknown
Sciurus pyrrhinus	Junin Red Squirrel	632433	Data deficient	Data deficient	Unknown	Population size and trend are unknown.
Sciurus richmondi	Richmond's Squirrel	632434	Near threatened	Near threatened	Unknown	Population size and trend are unknown according to IUCN.org
Sciurus sanborni	Sanborn's Squirrel	632435	Data deficient	Data deficient	Unknown	Population size and trend are unknown according to IUCN.org
Sciurus spadiceus	Southern Amazon Red Squirrel	632436	Least concern	Least concern	Unknown	Population size and trend are unknown according to IUCN.org
Sciurus stramineus	Guayaquil Squirrel	632437	Least concern	Least concern	Unknown	Population size and trend are unknown, still listed as "least concern" on IUCN.org
Sciurus variegatoides	Variegated Squirrel	632438	Least concern	Least concern	Stable	
Sciurus vulgaris	Eurasian Red Squirrel	632439	Least concern	Least concern	Decreasing	Population size is unknown and the population trend is listed as "decreasing", however, it is still listed as "least concern"
Sciurus yucatanensis	Yucatan Squirrel	632440	Least concern	Least concern	Stable	
Spermophilopsis leptodactylus	Long-clawed Ground Squirrel	632441	Least concern	Least concern	Unknown	Population size and trend are unknown, still listed as "least concern"
Spermophilus alashanicus	Alashan Ground Squirrel	632443	Least concern	Least concern	Decreasing	Population size is unknown and the population trend is listed as "decreasing", however, it is still listed as "least concern"
Spermophilus brevicauda	Brandt's Ground Squirrel	930266	Least concern	Least concern	Unknown	Population size and trend are unknown, still listed as "least concern"
Spermophilus citellus	European Ground Squirrel	632446	Endangered	Endangered	Decreasing	It is listed with a declining population (IUCN)

(continued)

Table 1.4 (continued)

Scientific name	Squirrel Species name	Taxonomic serial number	Conservation status (IUCN)	Conservation status (Illustrated Checklist of the mammals of the world)	Population trend (IUCN Red List)	Comments
Spermophilus dauricus	Daurian Ground Squirrel	632447	Least concern	Least concern	Unknown	Population size and trend are unknown, still listed as "least concern"
Spermophilus erythrogenys	Red-cheeked Ground Squirrel	632448	Least concern	Least concern	Stable	
Spermophilus fulvus	Yellow Ground Squirrel	632449	Least concern	Least concern	Unknown	Population size and trend are unknown according to IUCN.org
Spermophilus major	Russet Ground Squirrel	632451	Least concern	Least concern	Unknown	Population size and trend are unknown, still listed as "least concern" according to IUCN.org
Spermophilus musicus	Caucasian Mountain Ground Squirrel	632452	Near threatened	Near threatened	Unknown	Population size is unknown with pretty scattered habitat distribution according to IUCN.org
Spermophilus nilkaensis	Tien Shan Ground Squirrel	NA	NA	Least concern	NA	
Spermophilus pallidicauda	Pallid Ground Squirrel	930268	Least concern	Least concern	Unknown	Population size and trend are unknown, still listed as "least concern" according to IUCN.org
Spermophilus pygmaeus	Little Ground Squirrel	632454	Least concern	Least concern	Decreasing	Population size is unknown and the population trend is listed as "decreasing", however, it is still listed as "least concern"
Spermophilus ralli	Tien Shan Ground Squirrel	930269	Least concern	NA	Unknown	Population size and trend are unknown, still listed as "least concern" on IUCN.org, despite small and restricted habitat
Spermophilus relictus	Relict Ground Squirrel	632455	Least concern	Least concern	Unknown	Population size and trend are unknown, still listed as "least concern" according to IUCN.org
Spermophilus suslicus	Speckled Ground Squirrel	632456	Near threatened	Near threatened	Decreasing	Population size is unknown and the population trend is listed as "decreasing" according to IUCN.org
Spermophilus taurensis	Taurus Ground Squirrel	930306	Least concern	Least concern	Unknown	Population size and trend are unknown, still listed as "least concern" on IUCN.org
Spermophilus xanthoprymnus	Asia Minor Ground Squirrel	632458	Near threatened	Near threatened	Decreasing	Population size is unknown and the population trend is listed as "decreasing"
Sundasciurus altitudinis	Sumatran Mountain Squirrel	930331	Not evaluated	Not evaluated	NA	No data available on IUCN.org
Sundasciurus brookei	Brooke's Squirrel	632459	Least concern	Least concern	Unknown	Population size and trend are unknown, still listed as "least concern"
Sundasciurus davensis	Davao Squirrel	632460	Data deficient	Data deficient	Unknown	Population size and trend are unknown. Their habitat seems highly restricted and small on the distribution map of IUCN
Sundasciurus everetti	Bornean Mountain Ground Squirrel	NA	NA	Least concern	NA	

(continued)

Table 1.4 (continued)

Scientific name	Squirrel Species name	Taxonomic serial number	Conservation status (IUCN)	Conservation status (Illustrated Checklist of the mammals of the world)	Population trend (IUCN Red List)	Comments
Sundasciurus fraterculus	Fraternal Squirrel	632473	Vulnerable	Vulnerable	Decreasing	Population size is unknown and the population trend is decreasing. The habitat indicated by IUCN is very restricted to a few small islands
Sundasciurus hippurus	Horse-tailed Squirrel	632461	Near threatened	Near threatened	Decreasing	Population size is unknown and the population trend is listed as "decreasing"
Sundasciurus hoogstraali	Busuanga Squirrel	632462	Least concern	Least concern	Unknown	Population size and trend are unknown, still listed as "least concern". The distribution map on IUCN depicts a highly restricted island as its only habitat
Sundasciurus jentinki	Jentink's Squirrel	632463	Least concern	Least concern	Unknown	Population size and trend are unknown, still listed as "least concern", even though the depicted distribution map indicates a small restricted habitat
Sundasciurus juvencus	Northern Palawan Tree Squirrel	632464	Least concern	Least concern	Stable	Population size and trend are unknown, still listed as "least concern" according to IUCN.org
Sundasciurus lowii	Low's Squirrel	632465	Least concern	Least concern	Stable	
Sundasciurus mindanensis	Mindanao Squirrel	632466	Least concern	Least concern	Unknown	Population size and trend are unknown, still listed as "least concern" according to IUCN.org
Sundasciurus moellendorffi	Culion Tree Squirrel	632467	Near threatened	Near threatened	Decreasing	Population size is unknown and the population trend is listed as "decreasing". The distribution map on IUCN depicts highly restricted small islands as their only habitat
Sundasciurus philippinensis	Philippine Tree Squirrel	632468	Least concern	Least concern	Stable	
Sundasciurus rabori	Palawan Montane Squirrel	632469	Data deficient	Data deficient	Unknown	Population size and trend are unknown with highly restricted habitat for a small part of an island, according to IUCN.org
Sundasciurus samarensis	Samar Squirrel	632470	Least concern	Least concern	Unknown	Population size and trend are unknown, still listed as "least concern" according to IUCN.org
Sundasciurus steerii	Southern Palawan Tree Squirrel	632471	Least concern	Least concern	Stable	Population size and trend are unknown accordingly to IUCN.org, their habitat is highly restricted to one island and is despite that still listed as a "least concern" species
Sundasciurus tahan	Upland Squirrel	930332	Not evaluated	Not evaluated	NA	No data available on IUCN.org
Sundasciurus tenuis	Slender Squirrel	632472	Least concern	Least concern	Decreasing	Population size is unknown and the population trend is listed as "decreasing" according to IUCN.org, even though it is considered as "least concern"
Syntheosciurus brochus	Bangs's Mountain Squirrel	632474	Data deficient	Data deficient	Decreasing	Population size is unknown and the population trend is listed as "decreasing"
Tamias striatus	Eastern Chipmunk	180207	Least concern	Least concern	Stable	

(continued)

Table 1.4 (continued)

Scientific name	Squirrel Species name	Taxonomic serial number	Conservation status (IUCN)	Conservation status (Illustrated Checklist of the mammals of the world)	Population trend (IUCN Red List)	Comments
Tamiasciurus douglasii	Douglas's Squirrel	180167	Least concern	Least concern	Stable	
Tamiasciurus fremonti	Fremont's Squirrel	NA	Not evaluated	Not evaluated	NA	
Tamiasciurus hudsonicus	Red Squirrel	180166	Least concern	Least concern	Stable	
Tamiasciurus mearnsi	Mearns's Squirrel	632478	Endangered	NA	Decreasing	Population size is unknown and the population trend is listed as "decreasing". Habitat is highly restricted to its small ecological niche
Tamiops maritimus	Maritime Striped Squirrel	632479	Least concern	Least concern	Stable	
Tamiops mcclellandii	Himalayan Striped Squirrel	930264	Least concern	Least concern	Stable	
Tamiops rodolphii	Cambodian Striped Squirrel	930265	Least concern	Least concern	Stable	
Tamiops swinhoei	Swinhoe's Striped Squirrel	632482	Least concern	Least concern	Stable	
Trogopterus xanthipes	Complex-Toothed Flying Squirrel	632527	Near threatened	Near threatened	Decreasing	Population size is unknown and the population trend is listed as "decreasing"
Urocitellus armatus	Uinta Ground Squirrel	930314	Least concern	Least concern	Unknown	Population size and trend are unknown, still listed as "least concern" on IUCN.org
Urocitellus beldingi	Belding's Ground Squirrel	930315	Least concern	Least concern	Stable	
Urocitellus brunneus	Idaho Ground Squirrel	930316	Endangered	Critically Endangered	Increasing	This species is not listed on IUCN.org as species. There is only the subdivision available, which is Northern Idaho Ground Squirrel (*Urocitellus brunneus brunneus*) (Conservation status: Endangered, Population size: 1,000 adult individuals, Population trend: increasing) and Southern Idaho Ground Squirrel (*Spermophilus brunneus endemicus*) (Conservation status: Vulnerable, Population size and trend: unknown). The latter subspecies is not accounted for by ITIS.gov as subspecies. These two groups are considered "species" by IUCN.org
Urocitellus canus	Merriam's Ground Squirrel	930317	Least concern	Least concern	Unknown	Population size and trend are unknown, still listed as "least concern" according to IUCN.org
Urocitellus columbianus	Columbian Ground Squirrel	930318	Least concern	Least concern	Stable	
Urocitellus elegans	Wyoming Ground Squirrel	930319	Least concern	Least concern	Unknown	Population size and trend are unknown according to IUCN.org

(continued)

Table 1.4 (continued)

Scientific name	Squirrel Species name	Taxonomic serial number	Conservation status (IUCN)	Conservation status (Illustrated Checklist of the mammals of the world)	Population trend (IUCN Red List)	Comments
Urocitellus endemicus	Southern Idaho Ground Squirrel	NA	Vulnerable	NA	Unknown	
Urocitellus mollis	Piute Ground Squirrel	930320	Least concern	Least concern	Stable	
Urocitellus nancyae	Townsend's Ground Squirrel	NA	Vulnerable	NA	Decreasing	This species is commonly known as "*Urocitellus townsendii*". This is listed twice on IUCN Red List with two different scientific names
Urocitellus parryii	Arctic Ground Squirrel	180135	Least concern	Least concern	Unknown	Population size and trend are unknown, still listed as "least concern"
Urocitellus richardsonii	Richardson's Ground Squirrel	930322	Least concern	Least concern	Stable	
Urocitellus townsendii	Townsend's Ground Squirrel	930323	Vulnerable	Vulnerable	Decreasing	Population size is unknown and the population trend is listed as "decreasing" according to IUCN.org with small and restricted habitat
Urocitellus undulatus	Long-tailed Ground Squirrel	930324	Least concern	Least concern	Stable	
Urocitellus washingtoni	Washington Ground Squirrel	930325	Near threatened	Near threatened	Decreasing	Population size is unknown and the population trend is listed as "decreasing" according to IUCN.org
Xerospermophilus mohavensis	Mohave Ground Squirrel	930310	Near threatened	Near threatened	Decreasing	Population size is unknown and the population trend is decreasing according to IUCN.org
Xerospermophilus perotensis	Perote Ground Squirrel	930311	Endangered	Endangered	Decreasing	Population size is unknown and the population trend is listed as "decreasing" according to IUCN.org. Habitat is highly restricted to its small ecological niche
Xerospermophilus spilosoma	Spotted Ground Squirrel	930312	Least concern	Least concern	Stable	
Xerospermophilus tereticaudus	Round-tailed Ground Squirrel	930313	Least concern	Least concern	Stable	
Xerus erythropus	Striped Ground Squirrel	632483	Least concern	NA	Stable	
Xerus inauris	South African Ground Squirrel	632484	Least concern	NA	Stable	
Xerus princeps	Damara Ground Squirrel	632485	Least concern	NA	Stable	
Xerus rutilus	Unstriped Ground Squirrel	632486	Least concern	Least concern	Stable	
Conservation status count						
NE	Not evaluated		5	20	115	Unknown
DD	Data deficient		35	35	73	Decreasing
LC	Least concern		195	191	95	Stable
NT	Near threatened		23	24	4	Increasing

(continued)

Table 1.4 (continued)

Scientific name	Squirrel Species name	Taxonomic serial number	Conservation status (IUCN)	Conservation status (Illustrated Checklist of the mammals of the world)	Population trend (IUCN Red List)	Comments
VU	Vulnerable		17	15	10	Not available (NA)
EN	Endangered		15	14		
CR	Critically Endangered		2	3		
EW	Extinct in the wild		0	0		
EX	Extinct		0	0		
	Total		292	302		

Blue text indicates hyperlinks to the institution-specific website for the taxonomic records

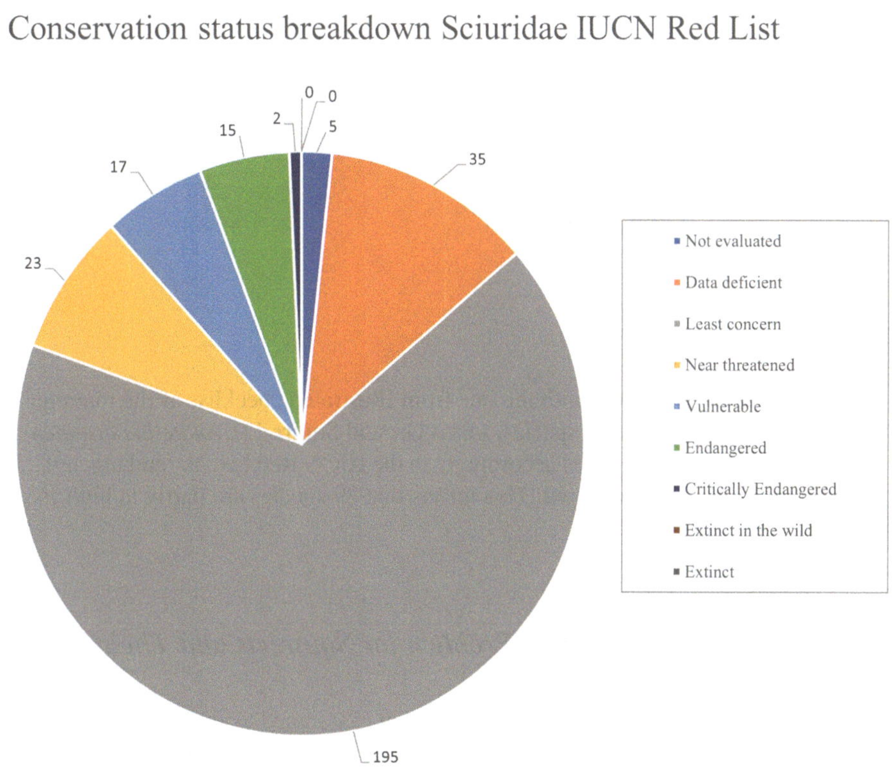

Fig. 1.1 Conservation status breakdown for all Sciuridae species in the IUCN Red List. Numbers indicate the number of species corresponding to the different classes. (Created by MS)

been released a few months before the cut-off date for this assessment (late 2020) and the website (www.iucnredlist.org) should be maintained constantly up to date. Looking at the presented numbers, it can be stated that the 'Illustrated checklist of the mammals of the world' considers fewer species listed as least concern (LC) and quite a few more as not evaluated (NE). Other than that, the numbers are pretty similar with only small deviations. One reason why the book has more species classified as NE compared to the IUCN Red List is that it contains many of the newly proposed species which are suggested to be elevated from the subspecies level to the species level, with 'reasonable' research findings to support the acceptance onto the taxonomic species list (see Sect. 1.5.2 for research findings and Sect. 1.2 and Table 1.3 for example species which are proposed to be elevated to species level). As many of these newly classified species have not yet been properly studied, information that is necessary for the conservation status classification is still lacking.

We find it is of crucial importance to further research these newly classified species as their conservation status is unknown and the worst-case scenarios cannot be excluded. Additionally, it seems noteworthy that the number of critically endangered species in the two lists presented above differs. From IUCN Red List two species are acknowledged to be critically endangered, Namdapha Flying Squirrel (*Biswamoyopterus biswasi*) and Vancouver Island Marmot (*Marmota vancouverensis*).

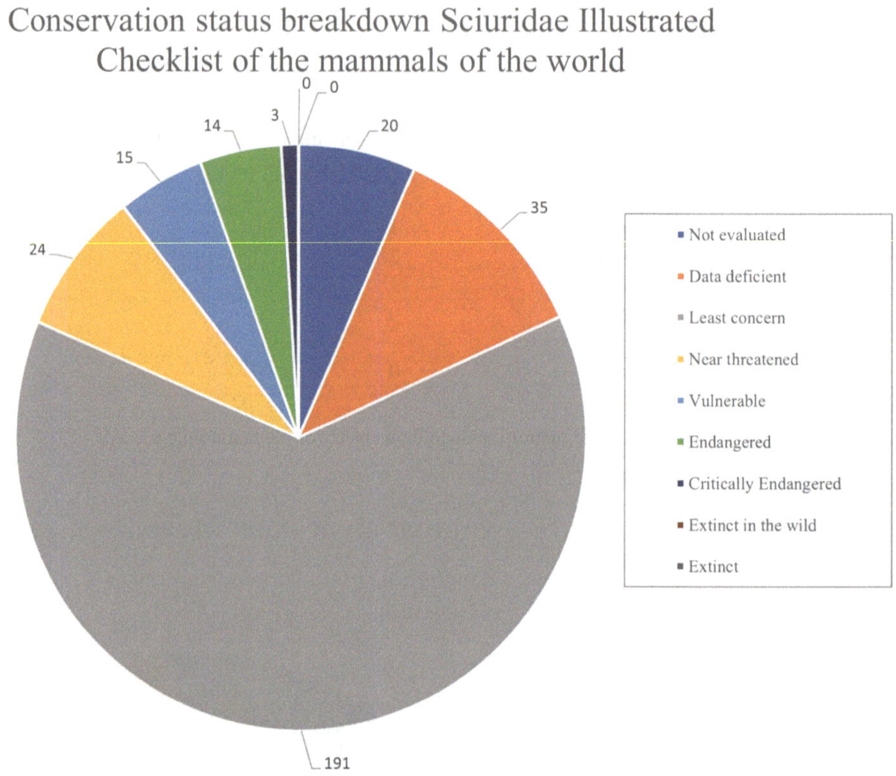

Fig. 1.2 Conservation status breakdown for all Sciuridae species in the book "Illustrated Checklist of the mammals of the world". Numbers indicate the number of species corresponding to the different classes. (Created by MS)

This classification is aligned with the one from Illustrated checklist of the mammals of the world, however, this book additionally acknowledges a third species, Idaho Ground Squirrel (*Urocitellus brunneus*), as critically endangered. These latter-mentioned species are classified accordingly to the IUCN Red List as "endangered". Both lists acknowledge 29 species being partly vulnerable and endangered. This means that 29 species are partly at high risk of extinction in the wild, or under high risk of endangerment.

1.6.4 The "Data Deficient" Problem for Squirrels and Their Conservation

"Data deficient" is a conservation status class that is assigned to species where not enough data is present to assess its risk of extinction. This can be understood in an administration as there is not always enough data for this species available to be able to say where the species lives, how many mature individuals of this species exist, whether the population trend is increasing or decreasing, and most importantly, whether the species is threatened to become extinct or not. Neither is it known how close this species is to be facing extinction. This class is evidently assigned to species with very low population sizes and highly restricted habitats. Additionally, it seems evident that the assigned research/conservation budget for range data research, range data analysis, etc. must be very low for these species. If the previous assumptions would be false, if the population size would be very large, with large and different habitats and with great conservation actions in execution, data would be present, and it could be classified properly. Therefore, to put it into perspective, more than 10% of the global squirrel species are classified as being "data deficient". Adding these 10% of unknown existence, to the other 10% of already endangered species mentioned in Sect. 1.6.3, over 20% of the global squirrel species seem to be in critical danger, at least their conservation management status can be considered rather dubious. This is far away from good management and best practice (Morden 2017; Silvy 2020). This message is a serious fact.

An additional aspect that is worth mentioning is that generally species that are classified as DD receive less public awareness regarding their extinction risk in contrast to species placed in the conservation classes such as "endangered", or "critically endangered". They are easily taken off the agenda that way and vanish from the public discussion. This is a big problem as they might even be in a worse situation than the species classified as EN, or CR. A great example of this is Bartels's Flying

Squirrel (*Hylopetes bartelsi*), which inhabits only one small volcano-like mountain (Pangrango) in West Java (Indonesia) – according to the IUCN Red List. This species is "DD" with an assumingly minuscule population size. Although this species' distribution is by far smaller than any CR-categorized squirrel species, it receives only very little attention for conservation actions and future research.

Only very few other studies address these issues, one we would like to cite is from Yu et al. (2014):

> Observations in the wild have been precluded by the rarity of some *Hylopetes* populations and practically nothing is known about their biological activities. For example, *H. bartelsi* Chasen, 1939 (the Bartel's flying squirrel) is found only from Mt. Pangrango, West Java, Indonesia; *H. sipora* Chasen, 1940 (the Sipora flying squirrel) occurs only in the Sipura island of the Mentawai Archipelago, southwest of Sumatra (Thorington and Hoffmann 2005) and *H. winstoni* (Sody 1949) (the Sumatran flying squirrel) is known only from a single specimen on the island of Sumatra. No specimens, either skin or skull, of these three species were publicly available for academic study.

The conservation statuses for the three upper-mentioned species are DD for *H. winstoni* and *H. bartelsi*, and EN for *H. sipora*. The corresponding population trends for these squirrels are "Unknown" for the first two species, and "Decreasing" for the last one, according to IUCN Red List.

1.6.5 Global Conservation Status Classes Maps and Their Problems

In order to visualize the data from Table 1.4 more globally and geographically, Figs. 1.3 and 1.4 have been created. These figures present the global distribution of squirrel species that have been categorized based on their assigned conservation status classes.

These presented figures aim to visualize the global squirrel distribution with their assigned conservation status classes by app. geographic center. The legends on the bottom left side in Figs. 1.3 and 1.4 indicate what the colors stand for and which conservation status class each color represents. By observing these tables, it is easily noticeable that the predominant color

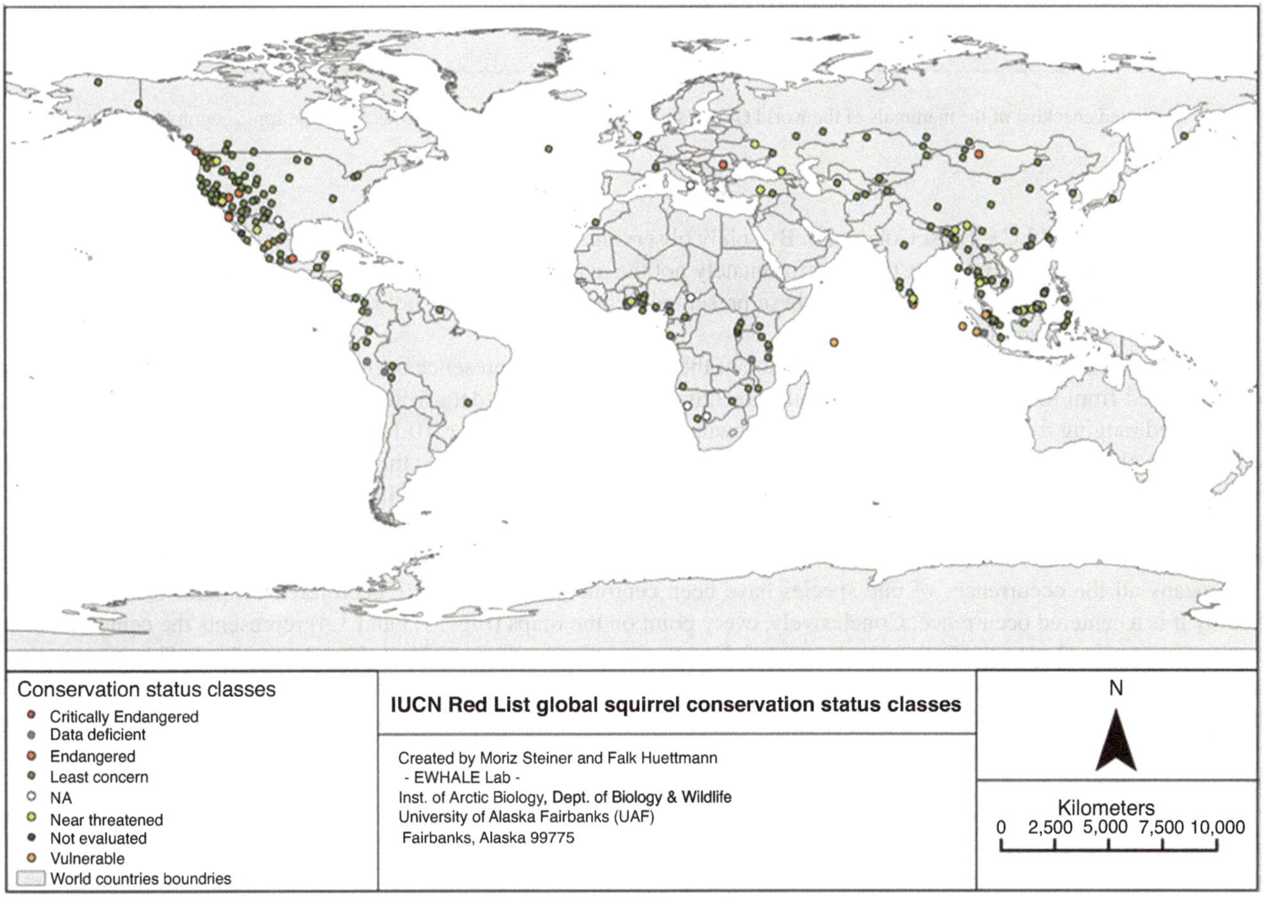

Fig. 1.3 IUCN Red List Global squirrel conservation status classes and their app. geographic centers

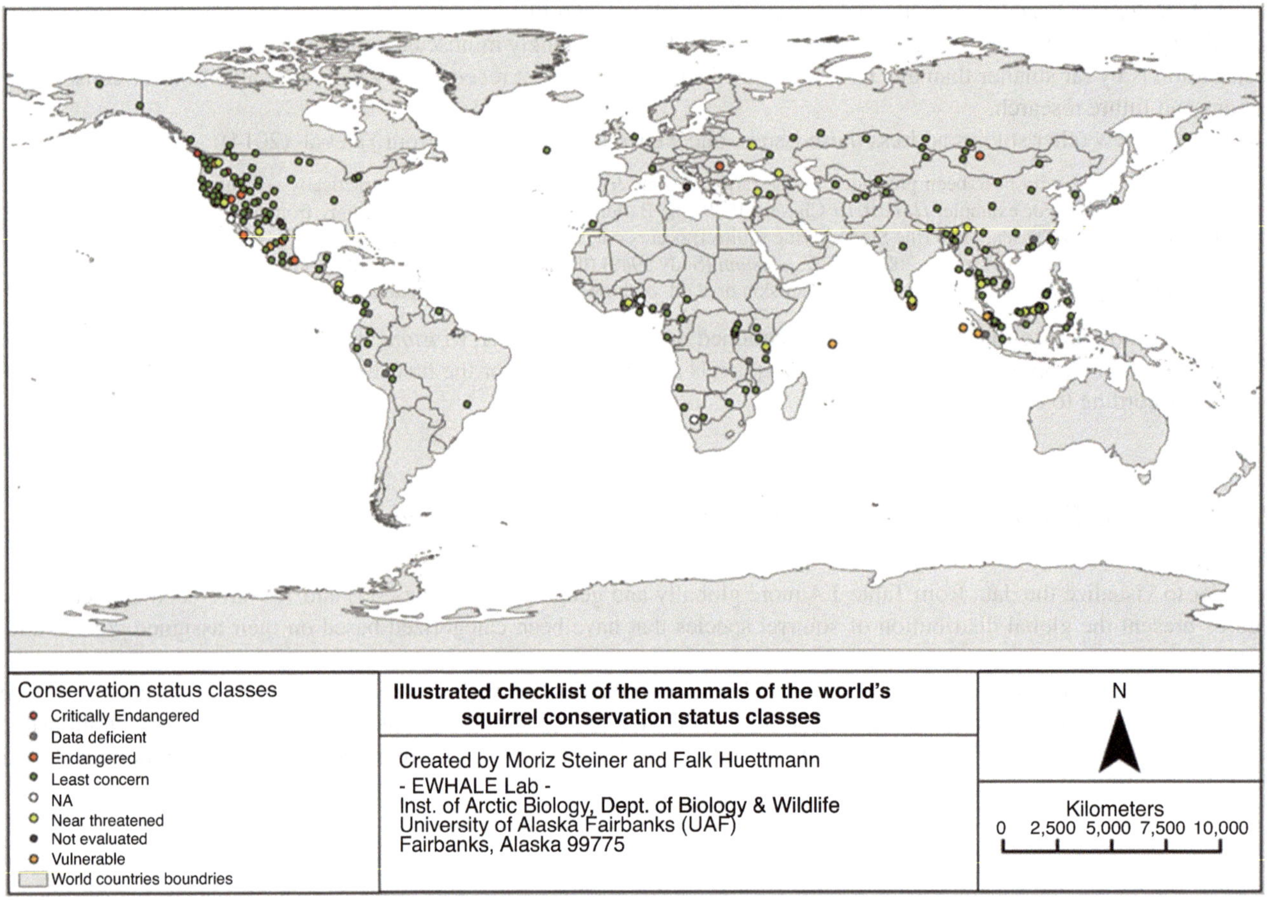

Fig. 1.4 Illustrated checklist of the mammals of the world Global squirrel conservation status classes and their app. geographic centers

is green (standing for "LC – Least Concern"). By solely observing these figures the global squirrel conservation seems to be in a pretty good situation. However, this is unfortunately not the reality. By explaining how these datasets have been created, we aim to provide a better insight into why these presented figures are in reality not representing the current conservation situation of the global squirrels at all.

Here we start with the georeferenced distribution of the squirrel species presence occurrences. These occurrence data have been obtained from the GBIF website as a bulk download where presence data from all global squirrel species have been downloaded ranging from the year 1960 to 2020 (download DOI: https://doi.org/10.15468/dl.665b59). This enormous dataset has been cleaned up, all subspecies have been removed as well as other species that are commonly not acknowledged to be part of the squirrel family (Sciuridae). Additionally, all occurrences containing spelling mistakes in the species name have been corrected and occurrences without geographical information have been removed. This cleaned dataset has been imported into QGIS as a .csv file. There, the vector analysis tool "Mean coordinate(s)…" has been used to center all the data points. This means all the occurrences of one species have been centralized to one single georeferenced point on the map, and thereby it is a centered occurrence. Conclusively, every point on the maps (Figs. 1.3 and 1.4) represents the centered occurrence of one squirrel species. To this newly created dataset, the information in Table 1.4 has been added. This means, every centered georeferenced squirrel presence point has been associated with the assigned conservation status class. So in QGIS, the data can be visualized based on the newly added data categories. This results in the data points being colored according to their assigned conservation status classes. To facilitate the interpretation of the graph, the official colors of the IUCN conservation status classes have been used when selecting the colors for the different classes. During this process, one major difficulty must be faced which leads to the conclusive statement above: these maps hardly represent the true conservation situation appropriately. This difficulty is created by the immense discrepancy between the GBIF taxonomic occurrence data set and the taxonomic lists from the IUCN Red List and the "Illustrated checklist of the mammals of the world" (Burgin et al. 2020). Already a very high number of species (79 out of around 300) are not found in the GBIF data set. Therefore, these species cannot be mapped, as geographic data is unavailable. If these 79 species could be added to the maps (Figs. 1.3 and 1.4), the global situation would look different. This is because many of these 79 species are DD (Data deficient – 22), NE (Not evaluated – 17), EN (Endangered – 6), VU (Vulnerable – 8), and only a few species are NT (Near threatened – 6), and LC (Least concern – 20). The numbers from NT and LC are – compared to the other categories – quite high, but in relation to the total number of these classes (see Table 1.4) they are actually very low. Therefore, in reality, the situation is much worse than depicted in those Figures above.

1.6.6 Over-Optimistic Conservation Classification of Species Dominates 'Authoritative' Sources Drive the Agenda

A general trend observed in Table 1.4 is that many species (195 and 191 for IUCN Red List and "Illustrated checklist of the mammals of the world" respectively) are considered to be Least concerned. The definition of "Least concern" according to IUCN Red List is "*A taxon is Least Concern when it has been evaluated against the criteria and does not qualify for Critically Endangered, Endangered, Vulnerable or Near Threatened. Widespread and abundant taxa are included in this category*". This pretty broad description allows a great range of species to satisfy these criteria and thus to be classified as such. It results in non-concrete classification boundaries and can also be observed in the distribution map that the IUCN Red List provides. Here, many habitats of these "Least concern species" are in fact very small (e.g. Olympic Marmot (*Marmota olympus*), Tien Shan Ground Squirrel (*Spermophilus nilkaensis*)), fragmented (e.g. Svynnerton's Bush Squirrel (*Paraxerus vexillarius*)), and sea-locked (e.g. Sanghir Squirrel (*Prosciurillus rosenbergii*), Busuanga Squirrel (*Sundasciurus hoogstraali*), Southern Palawan Tree Squirrel (*Sundasciurus steerii*)). If a species like *P. rosenbergii* is inhabiting such a small and remote island (Kepulauan Sangihe ~800 km^2), how can it be considered as a species with the least extinction concern, as it is completely locked to this singular minuscule island (for more detail see Steiner and Huettmann – Chap. 7, which deals with conservation issues of the world's squirrels on islands). Following the adage "One does not put all eggs in one basket."

And further, the percentage of species that are reported to have an increasing population size is minuscule (3–4 species out of around 300). This means that all other species (295+) must either be assigned to a stable, decreasing, or unknown population trend. Many, in fact, 100 squirrel species are considered to have the least extinction concern category, even though their population size is considered to be declining (25 species) or unknown (75 species) (see Table 1.4 and Fig. 1.5). These

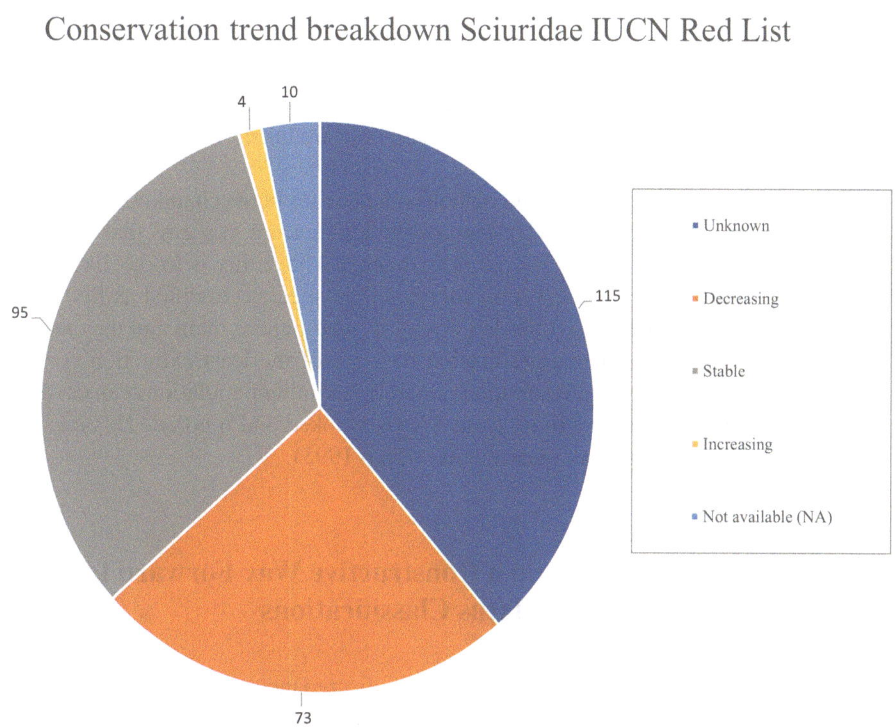

Fig. 1.5 Conservation trend breakdown Sciuridae Illustrated Checklist of the mammals of the world. Numbers indicate the number of species corresponding to the different classes. (Created by MS)

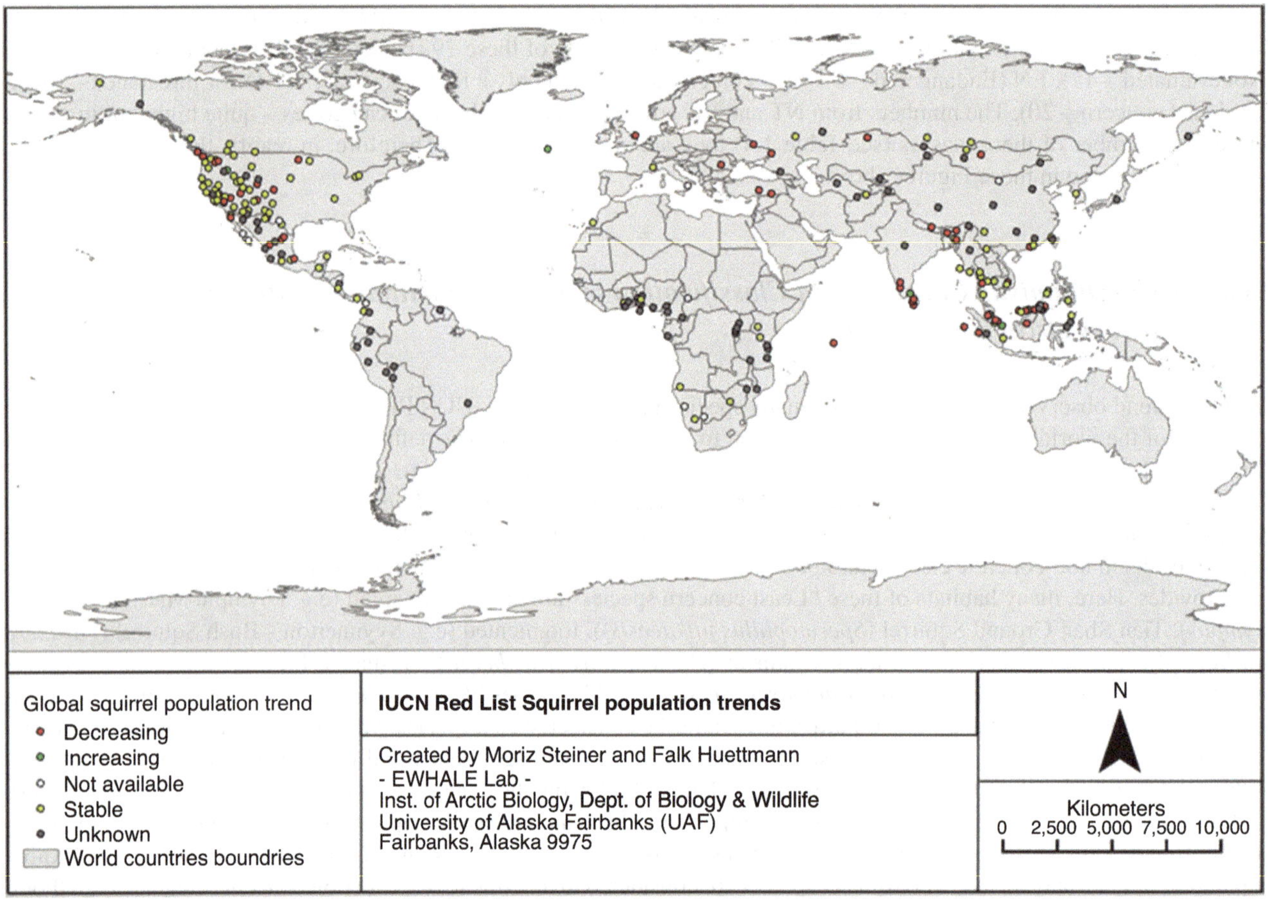

Fig. 1.6 IUCN Red List Global Squirrel population trends with app. geographic species centers

breakdowns are depicted in Fig. 1.5 as a pie chart to facilitate the interpretation and to better identify the magnitude of the presented issue (high number of "Unknown" and "Decreasing" population trends).

Figure 1.6 presented below depicts these mentioned population trends in a map form, using the same centered georeferencing method as described in Sect. 1.6.4. These data are taken from the IUCN Red List website as it is the only institution reporting this kind of data.

Considering all these observed and obvious mismatches between unknown and declining population trends, as well as the classification of these species to be of least extinction concern, it seems that many of these conservation status classes have been assigned to species in an over-optimistic manner. The reality is worse, likely much worse. Species that are classified as LC, receive generally less conservation attention than species classified as EN or VU, as they are not the focus of species conservation (see IUCN Red List for definition). It marginalizes them and they receive hardly any priority. This seems obvious because according to their classification they are of the least extinction concern. However, because of this artificial, man-made lowered conservation attention, possible declining populations can easily be overlooked by conservation management agencies/experts; and we think many are overlooked and ignored. This can conclusively easily result in a very small population size with a high risk of extinction (Raup 1995).

1.7 Proposed Solutions and a Constructive Way Forward for Global Squirrel Taxonomy and its Conservation Status Classifications

Greatly diminishing, or even erasing taxonomic discrepancies among institutions and generally, the field of taxonomy is commonly understood as a hardly feasible task. Here we presented several future pathways which aim to initiate these impossible-thought changes, with the greater ambition to improve institutional taxonomy and, thereby, facilitate species conservation and preservation.

1. A major revolution for many institutions would be the implementation of one mutually accepted and current species list. It would act as a reference for all globally important institutions, taxonomic, ecologic, and especially citizen-science-based ones. Citizen-science-based institutions have been highlighted here because this is a very common source of taxonomic errors, making taxonomic development and conservation management almost impossible. To better explain this, when citizens are logging taxonomy-based data online, it gets barely revised, and the citizens have the freedom to associate the recorded data to any species, that they considered appropriate. It can be imagined that thereby errors and obsolete, outdated classifications are imported and maintained in the database without a proper option for correction. The acceptance of a universally accepted species list could erase these occurring errors. By providing the opportunity/obligation to associate the logged data to an existing taxonomic species list, it can be avoided that those invalid species names are continuously logged and carried through – wrongly- in the database. To even further optimize this, it is proposed to establish collaborations between these mentioned citizen-science-based institutions, e.g. I-Naturalist, or GBIF, and institutions that are specifically focused on taxonomy, such as ITIS or MDD. If institutions like ITIS or MDD, are willing to provide citizen-science-based institutions with continuously updated, especially valid species lists (as their aim), and are willing to accept these lists and implement them, it would revolutionize institutional taxonomy for better management. If these kinds of collaborations could finally be established, the taxonomic discrepancies among institutions would assumingly drastically decrease. This is because most of the institutions would use the same species list and thereby enabling them to base all their data on A officially valid singular species list, ideally with a voucher specimen and documentation. This aligned taxonomy has a great influence on species conservation management and land/habitat questions, but not only that, because taxonomic chaos can also be avoided, and conservation projects can be executed more successfully.

1.7.1 Global Other Species/Genera/Families in Immediate Need of a Unified Taxonomic Revision

This final section aims to present additional species/genera/families that are not part of the Sciuridae family in need of a unified taxonomic revision. Table 1.5, presented below, aims to address these taxonomic issues for the species/genera/families to provide a bigger outlook, to support studies that pursue to change the taxonomy of these species. This list also aims to provide future research possibilities and supports our statements above that the taxonomy of some of our global species is still in major dispute.

Table 1.5 has been partly compiled and updated from a previously published table by the authors (Table 1 – Steiner and Huettmann 2021). It is understood that many of those examples provided in this table will not be addressed any time soon, and thus we add caution to a good and overly positive outlook that taxonomy-based species conservation would bring. Alternatives should be considered instead, e.g. to be more habitat and pixel-based.

1.8 Conclusion

We conclude that the squirrel species of the world and their taxonomies are not well aligned and poorly governed. With this chapter, the authors aim to spread awareness of this old western and still ongoing problem of small mammals and generic taxonomy for better conservation governance. It would be essential to consider these issues, re-review the species under major discrepancy, and strive therefore toward a more successful, realistic, precise, and effective species conservation management. This is relevant to effectively support squirrels overall with the far-reaching modern changes (e.g. caused by humans and nature – climate change, sea-level rise, habitat declines, etc.), and to conserve these beautiful creatures – a trusted public resource by law – and their habitats and wider ecosystems for future generations, just as many indigenous land management schemes did for millennia. Overall, a seemingly obvious trend has been observed throughout the data compilation for this study and the performed assessments. This trend consisted of an over-positive but unjustified conservation class assignation and agenda by some leading authorities and governance (e.g. IUCN Red List). This is especially true for species with a decreasing and unknown population trend. For a relevant conservation action, also, greater attention should be paid to the species that are classified as "Data Deficient" (DD). Because commonly, species classified as "Critically Endangered" (CR) or "Endangered" (EN), receive significantly higher conservation attention than DD species, even though those DD species are likely in even higher extinction danger than CR and EN species. The global realities like lack of resources, unabated poverty, governance failures, synergies, and finite resources must be fully considered. Conclusively, we suggest considering the mentioned wider concepts – including habitat and future perspectives for future generations – for the upcoming taxonomic revision of the squirrels in order to reach the highest possible conservation success and public governance trust possible.

Table 1.5 Global species/genera/families in immediate need of a taxonomic revision

Species	Short family/genus/species description	Description of the misclassification/taxonomic chaos
Brown Bear (*Ursus arctos*)	The brown bear (*Ursus arctos*) is found across Eurasia and North America and is divided into five different clades, some of which coexist in different regions (McLellan et al. 2017; Servheen et al. 1999)	For several bear species, the same set of subspecies is listed and causes large confusion problems in its conservation effort (Wilson and Ruff 1999). In addition, a recent hybridization with polar bears creates 'new species set ups' and management and legal problems in times of climate change
Red Wolfs (*Canis rufus*)	This species is native to the southeastern United States and its major characteristic is a reddish-tawny color to its fur. This species is currently recognized as endangered and grants protected status (Hinton et al. 2013)	The species status here is unclear, scientists do not agree whether it is a wolf species itself, a subspecies, or a hybrid between two subspecies (Phillips and Henry 1992)
Gorilla (*Gorilla gorilla*)	Inhabit the forest of central Sub-Saharan Africa, gorillas are ground-dwelling, predominantly herbivorous apes with a species division into the eastern gorillas and the western gorillas (both critically endangered)	Miscommunication between scientists can lead to taxonomic chaos and correspondingly to conservational interference (Cotterill et al. 2014"), shown by this species (see Groves 2002 for details)
Tree Kangaroos (*Dendrolagus*)	Tree-kangaroos are marsupials, inhabiting the tropical rainforests of New Guinea and far northeastern Australia, along with some of the islands in the region. All tree-kangaroos are considered threatened due to hunting and habitat destruction. They are the only true arboreal macropods (Procter-Gray and Ganslosser 1986)	Classification issues of Species and sub-species levels are broadly known and do not seem to be resolvable any time soon (Martin 2005)
Snow Petrels (*Pagodroma nivea*)	Is the only member of the genus *Pagodroma* and is divided into two subspecies (*P. nivea nivea* and *P. nivea confusa*), where individuals primarily differ in size (Henri and Schön 2017)	The taxonomic status of the Snow Petrel remains the subject of considerable controversy (Shirihai and Jarrett 2007; del Hoyo and Collar 2014). Few "pure" populations of both "subspecies" are known whereas most colonies consist of hybrids which can make it highly difficult for conservation management, e.g. during climate change with Antarctica as one of the strongest affected areas in man-made climate change
Birds of Paradise (*Paradisaeidae*)	The majority of species are found in Papua New Guinea and eastern Australia. The family has 42 species in 15 genera (Gill et al. 2020)	There is a long taxonomic debate about this species group for over 100 years. More taxonomic chaos has been created in the past years for this group by mixing instead of combining the genetic taxonomic approach and the morphologic one. It resulted in the classifications of several species and genera in different and multiple families (Irestedt et al. 2008). However, the conservation status of this group remains in peril
Canada Geese (Dusky Goose) (*Branta canadensis occidentalis*)	Dusky Canada geese represent one of the smallest populations of Canada goose in North America (William L. Finley – U.S. Fish and Wildlife Service)	The dusky Canada goose is occasionally merged with the Vancouver Canada goose which causes international misunderstandings with conservation management (Bromley 2003). It's of major legal concern for hunting
Lizards (*Lacertilia sp.*)	Lizards are a widespread group of reptiles with over 6000 species which are characterized by a small head and neck with a long body and tail	The taxonomy has always been in debate for this group and is currently outdated for the entire suborder. The latest update dates back to 1988 (Gauthier et al. 1988)
Arctic Cod (*Arctogadus glacialis*)	Arctic Cod is found circumpolar, also called Polar Cod	There are many genetic questions related to stocks and general biological characteristics, thus resulting in fishing questions about quotas and their disputes in the changing Arctic (as summarized by Cohen et al. 1990)
Patagonian Toothfish (*Dissostichus eleginoides*)	The toothfish is found in Antarctic waters and replaces the shark species there	This is a long-lived species and its life history details are known, as needed for valid fisheries management. Arguably, the Patagonian toothfish is a taxonomic group that is generally on the conservation decline and with Climate Change going on unabated (Huettmann 2012)
Charr (*Salvelinus*)	This genus contains primarily cold and fresh-water fish in the family Salmonidae, distributed circumpolar	The diversity within this genus makes speciation difficult and confusing. The fish inhabit different habitats, where food availabilities differ as well as other circumstantial factors, leading to high polytheism of several species and making it difficult for diversification of species (Reist et al. 2013)

(continued)

Table 1.5 (continued)

Species	Short family/genus/species description	Description of the misclassification/taxonomic chaos
Zooplankton (*No overarching scientific name available*)	This species group sits at the core of the global food chain and is now even harvested in Antarctic waters, causing probable food shortages for the entire fauna of the Antarctic region (Trathan and Reid 2009)	Already the life history of this species group is complex, with many life stages to be identified over years. However, the species' genetics remains to be resolved and with a matching morphological identification, key mutually accepted worldwide. Such an agreed classification and subsequent conservation management do not exist, and likely never will (Chiba et al. 2018) while zooplankton harvest is already practiced on an industrial scale
Rhodophyta (*Gracilaria*)	Rhodophyta also called red algae, are one of the oldest eukaryotic algae (Lee 2008)	According to Junfu and Bangmei (1984), the taxonomy of some species remains unsolved because of their extreme polymorphism and the lack of authentic specimens for comparative studies
(No common name available) *Lactobacillus acidophilus*	It is a species of gram-positive bacteria in the genus Lactobacillus and is a commercially significant bacterial probiotic	L. acidophilus has struggled with misidentification and misrepresentation, caused by limitations of differentiating phenotypically similar species by morphological and biochemical means (Bull et al. 2013), with negative influences on its use in the commercial field
Purple sulfur bacteria (*Allochromatium* and *Thiocapsa*)	These two bacteria genera occur mostly in dark environments, far from what is often considered a favorable living environment. Many members conduct anoxygenic photosynthesis	At the species level, several inconsistencies exist between the phenotypic and phylogenetic data, highlighting taxonomic problems within these genera (Herbert et al. 2005)

References

Ahmadi S, Maman S, Zoumenou R, Massougbodji A, Cot M, Glorennec P, Bodeau-Livinec F (2018) Hunting, sale, and consumption of bushmeat killed by Lead-Based ammunition in Benin. Int J Environ Res Public Health 15(6):1140

Álvarez-Castañeda ST (2007) Systematics of the antelope ground squirrel (Ammospermophilus) from islands adjacent to the Baja California peninsula. J Mammal 88(5):1160–1169

Arbogast BS, Schumacher KI, Kerhoulas NJ, Bidlack AL, Cook JA, Kenagy GJ (2017) Genetic data reveal a cryptic species of New World flying squirrel: glaucomys oregonensis. J Mammal 98(4):1027–1041

Beard J (2009) James Beard's American Cookery. Voracious

Benson E (2013) The urbanization of the eastern gray squirrel in the United States. J Am Hist 100(3):691–710

Bertolino S, Lurz PW (2013) Callosciurus squirrels: worldwide introductions, ecological impacts and recommendations to prevent the establishment of new invasive populations. Mammal Rev 43(1):22–33

Bertram BC, Moltu DP (1986) Reintroducing red squirrels into Regent's Park. Mammal Rev 16(2):81–88

Bizhanova N, Steiner M, Rametov N, Grachev A, Grachev Y, Bespalov M, Zhaparkulov T, Saparbayev S, Sailaukhanuly A, Bespalov S, Bolatuly A, Saparov K, Mohd Sah SA (2022) The elusive Turkestan lynx at the northwestern edge of geographic range: current suitable habitats and distribution forecast in the climate change. Sustainability MDPI

Boonstra R, Krebs CJ, Kanter M (1990) Arctic ground squirrel predation on collared lemmings. Can J Zool 68(4):757–760

Brehm JM, Maxted N, Ford-Lloyd BV, Martins-Louçao MA (2008) National inventories of crop wild relatives and wild harvested plants: case-study for Portugal. Genet Resour Crop Evol 55(6):779–796

Brockway LH (1979) Science and colonial expansion: the role of the British Royal Botanic Gardens. Am Ethnol 6(3):449–465

Bromley RGH (2003) Conservation assessment for the dusky Canada goose (Branta canadensis occidentalis Baird). Vol. 591. Portland, OR: US Department of Agriculture, Forest Service, Pacific Northwest Research Station

Brugière D, Fleury MC, Colyn M (2005) Structure of the squirrel community in the Forêt des Abeilles, central Gabon: rediscovery and revalidation of Funisciurus duchaillui Sanborn, 1953. Mammalia 69(2):223–231

Bull M, Plummer S, Marchesi J, Mahenthiralingam E (2013) The life history of Lactobacillus acidophilus as a probiotic: A tale of revisionary taxonomy, misidentification and commercial success. FEMS Microbiology Letters 349:77–87. https://doi.org/10.1111/1574-6968.12293

Burgin CJ, Wilson DE, Mittermeier RA, Rylands AB, Lacher TE, Sechrest W (2020) Illustrated checklist of the mammals of the world. Lynx Ediciones, Barcelona

Callahan JR (1993) Squirrels as predators. The Great Basin Naturalist, 137–144

Chavan VS, Rane NS, Watve A, Ruggiero M (2005) Resolving taxonmic discrepancies: role of electronic catalogues of known organisms. Biodivers Inform 2

Chiba S, Batten S, Martin CS, Ivory S, Miloslavich P, Weatherdon LV (2018) Zooplankton monitoring to contribute towards addressing global biodiversity conservation challenges. Journal of Plankton Research 40:509–518. https://doi.org/10.1093/plankt/fby030

Choudhury A (2002) Petaurista nobilis singhei-first record in India and a note on its taxonomy. J Bombay Nat Hist Soc 99(1):30–34

Choudhury AU (2007) A new gliding squirrel of the genus Petaurista Link from Arunachal Pradesh in north-east India. Newsl J Rhino Found Nat NE India 7:26–32

Choudhury AU (2009a) One more new species of Giant Flying Squirrel of the Genus Petaurista Link, 1795 from Arunachal Pradesh in north-east India. Newsl J Rhino Found Nat NE India 8:26–34

Choudhury AU (2009b) Replacement Latin name for Mechuka giant flying squirrels of the genus Petaurista Link, 1795. Newsl J Rhino Found Nat NE India 8:48

Choudhury AU (2013) Descriptions of a new species of giant flying squirrel of genus Petaurista Link, 1795 from Siang basin, Arunachal Pradesh in North East India. Newsl J Rhino Found Nat NE India 9:30–38

Cohen DM, Inada T, Iwamoto TN, Scialabba N (1990) FAO species catalogue. Vol. 10. Gadiform fishes of the world (Order Gadiformes). An annotated and illustrated catalogue of cods, hakes, grenadiers and other gadiform fishes known to date. FAO Fisheries Synopsis 125:1–442

Cotterill FP, Taylor PJ, Gippoliti S, Bishop JM, Groves CP (2014) Why one century of phenetics is enough: Response to "are there really twice as many bovid species as we thought?". Systematic Biology 63:819–832. https://doi.org/10.1093/sysbio/syu003

Czech B (2000) Economic growth as the limiting factor for wildlife conservation. Munich Personal RePEc Archive

D'Elía G, Fabre PH, Lessa EP (2019) Rodent systematics in an age of discovery: recent advances and prospects. J Mammal 100(3):852–871

Dalton JE (2017) Arctic ground squirrels as ecosystem engineers of Projected Tundra Shrub Encroachment. Doctoral dissertation, University of Illinois at Chicago

Del Hoyo J, Collar NJ (2014) HBW and birdlife international illustrated checklist of the birds of the world. Vol. 1: Non-passerines. Barcelona: Lynx Edicions

den Tex RJ, Thorington R, Maldonado JE, Leonard JA (2010) Speciation dynamics in the SE Asian tropics: putting a time perspective on the phylogeny and biogeography of Sundaland tree squirrels, Sundasciurus. Mol Phylogenet Evol 55(2):711–720

Dissanayake R, Oshida T (2012) The systematics of the dusky striped squirrel, Funambulus sublineatus (Waterhouse, 1838) (Rodentia: Sciuridae) and its relationships to Layard's squirrel, Funambulus layardi Blyth, 1849. J Nat Hist 46(1-2):91–116

Dobson FS (1990) Environmental influences on infanticide in Columbian ground squirrels. Ethology 84(1):3–14

Dubois A (2003) The relationships between taxonomy and conservation biology in the century of extinctions. Comptes Rendus Biologies 326:9–21

Ducarme F, Luque GM, Courchamp F (2013) What are "charismatic species" for conservation biologists. BioSciences Master Rev 10(2013):1–8

Ebach MC, Michaux B (2017) Establishing a framework for a natural area taxonomy. Acta Biotheor 65(3):167–177

Ebensperger LA, Blumstein DT (2007) Nonparental infanticide. In: Rodent societies: an ecological and evolutionary perspective, pp 267–279

Ewacha MV, Kaapehi C, Waterman JM, Roth JD (2016) Cape ground squirrels as ecosystem engineers: modifying habitat for plants, small mammals and beetles in Namib Desert grasslands. Afr J Ecol 54(1):68–75

Fa JE, Juste J, Burn RW, Broad G (2002) Bushmeat consumption and preferences of two ethnic groups in Bioko Island, West Africa. Hum Ecol 30(3):397–416

Figueiredo E, Smith GF (2010) The colonial legacy in African plant taxonomy. S Afr J Sci 106(3-4):1–4

Garofalo L, Mariacher A, Fanelli R, Fico R, Lorenzini R (2018) Hindering the illegal trade in dog and cat furs through a DNA-based protocol for species identification. PeerJ 6:e4902

Gauthier J, Kluge AG, Rowe T (1988) Amniote phylogeny and the importance of fossils. Cladistics 4:105–209. https://doi.org/10.1111/j.1096-0031.1988.tb00514.x

Gill F, Donsker D, Rasmussen P, Eds (2020) IOC world bird list (v10.1). https://doi.org/10.14344/IOC.ML.10.1

Griffiths JF, Griffiths AJ, Wessler SR, Lewontin RC, Gelbart WM, Suzuki DT, Miller JH (2005) An introduction to genetic analysis. Macmillan

Groves CP (2002) A history of gorilla taxonomy. In: Gorilla biology: A multidisciplinary perspective. Cambridge, UK: Cambridge University Press. pp. 15–34

Gündüz İ, Jaarola M, Tez C, Yeniyurt C, Polly PD, Searle JB (2007) Multigenic and morphometric differentiation of ground squirrels (Spermophilus, Scuiridae, Rodentia) in Turkey, with a description of a new species. Mol Phylogenet Evol 43(3):916–935

Hawkins MT, Helgen KM, Maldonado JE, Rockwood LL, Tsuchiya MT, Leonard JA (2016) Phylogeny, biogeography and systematic revision of plain long-nosed squirrels (genus Dremomys, Nannosciurinae). Mol Phylogenet Evol 94:752–764

Hebert PD, Gregory TR (2005) The promise of DNA barcoding for taxonomy. Syst Biol 54(5):852–859

Helgen KM, Cole FR, Helgen LE, Wilson DE (2009) Generic revision in the Holarctic ground squirrel genus Spermophilus. J Mammal 90(2):270–305

Henri R, Schön I (2017) Phylogeography and taxonomy of the snow petrel (Pagodroma nivea sl). In: Book of Abstracts. Brussels, Belgium: Royal Belgian Institute of Natural Sciences. pp. 82

Herbert RA, Ranchou-Peyruse A, Duran R, Guyoneaud R, Schwabe S (2005) Characterization of purple sulfur bacteria from the South Andros Black Hole cave system: Highlights taxonomic problems for ecological studies among the genera Allochromatium and Thiocapsa. Environmental Microbiology 7:1260–1268. https://doi.org/10.1111/j.1462-2920.2005.00815.x

Hinckley A, Hawkins MT, Achmadi AS, Maldonado JE, Leonard JA (2020) Ancient divergence driven by geographic isolation and ecological adaptation in forest dependent Sundaland tree squirrels. Front Ecol Evol 8:208

Hinton JW, Chamberlain MJ, Rabon DR (2013) Red wolf (Canis rufus) recovery: A review with suggestions for future research. Animals 3:722–744. https://doi.org/10.3390/ani3030722

Hope AG, Malaney JL, Bell KC, Salazar-Miralles F, Chavez AS, Barber BR, Cook JA (2016) Revision of widespread red squirrels (genus: Tamiasciurus) highlights the complexity of speciation within North American forests. Mol Phylogenet Evol 100:170–182

Huettmann F, Ed. (2012) Protection of the three poles. New York City, NY: Springer Science & Business Media

Irestedt M, Fuchs J, Jonsson KA, Ohlson JI, Pasquet E, Ericson PG (2008) The systematic affinity of the enigmatic Lamprolia victoriae (Aves: Passeriformes)? An example of avian dispersal between New Guinea and Fiji over Miocene intermittent land bridges? Molecular Phylogenetics and Evolution 48:1218–1222. https://doi.org/10.1016/j.ympev.2008.05.038

Jackson SM (2012) Gliding mammals of the world. Csiro Publishing

Jackson SM, Thorington RW Jr (2012) Gliding mammals: taxonomy of living and extinct species. Smithsonian Contributions to Zoology

Jansen L (2009) Classifications. In: Applied ontology: an introduction, pp 159–172

Jordan JS (1971) Yield from an intensively hunted population of Eastern Fox Squirrls, vol 186. Northeastern Forest Experiment Station, Forest Service, US Department of Agriculture

Junfu Z, Bangmei X (1984) Some problems in the taxonomy of Chinese species of Gracilaria (Rhodophyta). In: Eleventh international seaweed symposium. Dordrecht: Springer. pp. 59–62

Junqué de Fortuny E, Martens D, Provost F (2013) Predictive modeling with big data: is bigger really better? Big Data 1(4):215–226

Keuling O, Podgórski T, Monaco A, Melletti M, Merta D, Albrycht M, Gongora J et al (2018) Eurasian wild boar Sus scrofa (Linnaeus, 1758)
Kline PD (1964) Iowa squirrels: hunting statistics, sex and age ratios, and the influence of mast and agriculture. Proc Iowa Acad Sci 71(1):216–227
Koli VK, Bhatnagar C, Sharma SK (2013) Distribution and status of Indian giant flying squirrel (Petaurista philippensis Elliot) in Rajasthan, India. Natl Acad Sci Lett 36(1):27–33
Krishna MC, Kumar A, Tripathi OP, Koprowski JL (2016) Diversity, distribution and status of gliding squirrels in protected and non-protected areas of the Eastern Himalayas in India
Kryštufek B, Vohralík V (2012) Taxonomic revision of the Palaearctic rodents (Rodentia): Sciuridae: Xerinae 1 (Eutamias and Spermophilus)
Kryštufek B, Vohralík V (2013) Taxonomic revision of the Palaearctic rodents (Rodentia). Part 2. Sciuridae: Urocitellus, Marmota and Sciurotamias. Lynx, Series nova, 44
Kryštufek B, Mahmoudi A, Tesakov AS, Matějů J, Hutterer R (2016) A review of bristly ground squirrels Xerini and a generic revision in the African genus Xerus. Mammalia 80(5):521–540
Kwon D (2023) Highly cited genetics studies found to contain sequence errors. Nature
Lane RS, Mun J, Eisen RJ, Eisen L (2005) Western gray squirrel (Rodentia: Sciuridae): a primary reservoir host of Borrelia burgdorferi in Californian oak woodlands? J Med Entomol 42(3):388–396
Lee RE (2008) Phycology. Cambridge, UK: Cambridge university press
Li S, Feng Z (2017) Geographic variation of the large red flying squirrel, Petaurista albiventer (Gray, 1834) (Rodentia: Sciuridae), with a description of a new subspecies in Southwestern China. Pak J Zool 49(4)
Li S, He K, Yu FH, Yang QS (2013) Molecular phylogeny and biogeography of Petaurista inferred from the cytochrome b gene, with implications for the taxonomic status of P. caniceps, P. marica and P. sybilla. PLoS One 8(7):e70461
Li Q, Li XY, Jackson SM, Li F, Jiang M, Zhao W, Jiang XL et al (2019) Discovery and description of a mysterious Asian flying squirrel (Rodentia, Sciuridae, Biswamoyopterus) from Mount Gaoligong, southwest China. ZooKeys 864:147
Martin R (ed) (2005) Tree-kangaroos of Australia and New Guinea. CSIRO Publishing
May RM (1990) Taxonomy as destiny. Nature 347(6289):129–130
McLellan BN, Proctor MF, Huber D, Mitchel S (2017) Brown bear. IUCN Red List Threat Species, 2017–3
Mendes CP, Koprowski JL, Galetti M (2019) Neosquirrel: a data set of ecological knowledge on Neotropical squirrels. Mammal Rev 49(3):210–225
Morden T (2017) Principles of management. Routledge
Mori E, Menchetti M, Zozzoli R, Milanesi P (2019) The importance of taxonomy in species distribution models at a global scale: the case of an overlooked alien squirrel facing taxonomic revision. J Zool 307(1):43–52
Musser GA, Durden LA, Holden ME, Light JE (2010) Systematic review of endemic Sulawesi squirrels (Rodentia, Sciuridae), with descriptions of new species of associated sucking lice (Insecta, Anoplura) and phylogenetics and zoogeographic assessments of Sciurid lice. Bull Am Mus Nat Hist 339(1):1–260
Nguyen ST, Oshida T, Dang PH, Bui HT, Motokawa M (2018) A new species of squirrel (Sciuridae: Callosciurus) from an isolated island off the Indochina Peninsula in southern Vietnam. J Mammal 99(4):813–825
Ohl M (2007) 4 principles of taxonomy and classification: current procedures for naming and classifying organisms. In: Handbook of paleoanthropology. Springer, Berlin/Heidelberg. https://doi.org/10.1007/978-3-540-33761-4_4
Patterson BD, Norris RW (2016) Towards a uniform nomenclature for ground squirrels: the status of the Holarctic chipmunks. Mammalia 80(3):241–251
Payne J, Francis CM, Phillipps K (1985) Field guide to the mammals of Borneo. Sabah Society
Pelech SA, Smith JN, Boutin S (2010) A predator's perspective of nest predation: predation by red squirrels is learned, not incidental. Oikos 119(5):841–851
Phillips MK, Henry VG (1992) Comments on red wolf taxonomy. Conservation Biology 6(4):596–599
Potter B (2002) The tale of Timmy Tiptoes, vol 12. Penguin
Procter-Gray E, Ganslosser U (1986) The individual behaviors of Lumholtz's tree-kangaroo: repertoire and taxonomic implications. J Mammal 67(2):343–352
Pyle RL, Barik SK, Christidis L, Conix S, Costello MJ, van Dijk PP, Thiele KR et al (2021) Towards a global list of accepted species V. The devil is in the detail. Org Divers Evol:1–19
Raup DM (1995) Extinction rates. vol 11. In: Lawton JH, May RM (eds). Oxford University Press, Oxford
Reist JD, Power M, Dempson JB (2013) Arctic charr (Salvelinus alpinus): A case study of the importance of understanding biodiversity and taxonomic issues in northern fishes. Biodiversity 14:45–56. https://doi.org/10.1080/14888386.2012.725338
Rich B (2014) Mortgaging the Earth: World Bank, environmental impoverishment and the crisis of development. Routledge
Salkeld DJ, Leonhard S, Girard YA, Hahn N, Mun J, Padgett KA, Lane RS (2008) Identifying the reservoir hosts of the Lyme disease spirochete Borrelia burgdorferi in California: the role of the western gray squirrel (Sciurus griseus). Am J Trop Med Hyg 79(4):535
Sanamxay D, Douangboubpha B, Bumrungsri S, Xayavong S, Xayaphet V, Satasook C, Bates PJ (2013) Rediscovery of Biswamoyopterus (Mammalia: Rodentia: Sciuridae: Pteromyini) in Asia, with the description of a new species from Lao PDR. Zootaxa 3686(4):471–481
Schauffert CA, Koprowski JL, Greer VL, Alanen MI, Hutton KA, Young PJ (2002) Interactions between predators and Mt. Graham red squirrels (*Tamiasciurus hudsonicus grahamensis*). Southwest Nat 47(3):498–501
Schlesinger J (2017) A world trimmed with fur: wild things, pristine places, and the natural fringes of Qing rule. Stanford University Press
Servheen C, Herrero S, Peyton B, Pelletier K, Moll K, Moll J, eds. (1999) Bears: Status survey and conservation action plan (PDF). Missoula, MT: IUCN The World Conservation Union
Shirihai H, Jarrett B (2007) A complete guide to antarctic wildlife: A complete guide to the birds, mammals and natural history of the antarctic. London, UK: A & C Black London. pp. 544
Silvy NJ (ed) (2020) The wildlife techniques manual: volume 1: Research. Volume 2: management. JHU Press
Skibins JC, Powell RB, Hallo JC (2013) Charisma and conservation: charismatic megafauna's influence on safari and zoo tourists' pro-conservation behaviors. Biodivers Conserv 22(4):959–982
Smith CC (1970) The coevolution of pine squirrels (Tamiasciurus) and conifers. Ecol Monogr 40(3):349–371
Song LI, Fa-Hong YU (2013) Differentiation in cranial variables among six species of Hylopetes (Sciurinae: Pteromyini). Zool Res 34(E5):531120-E

Steiner M, Huettmann F (2021) Justification for a taxonomic conservation update of the rodent genus Tamiasciurus: addressing marginalization and mis-prioritization of research efforts and conservation laissez-faire for a sustainability outlook. Eur Zool J 88(1):86–116
Thorington RW Jr, Ferrell KE (2006) Squirrels: the animal answer guide. JHU Press
Thorington RW Jr, Schennum CE, Pappas LA, Pitassy D (2005) The difficulties of identifying flying squirrels (Sciuridae: Pteromyini) in the fossil record. J Vertebr Paleontol 25(4):950–961
Thorington RW Jr, Koprowski JL, Steele MA, Whatton JF (2012) Squirrels of the world. JHU Press
Trathan PN, Reid K (2009) Exploitation of the marine ecosystem in the sub-Antarctic: Historical impacts and current consequences. Papers and Proceedings of the Royal Society of Tasmania 143:9–14. https://doi.org/10.26749/rstpp.143.1.9
Trulio LA (1996) The functional significance of infanticide in a population of California ground squirrels (Spermophilus beecheyi). Behav Ecol Sociobiol 38(2):97–103
Vane-Wright RI, Humphries CJ, Williams PH (1991) What to protect? – Systematics and the agony of choice. Biol Conserv 55(3):235–254
Vanitharani J (2018) Conservation status and guidelines for the maintenance of endangered Grizzled Giant Squirrel Ratufa macroura in Srivilliputhur Wildlife Sanctuary. In: Indian hotspots. Springer, Singapore, pp 297–307
Wauters LA, Amori G, Aloise G, Gippoliti S, Agnelli P, Galimberti A et al (2017) New endemic mammal species for Europe: Sciurus meridionalis (Rodentia, Sciuridae). Hystrix 28(1)
Wikipedia (2021) Flora & Ulysses (film). https://en.wikipedia.org/wiki/Flora_%26_Ulysses_(film). Accessed 25 July 2022
Wikipedia (2022) List of development aid country donors. https://en.wikipedia.org/wiki/List_of_development_aid:country_donors. Accessed 25 July 2022
Wilson DE, Mittermeier RA (2011) Handbook of the mammals of the world, volume 2: hoofed mammals. Lynx Ediciones
Wilson DE, Ruff S (1999) The Smithsonian book of North American mammals. Washington, DC: Smithsonian Institution Press
Yu F, Yu F, Pang J, Kilpatrick CW, McGuire PM, Wang Y, Woods CA et al (2006) Phylogeny and biogeography of the Petaurista philippensis complex (Rodentia: Sciuridae), inter-and intraspecific relationships inferred from molecular and morphometric analysis. Mol Phylogenet Evol 38(3):755–766
Yu F, Lian X, Li Z, Xie M (2014) A molecular phylogenetic study of Hylopetes (Rodentia: *Sciuridae*) inferred from mitochondrial cytochrome-b gene. Biologia 69(12):1777–1783. https://doi.org/10.2478/s11756-014-0474-5

Chapter 2
Evolution, Extinction, and Extinction Rate Estimates of the World Squirrels

...big, slow, ancient and meaty, and you are done with...

Source unknown

Abstract Beyond Charles Darwnin, the taxonomy and subsequent phylogeny of squirrels, their biogeography, and their evolution explicit in time and space remain poorly understood, widely understudied, and controversial but relevant for land management and global well-being. Squirrels co-evolved with human societies for millennia too. Here we present a somewhat simplified but first and robust sketch of an evolutionary tracking model of the world squirrel's dispersal over time. The evolutionary tracking model in the first part of this chapter is based on the most recent phylogenetic studies (e.g. Menendez et al. 2020) and the information on the species differentiation which has been associated with the locations of fossil records (from e.g. www.fossilworks.org). This has resulted in a model which depicts the differentiation of squirrel species, mapped in GIS, to facilitate the understanding of the species dispersal across most of our planet. By identifying the habitats to which the squirrels dispersed in the present days and by analyzing the trend from the past, this conceptual model allows to predict future dispersal trends of the global squirrel species. An agreed-upon quantified global squirrel phylogeny is absent. But, by assuming the predictions (based on our presented tracking models) are a start, more accurate, and correct, conservation actions can be suggested for the future habitats of the squirrels. Here we also discuss recent extinctions, past and recent squirrel extinctions, as well as possible current and future extinction causes. We found that sudden anthropological-induced climate change, disease spreading, habitat destruction, and species invasion were the most predominant causes of species extinction. Additionally, by handling squirrel conservation with long-term sustainability approaches, it can help in a proactive fashion to turn around possible negative trends and improve future squirrel conservation and the world's well-being overall. Lastly, by aiming to estimate the actual extinction rate of the already extinct squirrel species and trying to associate it with the extinction cause, we attempt to put into perspective how much of the extinctions are caused by human, climatic – or geologic influence, by comparing this extinction rate to speciation rates. We found a speciation rate of 7.7 extant species/million years, and an extinction rate of 10.5 species/million years, indicating a long-term loss of species.

Keywords Squirrels · Evolution · Extinction rates · Speciation rate · Dispersal · Geographic Information System (GIS) models

2.1 Introduction

Evolution is usually referenced with Charles Darwin and his widely accepted textbook principles, but instead it remains a complex discipline (Bennett 2013; Churakov et al. 2010; Cohen and Lloyd 2014; McShea 2021; Xuesen et al. 1993). It is a field of study which generally includes the history and the development of species. It is officially defined as "*the way in which living things change and develop over millions of years*" (Cambridge Dictionary 2021a). The evolution of mammals and therefore also squirrels, has been a topic that has been debated for centuries and even today it is not really mutually agreed upon in the scientific world and outside (Carter and Mess 2007; Millien and Bovy 2010; Pennisi 2009; Smith and Peterson 2002). In the last few decades, genetic research and its communal applications have made major leaps forward (Grados et al. 2003; Houston and Macqueen 2019; Orr 2001; Owen et al. 2004; etc.), but not reduced tension or disagreement at all; rather vice versa. The evolutionary history of large and small mammals has been the main focus of many researchers for a long time, spilling into religion, spirituality, how to run the world, and how to perceive ourselves.

Supplementary Information The online version contains supplementary material available at https://doi.org/10.1007/978-3-031-23547-4_2.

M. Steiner, F. Huettmann, *Sustainable Squirrel Conservation*, https://doi.org/10.1007/978-3-031-23547-4_2

Depending on the source, the origins of what we nowadays define as squirrels are suspected to have started with the "Grande Coupure" also known as the Eocene-Oligocene extinction event (Lopez and Thaler 1974; Vianey-Liaud 1979; Hartenberger 1983). After this massive extinction event, new species 'spilled' and dispersed into newly vacant habitats and ecological niches, leading to great species diversity. This "Grande Coupure" was suspectively caused by a sudden climate change (Châteauneuf 1980). The ancestral species of the squirrels have likely used these newly vacant niches and have over time specialized to fit within them. This has led to the evolution of the first species which are nowadays considered squirrels. Nowadays, squirrels occur on all continents except Antarctica (natural distribution also excludes Australia & Oceania) and inhabit almost all global ecosystems (Bertolino and Lurz 2013; Thorington et al. 2012), except Greenland, Iceland, Papua New Guinea, very high altitudes, and extreme deserts (Steiner and Huettmann 2023, Chap. 3).

Additionally, the squirrels being a family of which its species occur virtually globally, also have a major contribution and play a major role in most global ecosystems. Squirrels can be found in the tropical forest, in parts of the tundra of the arctic, in the semi-dry deserts of the Sahel region, on remote islands, in temperate forests, and nowadays even in cities (Gibbs et al. 2019; Koprowski and Nandini 2008; Wheeler and Hik 2013). This highlights the importance of further studying these species and genera since for approximately 98–99% (302/307) of the squirrel species no basic life history is known, feeding behavior recordings are widely absent, distribution ranges are widely only assumed and often hand-drawn without underlying data presented, and population trends and data are practically non-existing. Knowing that these species occur in almost all ecosystems of this planet, and the most basic information for the majority of the species is unknown, it seems crucial to call for more action, research efforts, attention from the public, funding, and most importantly, a change in the attitude of abandoning *laissez-faire* (Steiner and Huettmann 2021), and moving towards a sustainable, science-based, conscious management. Such topics become increasingly more important and crucial in modern times with the rapid changes in the world (Gunn et al. 2021; Sih 2013; Sih et al. 2011). With the lack of basic information on these species, it seems virtually impossible to predict any outcomes/impacts for how the species will respond/are responding to the current human-induced climate change, biodiversity loss, and more. Pro-active management is then pretty much impossible in the public eye.

In the last few years, a couple of publications have been added to literature that describe the current, state-of-the-art genetic history of squirrels (Menéndez et al. 2020; Mercer and Roth 2003). These latest publications offer some great insights into the holistic evolution of squirrels. In this study, we follow the phylogenetic history published by Menéndez et al. (2020) as a reference for a geographical tracking map of the global squirrels' evolution. These created conceptual evolutionary tracking models represent the authors' spatial interpretation of the latest phylogenetic studies in this field. Since these models are conceptual, and rather an interpretation of the phylogenetic data, they should not be taken as the only and most perfect representations of the true evolution of squirrels. In these model narratives, the evolutionary path of each genus and the species have been presented in a rather simplistic manner, actions of moving back- and forwards in time and space have been widely excluded due to their complexity and lacking information sources. Consecutively, we want to open a discussion about this topic with this study and appreciate any form of further input, improvement, and research focused on this manner. It is noteworthy, that these first-time models have been created with the best-available open-access data to the date of publication. In order to illustrate the overall trend of the squirrel's evolution, Fig. 2.1 has been created.

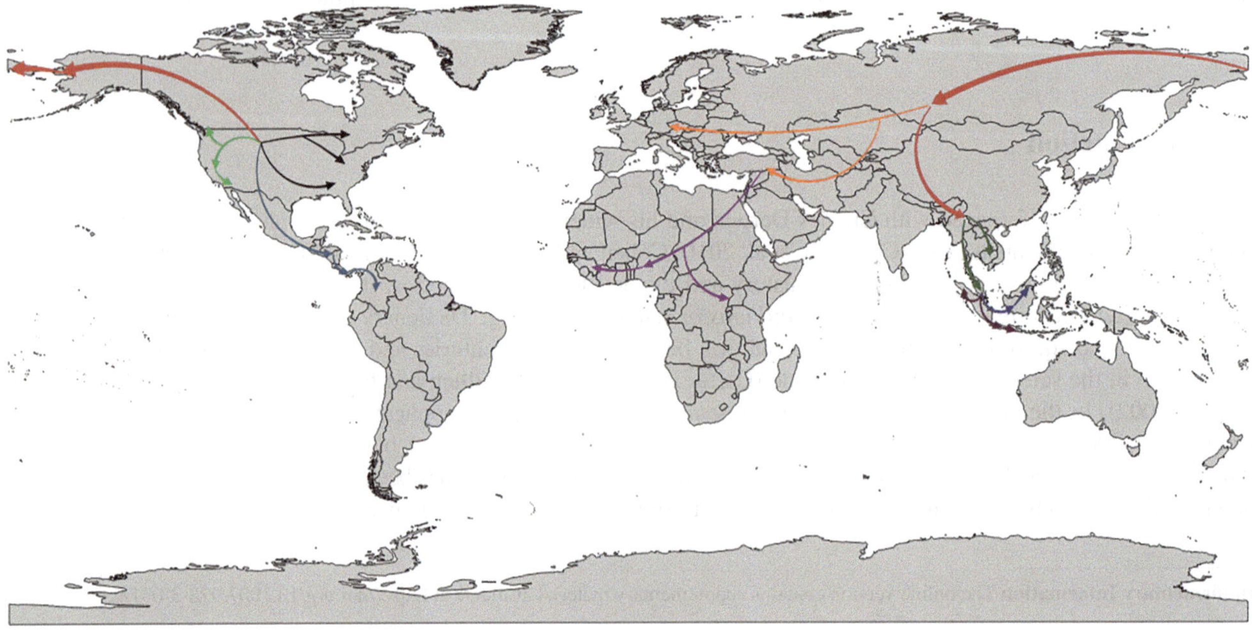

Fig. 2.1 Assumed global squirrel evolution range spread and overview over time

Figure 2.1 illustrates the overall trend that can be assumed for the evolution of the world's squirrels. An in-depth insight will be taken later on.

Apart from creating these first conceptual evolutionary tracking maps of all squirrel species, other evolutionary topics involving squirrels will be discussed here too. Namely, we will focus on some crucial evolutionary speciation events, the natural progress of evolution (e.g. the rise and fall of species), and topics such as extinction rates, and how these are related to modern issues (e.g. climate change).

Two factors that are commonly associated with the genetic and evolutionary speciation origins are presented in Textboxes 2.1 and 2.2 (Island dwarfism and gigantism, respectively).

Textbox 2.1: Island Dwarfism

Island dwarfism is a phenomenon related to genetic drift (Baeckens and Van Damme 2020; Flores-Manzanero et al. 2022); it is commonly observed in species and genera that have been isolated for a long time (several million years) and is expressed in particularly small body sizes and weights of species belonging to a family with commonly larger and heavier body sizes and weights (Baeckens and Van Damme 2020; Meiri et al. 2008; Prothero and Sereno 1982). Examples of this in the squirrel family can be observed for e.g. Least pygmy squirrel *(Exilisciurus exilis),* and African pygmy squirrel (*Myosciurus pumilio*). These mentioned species' total body lengths range from 10 to 14 cm with a weight of 12–26 grams (Payne et al. 1985; Thorington et al. 2012).

Textbox 2.2: Island Gigantism

Island gigantism is another phenomenon related to genetic drift. It demonstrates similarly to island dwarfism, but with species on the other end of the spectrum of body weights and sizes. This means that species that can be assigned to the phenomenon of island gigantism express a phenotype much larger and heavier than the closely related family or genus members (Baeckens and Van Damme 2020). This is similar to island dwarfism due to the complete isolation of this species for several million years and the full adaptation to the local environment and ecological niche. However, many changes come with body size e.g. metabolism or predator-prey relationships, and trophic positioning (Baxter-Gilbert et al. 2020; Köhler et al. 2010; Meiri et el. 2008; Vanek and Burke 2020). This phenomenon can similarly also be observed for species isolated on the mainland for several millions of years. An example of this can also be found in the squirrel family for e.g. Alpine marmot (*Marmota marmota*), and Bhutan giant flying squirrel (*Petaurista nobilis*). Several Marmot species (e.g. *M. marmota*) can reach a weight of up to 8 kg and more (Krystufek and Vohralik 2013). *P. nobilis* has been recorded to reach a total length of up to 1.3 m (Choudhury 2002; Jackson 2012).

This concept of island isolation and the evolution into new species has firstly been described by C. Darwin and his famous observations of finches (Galápagos finches) found on different islands (Marsh 2015), aka. Species radiation (Larsen 2014; Schluter 2000). The general concept follows the summary described below. "When a population of the same species inhabits a certain territory which is over time cleaved off from the mainland (due to abiotic factors), the population on this cleaved-off land piece remains in full isolation, leading over millions of years to the evolution of new species." (Source unknown). More details on these speciation-causing factors can be found throughout this study.

A cause of extinction, common to observe in many species of the global flora and fauna, especially in mammals and squirrels, is the rapid change in their surrounding environment (Bürger and Lynch 1997; Orr and Unckless 2008; Smith 1989). In modern times the environment is changing at one of the most rapid paces ever recorded (Hendry and Kinnison 1999). This is because many factors compose the environment in which all flora and fauna live in these environments are in rapid change. This includes the climate (Bradshaw and Holzapfel 2006; Franks and Weis 2008; Kelly and Goulden 2008), land use (Shi et al. 2018; Vessali 1996), water characteristics (salinity, pH, plastic content, temperature, etc.) (Hoegh-Guldberg et al. 2007; McPhee et al. 2009), advanced transportation (fast and broad-spectrum transportation of humans and animals around the world favoring disease transmissions and species invasion) (Groner et al. 2016), soil exploration & exploitation (Bouwman et al. 2013), destruction of pristine natural environments (Laurance 2010), etc.

All these mentioned components contribute to rapidly changing the environment humans, animals, and plants live in. Due to human activity and pressure worldwide (the Anthropocene), these changes have heavily accelerated in the past decades/

centuries, suggesting having led to a situation as severe as it has never been observed before (Otto 2018; Steffen et al. 2015). The risk of extinction is so high for many species that in recent years the situation has been referred to as the 6th mass extinction (Cowie et al. 2022; Pakistan 2017). This is likely due to the environmental changes occurring too rapidly to which the species cannot adapt fast enough, leading to vast human-driven extinctions.

In order to gain insight into the common causes of species extinction, a brief literature review has been conducted. This literature review also aimed to gain insights into the currently occurring trends of the environment and the world humans and squirrels live in, and how this is related to the possible extinction of squirrel species in the near future.

Additionally, in this study, we try to provide a first and somewhat coarse estimate of the recent extinction rate of squirrel species and the total rise and fall of species in the world of squirrels. In order to understand extinction, it must be said that it is defined as *"a situation in which something no longer exists"* (Cambridge Dictionary 2021b).

2.2 Methods

2.2.1 *Towards an Evolutionary Tracking Map of the World Squirrels*

The evolutionary tracking maps of the world's squirrels are a combination of state-of-the-art research works and newly created applications developed for this study. The genetic background of these maps has been published by Menéndez et al. (2020) which has been used as a reference for the evolutionary path of each squirrel species. In detail, Appendix S9 from Menéndez et al. (2020) has been used for the phylogenetic history of the squirrels. In order to use this publicly available data for this research, the individual moments in time at which subfamilies, tribes, genera, and species split off in the phylogenetic tree have been manually derived by MS. This has been done by manually associating each split-off with the time scale at the bottom of Appendix S9 (Menéndez et al. 2020). The information gathered from this has been described in the MS Excel document that can be found in Appendix 2.1 belonging to this study.

In order to geo-reference these phylogenetic records, the website www.fossilworks.org has been utilized to track fossil records. The website IUCN Red list (www.iucnredlist.org) has been used for the most recent and current distribution of each species. In cases where the IUCN Red list did not contain complete information on the squirrels' distribution, the book from Thorington et al. (2012) has been used to fill this gap of information. This complete dataset, the phylogenetic records linked with the approximate geographical location for each record, can be found in Appendix 2.1. In order to present this in a visual and more comprehendible manner, the georeferenced phylogenetic information has been mapped by MS using Open Source Geographic Information System OGIS, (QGIS – obtainable via https://www.qgis.org/en/site/forusers/download.html). These maps serve the aim of presenting the first-known maps of the evolutionary dispersal of squirrel genera and species on a relatively coarse level. It is meant to be used for further hypothesis testing. These maps do not serve the purpose of an in-depth and detailed visualization of each squirrel's evolution, as this is out of the scope of the study, and book.

2.2.2 *Extinction and New Speciation Rate in the Squirrel Family (The Rise and Fall of Squirrels)*

After understanding the general dispersion route of the global squirrel species and more insights into the past extinction causes of mammals and squirrels, now the extinction rate of the ancient (extinct) and present (extant) squirrel species is being calculated.

In order to estimate the rate of extinction and new speciation in the squirrel family (Sciuridae) very straightforward and rapid estimations have been performed.

For the extinction rate, Formulas 2.1 and 2.2 have been utilized. Those are based on simple calculations set up by MS. Formula 2.1 returns the percentage of species that have become extinct compared to all squirrel species that have ever been recorded.

$$\mathrm{ExR}_{\mathrm{Rel}} = \left(100 - \left(\text{Number of currently extant species / Total number of species ever recorded}\right) * 100\right) \tag{2.1}$$

In order to return the rate of species extinction per million years, Formula 2.2 has been utilized.

$$\mathrm{ExR}_{\mathrm{Abs}} = \left(\text{Number of species that got extinct / years since recording started}\right) \tag{2.2}$$

The rate of speciation in squirrels has been estimated using Formula 2.3. This includes all speciation events that have ever been recorded, starting from the first common ancestor of all squirrel species approx. 40 million years ago (Emry and Korth 2007).

$$SpR_A = \text{Total number of species recorded / Years since first ancestors have been recorded} \quad (2.3)$$

In order to generate the speciation rate of only currently extant species Formula 2.4 has been utilized. This formula only includes species that are currently considered extant.

$$SpR_E = \text{Number of species currently extant / Years since the first ancestor have been recorded} \quad (2.4)$$

The numbers of squirrel species that have been used for the calculations using the formulas above originate from two different sources. The number of currently extant species originates from Chap. 1 (Steiner and Huettmann 2023), and the number of species that have ever been recorded originates from the web portal www.fossilworks.org. The latter source has been used as it is considered to be the largest and most complete database of fossil records and collected data on extinct species. Another popular website to observe extinction numbers and trends is www.ourworldindata.org, where we see birds rank highest in terms of species extinct since 1500 (159 species extinct), mammals rank second (85 species extinct), and fish rank third (Ourworldindata 2020).

2.2.3 Literature Review

The literature review has been performed by using specific keywords in academic citation databases such as Google Scholar, Scopus, as well as the World Wide Web (www) using Google. Examples of these specific keywords are "Abiotic species extinction causes", "Recent species extinction", "Causes squirrel extinction", "Causes small mammal and rodent species extinction", etc. Generally, it has been focused on the first dozen records for each search term combination as we assumed that these literature records are the most recognized and globally accessible ones. For this literature review, only English language publications have been included.

2.3 Results

In Fig. 2.1 a great and early dispersal from the originating grounds of the first records can be observed. These records were found from the squirrels' ancestors in modern Wyoming (United States) migrating towards Europe (Luckett and Hartenberger 1985 and references within). This dispersal was possible thanks to the commonly known land bridge Beringia (Bering land bridge) that was open and allowed the dispersal of species between approximately 2–3 million years ago (Wen et al. 2016). This land bridge seems to play a key role in the dispersal of many species, including squirrels (Eddingsaas et al. 2004; Faerman et al. 2017; Galbreath et al. 2011; Oshida et al. 2009; Pečnerová and Martínková 2012), similarly as for many other species (Hopkins et al. 2013 and references within).

However, also a part of the species remained on the North American continent and started their dispersal path from there. This is assumed to have happened by the split off from the ancestral population into a diverging population finding new territories in the northern part and crossing Beringia, and another population moving more southwards, towards the remaining parts of North America. The reason why only two species (*Sciurus carolinensis*, and *Urocitellus parryii*) occur on multiple continents, especially the new and old world, can be due to the fact that the land bridge of Beringia limited the divergence of populations and did only allow divergence between limited times. This is also the case for islands (Chen and He 2009; Coyne 1992; Steinbauer et al. 2016). As it is commonly known, speciation is often caused by the isolation of a population without gene inflow and outflow (interaction) with other populations (Chen and He 2009; Coyne 1992; Steinbauer et al. 2016). This is the reason why many species, especially belonging to the squirrel family, occur solely on one or multiple neighboring islands (*Ratufa macroura* (Sri Lanka), *Funambulus layardi* (Sri Lanka), *Sciurus lis* (Japan), *Pteromys momonga* (Japan), *Petaurista leucogenys* (Japan), *Marmota vancouverensis* (Vancouver Island – Canada), *Hylopetes nigripes* (Palawan), *Exilisciurus concinnus* (Southeastern Philippines), *Exilisciurus exilis* (Borneo), *Sundasciurus jentinki* (Northern Borneo), *Callosciurus adamsi* (Northern Borneo), *Glyphotes simus* (Northern Borneo), *Sundasciurus fraterculus* (Western Sumatran Islands), etc.) (see Appendix 2.2 for more examples). Due to this isolation on islands, the species occurring only on one or a few small neighboring islands are generally considered more endangered (Jamieson et al. 2006; Johnson 1998; Sales 2005). However, this issue is barely recognized by any authorities, including the IUCN Red List. An example of this is Southern Palawan Tree Squirrel (*Sundasciurus steerii*), occurring on the southern tip of Palawan (Philippines) and a minor island nearby with a conservation status of "Least Concern" with an assumed "stable" population trend (Ong et al. 2016). The same can be observed for Busuanga Squirrel (*Sundasciurus hoogstraali*), occurring solely on the island Busuanga (Palawan – Philippines) with an area as small as 890 km² (Approximate size of Singapore or Kansas City for comparison) and being considered "Least Concern" with an unknown population trend (Balete and Ong 2016) (See Chap. 1 for details).

2.3.1 Evolutionary Tracking Maps

The visualization and implementation of the georeferenced phylogenetic records have been expressed in maps for each squirrel species, grouped by genera. Due to the extensive number of genera (52), only a few selected but representative subsets are included in this main section. The complete set of computed maps can be found in Appendix 2.2 of this study. The four handpicked genera that are presented here have been chosen based on the criteria:

1. globally known and relevant (*Sciurus*),
2. on the brink of extinction with available data (*Cynomys*),
3. representative for many similar genera (*Tamiops*),
4. currently still disagreed upon (*Tamiasciurus*).

The detailed subset of species within these genera that have been included in the creation of these conceptual evolutionary tracking models can be found in Table 2.1.

The figures of the *Sciurus* genius have been split into Fig. 2.2a, b in order to facilitate a clearer visualization of the evolutionary tracking path. Figure 2.2b illustrates the oldest divergence of three species of this genus, namely from the Japanese Squirrel (*Sciurus lis*), Eurasian Red Squirrel (*Sciurus vulgaris*), and Caucasian Squirrel (*Sciurus anomalus*). This earliest divergence of these three species and their common ancestor is assumed to have happened via the migration/divergence route using the land bridge Beringia approximately 20–15 million years ago (see Appendix 2.2). Once this common ancestor has bridged the land towards Eurasia, the ancestral species split off into three species; the first spit-off seems to have occurred for the Japanese Squirrel (*Sciurus lis*), currently endemic to Japan (Cassola 2016). The next species was the Eurasian Red Squirrel (*Sciurus vulgaris*) which is currently extant in most parts of Europe, Russia, and a few Asian countries (Shar et al. 2016). The last species of this first divergence event is the Caucasian Squirrel (*Sciurus anomalus*), which is currently inhabiting the Caucasian region (Yigit et al. 2016). Along with these three species belonging to the *Sciurus* genus, the Tufted Ground Squirrel (*Rheithrosciurus macrotis*) has also been assumed to have evolved from the common ancestor shared with the three previously discussed species. This species (*R. macrotis*) is currently occurring in Southeast Asia (Meijaard 2016) (see details in Appendixes 2.1 and 2.2, and Table 2.1).

In Fig. 2.2a the larger part of the genus *Sciurus* can be observed. These species never diverged towards Europe and Asia via Beringia, but rather remained in the North American continent. It is suspected that the evolution of these species happened in approximately sequential order, one after the other, meaning that once one species has found its ecological niche it remained there and specialized its characteristics to fit into this ecological niche (Peterson et al. 1999; Peterson et al. 2008).

In addition to the species mentioned above, the species belonging to this genus, and a few sister genera which have been included in the evolutionary tracking map of Fig. 2.2a are *S. aberti, S. griseus, S. niger, S. alleni, S. aureogaster, S. oculatus, S. carolinensis, S. colliaei, S. granatensis, S. colliaei, S. granatensis, S. deppei, S. variegatoides, S. stramineus, S. pyrrhinus, S. flammifer, S. igniventris, S. aestuans S. gilvigularis, S. ignites,* and *S. spadiceus*. In addition to these species from the genus *Sciurus*, the species *Syntheosciurus brochus, Microsciurus alfari*, and *Microsciurus flaviventer* have also been included in this evolutionary tracking map (Fig. 2.2a). All the references to the geographical occurrences can be found in Appendix 2.3 due to the extensive number of references, and all Taxonomic Serial Numbers (TSN's) can be found in Appendix 2.1.

How the species *Rheithrosciurus macrotis* and some species in sister genera are phylogenetically linked to the *Sciurus* genus can be found in Menéndez et al. (2020). Additionally, Table 2.1 summarizes the geographical divergence of all these species.

Figure 2.2c illustrates the evolutionary dispersal of the genus *Cynomys*. The species of this genus are among the most threatened species of this family for which data exists (2 out of 5 species are listed as "Endangered" on the IUCN Red List and all of the species show a decreasing population). The species that are included in this genus are Gunnison's prairie dog (*Cynomys gunnisoni*), White-tailed prairie dog (*Cynomys leucurus),* Black-tailed prairie dog (*Cynomys ludovicianus*), Mexican prairie dog (*Cynomys mexicanus*), and Utah prairie dog (*Cynomys parvidens*). These species occur nowadays all in a similar region, in Southwestern USA, and Mexico. It can also be seen from the evolutionary tracking map that the dispersion of these species did not occur over large distances, it seems like they rather remained on their original grounds. These species are also expected to have evolved all from one ancestor in the commonly occurring regions, where different populations have then specialized over time into the specific ecological niches, creating the five new species.

Table 2.1 Partial Evolutionary tracking sheet

Genus	Species	Tracking records ID	Tracking records	Record type	Time period	Location (today)	Record certainty	References
–	–	1	Start	Literature	40–36 mya	Wyoming (Central USA)	certain	Emry and Korth (2007), Luckett and Hartenberger (1985), and Mercer and Roth (2003)
Douglassciurus	*jeffersoni*	2	First known ancestor	Fossil record	40–36 mya	Wyoming (Central USA)	certain	Emry and Korth (2007), Luckett and Hartenberger (1985), and Mercer and Roth (2003)
–	*Sciurini-Pteromyini*	3	Subfamily split off into Tribes	Phylogenetic Tree derived	32 mya	USA	not fully certain	Menéndez et al. (2020)
Tamiasciurus and the other subfamily members	–	4	Genera split off	Phylogenetic Tree derived	20.5 mya	USA	not fully certain	Menéndez et al. (2020)
Tamiasciurus	*cf. yusheensis*	4a	First in lineage	Fossil record	5.3–3.6 mya	China	certain	Qiu and Storch (2000)
	hudsonicus & douglasii	4b	Species split off	Phylogenetic Tree derived	1 mya	USA	not fully certain	Menéndez et al. (2020)
		4c	Current Distribution	Fossil record & Current distribution	0.1 mya – Present	*T.h.* North and middle USA & Canada, *T.d.* Westcoast USA	certain	Burns (1991), Gidley and Gazin (1938), Hebda et al. (2008), Parmalee and Klippel (1981), Rasmussen (1974), and Schubert et al. (2003); etc.
	douglasii & mearnsi	4d	Species split off	Current distribution	2 mya – Present	*T.m.* (Baja California), *T.d.* (Western coast USA)	certain	Steiner and Huettmann (2021)
	hudsonicus & fremonti	4e	Species split off	Current distribution	0.4 mya – Present	Southern Central USA	certain	Steiner and Huettmann (2021)
Sciurus	–	5	Split off from North American and European tribe members	Phylogenetic Tree derived	13 mya	Beringia	certain	Mercer and Roth (2003)
Rheithrosciurus & Sciurus		5a	Genera split off	Phylogenetic Tree derived	11 mya	East Asia	not fully certain	Menéndez et al. (2020)
Rheithrosciurus	*macrotis*	5b	Isolation	Current distribution	11 mya	Borneo	certain	Meijaard (2016) and Menéndez et al. (2020)

(continued)

Table 2.1 (continued)

Genus	Species	Tracking records ID	Tracking records	Record type	Time period	Location (today)	Record certainty	References
Sciurus	*lis & vulgaris*	5c	Species split off	Phylogenetic Tree derived	2.5 mya	Japan and Europe	not fully certain	Menéndez et al. (2020)
		5d	Current Distribution	Phylogenetic Tree derived	Present	*S.l.* (Japan), *S.v.* (Europe)	not fully certain	Menéndez et al. (2020)
	anomalus	6	Species split off	Current distribution	10.6 mya – Present	Middle East	certain	Menéndez et al. (2020) and Yigit et al. (2016)
	aberti & griseus	7	Species split off	Current distribution	8.1 mya – Present	*S.a.* (South USA, Mexico), *S.g.* (USA west coast)	certain	Cassola (2017a), Lacher et al. (2016a), and Menéndez et al. (2020)
	niger	8	Species split off	Current distribution	8 mya – Present	Central-eastern USA	certain	Linzey et al. (2016a) and Menéndez et al. (2020)
	alleni	8a	Species split off	Current distribution	2 mya	Mexico	certain	de Grammont and Cuarón (2016a) and Menéndez et al. (2020)
	aureogaster & oculatus	8b	Species split off	Current distribution	0.1 mya	*S.a.* (Mexico & Guatemala), *S.o.* (Mexico)	certain	Álvarez-Castañeda et al. (2016a), Koprowski et al. (2017), and Menéndez et al. (2020)
	carolinensis & others	9	Species geographical division	Phylogenetic Tree derived	6 mya – Present	North America and South America	not fully certain	Menéndez et al. (2020)
	carolinensis	10	Species split off	Current distribution	6 mya – Present	Eastern North America	certain	Cassola (2016) and Menéndez et al. (2020)
Microsciurus	*alfari*	11	Species split off	Current distribution	4.1 mya	Colombia, Panama, Costa Rica	certain	Cassola (2016) and Menéndez et al. (2020)
Sciurus	*colliaei & granatensis* from other species	11aa	Species split off	Phylogenetic Tree derived	3.9 mya	Middle America	not fully certain	Menéndez et al. (2020)
	colliaei & granatensis	11ab	Species split off	Current distribution	1.9 mya – Present	*S.c.* (Western Mexico), *S.g.* (Panama, Costa Rica, North South America)	certain	de Grammont et al. (2016), Koprowski et al. (2016), and Menéndez et al. (2020)
	deppei	11ba	Species split off	Current distribution	3.4 mya – Present	Middle America	certain	Koprowski et al. (2016) and Menéndez et al. (2020)
	Sciurus variegatoides & Syntheosciurus brochus	11bb	Species split off	Current distribution	2.7 mya – Present	Middle America	certain	Koprowski et al. (2019), Menéndez et al. (2020), and Reid (2016)

(continued)

Table 2.1 (continued)

Genus	Species	Tracking records ID	Tracking records	Record type	Time period	Location (today)	Record certainty	References
	stramineus	12	Species split off	Current distribution	4.1 mya – Present	Ecuador, Peru	certain	Duckworth and Koprowski (2016) and Menéndez et al. (2020)
	Various *Sciurus* species	13	Species split off	Phylogenetic Tree derived	4 mya	Middle – South America	not fully certain	Menéndez et al. (2020)
	pyrrhinus	14aa	Species split off	Current distribution	2.7 mya – Present	Peru	certain	Amori et al. (2019a) and Menéndez et al. (2020)
	flammifer & *igniventris*	14ab	Species split off	Current distribution	1.6 mya – Present	*S. f.* (Venezuela), *S.i.* (Northwestern South America)	certain	Amori et al. (2016, 2019b), and Menéndez et al. (2020)
Microsciurus	*flaviventer*	14ba	Species split off	Current distribution	4 mya – Present	Northwestern South America	certain	Koprowski and Roach (2019a) and Menéndez et al. (2020)
Sciurus	various *Sciurus* species	14c	Species split off	Phylogenetic Tree derived	3 mya	South America	not fully certain	Menéndez et al. (2020)
	aestuans & *gilvigularis*	14da	Species split off	Current distribution	1.6 mya – Present	Northeastern South America	certain	Amori (2016), Amori et al. (2019c), and Menéndez et al. (2020)
	ignitus & *spadiceus*	14db	Species split off	Current distribution	1.5 mya – Present	Western South America	certain	Cassola (2016), Koprowski and Roach (2019b), and Menéndez et al. (2020)
Dremomys & *Tamiops* from others	–	6	Genera split off	Phylogenetic Tree derived	20.6 mya	Asia	not fully certain	Menéndez et al. (2020)
Dremomys & *Tamiops*	–	6a	Genus split off	Phylogenetic Tree derived	18.9 mya	Asia	not fully certain	Menéndez et al. (2020)
Tamiops	*rodolphii*	6ba	Species split off	Current distribution	17.8 mya – Present	Thailand, Vietnam, Laos, Cambodia	certain	Duckworth (2017a) and Menéndez et al. (2020)
	mcclellandii	6bb	Species split off	Current distribution	10.8 mya – Present	Southeast Asia	certain	Duckworth et al. (2017a) and Menéndez et al. (2020)
	maritimus & *swinhoei*	6bc	Species split off	Current distribution	5 mya – Present	Central – East Asia	certain	Duckworth (2017b, c) and Menéndez et al. (2020)

(continued)

Table 2.1 (continued)

Genus	Species	Tracking records ID	Tracking records	Record type	Time period	Location (today)	Record certainty	References
Cynomys	various species	8cdb	Species split off	Phylogenetic Tree derived	2.1 mya	USA	not fully certain	Menéndez et al. (2020)
	parvidens	8cdba	Species split off	Current distribution	1 mya – Present	Utah	certain	Menéndez et al. (2020) and Roach (2018)
	gunnisoni & *leucurus*	8cdbb	Species split off	Current distribution	0.6 mya – Present	*C.g.* (Utah, Colorado, Arizona, New Mexico), *C.l.* (Wyoming, Utah, Colorado)	certain	Cassola (2016) and Menéndez et al. (2020)
	ludovicianus & *mexicanus*	8cdbc	Species split off	Current distribution	0.6 mya – Present	*C.l.* (Central USA), *C.m.* (Mexico)	certain	Álvarez-Castañeda et al. (2019), Cassola (2016), and Menéndez et al. (2020)

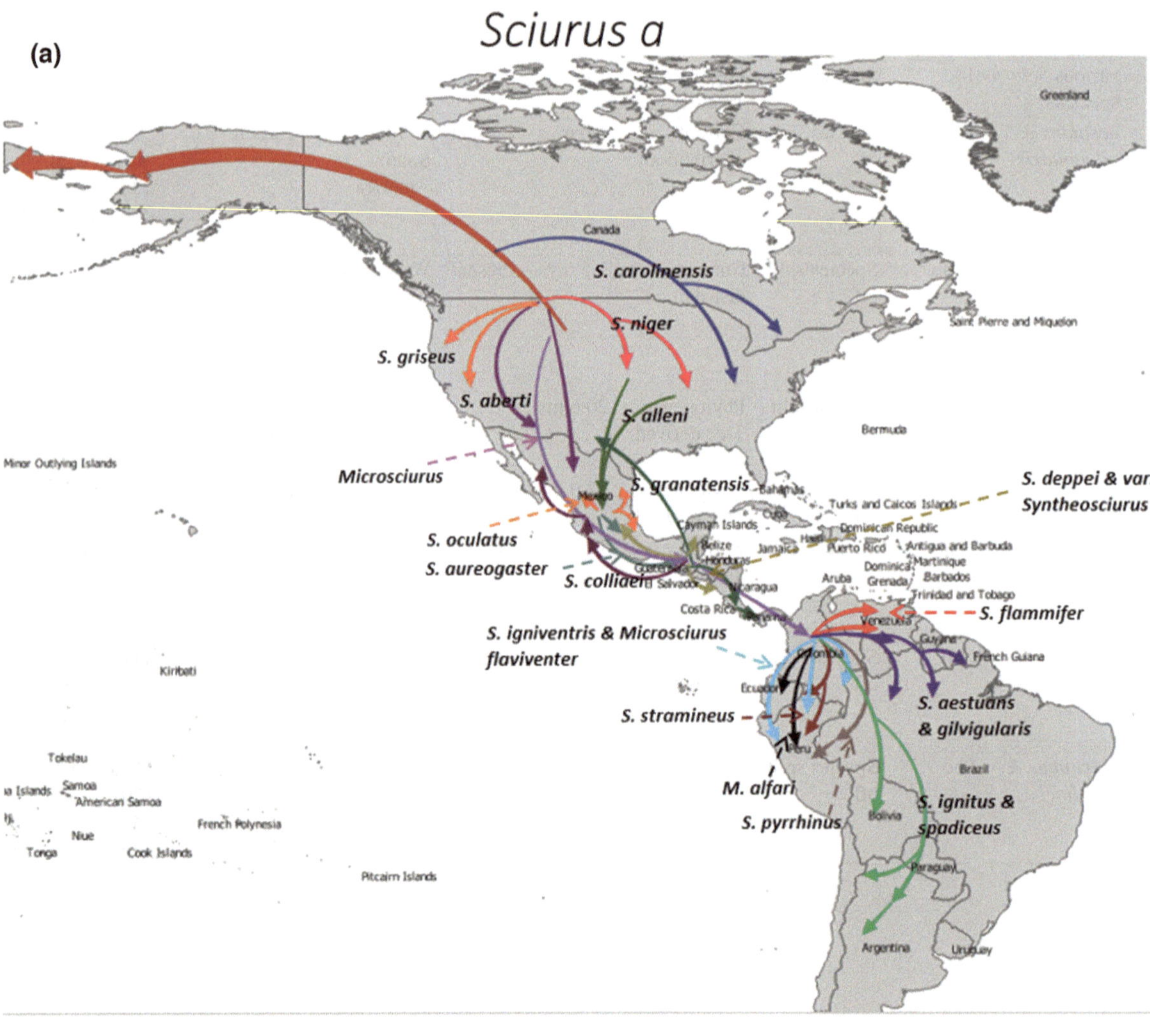

Figs. 2.2 Proposed evolutionary tracking map of the genus (**a**, **b**) *Sciurus*, (**c**) *Cynomys*, (**d**) *Tamiops*, (**e**) *Tamiasciurus*

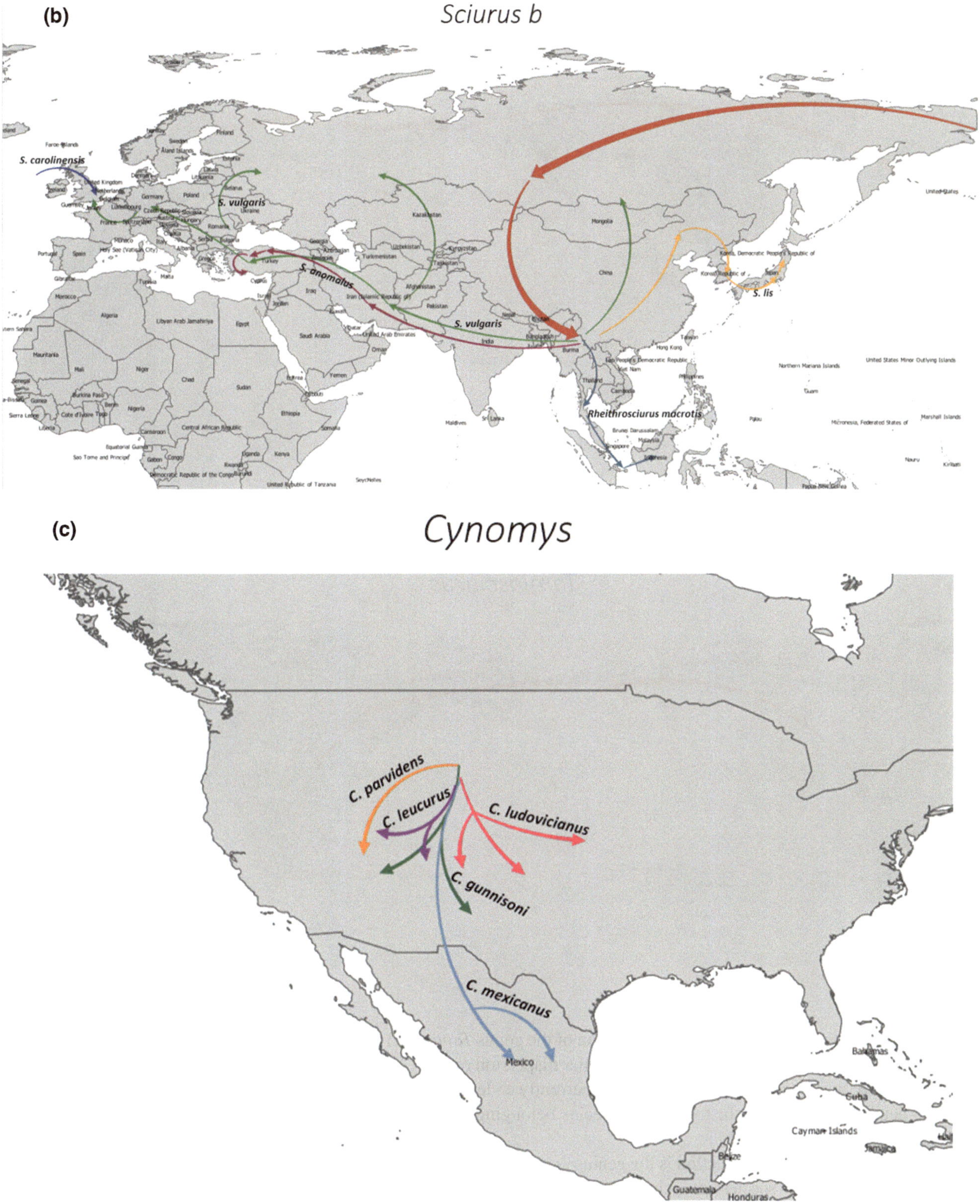

Fig. 2.2 (continued)

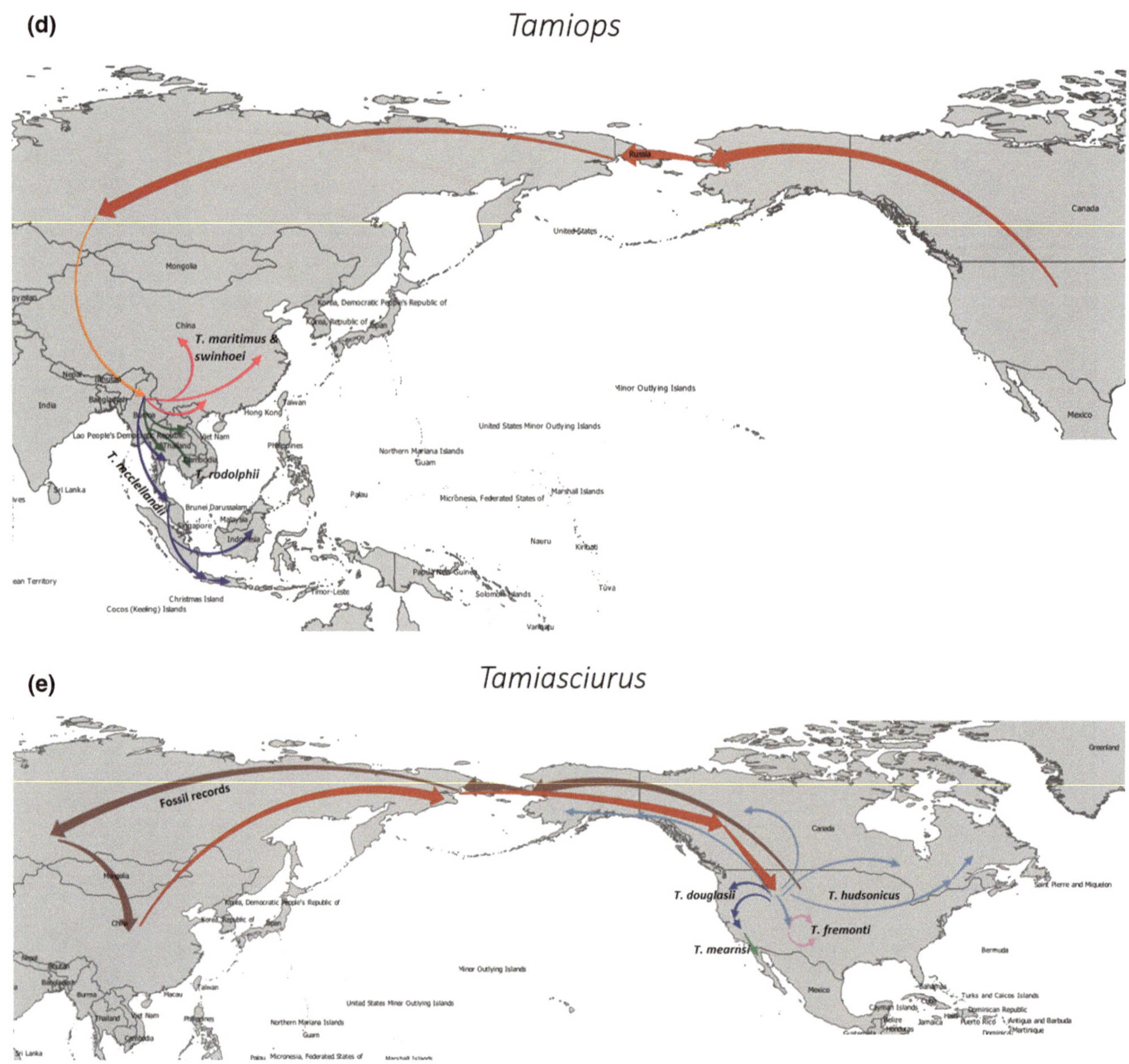

Fig. 2.2 (continued)

Figure 2.2d illustrates the evolutionary dispersion of the genus *Tamiops*. This genus has been included in this main section as it is an example of many other genera with a similar dispersion route. This genus contains the species *T. rodolphii, mcclellandii, maritimus,* and *swinhoei*. These species are currently endemic to Southeast Asia (see Table 2.1 and Duckworth 2017a, b, c). The detailed evolutionary history of the species belonging to this genus can be observed in Table 2.1, with included supporting references to literature.

The last genus included in the section is the genius *Tamiasciurus*. This genus has been included as it is to this day in dispute upon which and how many species are included in this genus. The number of including species ranges from 2 to 5 (Hope et al. 2016; Lindsay 1982; Steiner and Huettmann 2021). For the completion of this map, all five proposed species have been included. Not only the taxonomy but also the evolutionary history is under dispute according to several scientists (Churakov et al. 2010; Musser 2020). In literature, supported by fossil records, it is proposed that the common ancestor of this genus has diverged from the original grounds in Wyoming (USA) via Beringia towards Europe and Asia (Luckett and Hartenberger 1985 and references within). In Asia, apparent fossil records have been identified (Qiu and Storch 2000), however, in order to occur in the current distribution ranges, this species would have had to diverge back to North America via Beringia. This has been proposed by Mercer and Roth (2003). Thus far, the by far most commonly occurring species of this genus is the North American Red Squirrel (*Tamiasciurus hudsonicus*), it can be found from interior Alaska spreading all the way to eastern Canada and southwards into the states of New Mexico, and Arizona (Robold and Huettmann 2021; Savage et al. 2019; Steiner and Huettmann 2021; Thorington et al. 2012; Wilson and Reeder 2005). The second most occurring species, Douglas squirrel (*Tamiasciurus douglasii*) which has likely evolved from the former one, is occurring on the West Coast of the United States of America. The last three species cover a rather small distribution range and are often seen as the subspecies of the first 2 mentioned species. The Mearns's Squirrel (*Tamiasciurus mearnsi*) is commonly seen as a subspecies of the Douglas squirrel (Hope et al. 2016). However, there has also been significant evidence presented to consider this its own species (Koprowski et al. 2016; Steiner and Huettmann 2021 and references within). Currently, this species occurs only in a small region of Baja California (Mexico) (Koprowski et al. 2016; Steiner and Huettmann 2021 and references within). Southwestern Red Squirrel (*T. fremonti*) shares a similar story as *T. mearnsi*, however being labeled as a subspecies of *T. hudsonicus*. This species/proposed species currently occurs in the region around Texas (USA) (Hope et al. 2016). Lastly, Mount Graham Red Squirrel (*T. grahamensis)* is often considered as a subspecies of *T. fremonti*, however, also for this labeled subspecies evidence of an independent species has been published (Hope et al. 2016). *T. (h). grahamensis* occurs in modern days only in the Pinaleño Mountains (Mt. Graham) of Arizona (USA) (Audubon and Bachman 1847; Chen and Koprowski 2015).

Table 2.1 is a subsection of Appendix 2.2 which aims to summarize the evolutionary divergences of the species and genera in the focus of this main section.

In Table 2.1, for each evolutionary step, the corresponding record has been presented, describing for each record the recorded location supported by literature references. Additionally, for each record, the estimated point in time at which this event of split-off happened is noted, which has mainly been derived from Appendix S9 (Menéndez et al. 2020).

2.3.2 Extinction Rate and Speciation Rate Estimates

In order to calculate the overall extinction and speciation rate, historic data on the ancient squirrel species has been collected. MS compiled a dataset using the webpage www.fossilworks.org to gather all the information on ancient squirrel species. This compiled dataset contains all the extinct squirrel species that are expected to have lived on earth according to www.fossilworks.org, it is presented in Appendix 2.4. Using this best-available data set, the extinction rates and speciation rates have been estimated. This estimation includes all global extinct and extant squirrel species from the first known ancestor of the squirrel species until the present day.

Using Formulas 2.1 and 2.2 presented in Sect. 2.2.2, the percentage of extinct species and the estimated extinction rate of squirrels in the last 40 million years has been calculated.

$$\text{ExR}_{\text{Rel}} = \left(100 - \left(\text{Number of currently extant species / Total number of species ever recorded}\right) * 100\right)$$
$$\text{ExR}_{\text{Rel}} = \left(100 - \left(307/726\right) * 100\right) = 57.71\%$$

$$\text{ExR}_{\text{Abs}} = \left(\text{Number of species that got extinct / years since recording started}\right)$$
$$\text{ExR}_{\text{Abs}} = 419/40\,\text{my} = 10.5\,\text{species extinct / my}$$

These values used for the application of Formulas 2.1 and 2.2 have been obtained mainly from the web page www.fossilworks.org. In detail, the number of currently extant species (307) has been taken from (Steiner and Huettmann 2023, Chap. 1) as it is assumed to be the most updated species number of the squirrel family. The total number of squirrel species ever recorded has been calculated using the total number of currently extant species (307) plus the total number of extinct species, retrieved from Appendix 2.4 created by MS for this study. In Formula 2.2, the number of species that got extinct in the past (419) was obtained as described before from Appendix 2.4, and the number for "years since recording started" has been obtained from literature records (Emry and Korth 2007), as it is the commonly agreed start of the evolution of squirrels.

Similarly, using Formula 2.3, the speciation rate of all ever-recorded squirrel species can be calculated.

$$\text{SpR}_{\text{A}} = \text{Total number of species recorded / Years since first ancestors have been recorded}$$
$$\text{SpR}_{\text{A}} = 726/40\,\text{my} = 18.15\,\text{new species / my}$$

On the other hand, the speciation rate of the currently extant species has been calculated by applying Formula 2.4.

$$\text{SpR}_{\text{E}} = \text{Number of species currently extant / Years since the first ancestor have been recorded}$$
$$\text{SpR}_{\text{E}} = 307/40 = 7.7\,\text{extant species / my}$$

This latter speciation rate indicates how many squirrel species have successfully evolved in the last 40 million years and survived until this day.

For this calculation, the origin of the values used is similar to the one described above. The total number of species ever recorded is again the sum of the currently extant species (307) and the species that have gone extinct to this day (419), and the "years since the first ancestors have been recorded" is the same as above (40 my). In Formula 2.4, the number of species currently extant has been retrieved from Steiner and Huettmann (2023, Chap. 1), and the time since the first ancestors have been recorded has been retrieved from literature (Emry and Korth 2007).

The estimations presented above aim to illustrate the minimum estimate, meaning that the number of species extinct is assumed to be a minimum since the records on which it is based are solely obtained from fossil records which cannot be assumed to be entirely complete for the last 40 million years. Also, the amount of research conducted on this topic is very limited and outdated, and subsequently, modern analysis, data, and updates are widely lacking. To a very large extent, the records of extinct species are dating back to times before genetic analysis and phylogenetic approaches have been used in science on a broad scale, and therefore makes these records seem quite outdated in the modern world. Further research, utilizing modern tools, would be necessary to come to more accurate results and better estimates.

2.3.3 Literature Review

This literature review mainly fulfills the aim of providing literature records that can indicate the causes that have led to the high number of extinct squirrel species, and the high extinction rate during the last millions of years (calculated in Sect. 2.3.2).

The results of the literature review have been presented using a meta-analysis approach in Table 2.2.

Table 2.2 Meta-analysis results of the literature review on common causes of species extinction

Extinction cause	Extinction animal	Age of concern	References
Anthropogenic climate change	All global biota	Present & upcoming 100+ years	Cahill et al. (2013), Maclean and Wilson (2011), Thomas et al. (2006), and Román-Palacios and Wiens (2020)
Infectious diseases	All global biota	Since at aleast 1500	Pedersen et al. (2007) and Smith et al. (2006)
Human-driven selection pressure	All global biota	Present	Otto (2018)
Distribution and abundance relationship	Australian marsupials	Past and present	Johnson (1998)
Anthropogenic pressure	All global biota	Present & near future	Ceballos et al. (2015) and Pimm et al. (2014)
Increased tropical forest destruction	Tropical species	Present	Sayer and Whitmore (1991)
Increasing human population	World's carnivores (particularly African viverrid species)	Present & near future	Cardillo et al. (2004)
Increasing human footprint	All global biota	Present & near future	Di Marco et al. (2018)
Widespread Modification of Lakes and Rivers	North American Freshwater Fauna	Since at least 1900	Ricciardi and Rasmussen (1999)
Fragmentation, decrease in abundance, and host plant species vanishment	Birds and Butterflies on Barro Colorado Island, Panama	Since at least 1923	Basset et al. (2015)
Invasive species dominance	Eurasian Red Squirrel	Present & near future	Bertolino et al. (2014) and Gurnell et al. (2004)
Deforestation and habitat fragmentation for agriculture & tourism	Squirrel monkeys in Costa Rica	Present	Boinski et al. (1998)
Hyper-predation	Australian conilurine rodents	Recent past	Smith and Quin (1996)
Unbalanced ecosystems	Rodents in the Chihuahuan desert	Present	Valone and Brown (1995)
Habitat fragmentation, Pathogens & proliferation of human-adapted competitors	Allegheny woodrat (*Neotoma magister*)	Recent past	LoGiudice (2006)
Introduction of diseases	Small mammals of Mallorca	Past 5 millennia	Bover and Alcover (2008)
Climatic changes	Small mammals	Late-Pleistocene	Blois et al. (2010)
Contemporary fire patterns	Brush-tailed Rabbit-rat (*Conilurus penicillatus*)	Present & near future	Firth et al. (2010)
Forest fragmentation	Small mammals in Thailand	Recent past & present	Gibson et al. (2013)
Uncontrolled hunting and predation (especially on islands)	Birds and mammals	Since 1500	Loehle and Eschenbach (2012)

From this table, it can be observed that issues and causes leading to the extinction of species are not just something that has been recorded in the past but similar issues are forecasted and already happening in the modern world, warranting us of the upcoming near future. Arguably, any human-driven impacts are to be reduced; How that is to be done remains unclear though while human-driven habitat loss, consumption, and climate change continue virtually unabated (Coverdale et al. 2013; Lynas et al. 2021; Monteiro et al. 2010; Salmona et al. 2017).

The causes that have predominately been observed are sudden anthropological-induced climate change, disease spreading, habitat destruction, and species invasion (Cahill et al. 2013; Case 1990; McCallum 2012; Román-Palacios and Wiens 2020; Zhenshan and Shuguang 2002).

2.4 Discussion

Species generally evolve over time (evolution). Throughout evolution, species tend to rise and fall, and the rate at which these old species become extinct and new species arise is commonly depending on the type of animal and plant, and in which era they live(d). The interaction, ecology, is the main driver and in response to climate and many other environmental factors, including universe ones (e.g. moon and sun cycles – Tran et al. 2011; Wuitchik et al. 2019). Certain families and types of animals generally have a longer species lifespan compared to others (Levin 2013). In this study, we considered the evolution of the species belonging to the family of squirrels. In order to discuss the evolution of the squirrels, evolutionary tracking maps have been created that aim to visualize the dispersal route of all the currently extant species. These tracking maps aimed to provide a coarse overview of where the origin of each genus and species was and how it dispersed to the current territories of its occurrence. The full data set of these evolutionary tracking maps can be found in Appendix 2.2, with ISO-compliant metadata. In these maps, one major trend was able to be observed, namely the divergence from the earliest ancestral grounds in the modern USA towards Eurasia via the land bridge Beringia. This land bridge is commonly known as a hotspot of species divergence between the old and the new world. An example species which managed to find suitable habitats in both "worlds" is the Pika (Ge et al. 2013). Often it is recorded that species depart their evolutionary migration on the Eurasian side (Old World) in order to end up on the North American side (New World). However, for squirrels, it seems to be the opposite. Once their ancestral species crossed the Bering land bridge, it is commonly observed that populations split up in eastern Russia and from there diverged into Asia and Southeast Asia in one direction and towards Europe and Africa in the other direction. This main dispersal route can be observed for several genera, actually, all squirrel genera and species currently occurring in Europe, Africa, and Asia. In the new world, species mainly remained in the northern hemisphere, and only a few species diverged into middle America and consecutively into southern America. The evolutionary tracking maps, presented in the main section, illustrate this commonly observed trend of the squirrels' evolution very well. Throughout the upcoming chapters, an in-depth analysis of the individual habitats, which squirrels inhabit in the modern world, is presented with the greater intention to assess this widely marginalized family using state-of-the-art assessment tools and open-data sharing approaches. In order to better understand this evolution of squirrels, a literature review on common extinction causes for squirrels and related species has been performed in the study. This literature review yielded great insights into the common extinction causes of the recent past, presence, and the near future. The main causes for species extinction are according to this brief literature review, closely bound with the increased presence of humans on this earth and our heavy impact on nature. In detail, anthropologically induced climate change has been described by several literature references as the main reason for species extinction. In addition to this, human practices such as habitat fragmentation, habitat destruction (e.g. forest destruction), and unsustainable nature management have been described as the main reasons and causes for species extinction. Other often observed causes for extinction include sea-level rise, coastline flooding, and disease outbreaks within isolated populations (Gilbert et al. 2020; Field et al. 2017; Keller et al. 1993).

In order to determine whether this recently increased extinction of species is also true for squirrels, a first attempt has been to compute the extinction rate of the squirrel family in this study. The results presented in Sect. 2.3.2 demonstrate a very high rate at which species have gone extinct in the last 40 million years (similar patterns have been observed e.g. Birds with a 1,000–10,000 times higher observed extinction rate compared to background extinction rates (which is already very high for birds) see Birdlife International 2022). We observed that extinction rates are higher than speciation rates (speciation rate of 7.7 extant species/million years, and an extinction rate of 10.5 species/million years). This indicates that this species group is in great danger long-term due to its derived negative species number trend. This again indicates an immediate need for long-term conservation projects to be set up and to invest resources into conservation research. Not all of these mentioned extinctions of squirrels are caused by human influence, however, it demonstrates that squirrels, as a family, are rather short-living species, and thus, under greater threat compared to other species. The web page www.recentlyextinctspecies.com lists approximately 12 recently extinct species from the squirrel family. Over the past 40 million years, where squirrels have been present on this earth, the main extinction and speciation causes for squirrels have been assumed to be climate change, changes in sea levels, terrestrial land drift off, and similar abiotic factors (e.g. see Châteauneuf 1980; Flenner and Sahlén 2008; Haq et al. 1987; Li and Li 2018; Ma et al. 2016). The latter factors are more commonly observed as causes for speciation rather than extinction, because as discussed in the introduction, once a population becomes isolated, due to terrestrial land split (often also supported by changes in the sea level) new species arise over time. This can be observed on a bigger scale for the tribes Sciurini and Marmotini in response to the changes in sea levels in that era when their speciation is estimated to have happened approx. 32 million years ago (Menendez et al. 2020). Some environmental events can lead to both extinction and speciation. This is because once a species is eradicated, the ecological niche becomes vacant and new speciation is possible and commonly observed in such scenarios.

Conclusively, it can be stated that squirrels are among the species with a rather short species lifespan, making them even more vulnerable compared to other species groups. The fact that many species only occur on one or a few neighboring small islands does not aid the preservation of the global squirrel species. Nowadays, the Anthropocene acting in those evolutionary cradles makes for peculiar conservation management and subsequent threats. However, with modern, sustainable, and science-based conservation (Dinerstein et al. 2019), further conservation management failures can hopefully widely be avoided and long-term success in the conservation and preservation of the squirrels on this earth can be achieved.

As can be seen in Appendix 2.4, the records for the ancient squirrel species are widely outdated. Those records have most likely been established without any modern approaches e.g. genetic analysis, Machine Learning, AI, etc. Additionally, it seems like the financial efforts for determining the evolution of this species group are kept very limited, only available for museum-based research of often old and single specimens (often repatriation specimens (Dias et al. 2017; Turner 2005)). This suggests that in current days there is still an imbalanced, somewhat unfair, and uneven allocation of financial resources – public and private – especially in small mammal and squirrel research. This failed allocation of money for the wider public good can be observed even more severely with the lack of any conservation efforts for the understudied squirrel species (302/307), as virtually, no scientific and conservational efforts are known for this majority of squirrel species. This topic of unequal and somewhat failed allocation of financial resources, in combination with governance failure, is widely known now, publicly debated for years already (e.g. see Czech 2000), and it gets summarized and discussed more thoroughly in Chap. 13 (Steiner and Huettmann 2023).

Another crucial and noteworthy topic that must be discussed here is the ecological fate of squirrels in cities and general anthropological settings. In the last century, especially the last few decades, squirrels are more and more observed and held within urban settings (Bowers and Breland 1996; Van Der Merwe et al. 2005). In the Anthropocene, this raises the question of how life in cities affects the squirrels themselves. Does urban life affect the squirrels' fertility; their life history, their behavior, their longevity, and their stress level? etc. These are all situations that are already happening and have a negative influence on many squirrel species (e.g. reviews and studies by Iglesias-Carrasco et al. 2020; Lair 1985; Mccleery 2009; Sarno et al. 2015). However, it is widely ignored by the public and the research sector. For the public, politicians, and institutions squirrels are interesting because they are considered "cute" and an attraction for tourists. However, most of the time it is not looked behind the scenes of how a life of a squirrel changes by living in cities, parks, suburban areas, etc. Urban encroachment has major effects on habitats, and squirrels. This cuteness of wildlife is sometimes being used for financial and economic development. For instance, squirrels and other animals considered "cute" are actually used in advertisements for city parks, natural parks (ecotourism), institutions such as universities, zoos, etc. (e.g. see National Park Foundation 2022; Prague Zoo 2022; YouTube 2021). However, despite this use for advertisement, the conservation of squirrels is still widely ignored and marginalized on management agendas. One would think that with all these advertisements, and economic interest, the species would receive increased grants, funding, research efforts, etc. However, this is not true except for the 4–5 over-studied cosmopolitan species (Eastern Gray Squirrel (*Sciurus carolinensis*), Eurasian Red Squirrel (*Sciurus vulgaris*), North American Red Squirrel (*Tamiasciurus hudsonicus*), Fox squirrel (*Sciurus niger*), and Eastern Chipmunk (*Tamias striatus*)).

It is noteworthy to have thought about how these factors that are likely influencing the squirrel's lives and evolution in cities, can also possibly lead to the extinction of whole species. This is a valid thought since many species e.g. tropical forest species, e.g. Indian giant squirrels are feeding predominantly on fruits and seeds found within those forests, however, if these food sources are lacking in the urbanized environments, the species have to either adjust immediately within one generation or two, to the sudden change in diet, or it will result in drastic population declines (Fuller et al. 1995; Pomfret et al. 2012). This was so far likely avoided by the species remaining on the original habitat grounds or moving to the remaining wilderness areas (tropical forests, steppe, boreal forests, tempered forests, and more). However, as discussed in Chap. 5 (Steiner and Huettmann 2023), with the increased pressure of humans on the environment, the rapid change in land use, and the high destruction of natural lands for the expansion of urban environments, these original grounds inhabited by the squirrels drastically decrease and become fragmented or completely lost (Cielemęcka 2020; GhulamRabbany et al. 2013; Johnson and Lewis 2007; Oka and Majuk 2016), likely in the upcoming decades. This leads to the final question, where do the squirrels go in the near future with the rapid growth of the human species? Does it lead to a flourishing of species, declines, or extinctions? Only science-based research and increased efforts will provide better predictions to answer these questions, by attempting not to lose species that we do not even know about. With these open questions and "food for thought", this study also aimed to act as an initial milestone for opening a platform to discuss the topics presented here and simultaneously calls for future initiatives to improve the current lack of information for the large majority and heavily understudied squirrel species.

References

Amori G, Koprowski J, Roth L (2016) Sciurus igniventris (errata version published in 2017). The IUCN Red List of threatened species 2016: e.T20013A115154863. https://doi.org/10.2305/IUCN.UK.2016-3

Audubon JJ, Bachman J (1847) The viviparous quadrupeds of North America, vol 1. Wiley and Putnam

Baeckens S, Van Damme R (2020) The island syndrome. Curr Biol 30(8):R338–R339

Balete DS, Ong P (2016) Sundasciurus hoogstraali. The IUCN Red List of Threatened Species 2016: e.T42566A22249540. https://doi.org/10.2305/IUCN.UK.2016-2.RLTS.T42566A22249540.en. Accessed 10 Oct 2021

Basset Y, Barrios H, Segar S, Srygley RB, Aiello A, Warren AD, et al., Ramirez JA (2015) The butterflies of Barro Colorado Island, Panama: local extinction since the 1930s. PLoS One, 10(8), e0136623

Baxter-Gilbert J, Riley JL, Wagener C, Mohanty NP, Measey J (2020) Shrinking before our isles: the rapid expression of insular dwarfism in two invasive populations of guttural toad (Sclerophrys gutturalis). Biol Lett 16(11):20200651

Bennett KD (2013) Is the number of species on Earth increasing or decreasing? Time, chaos and the origin of species. Palaeontology 56(6):1305–1325

Bertolino S, Lurz PW (2013) Callosciurus squirrels: worldwide introductions, ecological impacts and recommendations to prevent the establishment of new invasive populations. Mammal Rev 43(1):22–33

Bertolino S, Di Montezemolo NC, Preatoni DG, Wauters LA, Martinoli A (2014) A grey future for Europe: Sciurus carolinensis is replacing native red squirrels in Italy. Biol Invasions 16(1):53–62

Birdlife International (2022) Numerous bird species have been driven extinct. http://datazone.birdlife.org/sowb/state/theme1#:~:text=In%20modern%20times%2C%20birds%20have,extinctions%20on%20continents%20is%20increasing. Accessed 28 July 2022

Blois JL, McGuire JL, Hadly EA (2010) Small mammal diversity loss in response to late-Pleistocene climatic change. Nature 465(7299):771–774

Boinski S, Jack K, Lamarsh C, Coltrane JA (1998) Squirrel monkeys in Costa Rica: drifting to extinction. Oryx 32(1):45–58

Bouwman L, Goldewijk KK, Van Der Hoek KW, Beusen AH, Van Vuuren DP, Willems J, Stehfest E et al (2013) Exploring global changes in nitrogen and phosphorus cycles in agriculture induced by livestock production over the 1900–2050 period. Proc Natl Acad Sci 110(52):20882–20887

Bover P, Alcover JA (2008) Extinction of the autochthonous small mammals of Mallorca (Gymnesic Islands, Western Mediterranean) and its ecological consequences. J Biogeogr 35(6):1112–1122

Bowers MA, Breland B (1996) Foraging of gray squirrels on an urban-rural gradient: use of the GUD to assess anthropogenic impact. Ecol Appl 6(4):1135–1142

Bradshaw WE, Holzapfel CM (2006) Evolutionary response to rapid climate change. Science, 312(5779), 1477–1478

Buerger R, Lynch M (1997) Adaptation and extinction in changing environments. In: Environmental stress, adaptation and evolution, Springer Basel AG (Basel - Switzerland), pp 209–239

Cahill AE, Aiello-Lammens ME, Fisher-Reid MC, Hua X, Karanewsky CJ, Yeong Ryu H, Wiens JJ et al (2013) How does climate change cause extinction? Proc R Soc B Biol Sci 280(1750):20121890

Cambridge Dictionary (2021a) Evolution. https://dictionary.cambridge.org/dictionary/english/evolution. Accessed 10 Oct 2021

Cambridge Dictionary (2021b) Extinction. https://dictionary.cambridge.org/dictionary/english/extinction. Accessed 10 Oct 2021

Cardillo M, Purvis A, Sechrest W, Gittleman JL, Bielby J, Mace GM, Moritz C (2004) Human population density and extinction risk in the world's carnivores. PLoS Biol 2(7):e197

Carter AM, Mess A (2007) Evolution of the placenta in eutherian mammals. Placenta 28(4):259–262

Case TJ (1990) Invasion resistance arises in strongly interacting species-rich model competition communities. Proc Natl Acad Sci 87(24):9610–9614

Cassola F (2016) Sciurus lis (errata version published in 2017). The IUCN Red List of Threatened Species 2016: e.T20014A115155004. https://doi.org/10.2305/IUCN.UK.2016-3.RLTS.T20014A22247065.en. Accessed 10 Oct 2021

Ceballos G, Ehrlich PR, Barnosky AD, García A, Pringle RM, Palmer TM (2015) Accelerated modern human–induced species losses: Entering the sixth mass extinction. Sci Adv 1(5):e1400253

Châteauneuf JJ (1980) Palynostratigraphie et paléoclimatologie de l'Eocène supérieur et de l'Oligocène du Bassin de Paris (France)

Chen XY, He F (2009) Speciation and endemism under the model of island biogeography. Ecology 90(1):39–45

Chen HL, Koprowski JL (2015) Animal occurrence and space use change in the landscape of anthropogenic noise. Biol Conserv 192:315–322

Choudhury A (2002) Petaurista nobilis singhei-first record in India and a note on its taxonomy. J Bombay Nat Hist Soc 99(1):30–34

Churakov G, Sadasivuni MK, Rosenbloom KR, Huchon D, Brosius J, Schmitz J (2010) Rodent evolution: back to the root. Mol Biol Evol 27(6):1315–1326

Cielemęcka O (2020) Forest futures: biopolitics, purity, and extinction in Europe's last 'pristine' forest. J Gend Stud 29(1):63–75

Cohen E, Lloyd S (2014) Disciplinary evolution and the rise of the transdiscipline. Inform Sci 17:189–215

Coverdale TC, Herrmann NC, Altieri AH, Bertness MD (2013) Latent impacts: the role of historical human activity in coastal habitat loss. Front Ecol Environ 11(2):69–74

Cowie RH, Bouchet P, Fontaine B (2022) The sixth mass extinction: fact, fiction or speculation? Biol Rev 97:640–663

Coyne JA (1992) Genetics and speciation. Nature 355(6360):511–515

Czech B (2000) Shoveling fuel for a runaway train: errant economists, shameful spenders, and a plan to stop them all. University of California Press

Di Marco M, Venter O, Possingham HP, Watson JE (2018) Changes in human footprint drive changes in species extinction risk. Nat Commun 9(1):1–9

Dias D, Fonseca CB, Correa L, Soto N, Portela A, Juarez K, Junior J et al (2017) Repatriation Data: More than two million species occurrence records added to the Brazilian Biodiversity Information Facility Repository (SiBBr). Biodivers Data J 5:e12012

Dinerstein E, Vynne C, Sala E, Joshi AR, Fernando S, Lovejoy TE, et al., Wikramanayake E (2019) A global deal for nature: guiding principles, milestones, and targets. Science advances, 5(4), eaaw2869

Eddingsaas AA, Jacobsen BK, Lessa EP, Cook JA (2004) Evolutionary history of the arctic ground squirrel (Spermophilus parryii) in Nearctic Beringia. J Mammal 85(4):601–610

Emry RJ, Korth WW (2007) A new genus of squirrel (Rodentia, Sciuridae) from the mid-Cenozoic of North America. J Vertebr Paleontol 27(3):693–698

Faerman M, Bar-Gal GK, Boaretto E, Boeskorov GG, Dokuchaev NE, Ermakov OA, Formozov NA et al (2017) DNA analysis of a 30,000-year-old Urocitellus glacialis from northeastern Siberia reveals phylogenetic relationships between ancient and present-day Arctic ground squirrels. Sci Rep 7(1):1–10

Field CR, Bayard TS, Gjerdrum C, Hill JM, Meiman S, Elphick CS (2017) High-resolution tide projections reveal extinction threshold in response to sea-level rise. Glob Chang Biol 23(5):2058–2070

Firth RS, Brook BW, Woinarski JC, Fordham DA (2010) Decline and likely extinction of a northern Australian native rodent, the Brush-tailed Rabbit-rat Conilurus penicillatus. Biol Conserv 143(5):1193–1201

Flenner IDA, Sahlén G (2008) Dragonfly community re-organisation in boreal forest lakes: rapid species turnover driven by climate change? Insect Conserv Diver 1(3):169–179

Flores-Manzanero A, Valenzuela-Galván D, Cuarón AD, Vázquez-Domínguez E (2022) Conservation genetics of two critically endangered island dwarf carnivores. Conserv Genet 23(1):35–49

Franks SJ, Weis AE (2008) A change in climate causes rapid evolution of multiple life-history traits and their interactions in an annual plant. J Evol Biol 21(5):1321–1334

Fuller RJ, Gregory RD, Gibbon DW, Marchant JH, Wilson JD, Baillie SR, Carter N (1995) Population declines and range contractions among lowland farmland birds in Britain. Conserv Biol 9(6):1425–1441

Galbreath KE, Cook JA, Eddingsaas AA, DeChaine EG (2011) Diversity and demography in Beringia: multilocus tests of paleodistribution models reveal the complex history of arctic ground squirrels. Evolution: International Journal of Organic. Evolution 65(7):1879–1896

Ge D, Wen Z, Xia L, Zhang Z, Erbajeva M, Huang C, Yang Q (2013) Evolutionary history of lagomorphs in response to global environmental change. PLoS One 8(4):e59668

GhulamRabbany M, Afrin S, Rahman A, Islam F, Hoque F (2013) Environmental effects of tourism. Am J Environ Energy Power Res 1(7):117–130

Gibbs JP, Buff MF, Cosentino BJ (2019) The biological system—Urban wildlife, adaptation, and evolution: Urbanization as a driver of contemporary evolution in Gray Squirrels (Sciurus carolinensis). In: Understanding urban ecology. Springer, Cham, pp 269–286

Gibson L, Lynam AJ, Bradshaw CJ, He F, Bickford DP, Woodruff DS, et al., Laurance WF (2013) Near-complete extinction of native small mammal fauna 25 years after forest fragmentation. Science, 341(6153), 1508–1510

Gilbert M, Sulikhan N, Uphyrkina O, Goncharuk M, Kerley L, Castro EH, et al., Cleaveland S (2020) Distemper, extinction, and vaccination of the Amur tiger. Proceedings of the National Academy of Sciences, 117(50), 31954–31962

Grados MA, Walkup J, Walford S (2003) Genetics of obsessive-compulsive disorders: new findings and challenges. Brain Dev 25:S55–S61

Groner ML, Maynard J, Breyta R, Carnegie RB, Dobson A, Friedman CS, Harvell CD et al (2016) Managing marine disease emergencies in an era of rapid change. Philos Trans R Soc B Biol Sci 371(1689):20150364

Gunn RL, Hartley IR, Algar AC, Niemelä PT, Keith SA (2021) Understanding behavioural responses to human-induced rapid environmental change: a meta-analysis. Oikos. https://doi.org/10.1111/oik.08366

Gurnell J, Wauters LA, Lurz PW, Tosi G (2004) Alien species and interspecific competition: effects of introduced eastern grey squirrels on red squirrel population dynamics. J Anim Ecol 73(1):26–35

Haq BU, Hardenbol JAN, Vail PR (1987) Chronology of fluctuating sea levels since the Triassic. Science 235(4793):1156–1167

Hartenberger JL (1983) La Grande Coupure. Pour La Science 67:f26–f38

Hendry AP, Kinnison MT (1999) Perspective: the pace of modern life: measuring rates of contemporary microevolution. Evolution 53(6):1637–1653

Hoegh-Guldberg O, Mumby PJ, Hooten AJ, Steneck RS, Greenfield P, Gomez E, Hatziolos ME et al (2007) Coral reefs under rapid climate change and ocean acidification. Science 318(5857):1737–1742

Hope AG, Malaney JL, Bell KC, Salazar-Miralles F, Chavez AS, Barber BR, Cook JA (2016) Revision of widespread red squirrels (genus: Tamiasciurus) highlights the complexity of speciation within North American forests. Mol Phylogenet Evol 100:170–182

Hopkins DM, Matthews JV, Schweger CE (eds) (2013) Paleoecology of Beringia. Elsevier

Houston RD, Macqueen DJ (2019) Atlantic salmon (Salmo salar L.) genetics in the 21st century: taking leaps forward in aquaculture and biological understanding. Anim Genet 50(1):3–14

Iglesias-Carrasco M, Aich U, Jennions MD, Head ML (2020) Stress in the city: meta-analysis indicates no overall evidence for stress in urban vertebrates. Proc R Soc B 287(1936):20201754

Jackson SM (2012) Gliding mammals of the world. Csiro Publishing

Jamieson IG, Wallis GP, Briskie JV (2006) Inbreeding and endangered species management: is New Zealand out of step with the rest of the world? Conserv Biol 20(1):38–47

Johnson CN (1998) Species extinction and the relationship between distribution and abundance. Nature 394(6690):272–274

Johnson DL, Lewis LA (2007) Land degradation: creation and destruction. Rowman & Littlefield

Keller G, Barrera E, Schmitz B, Mattson E (1993) Gradual mass extinction, species survivorship, and long-term environmental changes across the Cretaceous-Tertiary boundary in high latitudes. Geol Soc Am Bull 105(8):979–997

Kelly AE, Goulden ML (2008) Rapid shifts in plant distribution with recent climate change. Proc Natl Acad Sci 105(33):11823–11826

Köhler M, Pérez-Mellado V, Ramon MM (2010) Fast or slow? The evolution of life history traits associated with insular dwarfing. Islands Evol 19:261–280

Koprowski JL, Nandini R (2008) Global hotspots and knowledge gaps for tree and flying squirrels, vol 95. Curr Sci, pp 851–856

Koprowski JL, Steele MA, Ramos-Lara N (2016) Tamiasciurus mearnsi (Rodentia: Sciuridae). Mamm Species 48(936):66–72
Krystufek B, Vohralik V (2013) Taxonomic revision of the Palaearctic rodents (Rodentia). Part 2. Sciuridae: Urocitellus, Marmota and Sciurotamias. Lynx, series nova, 44
Lair H (1985) Mating seasons and fertility of red squirrels in southern Quebec. Can J Zool 63(10):2323–2327
Larsen CS (2014) Our origins: Discovering physical anthropology. WW Norton & Company
Laurance WF (2010) Habitat destruction: death by a thousand cuts. Conserv Biol All 1(9):73–88
Levin SA (2013) Encyclopedia of biodiversity. Elsevier
Li F, Li S (2018) Paleocene–Eocene and Plio–Pleistocene sea-level changes as "species pumps" in Southeast Asia: evidence from Althepus spiders. Mol Phylogenet Evol 127:545–555
Lindsay SL (1982) Systematic relationship of parapatric tree squirrel species (Tamiasciurus) in the Pacific Northwest. Can J Zool 60(9):2149–2156
Loehle C, Eschenbach W (2012) Historical bird and terrestrial mammal extinction rates and causes. Divers Distrib 18(1):84–91
LoGiudice K (2006) Toward a synthetic view of extinction: a history lesson from a North American rodent. Bioscience 56(8):687–693
Lopez N, Thaler L (1974) Sur Ie plus ancien Lagomorphe europeen et la "Grande Coupure" Oligocene de Stehlin. Palaeovertebrata 6:243–251
Luckett WP, Hartenberger JL (1985) Evolutionary relationships among rodents: comments and conclusions. In: Evolutionary relationships among rodents. Springer, Boston, pp 685–712
Lynas M, Houlton BZ, Perry S (2021) Greater than 99% consensus on human caused climate change in the peer-reviewed scientific literature. Environ Res Lett 16(11):114005
Ma X, Gong Y, Chen D, Racki G, Chen X, Liao W (2016) The Late Devonian Frasnian–Famennian Event in South China—patterns and causes of extinctions, sea level changes, and isotope variations. Palaeogeogr Palaeoclimatol Palaeoecol 448:224–244
Maclean IM, Wilson RJ (2011) Recent ecological responses to climate change support predictions of high extinction risk. Proc Natl Acad Sci 108(30):12337–12342
Marsh G (2015) Darwin's iconic finches join genome club. Nat News 518(7538):147
McCallum H (2012) Disease and the dynamics of extinction. Philos Trans R Soc B Biol Sci 367(1604):2828–2839
Mccleery RA (2009) Changes in fox squirrel anti-predator behaviors across the urban–rural gradient. Landsc Ecol 24(4):483–493
McPhee MG, Proshutinsky A, Morison JH, Steele M, Alkire MB (2009) Rapid change in freshwater content of the Arctic Ocean. Geophys Res Lett 36(10):L10602
McShea DW (2021) Evolution of complexity. In: Evolutionary developmental biology: a reference guide. Springer Cham (Cham - Switzerland), pp 169–179
Meijaard E (2016) Rheithrosciurus macrotis. The IUCN Red List of Threatened Species 2016: e.T19474A22248783. https://doi.org/10.2305/IUCN.UK.2016-2.RLTS.T19474A22248783.en. Accessed 10 Oct 2021
Meiri S, Cooper N, Purvis A (2008) The island rule: made to be broken? Proc R Soc B Biol Sci 275(1631):141–148
Menéndez I, Gómez Cano AR, Cantalapiedra JL, Peláez-Campomanes P, Álvarez-Sierra MÁ, Hernández Fernández M (2020) A multi-layered approach to the diversification of squirrels. Mammal Rev 51(1):66–81
Mercer JM, Roth VL (2003) The effects of Cenozoic global change on squirrel phylogeny. Science 299(5612):1568–1572
Millien V, Bovy H (2010) When teeth and bones disagree: body mass estimation of a giant extinct rodent. J Mammal 91(1):11–18
Monteiro CA, Levy RB, Claro RM, de Castro IRR, Cannon G (2010) Increasing consumption of ultra-processed foods and likely impact on human health: evidence from Brazil. Public Health Nutr 14(1):5–13
Musser G (2020) "Rodent". Encyclopedia Britannica. https://www.britannica.com/animal/rodent. Accessed 28 July 2022
National Park Foundation (2022) The Squirrel Family: America's most endearing rodents. https://www.nationalparks.org/connect/blog/squirrel-family-americas-most-endearing-rodents. Accessed 28 July 2022
Oka PO, Majuk SM (2016) Collaborating with Nature: The Pristine Habitats, in Ugep, Yakurr LGA, Nigeria. J Geosci Environ Protect 4(02):46
Ong P, Balete D, Heaney L (2016) Sundasciurus steerii. The IUCN red list of threatened species 2016: e.T21163A22249713. https://doi.org/10.2305/IUCN.UK.2016-2.RLTS.T21163A22249713.en. Accessed 10 Oct 2021
Orr HA (2001) The genetics of species differences. Trends Ecol Evol 16(7):343–350
Orr HA, Unckless RL (2008) Population extinction and the genetics of adaptation. Am Nat 172(2):160–169
Oshida T, Arslan A, Noda M (2009) Phylogenetic relationships among the Old World Sciurus squirrels. Folia Zool 58(1):14
Otto SP (2018) Adaptation, speciation and extinction in the Anthropocene. Proc R Soc B 285(1891):20182047
Ourworldindata (2020) How many bird species have gone extinct?. https://ourworldindata.org/birds#how-many-bird-species-have-gone-extinct. Accessed 28 July 2022
Owen MJ, Williams NM, O'Donovan MC (2004) The molecular genetics of schizophrenia: new findings promise new insights. Mol Psychiatry 9(1):14–27
Pakistan R (2017) Are we witnessing the world's 6th mass extinction? A new study based on the range contractions of 27,000 terrestrial vertebrate species and a more detailed analysis of population. Oryx 51(4):571–580
Payne J, Francis CM, Phillipps K (1985) Field guide to the mammals of Borneo. Sabah Society
Pečnerová P, Martínková N (2012) Evolutionary history of tree squirrels (Rodentia, Sciurini) based on multilocus phylogeny reconstruction. Zool Scr 41(3):211–219
Pedersen AB, Jones KE, Nunn CL, Altizer S (2007) Infectious diseases and extinction risk in wild mammals. Conserv Biol 21(5):1269–1279
Pennisi E (2009) Agreeing to disagree. Science 323(5915):706–708
Peterson AT, Soberón J, Sánchez-Cordero V (1999) Conservatism of ecological niches in evolutionary time. Science 285(5431):1265–1267
Peterson AT, Papeş M, Soberón J (2008) Rethinking receiver operating characteristic analysis applications in ecological niche modeling. Ecological modelling, 213(1), 63–72
Pimm SL, Jenkins CN, Abell R, Brooks TM, Gittleman JL, Joppa LN, Sexton JO et al (2014) The biodiversity of species and their rates of extinction, distribution, and protection. Science 344(6187):1246752
Pomfret JK, Nocera JJ, Kyser TK, Reudink MW (2012) Linking population declines with diet quality in Vaux's swifts. Northwest Sci 88(4):305–313
Prague Zoo (2022) Animal of the week: European ground squirrel. https://www.zoopraha.cz/en/about-zoo/news/director-s-view/11889-animal-of-the-week-european-ground-squirrel. Accessed 28 July 2022

Prothero DR, Sereno PC (1982) Allometry and paleoecology of medial Miocene dwarf rhinoceroses from the Texas Gulf Coastal Plain. Paleobiology 8(1):16–30

Qiu Z, Storch G (2000) The early Pliocene micromammalian fauna of Bilike, Inner Mongolia, China (Mammalia: Lipotyphla, Chiroptera, Rodentia, Lagomorpha). Senckenb Lethaea 80(1):173–229

Ricciardi A, Rasmussen JB (1999) Extinction rates of North American freshwater fauna. Conserv Biol 13(5):1220–1222

Robold RB, Huettmann F (2021) High-resolution prediction of American red squirrel in Interior Alaska: a role model for conservation using open access data, machine learning, GIS and LIDAR. PeerJ 9:e11830

Román-Palacios C, Wiens JJ (2020) Recent responses to climate change reveal the drivers of species extinction and survival. Proc Natl Acad Sci 117(8):4211–4217

Sales J (2005) The endangered kiwi: a review. Folia Zoologica-Praha 54(1/2):1

Salmona J, Heller R, Quéméré E, Chikhi L (2017) Climate change and human colonization triggered habitat loss and fragmentation in Madagascar. Mol Ecol 26(19):5203–5222

Sarno RJ, Parsons M, Ferris A (2015) Differing vigilance among gray squirrels (Sciuridae carolinensis) along an urban–rural gradient on Long Island. Urban Ecosyst 18(2):517–523

Savage AR, Manlick PJ, Romanski MC, Pauli JN (2019) A reclassification of red squirrels, Tamiasciurus hudsonicus (Rodentia: Sciuridae), on Isle Royale. Biol J Linn Soc 127(2):213–223

Sayer JA, Whitmore TC (1991) Tropical moist forests: destruction and species extinction. Biol Conserv 55(2):199–213

Schluter D (2000) The ecology of adaptive radiation. Oxford University Press, Oxford

Shar S, Lkhagvasure D, Bertolino S, Henttonen H, Kryštufek B, Meinig H (2016) Sciurus vulgaris (errata version published in 2017). The IUCN Red List of Threatened Species 2016: e.T20025A115155900. https://doi.org/10.2305/IUCN.UK.2016-3.RLTS.T20025A22245887.en. Accessed 10 Oct 2021

Shi G, Jiang N, Yao L (2018) Land use and cover change during the rapid economic growth period from 1990 to 2010: A case study of shanghai. Sustainability 10(2):426

Sih A (2013) Understanding variation in behavioural responses to human-induced rapid environmental change: a conceptual overview. Anim Behav 85(5):1077–1088

Sih A, Ferrari MC, Harris DJ (2011) Evolution and behavioural responses to human-induced rapid environmental change. Evol Appl 4(2):367–387

Smith JM (1989) The causes of extinction. Philos Trans R Soc London B Biol Sci 325(1228):241–252

Smith AB, Peterson KJ (2002) Dating the time of origin of major clades: molecular clocks and the fossil record. Annu Rev Earth Planet Sci 30(1):65–88

Smith AP, Quin DG (1996) Patterns and causes of extinction and decline in Australian conilurine rodents. Biol Conserv 77(2–3):243–267

Smith KF, Sax DF, Lafferty KD (2006) Evidence for the role of infectious disease in species extinction and endangerment. Conserv Biol 20(5):1349–1357

Steffen W, Broadgate W, Deutsch L, Gaffney O, Ludwig C (2015) The trajectory of the Anthropocene: the great acceleration. Anthropocene Rev 2(1):81–98

Steinbauer MJ, Field R, Grytnes JA, Trigas P, Ah-Peng C, Attorre F, Beierkuhnlein C et al (2016) Topography-driven isolation, speciation and a global increase of endemism with elevation. Glob Ecol Biogeogr 25(9):1097–1107

Steiner M, Huettmann F (2021) Justification for a taxonomic conservation update of the rodent genus Tamiasciurus: addressing marginalization and mis-prioritization of research efforts and conservation laissez-faire for a sustainability outlook. Eur Zool J 88(1):86–116

Thomas CD, Franco AM, Hill JK (2006) Range retractions and extinction in the face of climate warming. Trends Ecol Evol 21(8):415–416

Thorington RW Jr, Koprowski JL, Steele MA, Whatton JF (2012) Squirrels of the world. JHU Press

Tran D, Nadau A, Durrieu G, Ciret P, Parisot JP, Massabuau JC (2011) Field chronobiology of a molluscan bivalve: how the moon and sun cycles interact to drive oyster activity rhythms. Chronobiol Int 28(4):307–317

Turner TR (ed) (2005) Biological anthropology and ethics: From repatriation to genetic identity. SUNY Press

Valone TJ, Brown JH (1995) Effects of competition, colonization, and extinction on rodent species diversity. Science 267(5199):880–883

Van Der Merwe M, Brown JS, Jackson WM (2005) The coexistence of fox (Sciurus niger) and gray (S. caroliniensis) squirrels in the Chicago metropolitan area. Urban Ecosyst 8(3):335–347

Vanek JP, Burke RL (2020) Insular dwarfism in female Eastern Hog-nosed Snakes (Heterodon platirhinos; Dipsadidae) on a barrier island. Can J Zool 98(3):157–164

Vessali KV (1996) Land use impacts of rapid transit: a review of the empirical literature. Berkeley Plann J 11(1):71–105

Vianey-Liaud M (1979) L'evolution des Rongeurs a l'Oligocene en Europe occidentale. Paléo 166:136–236

Wen J, Nie ZL, Ickert-Bond SM (2016) Intercontinental disjunctions between eastern Asia and western North America in vascular plants highlight the biogeographic importance of the Bering land bridge from late Cretaceous to Neogene. J Syst Evol 54(5):469–490

Wheeler HC, Hik DS (2013) Arctic ground squirrels U rocitellus parryii as drivers and indicators of change in northern ecosystems. Mammal Rev 43(3):238–255

Wilson DE, Reeder DM (eds) (2005) Mammal species of the world: a taxonomic and geographic reference, vol 1. JHU Press

Wuitchik DM, Wang D, Pells TJ, Karimi K, Ward S, Vize PD (2019) Seasonal temperature, the lunar cycle and diurnal rhythms interact in a combinatorial manner to modulate genomic responses to the environment in a reef-building coral. Mol Ecol 28(16):3629–3641

Xuesen Q, Jingyuan Y, Ruwei D (1993) A new discipline of science—The study of open complex giant system and its methodology. J Syst Eng Electron 4(2):2–12

Yigit N, Kryštufek B, Sozen M, Bukhnikashvili A, Shenbrot G (2016) Sciurus anomalus (errata version published in 2017). The IUCN Red List of Threatened Species 2016: e.T20000A115154256. https://doi.org/10.2305/IUCN.UK.2016-3.RLTS.T20000A22245460.en. Accessed 10 Oct 2021

YouTube (2021) Squirrels at the University of Oklahoma (4K Mini-Doc). https://www.youtube.com/watch?v=Z3pj8SkYPQY. Accessed 28 July 2022

Zhenshan L, Shuguang W (2002) Study on the relations between the animal species extinction and habitat destruction. Acta Ecol Sin 22(4):535–540

Part II
Squirrels of the World in the Anthropocene: A Data-Driven Digital Assessment of the Global Squirrel Species

Chapter 3
Habitat Trends of the World's Squirrels and Their Interactions with the Modern World: Relevance for a New Digital Model-Based Conservation Management

Forgiveness says you are given another chance to make a new beginning

Desmond Tutu

Abstract This study investigates and quantifies the preferred ecological and climatic niche for all extant global squirrel species with available data. That is done by using open-access GBIF.org point data, and 132 Geographic Information System (GIS) environmental predictor maps we compiled. We make it publicly available as a value-added open-access data set (including temperature, precipitation, and other factors e.g. altitude, slope, forest cover, soil characteristics, human influence index, proximity to roads, protected areas, etc.). These environmental layers link with the squirrels' distribution across the globe. These best-available predicted squirrel distribution maps for 233 species are then used to identify possible current and future trends to which squirrels diverged during their evolution (= a more detailed outcome of Chapter two's evolutionary dispersion). This has the primary aim to identify whether species tended to diverge to certain regions around the globe, e.g. whether hotspot regions exist where more species occur, in terms of population numbers and species diversity when compared to other areas. Additionally, it aims to identify "regions of high conservation risk" allowing us to see regions where the present species are threatened, due to habitat loss or/and human influence, even warfare, poor governance, and law enforcement. These "regions of/ under high risk" include cities, old-growth forests (primarily for tree squirrels), tropics, and islands. Cities have been considered as regions of/under risk since it has been identified that many squirrel hotspots are near or in cities with high human densities and impacts, which can possibly lead to disease transmission between humans and invasive mammal species (zoonosis – recent examples: Covid-19, rabies, and bubonic plague). Old-growth forests, islands, and the tropics have also been considered as regions of/under high risk since these are all habitats that are affected and threatened by climatic, geologic, or/and human influence. This work sets the baseline for upcoming chapters and includes studies assessing all these regions of/ under high risk in detail. This is done together with the associated specific problems of each habitat/ region, trying to seek greater conservation success for the threatened species at stake, on a global scale.

Keywords Squirrels · Sciuridae · Habitat identification · Ecological niche · GIS · Climate model · Regions of/under high conservational risk

3.1 Introduction

This section aims to briefly introduce the underlying data and theory behind species distribution models (SDMs) and subsequentially presents the background information for the extended SDMs created for the entire squirrel family (Sciuridae) on a global scale. Additionally, the data for the SDMs that have been used are presented here with a brief explanation of where and how it has been retrieved and why this data has been used.

Supplementary Information The online version contains supplementary material available at https://doi.org/10.1007/978-3-031-23547-4_3.

M. Steiner, F. Huettmann, *Sustainable Squirrel Conservation*, https://doi.org/10.1007/978-3-031-23547-4_3

Textbox 1: What Are Species Distribution Models (SDM)?

Species distribution models (SDMs) are models created by technological computer software that combine presence occurrence data of the species with environmental characteristics that describe the surrounding habitat or better, their ecological niche. Species distribution models are defined by Elith and Leathwick (2009) as "*Species distribution models (SDMs) are numerical tools that combine observations of species occurrence or abundance with environmental estimates.*" Usually, those are performed nowadays with machine learning tools as those have been shown to perform superior e.g. Elith et al. (2006), Humphries and Huettmann (2018), and Humphries et al. (2018). If maxent is used instead of more complex and deeper Machine Learning (ML) ensembles and Artificial Intelligence (AI), those still can serve as a rapid assessment tool. If done in an open access framework, those models and data can then better be used and extended by the community at large (*sensu* Huettmann 2015; see also Zuckerberg et al. 2011; Humphries et al. 2018).

Textbox 2: What Are Species Distribution Models Used for?

These models aim to quantify the information of the imported occurrence data with the environmental characteristics/variables. This can result in a statistical relative index prediction of the species of interest. This means that for every map unit the prediction will be calculated, resulting in a, in our case, global model of the possible occurrence index of the species at stake. It is essential that those models predict and generalize well for valid inference (Breiman 2001). Most importantly, these SDMs often can act as the baseline for species conservation assessments and finally aid to set up conservation management projects/initiatives.

Textbox 3: How Are Species Distribution Models Created?

There are many ways to create SDMs, from correlations or linear regression to multiple regressions, deep learning, machine learning, and neural networks (e.g. Deneu et al. 2021; Hegel et al. 2010; Jain et al. 2020; Phillips 2005). Often they are created by technical software such as Maxent or R packages and underlying algorithms. For the SDMs created and discussed throughout this chapter and book, this previously mentioned software Maxent has been used (www.biodiversityinformatics.amnh.org/open_source/maxent/), e.g. for rapid assessments in an automated fashion. Details on the practical creation can be found in Appendix 3.1. For a wider Machine Learning context, see for instance Humphries et al. (2018).

Textbox 4: What Are Rasters and Pixels?

Pixels are components and building blocks for rasters representing landscapes. A raster is built up of multiple pixels which have a pre-definable size. Usually, the more pixels a raster contains, the bigger it is. However, the pixel size can also be adjusted/changed, and thereby the raster can be adjusted to be more, or less sophisticated/detailed (increased/decreased resolution). The number of pixels one certain area of a raster contains determines how sophisticated the raster is. For instance, for a region of 100 square meters, one single pixel can be used which means that for this whole area there is one information entry available to describe its landscape/habitat characteristics. However, this area of 100 square meters can also be split into many different pixels (e.g. 100 = 10 × 10). This means that each square meter contains individually one information entry for this certain characteristic that the raster is representing. This leads to a more sophisticated raster, describing the habitat. A raster has been described as "*a type of computer image that is based on a rectangular grid (= pattern of lines and columns) of pixels (= small dots)*" by the Cambridge Dictionary (2021a). It is a flat but quite powerful representation of the world and includes a geographic projection to account for the earth's shape and surface.

Textbox 5: What Are Environmental Layers?
Environmental layers, also called, environmental predictors, are data files that are usually raster files. These rasters contain for each pixel an environmental variable (e.g. long-term average temperature per month, minimum precipitation per month, etc.) used in the drill down. If such a raster is used for statistical modeling such as SDMs, they are called environmental predictors.

Textbox 6: What Is the Outcome of These Species Distribution Models (SDMs)?
The most important outcomes of these models are predictions to provide generalized inference; it's a predicted Relative Index of Ocurrence (RIO; not a probability). These files from Maxent are represented in the output directory, usually as the HTML report file and the ASC raster data file. The HTML file presents all the findings summarized as a report in a clear fashion. This file includes a map where the predicted occurrences are modeled, which facilitates the interpretation by using different colors to identify the different stages of the relative probability index of the species' occurrence. In addition, the HTML file also contains all the statistical summaries of the model run. This entails the ranking of the most significant and most contributing predictors for each model. Additionally, it also provides the response curves of each predictor, representing its performance during the model run. Last but not least, this HTML file also contains an explanation of how the model has been created and contains information on the statistical measures used for the models, such as ROCs, AUC, P-values, etc.

The other important output file is the geo-referenced ASC file (generic ASCII format). This file is most important if the user wants to use the data for mapping purposes and present it as a GIS map to increase the understanding of the species distribution model. This ASC file can then be further imported into mapping software such as QGIS or ArcGIS, where further use can be made of these predictive models. This ASC file is a raster file, and each raster pixel contains the specific relative probability of the occurrence of the species in question. By using this visualized result from the models it is possible to graphically present hotspots of the species as well as their spatial distribution. For the identification of their regions of/under high risk, this feature of identifying hotspots has been utilized here. These regions of/under high risk will be a topic of high importance throughout this book, and more will be discussed about them later on.

For this study, 233 SDMs have been created using 132 environmental layers each. This means that for all global squirrel species, for which distribution data is available on the online platform Global Biodiversity Information Facility (GBIF – www.gbif.org), a SDM has been created. In order to create the most complete and highest quality SDMs possible, a high number (132) of environmental layers/predictors have been compiled and utilized for these SDMs. The other reason why we utilized this high number of predictors is to provide the software with all possible environmental descriptors so the ecological niche can be described as best as possible. Another reason for performing the SDM with so many environmental predictors is that it aims to eliminate the bias related to data deficiency in a certain location or poor data in some regions. The presence distribution data has been obtained from GBIF by using an R package called RGBIF. Information and details about this can be found in Appendix 3.1 and via the following link (www.cran.r--project.org/web/packages/rgbif/index.html). The cleaned-up data set can be found in the appendix of this study. For details on how this dataset has been cleaned up, please see Appendix 3.1. For a correct and unbiased scientific approach, conceptually, such scientific data should not really be cleaned up, because data is changed and removed from the raw dataset due to an underlying 'opinion' of the operator. However, this is unfortunately necessary for the SDM process with Maxent as the software does not accept any other file type and is not so robust. The negative impacts of this data clean-up and its pitfalls are discussed in more detail in Sect. 3.5.2 below. Machine Learning can usually handle outliers well and is robust so that a heavy data cleaning (to meet parametric assumptions) is barely needed, except for Maxent; *sensu* Humphries et al. (2018). A list of all the species for which the SDMs have been performed can be seen in Table 3.1. This table includes additionally an occurrence count of all the included species. This number represents how many distribution data points are available for each species and utilized for the SDMs.

Table 3.1 SDM species list with occurrence data count

Original species name	Corrected species name	GBIF occurrence points count	Success of the SDM	Comments
Aeromys tephromelas	*Aeromys tephromelas*	7	Yes	
Aeromys thomasi	*Aeromys thomasi*	11	Yes	
Ammospermophilus fossilis	*NA*	3	No data	This species is commonly known to be extinct (for approx. 10.3 Ma), according to fossilworks.org.
Ammospermophilus harrisii	*Ammospermophilus harrisii*	534	Yes	
Ammospermophilus insularis	*Ammospermophilus insularis*	18	Yes	This species is usually mentioned as subspecies of "*Ammospermophilus leucurus*".
Ammospermophilus interpres	*Ammospermophilus interpres*	152	Yes	
Ammospermophilus leucurus	*Ammospermophilus leucurus*	1487	Yes	
Ammospermophilus nelsoni	*Ammospermophilus nelsoni*	131	Yes	
Atlantoxerus getulus	*Atlantoxerus getulus*	135	Yes	
Atlantoxerus rhodius	*NA*	2	Yes	This species is commonly known to be extinct (for approx. 4.9 Ma), according to fossilworks.org.
Belomys pearsonii	*Belomys pearsonii*	4	Yes	
Blackia miocaenica	*NA*	4	Yes	This species is commonly known to be extinct (for approx. 2.588 Ma), according to fossilworks.org.
Callosciurus adamsi	*Callosciurus adamsi*	1	100% BIO3_2_5min	
Callosciurus baluensis	*Callosciurus baluensis*	3	Yes	
Callosciurus caniceps	*Callosciurus caniceps*	116	Yes	
Callosciurus erythraeus	*Callosciurus erythraeus*	784	Yes	
Callosciurus finlaysonii	*Callosciurus finlaysonii*	173	Yes	
Callosciurus inornatus	*Callosciurus inornatus*	1	No data	
Callosciurus nigrovittatus	*Callosciurus nigrovittatus*	12	Yes	
Callosciurus notatus	*Callosciurus notatus*	641	Yes	
Callosciurus orestes	*Callosciurus orestes*	23	Yes	
Callosciurus phayrei	*Callosciurus phayrei*	1	No data	
Callosciurus prevostii	*Callosciurus prevostii*	125	Yes	
Callosciurus pygerythrus	*Callosciurus pygerythrus*	74	Yes	
Callospermophilus lateralis	*Callospermophilus lateralis*	3269	Yes	
Callospermophilus madrensis	*Callospermophilus madrensis*	7	Yes	
Callospermophilus saturatus	*Callospermophilus saturatus*	293	Yes	
Cedromus savannae	*NA*	5	Yes	This species is commonly known to be extinct (for approx. 26.3 Ma), according to fossilworks.org.

(continued)

Table 3.1 (continued)

Original species name	Corrected species name	GBIF occurrence points count	Success of the SDM	Comments
Citellus adocetus	*Notocitellus adocetus*	1	100% World_MAX_RH_MAR	
Citellus beecheyi	*Otospermophilus beecheyi*	2	No data	
Citellus finlayensis	*NA*	1	No data	This species is commonly known to be extinct (for approx. 1.8Ma), according to mindat.org.
Citellus junturensis	*NA*	7	Yes	This species is commonly known to be extinct (for approx. 4.9 Ma), according to fossilworks.org.
Citellus matthewi	*NA*	3	Yes	This species is commonly known to be extinct (for approx. 10.3 Ma), according to fossilworks.org.
Citellus mcgheei	*NA*	1	No data	This species is commonly known to be extinct (for approx. 1.8 Ma), according to fossilworks.org.
Citellus parryii	*Urocitellus parryii*	6	Yes	
Citellus shotwelli	*NA*	7	Yes	This species is commonly known to be extinct (for approx. 5.332 Ma), according to fossilworks.org.
Cryptopterus webbi	*NA*	3	Yes	This species is commonly known to be extinct (for approx. 3.6 Ma), according to fossilworks.org.
Cynomys gunnisoni	*Cynomys gunnisoni*	427	Yes	
Cynomys hibbardi	*NA*	2	Yes	This species is commonly known to be extinct (for approx. 0.3 Ma), according to fossilworks.org.
Cynomys leucurus	*Cynomys leucurus*	400	Yes	
Cynomys ludovicianus	*Cynomys ludovicianus*	2290	Yes	
Cynomys mexicanus	*Cynomys mexicanus*	173	Yes	
Cynomys parvidens	*Cynomys parvidens*	113	Yes	
Dremomys everetti	*Dremomys everetti*	3	Yes	
Dremomys lokriah	*Dremomys lokriah*	15	Yes	
Dremomys pernyi	*Dremomys pernyi*	26	Yes	
Dremomys pyrrhomerus	*Dremomys pyrrhomerus*	4	Yes	
Dremomys rufigenis	*Dremomys rufigenis*	18	Yes	
Echinosciurus aureogaster	*Sciurus aureogaster*	8	Yes	Has been proposed to move from the genus *Sciurus* to *Echinosciurus*, this is however not mutually accepted among institutions (de Abreu-Jr et al. 2020).
Echinosciurus cinereus	*Eupetaurus cinereus*	2	Yes	
Echinosciurus colliaei	*Sciurus colliaei*	2	No data	Has been proposed to move from the genus *Sciurus* to *Echinosciurus*, this is however not mutually accepted among institutions (de Abreu-Jr et al. 2020).
Echinosciurus deppei	*Sciurus deppei*	1	No data	Has been proposed to move from the genus *Sciurus* to *Echinosciurus*, this is however not mutually accepted among institutions (de Abreu-Jr et al. 2020).
Eutamias minimus	*Neotamias minimus*	67	Yes	
Eutamias canipes	*Neotamias canipes*	2	Yes	
Eutamias dorsalis	*Neotamias dorsalis*	1	Yes	
Eutamias merriami	*Neotamias merriami*	1	No data	

(continued)

Table 3.1 (continued)

Original species name	Corrected species name	GBIF occurrence points count	Success of the SDM	Comments
Eutamias ochrogenys	*Neotamias ochrogenys*	56	Yes	
Eutamias quadrivitatus	*Neotamias quadrivitatus*	2	No data	
Eutamias sibiricus	*Eutamias sibiricus*	452	Yes	
Eutamias siskiyou	*Neotamias siskiyou*	30	Yes	
Eutamias speciosus	*Neotamias speciosus*	1	No data	
Eutamias townsendii	*Neotamias townsendii*	3	Yes	
Euxerus erythropus	*Euxerus erythropus*	80	Yes	
Exilisciurus concinnus	*Exilisciurus concinnus*	2	Yes	
Exilisciurus exilis	*Exilisciurus exilis*	53	Yes	
Exilisciurus whiteheadi	*Exilisciurus whiteheadi*	12	Yes	
Funambulus layardi	*Funambulus layardi*	4	Yes	
Funambulus palmarum	*Funambulus palmarum*	301	Yes	
Funambulus pennantii	*Funambulus pennantii*	359	Yes	
Funambulus sublineatus	*Funambulus sublineatus*	28	Yes	
Funambulus tristriatus	*Funambulus tristriatus*	23	Yes	
Funisciurus anerythrus	*Funisciurus anerythrus*	60	Yes	
Funisciurus carruthersi	*Funisciurus carruthersi*	18	Yes	
Funisciurus congicus	*Funisciurus congicus*	17	Yes	
Funisciurus isabella	*Funisciurus isabella*	4	Yes	
Funisciurus lemniscatus	*Funisciurus lemniscatus*	1	100% VE4	
Funisciurus leucogenys	*Funisciurus leucogenys*	75	Yes	
Funisciurus pyrropus	*Funisciurus pyrropus*	29	Yes	
Funisciurus substriatus	*Funisciurus substriatus*	68	Yes	
Geosciurus inauris	*Geosciurus inauris*	248	Yes	
Geosciurus princeps	*Geosciurus princeps*	13	Yes	
Glaucomys oregonensis	*Glaucomys oregonensis*	95	Yes	
Glaucomys sabrinus	*Glaucomys sabrinus*	888	Yes	
Glaucomys volans	*Glaucomys volans*	1083	Yes	
Glyphotes simus	*Glyphotes simus*	1	100% BIO3_2_5min	
Guerlinguetus ingrami	*Sciurus aestuans*	18	Yes	Is commonly seen as subspecies of *Sciurus aestuans*.
Heliosciurus gambianus	*Heliosciurus gambianus*	125	Yes	
Heliosciurus mutabilis	*Heliosciurus mutabilis*	26	Yes	
Heliosciurus punctatus	*Heliosciurus punctatus*	28	Yes	

(continued)

Table 3.1 (continued)

Original species name	Corrected species name	GBIF occurrence points count	Success of the SDM	Comments
Heliosciurus rufobrachium	*Heliosciurus rufobrachium*	188	Yes	
Heliosciurus ruwenzorii	*Heliosciurus ruwenzorii*	25	Yes	
Heliosciurus undulatus	*Heliosciurus undulatus*	17	Yes	
Hesperopetes blacki	*NA*	1	No data	This species is commonly known to be extinct (for approx. 20.43 Ma), according to fossilworks.org.
Hesperopetes jamesi	*NA*	1	No data	This species is commonly known to be extinct (for approx. 20.43 Ma), according to fossilworks.org.
Hesperopetes thoringtoni	*NA*	1	No data	This species is commonly known to be extinct (for approx. 33.9 Ma), according to fossilworks.org.
Hesperosciurus aberti	*Sciurus aberti*	4	Yes	Has been proposed to move from the genus *Sciurus* to *Hesperosciurus*, this is however not mutually accepted among institutions (de Abreu-Jr et al. 2020).
Hylopetes alboniger	*Hylopetes alboniger*	2	Yes	
Hylopetes fimbriatus	*Eoglaucomys fimbriatus*	6	Yes	
Hylopetes nigripes	*Hylopetes nigripes*	3	Yes	
Hylopetes phayrei	*Hylopetes phayrei*	1	100% VE4	
Hylopetes spadiceus	*Hylopetes spadiceus*	3	68.2% VE4, 31.8% BIO3_2_5min	
Hyosciurus ileile	*Hyosciurus ileile*	3	Yes	
Ictidomys mexicanus	*Ictidomys mexicanus*	319	Yes	
Ictidomys parvidens	*Ictidomys parvidens*	454	Yes	
Ictidomys tridecemlineatus	*Ictidomys tridecemlineatus*	1310	Yes	
Iomys horsfieldii	*Iomys horsfieldii*	1	93.3% HII, 4.7% LC12asc2, 2% BIO3_2_5min	
Lariscus hosei	*Lariscus hosei*	1	100% BIO3_2_5min	
Lariscus insignis	*Lariscus insignis*	13	Yes	
Lariscus niobe	*Lariscus niobe*	4	Yes	
Marmota baibacina	*Marmota baibacina*	75	Yes	
Marmota bobak	*Marmota bobak*	84	Yes	
Marmota broweri	*Marmota broweri*	38	Yes	
Marmota caligata	*Marmota caligata*	915	Yes	
Marmota camtschatica	*Marmota camtschatica*	1	No data	
Marmota caudata	*Marmota caudata*	15	Yes	
Marmota flaviventris	*Marmota flaviventris*	2503	Yes	
Marmota himalayana	*Marmota himalayana*	55	Yes	
Marmota korthi	*NA*	1	No data	This species is commonly known to be extinct (for approx. 4.9 Ma), according to fossilworks.org.
Marmota marmota	*Marmota marmota*	1010	Yes	
Marmota monax	*Marmota monax*	5748	Yes	
Marmota olympus	*Marmota olympus*	100	Yes	

(continued)

Table 3.1 (continued)

Original species name	Corrected species name	GBIF occurrence points count	Success of the SDM	Comments
Marmota sawrockensis	*NA*	6	Yes	This species is commonly known to be extinct (for approx. 1.9 Ma), according to fossilworks.org.
Marmota sibirica	*Marmota sibirica*	16	Yes	
Marmota vancouverensis	*Marmota vancouverensis*	21	Yes	
Menetes berdmorei	*Menetes berdmorei*	34	Yes	
Mesosciurus granatensis	*Sciurus granatensis*	3	Yes	
Microsciurus alfari	*Microsciurus alfari*	53	Yes	
Microsciurus flaviventer	*Microsciurus flaviventer*	29	Yes	
Microsciurus mimulus	*Microsciurus mimulus*	38	Yes	
Microsciurus pucheranii	*Sciurus pucheranii*	2	Yes	Has been proposed to move from the genus *Sciurus* to *Leptosciurus*, this is however not mutually accepted among institutions (de Abreu-Jr et al. 2020).
Miopetaurista thaleri	*NA*	2	Yes	This species is commonly known to be extinct (for approx. 4.2 Ma), according to fossilworks.org.
Miospermophilus wyomingensis	*NA*	11	Yes	This species is commonly known to be extinct (for approx. 13.6 Ma), according to fossilworks.org.
Nannosciurus melanotis	*Nannosciurus melanotis*	3	Yes	
Neosciurus carolinensis	*Sciurus carolinensis*	254	Yes	
Neotamias alpinus	*Neotamias alpinus*	17	Yes	
Neotamias amoenus	*Neotamias amoenus*	383	Yes	
Neotamias bulleri	*Neotamias bulleri*	3	Yes	
Neotamias canipes	*Neotamias canipes*	67	Yes	
Neotamias cinereicollis	*Neotamias cinereicollis*	19	Yes	
Neotamias dorsalis	*Neotamias dorsalis*	280	Yes	
Neotamias durangae	*Neotamias durangae*	19	Yes	
Neotamias merriami	*Neotamias merriami*	419	Yes	
Neotamias minimus	*Neotamias minimus*	783	Yes	
Neotamias obscurus	*Neotamias obscurus*	41	Yes	
Neotamias ochrogenys	*Neotamias ochrogenys*	15	Yes	
Neotamias palmeri	*Neotamias palmeri*	21	Yes	
Neotamias panamintinus	*Neotamias panamintinus*	12	Yes	
Neotamias quadrimaculatus	*Neotamias quadrimaculatus*	65	Yes	
Neotamias quadrivittatus	*Neotamias quadrivittatus*	60	Yes	
Neotamias ruficaudus	*Neotamias ruficaudus*	15	Yes	
Neotamias rufus	*Neotamias rufus*	40	Yes	
Neotamias senex	*Neotamias senex*	16	Yes	
Neotamias siskiyou	*Neotamias siskiyou*	9	Yes	
Neotamias sonomae	*Neotamias sonomae*	137	Yes	
Neotamias speciosus	*Neotamias speciosus*	153	Yes	

(continued)

Table 3.1 (continued)

Original species name	Corrected species name	GBIF occurrence points count	Success of the SDM	Comments
Neotamias townsendii	*Neotamias townsendii*	366	Yes	
Neotamias umbrinus	*Neotamias umbrinus*	151	Yes	
Notocitellus adocetus	*Notocitellus adocetus*	17	Yes	
Notocitellus annulatus	*Notocitellus annulatus*	64	Yes	
Nototamias hulberti	*NA*	71	No data	This species is commonly known to be extinct (for approx. 15.97 Ma), according to fossilworks.org.
Otospermophilus atricapillus	*Otospermophilus atricapillus*	22	Yes	
Otospermophilus beecheyi	*Otospermophilus beecheyi*	8317	Yes	
Otospermophilus variegatus	*Otospermophilus variegatus*	3968	Yes	
Palaeosciurus goti	*NA*	8	Yes	This species is commonly known to be extinct (for approx. 23.03 Ma), according to fossilworks.org.
Parasciurus niger	*Sciurus niger*	130	Yes	Has been proposed to move from the genus *Sciurus* to *Parasciurus*, this is however not mutually accepted among institutions (de Abreu-Jr et al. 2020).
Paraxerus alexandri	*Paraxerus alexandri*	18	Yes	
Paraxerus boehmi	*Paraxerus boehmi*	55	Yes	
Paraxerus cepapi	*Paraxerus cepapi*	378	Yes	
Paraxerus cooperi	*Paraxerus cooperi*	1	80.8% VE4, 19.2% Prec09	
Paraxerus flavovittis	*Paraxerus flavovittis*	6	Yes	
Paraxerus ipoensis	*NA*	34	Yes	
Paraxerus lucifer	*Paraxerus lucifer*	2	54% WorldSoil2, 46% VE4	
Paraxerus ochraceus	*Paraxerus ochraceus*	57	Yes	
Paraxerus palliatus	*Paraxerus palliatus*	38	Yes	
Paraxerus poensis	*Paraxerus poensis*	19	Yes	
Paraxerus vexillarius	*Paraxerus vexillarius*	3	Yes	
Petaurista alborufus	*Petaurista alborufus*	32	Yes	
Petaurista elegans	*Petaurista elegans*	19	Yes	
Petaurista leucogenys	*Petaurista leucogenys*	5	Yes	
Petaurista magnificus	*Petaurista magnificus*	7	Yes	
Petaurista mishmiensis	*Petaurista mishmiensis*	1	No data	
Petaurista nobilis	*Petaurista nobilis*	1	100% VE4	
Petaurista petaurista	*Petaurista petaurista*	62	Yes	
Petaurista philippensis	*Petaurista philippensis*	51	Yes	
Petaurista xanthotis	*Petaurista xanthotis*	1	100% WCaltitude	
Petauristodon pattersoni	*NA*	40	No data	This species is commonly known to be extinct (for approx. 15.97 Ma), according to fossilworks.org.
Petinomys genibarbis	*Petinomys genibarbis*	1	100% BIO3_2_5min	
Petinomys lugens	*Petinomys lugens*	1	100% Prec10	

(continued)

Table 3.1 (continued)

Original species name	Corrected species name	GBIF occurrence points count	Success of the SDM	Comments
Petinomys setosus	*Petinomys setosus*	1	100% BIO3_2_5min	
Poliocitellus franklinii	*Poliocitellus franklinii*	173	Yes	
Prosciurillus leucomus	*Prosciurillus leucomus*	2	98.6% VE4, 1.4% BIO3_2_5min	
Prosciurillus murinus	*Prosciurillus murinus*	27	Yes	
Prosciurillus rosenbergii	*Prosciurillus rosenbergii*	1	Yes	
Protosciurus douglassi	*NA*	5	No data	This species is commonly known to be extinct (for approx. 33.3 Ma), according to fossilworks.org.
Protosciurus mengi	*NA*	4	No data	This species is commonly known to be extinct (for approx. 33.3 Ma), according to fossilworks.org.
Protospermophilus		1	No data	No species data available, the entry should be removed
Protospermophilus kelloggi	*NA*	10	yes	This species is commonly known to be extinct (for approx. 15.97 Ma), according to fossilworks.org.
Protospermophilus protospermophilus	*NA*	1	No data	This species is commonly known to be extinct (for approx. 5.332 Ma), according to fossilworks.org.
Protoxerus aubinnii	*Protoxerus aubinnii*	5	Yes	
Protoxerus stangeri	*Protoxerus stangeri*	98	Yes	
Pteromys volans	*Pteromys volans*	112	Yes	
Ratufa affinis	*Ratufa affinis*	54	Yes	
Ratufa bicolor	*Ratufa bicolor*	196	Yes	
Ratufa indica	*Ratufa indica*	129	Yes	
Ratufa macroura	*Ratufa macroura*	86	Yes	
Rubrisciurus rubriventer	*Rubrisciurus rubriventer*	5	Yes	
Sciurillus pusillus	*Sciurillus pusillus*	1	94.8% VE, 5.2% WorldSoil2	
Sciuropterus jamesi	*NA*	5	Yes	This species is commonly known to be extinct (for approx. 10.3 Ma), according to fossilworks.org.
Sciuropterus mathewsi	*NA*	6	Yes	This species is commonly known to be extinct (for approx. 10.3 Ma), according to fossilworks.org.
Sciuropterus minimus	*NA*	3	Yes	This species is commonly known to be extinct (for approx. 13.6 Ma), according to fossilworks.org.
Sciuropterus uphami	*NA*	3	Yes	This species is commonly known to be extinct (for approx. 13.6 Ma), according to fossilworks.org.
Sciurotamias davidianus	*Sciurotamias davidianus*	17	Yes	
Sciurus aberti	*Sciurus aberti*	765	Yes	
Sciurus aestuans	*Sciurus aestuans*	149	Yes	
Sciurus alleni	*Sciurus alleni*	275	Yes	
Sciurus anomalus	*Sciurus anomalus*	36	Yes	
Sciurus arizonensis	*Sciurus arizonensis*	306	Yes	
Sciurus aureogaster	*Sciurus aureogaster*	2026	Yes	

(continued)

Table 3.1 (continued)

Original species name	Corrected species name	GBIF occurrence points count	Success of the SDM	Comments
Sciurus carolinensis	*Sciurus carolinensis*	64463	Yes	Duplicates have been removed. Maximum iteration decreased to 50. Occurrence points are rounded to the nearest paired number for the 5th decimal, duplicates removed again.
Sciurus colliaei	*Sciurus colliaei*	219	Yes	
Sciurus deppei	*Sciurus deppei*	200	Yes	
Sciurus granatensis	*Sciurus granatensis*	936	Yes	
Sciurus griseus	*Sciurus griseus*	2864	Yes	
Sciurus ignitus	*Sciurus ignitus*	35	Yes	
Sciurus igniventris	*Sciurus igniventris*	11	Yes	
Sciurus ingrami	*Sciurus aestuans*	18	Yes	
Sciurus lis	*Sciurus lis*	12	Yes	
Sciurus meridionalis	*Sciurus meridionalis*	13	Yes	
Sciurus nayaritensis	*Sciurus nayaritensis*	123	Yes	
Sciurus niger	*Sciurus niger*	23157	Yes	
Sciurus oculatus	*Sciurus oculatus*	64	Yes	
Sciurus pucheranii	*Sciurus pucheranii*	47	Yes	
Sciurus pyrrhinus	*Sciurus pyrrhinus*	2	Yes	
Sciurus richmondi	*Sciurus richmondi*	46	Yes	
Sciurus sabrinus	*Glaucomys sabrinus*	1	No data	
Sciurus sanborni	*Sciurus sanborni*	1	76.3% Prec02, 23.7% VE4	
Sciurus spadiceus	*Sciurus spadiceus*	48	Yes	
Sciurus stramineus	*Sciurus stramineus*	146	Yes	
Sciurus variegatoides	*Sciurus variegatoides*	929	Yes	
Sciurus vulgaris	*Sciurus vulgaris*	39083	Yes	Duplicates have been removed, and 500 iterations reduced to 150 due to the very high number of presence points
Sciurus yucatanensis	*Sciurus yucatanensis*	341	Yes	
Spermophilinus besanus	*NA*	2	Yes	This species is commonly known to be extinct (for approx. 13.65 Ma), according to fossilworks.org.
Spermophilinus giganteus	*NA*	1	No data	This species is commonly known to be extinct (for approx. 4.9 Ma), according to fossilworks.org.
Spermophilopsis leptodactylus	*Spermophilopsis leptodactylus*	17	Yes	
Spermophilus adocetus	*Notocitellus adocetus*	197	Yes	Solely duplicates, have been removed and merged.
Spermophilus alashanicus	*Spermophilus alashanicus*	16	Yes	
Spermophilus annulatus	*Notocitellus annulatus*	4	Yes	
Spermophilus armatus	*Urocitellus armatus*	49	Yes	
Spermophilus atricapillus	*Otospermophilus atricapillus*	8	Yes	
Spermophilus aureogaster	*Sciurus aureogaster*	2	No data	
Spermophilus beecheyi	*Otospermophilus beecheyi*	132	Yes	
Spermophilus beldingi	*Urocitellus beldingi*	67	Yes	

(continued)

Table 3.1 (continued)

Original species name	Corrected species name	GBIF occurrence points count	Success of the SDM	Comments
Spermophilus boothi	*NA*	7	Yes	This species is commonly known to be extinct (for approx. 1.3 Ma), according to fossilworks.org.
Spermophilus brevicauda	*Spermophilus brevicauda*	1	Yes	
Spermophilus brunneus	*Urocitellus brunneus*	1	No data	
Spermophilus carolinensis	*Sciurus carolinensis*	1	100% VE4	
Spermophilus citellus	*Spermophilus citellus*	267	Yes	
Spermophilus columbianus	*Urocitellus columbianus*	84	Yes	
Spermophilus dauricus	*Spermophilus dauricus*	28	Yes	
Spermophilus elegans	*Urocitellus elegans*	9	Yes	
Spermophilus erythrogenys	*Spermophilus erythrogenys*	19	Yes	
Spermophilus franklinii	*Poliocitellus franklinii*	50	Yes	
Spermophilus fulvus	*Spermophilus fulvus*	17	Yes	
Spermophilus johnstoni	*NA*	1	No data	This species is commonly known to be extinct (for approx. 1.8 Ma), according to fossilworks.org.
Spermophilus lateralis	*Callospermophilus lateralis*	259	Yes	Solely duplicates, have been removed and merged.
Spermophilus leucurus	*Ammospermophilus leucurus*	1	100% HII	
Spermophilus lorisrusselli	*NA*	3	Yes	This species is commonly known to be extinct (for approx. 0.3 Ma), according to fossilworks.org.
Spermophilus madrensis	*Callospermophilus madrensis*	5	Yes	
Spermophilus major	*Spermophilus major*	38	Yes	
Spermophilus meltoni	*NA*	1	No data	This species is commonly known to be extinct (for approx. 1.8 Ma), according to fossilworks.org.
Spermophilus mexicanus	*Ictidomys mexicanus*	299	Yes	
Spermophilus mohavensis	*Xerospermophilus mohavensis*	10	Yes	
Spermophilus mollis	*Urocitellus mollis*	15	Yes	
Spermophilus musicus	*Spermophilus musicus*	5	Yes	
Spermophilus navigator	*NA*	14	No data	Species is generally not considered valid.
Spermophilus pallidicauda	*Spermophilus pallidicauda*	1	Yes	
Spermophilus parryii	*Urocitellus parryii*	341	Yes	
Spermophilus parvidens	*Ictidomys parvidens*	9	Yes	
Spermophilus perotensis	*Xerospermophilus perotensis*	2	No data	
Spermophilus pygmaeus	*Spermophilus pygmaeus*	14	Yes	
Spermophilus relictus	*Spermophilus relictus*	3	Yes	

(continued)

Table 3.1 (continued)

Original species name	Corrected species name	GBIF occurrence points count	Success of the SDM	Comments
Spermophilus richardsonii	*Urocitellus richardsonii*	98	Yes	
Spermophilus saturatus	*Callospermophilus saturatus*	163	Yes	
Spermophilus spilosoma	*Xerospermophilus spilosoma*	232	Yes	
Spermophilus suslicus	*Spermophilus suslicus*	2	Yes	
Spermophilus tereticaudus	*Xerospermophilus tereticaudus*	65	Yes	
Spermophilus townsendii	*Urocitellus townsendii*	27	Yes	
Spermophilus tridecemlineatus	*Ictidomys tridecemlineatus*	937	Yes	
Spermophilus undulatus	*Urocitellus undulatus*	1	Yes	
Spermophilus variegatus	*Otospermophilus variegatus*	210	Yes	
Spermophilus washingtoni	*Urocitellus washingtoni*	44	Yes	
Spermophilus xanthoprymnus	*Spermophilus xanthoprymnus*	8	Yes	
Sundasciurus everetti	*Sundasciurus everetti*	26	Yes	
Sundasciurus hippurus	*Sundasciurus hippurus*	3	Yes	
Sundasciurus jentinki	*Sundasciurus jentinki*	14	Yes	
Sundasciurus juvencus	*Sundasciurus juvencus*	4	No data	
Sundasciurus lowii	*Sundasciurus lowii*	17	Yes	
Sundasciurus philippinensis	*Sundasciurus philippinensis*	1	No data	
Sundasciurus tenuis	*Sundasciurus tenuis*	71	Yes	
Tamias alpinus	*Neotamias alpinus*	9	Yes	
Tamias amoenus	*Neotamias amoenus*	954	Yes	
Tamias aristus	*NA*	2	Yes	This species is commonly known to be extinct (for approx. 0.012 Ma), according to fossilworks.org.
Tamias bulleri	*Neotamias bulleri*	8	Yes	
Tamias canipes	*Neotamias canipes*	75	Yes	
Tamias cinereicollis	*Neotamias cinereicollis*	117	Yes	
Tamias dorsalis	*Neotamias dorsalis*	456	Yes	
Tamias durangae	*Neotamias durangae*	1	No data	
Tamias merriami	*Neotamias merriami*	39	Yes	
Tamias minimus	*Neotamias minimus*	2392	Yes	
Tamias obscurus	*Neotamias obscurus*	16	No data	
Tamias ochrogenys	*Neotamias ochrogenys*	33	Yes	
Tamias palmeri	*Neotamias palmeri*	4	No data	
Tamias panamintinus	*Neotamias panamintinus*	14	Yes	
Tamias quadrimaculatus	*Neotamias quadrimaculatus*	45	Yes	

(continued)

Table 3.1 (continued)

Original species name	Corrected species name	GBIF occurrence points count	Success of the SDM	Comments
Tamias quadrivittatus	*Neotamias quadrivittatus*	305	Yes	
Tamias ruficaudus	*Neotamias ruficaudus*	67	Yes	
Tamias rufus	*Neotamias rufus*	185	Yes	
Tamias senex	*Neotamias senex*	271	Yes	
Tamias sibiricus	*Eutamias sibiricus*	638	Yes	
Tamias siskiyou	*Neotamias siskiyou*	163	Yes	
Tamias sonomae	*Neotamias sonomae*	26	Yes	
Tamias speciosus	*Neotamias speciosus*	278	Yes	
Tamias striatus	*Tamias striatus*	12987	Yes	
Tamias townsendii	*Neotamias townsendii*	574	Yes	
Tamias umbrinus	*Neotamias umbrinus*	336	Yes	
Tamiasciurus douglasii	*Tamiasciurus douglasii*	1968	Yes	
Tamiasciurus mearnsi	*Tamiasciurus mearnsi*	56	Yes	
Tamiasciurus hudsonicus	*Tamiasciurus hudsonicus*	12227	Yes	
Tamiasciurus pennsylvanicus	*NA*	1	100% VE4	Unvalid species.
Tamiops macclellandii	*Tamiops mcclellandii*	107	Yes	
Tamiops maritimus	*Tamiops maritimus*	47	Yes	
Tamiops rodolphii	*Tamiops rodolphii*	15	Yes	
Tamiops swinhoei	*Tamiops swinhoei*	25	Yes	
Urocitellus armatus	*Urocitellus armatus*	559	Yes	
Urocitellus beldingi	*Urocitellus beldingi*	349	Yes	
Urocitellus brunneus	*Urocitellus brunneus*	8	Yes	
Urocitellus canus	*Urocitellus canus*	132	Yes	
Urocitellus columbianus	*Urocitellus columbianus*	822	Yes	
Urocitellus elegans	*Urocitellus elegans*	405	Yes	
Urocitellus mollis	*Urocitellus mollis*	132	Yes	
Urocitellus parryii	*Urocitellus parryii*	1386	Yes	
Urocitellus richardsonii	*Urocitellus richardsonii*	585	Yes	
Urocitellus townsendii	*Urocitellus townsendii*	106	Yes	
Urocitellus tridecemlineatus	*Ictidomys tridecemlineatus*	60	Yes	
Urocitellus undulatus	*Urocitellus undulatus*	157	Yes	
Urocitellus washingtoni	*Urocitellus washingtoni*	35	Yes	
Xenospermophilus mohavensis	*Xerospermophilus mohavensis*	1	No data	
Xenospermophilus spilosoma	*Xerospermophilus spilosoma*	2	Yes	
Xerospermophilus mohavensis	*Xerospermophilus mohavensis*	16	Yes	
Xerospermophilus perotensis	*Xerospermophilus perotensis*	2	Yes	
Xerospermophilus spilosoma	*Xerospermophilus spilosoma*	439	Yes	
Xerospermophilus tereticaudus	*Xerospermophilus tereticaudus*	660	Yes	

(continued)

Table 3.1 (continued)

Original species name	Corrected species name	GBIF occurrence points count	Success of the SDM	Comments
Xerus erythropus	*Geosciurus erythropus*	103	Yes	
Xerus inauris	*Geosciurus inauris*	5	Yes	
Xerus rutilus	*Xerus rutilus*	53	Yes	
		Sum		
		227341		
	Taxonomic changes count			
	83			
	Extinct specimen			
	38			
		Duplicates have been removed from the data set. The data assigned to the old classification is merged with the data of the new classifications due to taxonomic updates and the species name has been changed. Models have then been performed using the updated dataset.		

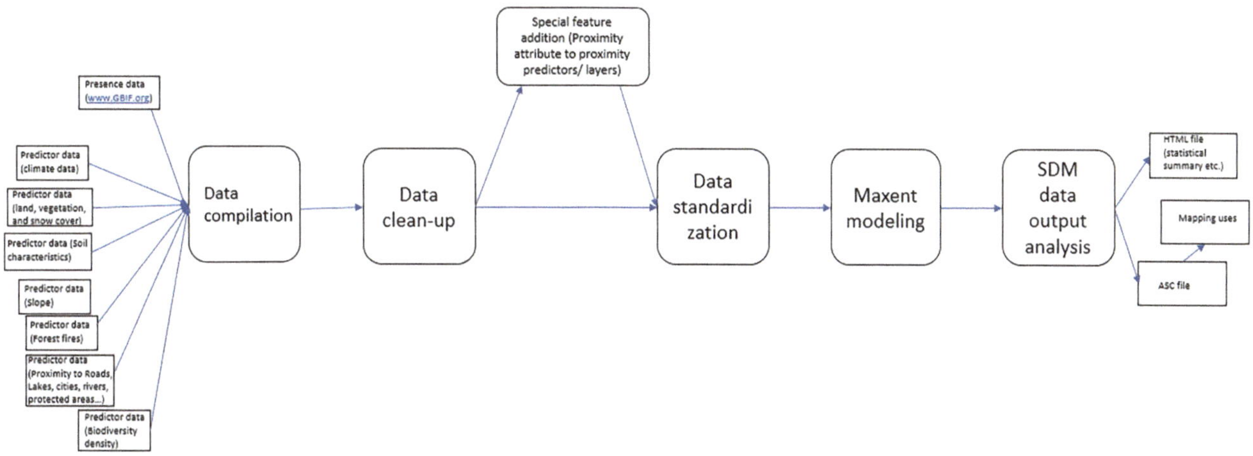

Fig. 3.1 Workflow overview of the SDM creation

For the column in Fig. 3.1, entitled "GBIF occurrence points count" it is noteworthy to shortly discuss the observed trend. ecause the occurrence point counts seemingly vary strongly, they range from 1 to almost 65.000. For species that occur in very high numbers (e.g. Eastern gray squirrel (*Sciurus carolinensis* (64,463)), Eurasian red squirrel (*Sciurus vulgaris* (39,083)), Fox squirrel *(Sciurus niger* (23,157))), it likely indicates that the population is big and distributed across large land areas. Another observed pattern is the spatial locations of these recorded data points. All of these species with very high point data numbers occur in regions where species distribution data logging/collection is very prominent and executed for decades already (e.g. USA, or Western Europe). If these occurrence numbers are compared with South Asian species, or species occurring in the tropics and on islands, the occurrence numbers drastically decrease. This however does not necessarily mean that the squirrel populations are significantly smaller. This rather suggests that the number of recorded occurrence points indicates the presence of data logging surveyors and data submissions, rather than squirrel individuals themselves. This observation has been identified and discussed previously and not only for squirrels e.g. see Raup (1995) and Humphries et al. (2018) for such data types. It suggests that a sampling bias for such "over-sampled" species is easily created and the quality of the predicted occurrence of the SDMs is different than for other species. Further details on this topic, the influences on this study, and holistic conservation are discussed in Sect. 3.5.3.

To compile the SDMs, next to occurrence data, also environmental layers/predictors are needed. These environmental predictors have been compiled from a series of public open-access websites. An overview of all predictors and their sources can be observed in Table 3.2.

Table 3.2 Predictor overview with sources

Predictor name	Source	Explanation	Citation
BIO1_2_5min – BIO19_2_5min; tmin1 – tmin12; tmax1 – tmax12; tavg1 – tavg12; srad1 – srad12; prec1 – prec12; Wcaltitude	Worldclim (https://www.worldclim.org/data/worldclim21.html)	These datasets represent most of the climate data utilized for the SDM.	Fick and Hijmans (2017)
FAOCC	FAO Geonetwork (http://www.fao.org/geonetwork/)	This predictor represents the global climate classes.	
LC12asc2 ; VE4	Geospatial Information Authority of Japan (https://www.gsi.go.jp/kankyochiri/gm_global_e.html)	These predictors represent the global land cover (LC12asc2), and the global vegetation cover (VE4).	
GlobalRiversProxy2	A Simple Global River Bank full Width & Depth Database (http://gaia.geosci.unc.edu/rivers/)	This predictor represents all global mid-size and large rivers.	Andreadis et al. (2013)
GlobalBigRivers11	Global major rivers (https://www.arcgis.com/home/item.html?id=44e8358cf83a4b43bc863646cd695945)	This predictor represents all global large rivers.	
GlobalCities2	Global Cities (https://hub.arcgis.com/datasets/6996f03a1b364dbab4008d99380370ed_0?geometry=-65.394%2C25.931%2C73.737%2C49.818)	This predictor represents all global cities.	
GlobalLakes2	Global Lakes and Wetlands Database (GLWD) (http://www.fao.org/land-water/land/land-governance/land-resources-planning-toolbox/category/details/es/c/1043160/)	This predictor represents all global lakes and wetlands.	
GlobalSnowCoverMonthJan2021_7; FFJan2020_3; FFFeb2020_3; FFMar2020_3; FFMay2020_3,FFJun2020_3; FFJul2020_3; FFAug2020_3; FFSep2020_3; FFOct2020_3; FFNov2020_3; FFJan2021_3	Global Snow Cover and Forest fires (https://neo.sci.gsfc.nasa.gov/view.php?datasetId=MOD10C1_M_SNOW and https://neo.sci.gsfc.nasa.gov/view.php?datasetId=MOD14A1_M_FIRE&year=2020)	These predictors mainly represent the Global snow cover in the month of January, and the forest fire information for nearly all months of 2020 with exception of April and December as these months were not available.	
WorldSoil2	Global Soil characteristics map (https://webarchive.iiasa.ac.at/Research/LUC/External-World-soil-database/HTML/HWSD_Data.html?sb=4)	This predictor represents all global soil types and its characteristics.	
WorldProtectedAreasMerged4	Global Protected areas. (https://www.protectedplanet.net/en/search-areas?geo_type=region&filters%5Bis_type%5D%5B%5D=terrestrial)	This predictor represents all global protected areas merged into one shapefile.	UNEP-WCMC and IUCN (2020)
WorldMammaldensity4; WorldRodentDensity3; WorldThreatenedMammalDensity3; GlobalBirdDensity2	Global Mammal density. Proximity maps for the world mammal density, world rodent density, world threatened mammal density (https://biodiversitymapping.org/index.php/mammals/)	These predictors mainly represent global biodiversity densities. In detail, they contain the world mammal density, world rodent density, world bird density, and the world's threatened mammal density.	Jenkins et al. (2013); Pimm et al. (2014)

GlobalRoadsProxy2	Global Roads - Socioeconomic data and applications center (SEDAC) - Data center in NASA's Earth Observatory System Data and Information System (EOSDIS) (https://sedac.ciesin.columbia.edu/data/set/groads-global-roads-open-access-v1/data-download)	This predictor represents the global proximity to all world's roads. Minor roads may not be included.	
HII1	Human Influence Index (HII). (https://sedac.ciesin.columbia.edu/data/set/wildareas-v2-human-influence-index-geographic/data-download)	This predictor represents the global Human Influence index	
WorldSlope1	Slope. (https://scholarworks.alaska.edu/handle/11122/7151)	This predictor represents the global terrestrial and aquatic slope.	Sriram and Huettmann (unpublished)
World_MAX_RH_JAN – World_MAX_RH_DEC; World_MIN_RH_JAN – World_MIN_RH_DEC	Global Monthly Relative Humidity. (http://palebludata.com/?q=data)	This predictor set represents the global maximum and minimum relative humidity for the months January to December of the year 2020.	Jones and Wint (2015)

Land cover index code (source: https://globalmaps.github.io/glcnmo.html)

The following table shows 20 classifications.

Code	Class name	Code	Class name
1	Broadleaf Evergreen Forest	11	Cropland
2	Broadleaf Deciduous Forest	12	Paddy field
3	Needleleaf Evergreen Forest	13	Cropland / Other Vegetation Mosaic
4	Needleleaf Deciduous Forest	14	Mangrove
5	Mixed Forest	15	Wetland
6	Tree Open	16	Bare area,consolidated(gravel,rock)
7	Shrub	17	Bare area,unconsolidated (sand)
8	Herbaceous	18	Urban
9	Herbaceous with Sparse Tree/Shrub	19	Snow / Ice
10	Sparse vegetation	20	Water bodies

Vegetation cover index code (source: https://globalmaps.github.io/ptc.html)

Percent tree cover is classified into 0% to 100%.

Code	Class name
0–100	Percent Tree Cover (%)
254	Water bodies
255	No data

How these environmental predictors have been compiled, standardized, and improved can be found in Appendix 3.1.

3.2 Methods

In this section, two main methods for this chapter are discussed. To begin with, the methods for creating SDMs are shortly explained. Consecutively, it will be discussed how the outcomes of these SDMs can be interpreted for a science-based conservation assessment of the world's squirrels.

3.2.1 Workflow of Creating Species Distribution Models (SDM)

Section 3.1, it has been explained what SDMs are, what they can be used for, and what they result in. Now, it will be discussed how to actually perform a SDM. Figure 3.1 depicts an overview of all the steps that must be taken to run a SDM successfully and to be able to use the output for further mapping purposes.

The generic process of creating a SDM, which is illustrated in Fig. 3.1, consists of five main elements

1. Data compilation,
2. Data clean-up,
3. Data standardization,
4. Maxent modeling,
5. SDM Data output analysis

Next to these main steps, some additional steps can further be taken to increase the quality and outcome of the SDMs. As seen in Fig. 3.1, these additional steps include an extensive data compilation (presented in detail in Table 3.2 in Sect. 3.1). Additionally, some predictors can be adjusted to obtain high-quality predictors, an example of this is the modification of predictors into proximity predictors.

Proximity predictors have a special feature, where each pixel contains the information of the distance between each pixel and the environmental, or anthropological feature of interest (e.g. lakes, rivers, roads, cities, protected areas, etc.). These proximity predictors are commonly created with the Euclidean Distance analysis (in ArcGIS in our case). Adding proximity predictors to a SDM has the advantage of adding the distance factor to the predictions. The final additional feature that can be added to the process of creating and utilizing SDMs is to improve the interpretation of the output data. This can be achieved by mapping the predicted occurrence of the study species.

The data documentation is crucial to document and present each processing step. Those metadata are provided for the main products in the Appendix.

An in-depth guide on how to perform all these major and minor steps for a SDM can be found in Appendix 3.1 (entitled "Maxent species distribution model creation"). In that appendix, it is discussed how the data should ideally be compiled, how and why the data must be cleaned up, how this data is standardized, and how these special features (Euclidean Distance) are added to the layers of interest. Additionally, the guide explains how SDMs are successfully performed in Maxent. The output data types are being discussed and additional explanation is given on how to increase the interpretation of the output data. Once all these steps have been performed, it can be investigated what these SDMs depict and indicate. This SDM analysis is described in Sect. 3.2.2.

An example of how such an improved mapped SDM for the squirrel species e.g. Eurasian Red Squirrel (*Sciurus vulgaris*) can look like, after following the steps in the appendix, can be observed in Fig. 3.2.

3.2.2 SDM Analysis and Conservation Threat Identification for All Global Squirrels

After creating the SDMs for 233 squirrel species (details on how performed see Fig. 3.1), each model has been mapped in GIS and assessed there. This assessment focused on two main characteristics, the primary spatial distribution, as well as the identification of possible distribution/occurrence hotspots, and thus, major conservation hotspots. This assessment has been performed for all global squirrel species. The outcome of this assessment can be observed in Table 3.3.

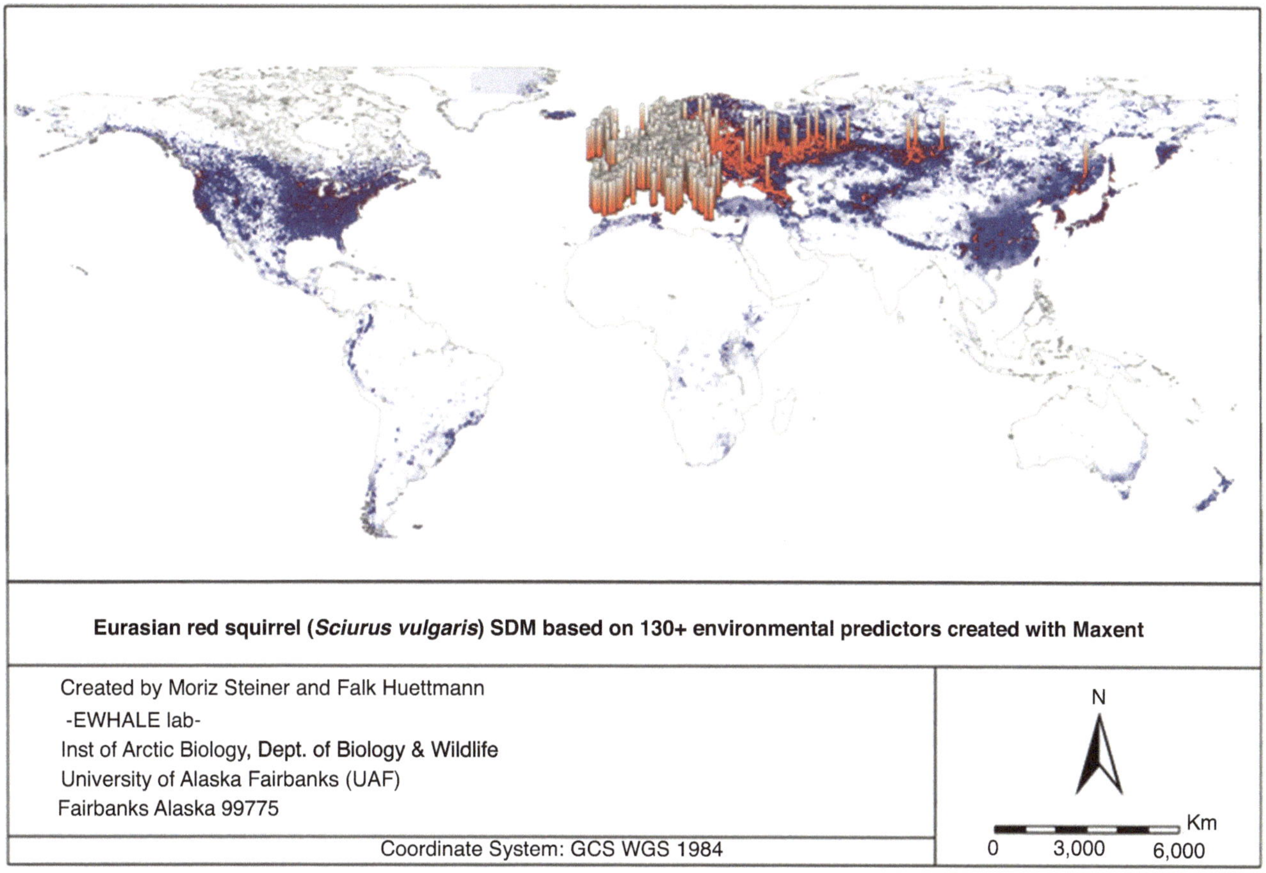

Fig. 3.2 Improved SDM map of the squirrel species Eurasian Red Squirrel (*Sciurus vulgaris*)

3.3 Results

3.3.1 Global Squirrel Hot–/Coldspots: A First (Digital) Global Overview

This section aims to present a macro-overview of the global hot–/coldspots of all the world's squirrels for which sufficient distribution data is publicly available. This is the first time such work has ever been attempted. In order to provide this macro-overview of all global squirrel species, Fig. 3.3 has been created that summarizes all the created SDMs and aims to illustrate the global distribution of hot–/coldspots.

Figure 3.3 provides a global macro-overview of the regions to which squirrels diverged over time and have established their ecological niche. This figure has been created by overlaying all available SMDs that have been created for this study. Following this illustration, it can be identified that the global hotspots of squirrels (family Sciuridae) are Southern Africa, Papua New Guinea (PNG), Western tropical Africa, Eastern Europe, Asia, North-western South America, and North America except the Canadian higher latitudes. The coldspots of the global squirrel distribution have been identified to be Antarctica, Greenland, higher latitudes of Canada, Northern and Central Africa, Western and southern South America, Australia, Northern Russia, and the southern Middle East. Ecologically, these coldsposts are unified by cold/glacier characteristics, desert-like environments, and highly remote islands, as well as spatial limitations (e.g. Wallace line). This indicates that squirrels inhabit all the global regions that are considered to be commonly inhabited by terrestrial mammals.

In the upcoming (Sect. 3.4), this macro-overview will be further broken down into specific regions to provide a detailed overview, towards a more micro/local, species-specific overview.

Table 3.3 Species distribution model analysis and threat identification for all global squirrels

Species	SDM available	Range description	Region of/under high risk participant	Allocated region of/under high risk	Comments
Aeretes melanopterus	No	NA	NA	NA	
Aeromys tephromelas	Yes	Malaysia; Indonesia	Yes	Tropics, Islands	
Aeromys thomasi	Yes	Sabah, Malaysia	Yes	Tropics, Islands	
Ammospermophilus harrisii	Yes	Arizona, Nevada USA; Baja California, Mexico	No	NA	
Ammospermophilus insularis	No	NA	NA	NA	
Ammospermophilus interpres	Yes	Colorado, New Mexico, USA; Chihuahua, Mexico	NA	NA	
Ammospermophilus leucurus	Yes	Southern Westcoast USA; Baja California Mexico	NA	NA	
Ammospermophilus nelsoni	Yes	California, USA	NA	NA	
Atlantoxerus getulus	Yes	Morocco; Tenerife	Yes	Islands	
Belomys pearsonii	Yes	Bangladesh; Taiwan; Coast of China and Vietnam; Southern coast of Japan	Yes	Tropics, Islands	
Biswamoyopterus biswasi	No	NA	NA	NA	
Biswamoyopterus gaoligongensis	No	NA	NA	NA	
Biswamoyopterus laoensis	No	NA	NA	NA	
Callosciurus adamsi	Yes	NA	NA	NA	Seems to have an incorrect SDM
Callosciurus baluensis	Yes	Coastal regions of Malaysia	Yes	Tropics, Islands	Check how many presence points. Seems extremely restricted
Callosciurus caniceps	Yes	Myanmar; Thailand; Cambodia; Malaysia	Yes	Tropics, Islands, Cities	
Callosciurus erythraeus	Yes	Nepal; Bhutan; Bangladesh; Myanmar; China; Taiwan	Yes	Tropics, Islands, Cities	The Netherlands and West Germany also high presence.
Callosciurus finlaysonii	Yes	Myanmar; Thailand; Cambodia; Laos	Yes	Tropics, Islands, Cities	
Callosciurus honkhoaiensis	No	NA	NA	NA	
Callosciurus inornatus	Yes	NA	NA	NA	Seems to have an incorrect SDM
Callosciurus melanogaster	No	NA	NA	NA	
Callosciurus nigrovittatus	Yes	Sumatra, West Java, Indonesia; Peninsula Malaysia	Yes	Tropics, Islands, Cities, Old-growth forests	
Callosciurus notatus	Yes	Sumatra, Java, Indonesia; Malaysia; Brunei	Yes	Tropics, Islands, Cities	
Callosciurus orestes	Yes	The southern tip of Vietnam; Sabah, Malaysia	Yes	Tropics, Islands	
Callosciurus phayrei	Yes	NA	NA	NA	Seems to have an incorrect SDM
Callosciurus prevostii	Yes	Malaysia; Sumatra; West Javal; Brunei; Sri Lanka	Yes	Tropics, Islands, Cities	A very good example for a species in the tropics and on islands in cities, especially cities.
Callosciurus pygerythrus	Yes	Nepal; Bhutan; Bangladesh; North-eastern India; Myanmar	NA	NA	
Callosciurus quinquestriatus	No	NA	NA	NA	
Callospermophilus lateralis	Yes	North-western USA; South-western Canada	NA	NA	

(continued)

Table 3.3 (continued)

Species	SDM available	Range description	Region of/under high risk participant	Allocated region of/under high risk	Comments
Callospermophilus madrensis	Yes	Chihuahua, Durnago, Mexico	NA	NA	
Callospermophilus saturatus	Yes	Washington State, USA; British Colombia, Canada	Yes	Cities, Old-growth forests	
Cynomys gunnisoni	Yes	New Mexico, Colorado, Utah, Arizona, USA	NA	NA	
Cynomys leucurus	Yes	Colorado, Wyoming, Utah, USA	NA	NA	
Cynomys ludovicianus	Yes	Central USA	NA	NA	
Cynomys mexicanus	Yes	Central Mexico	NA	NA	
Cynomys parvidens	Yes	Utah, USA	NA	NA	
Dremomys everetti	Yes	Sabah, Malaysia	NA	NA	
Dremomys gularis	No	NA	NA	NA	
Dremomys lokriah	Yes	Nepal; Bhutan; Bangladesh	Yes	Tropics	
Dremomys pernyi	Yes	Taiwan; China	Yes	Islands, Cities, Islands	
Dremomys pyrrhomerus	Yes	Taiwan; Hainan, Hunan, China	Yes	Islands, Cities	
Dremomys rufigenis	Yes	Myanmar; North-eastern and Southern India; Yunnan, China; Malaysian Peninsula	Yes	Tropics	
Eoglaucomys fimbriatus	Yes	Taiwan; North-western India	NA	NA	
Epixerus ebii	No	NA	NA	NA	
Eupetaurus cinereus	No	NA	NA	NA	
Eutamias sibiricus	Yes	Russia; Japan; South Korea	Yes	Islands	Some weird hotspots also around Europe, maybe confused with Sciurus vulgaris
Euxerus erythropus	Yes	Sub-Saharan and tropical regions Africa	Yes	Tropics, Old-growth forests	Very clear in most tropics regions
Exilisciurus concinnus	Yes	Philippines; Solomon Islands; Papua New Guinea	Yes	Tropics, Islands, Old-growth forests	Good for islands in tropics
Exilisciurus exilis	Yes	Brunei; Sabah, Malaysia	Yes	Tropics, Islands	
Exilisciurus whiteheadi	Yes	Brunei; Sabah, Malaysia; Sri Lankan coasts; Banda Aceh, Balikpapan, Indonesia	Yes	Tropics, Islands, Cities	Also quite clear for some cities
Funambulus layardi	Yes	Sri Lanka; Fiji	Yes	Tropics, Islands	Good for islands in the tropics
Funambulus obscurus	No	NA	NA	NA	
Funambulus palmarum	Yes	Southern India; Sri Lanka	Yes	Tropics, Islands	
Funambulus pennantii	Yes	India; UAE	NA	NA	
Funambulus sublineatus	Yes	Southern India; Sri Lanka	Yes	Tropics, Islands	
Funambulus tristriatus	Yes	South-western India; Sri Lanka	Yes	Tropics, Islands	
Funisciurus anerythrus	Yes	Western tropical regions, Africa	Yes	Tropics	

(continued)

Table 3.3 (continued)

Species	SDM available	Range description	Region of/under high risk participant	Allocated region of/under high risk	Comments
Funisciurus bayonii	No	NA	NA	NA	
Funisciurus carruthersi	Yes	Rwanda; Burundi; DRC	Yes	Tropics	
Funisciurus congicus	Yes	Angola; Namibia; Ethiopia; Paita, Peru	Yes	Tropics	
Funisciurus duchaillui	No	NA	NA	NA	
Funisciurus isabella	Yes	Cameroon; Central African Republic; Equatorial Guinea; Gabon	Yes	Tropics, Old-growth forests	
Funisciurus lemniscatus	Yes	NA	NA	NA	Seems to have an incorrect SDM
Funisciurus leucogenys	Yes	Togo; Benin	Yes	Tropics	
Funisciurus pyrropus	Yes	Western tropical regions, Africa	Yes	Tropics	
Funisciurus substriatus	Yes	Togo; Benin	Yes	Tropics	
Geosciurus erythropus	Yes	Western tropical regions, Africa; Kenya	Yes	Tropics	
Geosciurus inauris	Yes	Botswana; Namibia; South Africa	NA	NA	
Geosciurus princeps	Yes	Namibia; South Africa	NA	NA	
Glaucomys oregonensis	Yes	West Coast of the USA	NA	NA	
Glaucomys sabrinus	Yes	Eastern Canada; Western USA, Alaska	Yes	Islands, Old-growth forests	
Glaucomys volans	Yes	Eastern USA	Yes	Cities	
Glyphotes simus	Yes	NA	NA	NA	Seems to have an incorrect SDM
Heliosciurus gambianus	Yes	Western tropical regions, Africa	Yes	Tropics	
Heliosciurus mutabilis	Yes	Zimbabwe; Mozambique; Madagascar; North-east Queensland, Australia	Yes	Tropics	
Heliosciurus punctatus	Yes	Côte d'Ivoire; Ghana; Togo; Benin	Yes	Tropics	
Heliosciurus rufobrachium	Yes	Tropical regions, Africa	Yes	Tropics	
Heliosciurus ruwenzorii	Yes	Rwanda; Burundi	Yes	Tropics	Very small region in the tropics
Heliosciurus undulatus	Yes	Mozambique	Yes	Tropics, Cities	
Hylopetes alboniger	Yes	Tripura, India	Yes	Tropics	Very small region in the tropics
Hylopetes bartelsi	No	NA	NA	NA	
Hylopetes electilis	No	NA	NA	NA	
Hylopetes nigripes	Yes	Palawan, Philippines	Yes	Tropics, Islands	Very small region in the tropics
Hylopetes phayrei	Yes	NA	NA	NA	Seems to have an incorrect SDM
Hylopetes platyurus	No	NA	NA	NA	
Hylopetes sagitta	No	NA	NA	NA	
Hylopetes sipora	No	NA	NA	NA	
Hylopetes spadiceus	Yes	NA	NA	NA	Seems to have an incorrect SDM
Hylopetes winstoni	No	NA	NA	NA	
Hyosciurus heinrichi	No	NA	NA	NA	
Hyosciurus ileile	Yes	Central Indonesia; Papua; Papua New Guinea; Taiwan	Yes	Tropics, Islands, Old-growth forests	
Ictidomys mexicanus	Yes	Mexico; Texas, New Mexico, USA; Guatemala	Yes	Tropics	

(continued)

Table 3.3 (continued)

Species	SDM available	Range description	Region of/under high risk participant	Allocated region of/under high risk	Comments
Ictidomys parvidens	Yes	North-eastern Mexico; Texas, USA	NA	NA	
Ictidomys tridecemlineatus	Yes	Central North America	Yes	Cities	
Iomys horsfieldii	Yes	NA	NA	NA	Seems to have an incorrect SDM
Iomys sipora	No	NA	NA	NA	
Lariscus hosei	Yes	NA	NA	NA	Seems to have an incorrect SDM
Lariscus insignis	Yes	Singapore; Malaysia; Sumatra, Sabah, Java, Indonesia; Brunei; Papua New Guinea; Sri Lanka	Yes	Tropics, Islands, Cities	Also quite clear for some cities
Lariscus niobe	Yes	Bengkulu, Indonesia; Papua New Guinea	Yes	Tropics, Islands, Old-growth forests	very small regions
Lariscus obscurus	No	NA	NA	NA	
Marmota baibacina	Yes	Central Russia	NA	NA	
Marmota bobak	Yes	Western Russia; Ukraine	NA	NA	
Marmota broweri	Yes	Northern Alaska, USA	NA	NA	
Marmota caligata	Yes	Alaska, Washington State, Montana, USA; Yukon, British Colombia, Canada	Yes	Old-growth forests	
Marmota camtschatica	Yes	NA	NA	NA	Seems to have an incorrect SDM
Marmota caudata	Yes	Kyrgyzstan; Tajikistan; Uzbekistan	NA	NA	
Marmota flaviventris	Yes	Westeren North America	NA	NA	
Marmota himalayana	Yes	Himalayan region	NA	NA	
Marmota kastschenkoi	No	NA	NA	NA	
Marmota marmota	Yes	Central Europe, Alps and Pyrenees	NA	NA	
Marmota menzbieri	No	NA	NA	NA	
Marmota monax	Yes	Eastern North America; Alaska USA; Japan; Korea	Yes	Cities	
Marmota olympus	Yes	Washington State, USA; British Colombia, Canada	Yes	Old-growth forests	Restricted to one national park which is an old-growth forest
Marmota sibirica	Yes	Mongolia	NA	NA	
Marmota vancouverensis	Yes	Vancouver Island, Canada	Yes	Islands	
Menetes berdmorei	Yes	Thailand; Cambodia; Laos; Vietnam	Yes	Tropics	
Microsciurus alfari	Yes	Guatemala; Nicaragua; Costa Rica; Panama; Colombia; Jamaica	Yes	Tropics, Islands, Old-growth forests	
Microsciurus flaviventer	Yes	Colombia; Ecuador; Peru; Western Brazil	Yes	Tropics, Old-growth forests	
Microsciurus mimulus	Yes	Costa Rica; Panama; Colombia; Ecuador	Yes	Tropics	
Microsciurus santanderensis	No	NA	NA	NA	
Myosciurus pumilio	No	NA	NA	NA	
Nannosciurus melanotis	Yes	Coastal Indonesia	Yes	Tropics, Islands	A good example of a sea-level rise threat
Neotamias alpinus	Yes	Sierra Nevada, California, USA	NA	NA	

(continued)

Table 3.3 (continued)

Species	SDM available	Range description	Region of/under high risk participant	Allocated region of/under high risk	Comments
Neotamias amoenus	Yes	North-western USA; South-western Canada	NA	NA	
Neotamias bulleri	Yes	Mexico	NA	NA	
Neotamias canipes	Yes	New Mexico, USA; Eastern Algeria	NA	NA	Weird distribution
Neotamias cinereicollis	Yes	New Mexico, Arizona, USA	NA	NA	
Neotamias dorsalis	Yes	New Mexico, Colorado, Utah, Arizona, USA; Western Mexico	NA	NA	
Neotamias durangae	Yes	Northern Mexico	NA	NA	
Neotamias merriami	Yes	California, USA	Yes	Cities	
Neotamias minimus	Yes	North-eastern North America	NA	NA	
Neotamias obscurus	Yes	Baja California, Mexico	NA	NA	
Neotamias ochrogenys	Yes	West coast USA	NA	NA	
Neotamias palmeri	Yes	Las Vegas, Nevada, USA	Yes	Cities	Very small distribution range
Neotamias panamintinus	Yes	Arizona, Nevada, USA	NA	NA	
Neotamias quadrimaculatus	Yes	Sierra Nevada, California, USA	NA	NA	
Neotamias quadrivittatus	Yes	New Mexico, Colorado, Utah, Arizona, USA	NA	NA	
Neotamias ruficaudus	Yes	Montana, Washington State, USA	NA	NA	
Neotamias rufus	Yes	Utah, Colorado, USA	NA	NA	
Neotamias senex	Yes	California, Oregon, Washington State, USA	NA	NA	
Neotamias siskiyou	Yes	California, Oregon, USA	NA	NA	
Neotamias solivagus	No	NA	NA	NA	
Neotamias sonomae	Yes	California, USA	NA	NA	
Neotamias speciosus	Yes	California, USA	NA	NA	
Neotamias townsendii	Yes	West coast USA; Vancouver Island, Canada	Yes	Islands	
Neotamias umbrinus	Yes	Western USA	NA	NA	
Notocitellus adocetus	Yes	Southern Mexico	Yes	Tropics	
Notocitellus annulatus	Yes	Western coast Mexico	NA	NA	
Otospermophilus atricapillus	Yes	Baja California, Mexico	NA	NA	
Otospermophilus beecheyi	Yes	West coast USA; Baja California, Mexico	NA	NA	Also quite high predicted occurrence in Spain, Portugal, and Cyprus
Otospermophilus variegatus	Yes	Central-south USA; Mexico	NA	NA	
Paraxerus alexandri	Yes	Rwanda,: Burundi; Uganda	Yes	Tropics	
Paraxerus boehmi	Yes	Rwanda,: Burundi; DRC; Uganda	Yes	Tropics	
Paraxerus cepapi	Yes	Zambia; Zimbabwe; Botswana; South Africa	Yes	Tropics, Old-growth forests	
Paraxerus cooperi	Yes	NA	NA	NA	Seems to have an incorrect SDM
Paraxerus flavovittis	Yes	Tanzania	Yes	Tropics	Very small distribution range
Paraxerus lucifer	Yes	NA	NA	NA	Seems to have an incorrect SDM
Paraxerus ochraceus	Yes	Tanzania; Kenya	Yes	Tropics	

(continued)

Table 3.3 (continued)

Species	SDM available	Range description	Region of/under high risk participant	Allocated region of/under high risk	Comments
Paraxerus palliatus	Yes	Tanzania; Luanda, Angola; Gabon	Yes	Tropics, Cities	
Paraxerus poensis	No	NA	NA	NA	
Paraxerus vexillarius	Yes	Luanda, Angola	Yes	Tropics, Cities	
Paraxerus vincenti	No	NA	NA	NA	
Petaurillus emiliae	No	NA	NA	NA	
Petaurillus hosei	No	NA	NA	NA	
Petaurillus kinlochii	No	NA	NA	NA	
Petaurista albiventer	No	NA	NA	NA	
Petaurista alborufus	Yes	Taiwan	Yes	Islands	
Petaurista caniceps	No	NA	NA	NA	
Petaurista elegans	Yes	Taiwan; Indonesia, Papua New Guinea; Laos; Vietnam; Malaysia	Yes	Tropics, Islands, Old-growth forests	
Petaurista grandis	No	NA	NA	NA	
Petaurista hainana	No	NA	NA	NA	
Petaurista lena	No	NA	NA	NA	
Petaurista leucogenys	Yes	Japan, South Korea	Yes	Islands	
Petaurista magnificus	Yes	Vietnam; Nepal; Bhutan; Arunachal Pradesh, India	Yes	Old-growth forests	
Petaurista marica	No	NA	NA	NA	
Petaurista mechukaensis	No	NA	NA	NA	
Petaurista mishmiensis	Yes	NA	NA	NA	Seems to have an incorrect SDM
Petaurista nobilis	Yes	NA	NA	NA	Seems to have an incorrect SDM
Petaurista petaurista	Yes	Malaysia; Indonesia; China; Taiwan	Yes	Tropics, Islands	
Petaurista philippensis	Yes	Taiwan, West coast India	Yes	Tropics, Islands	
Petaurista siangensis	No	NA	NA	NA	
Petaurista sybilla	No	NA	NA	NA	
Petaurista xanthotis	Yes	NA	NA	NA	Seems to have an incorrect SDM
Petaurista yunanensis	No	NA	NA	NA	
Petinomys crinitus	No	NA	NA	NA	
Petinomys fuscocapillus	No	NA	NA	NA	
Petinomys genibarbis	Yes	NA	NA	NA	Seems to have an incorrect SDM
Petinomys hageni	No	NA	NA	NA	
Petinomys lugens	Yes	NA	NA	NA	Seems to have an incorrect SDM
Petinomys mindanensis	No	NA	NA	NA	
Petinomys setosus	Yes	NA	NA	NA	Seems to have an incorrect SDM
Petinomys vordermanni	No	NA	NA	NA	
Poliocitellus franklinii	Yes	Northern Nord America	NA	NA	
Prosciurillus abstrusus	No	NA	NA	NA	
Prosciurillus alstoni	No	NA	NA	NA	
Prosciurillus leucomus	Yes	NA	NA	NA	Seems to have an incorrect SDM
Prosciurillus murinus	Yes	Indonesia; Papua	Yes	Tropics, Islands, Old-growth forests	
Prosciurillus rosenbergii	No	NA	NA	NA	
Prosciurillus topapuensis	No	NA	NA	NA	
Prosciurillus weberi	No	NA	NA	NA	

(continued)

Table 3.3 (continued)

Species	SDM available	Range description	Region of/under high risk participant	Allocated region of/under high risk	Comments
Protoxerus aubinnii	Yes	Côte d'Ivoire; Ghana; Togo; Benin	Yes	Tropics, Cities	
Protoxerus stangeri	Yes	Tropical regions, Africa	Yes	Tropics, Old-growth forests	
Pteromys momonga	No	NA	NA	NA	
Pteromys volans	Yes	Sweden; Finland; Japan; Western Russia	NA	NA	
Pteromyscus pulverulentus	No	NA	NA	NA	
Ratufa affinis	Yes	Malaysia	Yes	Tropics, Islands	
Ratufa bicolor	Yes	Nepal; Bhutan; Malaysia; Myanmar; Cambodia; Papua New Guinea, Sumatra, Indonesia	Yes	Tropics, Islands, Old-growth forests	
Ratufa indica	Yes	India; Sri Lanka	Yes	Tropics, Islands	
Ratufa macroura	Yes	Southern India; Sri Lanka	Yes	Tropics, Islands	
Rheithrosciurus macrotis	No	NA	NA	NA	
Rhinosciurus laticaudatus	No	NA	NA	NA	
Rubrisciurus rubriventer	Yes	Brunei; Indonesia; Malaysia; Buinea; Papua New Guinea; Sri Lanka	Yes	Tropics, Islands, Old-growth forests	
Sciurillus pusillus	Yes	NA	NA	NA	Seems to have an incorrect SDM
Sciurotamias davidianus	Yes	China; Bangladesh	NA	NA	
Sciurotamias forresti	No	NA	NA	NA	
Sciurus aberti	Yes	New Mexico, Colorado, Utah, Arizona, USA; Mexico	NA	NA	
Sciurus aestuans	Yes	Brazil; Bolivia	Yes	Tropics, Cities, Old-growth forests	
Sciurus alleni	Yes	Mexico	Yes	Tropics	
Sciurus anomalus	Yes	Turkey; Syria; Lebanon	NA	NA	
Sciurus arizonensis	Yes	Arizona, New Mexico, USA	NA	NA	
Sciurus aureogaster	Yes	Mexico; Guatemala	Yes	Tropics	
Sciurus carolinensis	Yes	Europe; Eastern USA; Japan; China	Yes	Cities	
Sciurus colliaei	Yes	West Coast Mexico	Yes	Tropics	
Sciurus deppei	Yes	Middle America	Yes	Tropics	
Sciurus flammifer	No	NA	NA	NA	
Sciurus gilvigularis	No	NA	NA	NA	
Sciurus granatensis	Yes	Guatemala; Costa Rica; Panama; Colombia; Ecuador	Yes	Tropics	
Sciurus griseus	Yes	West coast USA	NA	NA	
Sciurus ignitus	Yes	Peru; Bolivia; Brazil	Yes	Tropics, Old-growth forests	
Sciurus igniventris	Yes	Peru; Ecuador; Brazil; Colombia	Yes	Tropics	Possible example for tropics
Sciurus lis	Yes	Japan, South Korea	NA	NA	Interestingly also predicted occurrence in Tasmania, Australia

(continued)

Table 3.3 (continued)

Species	SDM available	Range description	Region of/under high risk participant	Allocated region of/under high risk	Comments
Sciurus meridionalis	Yes	Sardegna, Sicilia, Southern Italy; Corsica, France; Crete, Greece	Yes	Islands	
Sciurus nayaritensis	Yes	Mexico	Yes	Tropics	
Sciurus niger	Yes	USA	Yes	Cities	
Sciurus oculatus	Yes	Southern Mexico	Yes	Tropics, Cities	
Sciurus pucheranii	Yes	Colombia; Ecuador	Yes	Tropics	
Sciurus pyrrhinus	Yes	Peru	Yes	Tropics, Old-growth forests	
Sciurus richmondi	Yes	Nicaragua	Yes	Tropics	Good for a small region in the tropics
Sciurus sanborni	Yes	NA	NA	NA	Seems to have an incorrect SDM
Sciurus spadiceus	Yes	Ecuador; Peru; Brazil; Bolivia	Yes	Tropics, Old-growth forests	
Sciurus stramineus	Yes	Ecuador; Northern Peru	Yes	Tropics	
Sciurus variegatoides	Yes	Guatemala; Honduras; El Salvador; Nicaragua; Costa Rica; Panama	Yes	Tropics, Cities	
Sciurus vulgaris	Yes	Europe	Yes	Cities	
Sciurus yucatanensis	Yes	Southern Mexico; Belize; Guatemala	Yes	Tropics	
Spermophilopsis leptodactylus	Yes	Iran; Turkmenistan	NA	NA	
Spermophilus alashanicus	Yes	Mongolia; China	NA	NA	
Spermophilus brevicauda	No	NA	NA	NA	
Spermophilus citellus	Yes	Eastern Europe	NA	NA	
Spermophilus dauricus	Yes	Mongolia, Northern China; Eastern Russia	NA	NA	
Spermophilus erythrogenys	Yes	Russia	NA	NA	
Spermophilus fulvus	Yes	Turkey; Kazakhstan; Uzbekistan; Kyrgystan; Tajikistan	NA	NA	
Spermophilus major	Yes	Republic of Bashkortostan, Russia	NA	NA	
Spermophilus musicus	Yes	Giorgia; Armenia	Yes	Old-growth forests	Very small distribution range
Spermophilus nilkaensis	No	NA	NA	NA	
Spermophilus pallidicauda	Yes	Western Mongolia	NA	NA	
Spermophilus perotensis	Yes	NA	NA	NA	Seems to have an incorrect SDM
Spermophilus pygmaeus	Yes	South-western Russia; Ukraine	NA	NA	
Spermophilus relictus	Yes	Kazakhstan	NA	NA	Very small distribution patches
Spermophilus suslicus	Yes	Ukraine; Russia; Kazakhstan	Yes	Cities	Very small distribution patches
Spermophilus taurensis	No	NA	NA	NA	
Spermophilus xanthoprymnus	Yes	Turkey; Iran	NA	NA	
Sundasciurus altitudinis	No	NA	NA	NA	
Sundasciurus brookei	No	NA	NA	NA	
Sundasciurus davensis	No	NA	NA	NA	
Sundasciurus everetti	Yes	Brunei; Sabah, Banda Aceh, Malaysia	Yes	Tropics, Islands, Cities, Old-growth forests	
Sundasciurus fraterculus	No	NA	NA	NA	

(continued)

Table 3.3 (continued)

Species	SDM available	Range description	Region of/under high risk participant	Allocated region of/under high risk	Comments
Sundasciurus hippurus	Yes	Malaysia; Brunei; Sumatra, Indonesia	Yes	Tropics, Islands	
Sundasciurus hoogstraali	No	NA	NA	NA	
Sundasciurus jentinki	Yes	Northern Borneo; Brunei	Yes	Tropics, Islands, Old-growth forests	
Sundasciurus juvencus	Yes	NA	NA	NA	Seems to have an incorrect SDM
Sundasciurus lowii	Yes	Borneo, Malaysian Peninsula; Brunei; Sumatra, Indonesia; Papua; Papua New Guinea	Yes	Tropics, Islands, Old-growth forests	
Sundasciurus mindanensis	No	NA	NA	NA	
Sundasciurus moellendorffi	No	NA	NA	NA	
Sundasciurus natunensis	No	NA	NA	NA	
Sundasciurus philippinensis	Yes	NA	NA	NA	Seems to have an incorrect SDM
Sundasciurus rabori	No	NA	NA	NA	
Sundasciurus robinsoni	No	NA	NA	NA	
Sundasciurus samarensis	No	NA	NA	NA	
Sundasciurus steerii	No	NA	NA	NA	
Sundasciurus tahan	No	NA	NA	NA	
Sundasciurus tenuis	Yes	Malaysian Peninsula; Sumatra, Indonesia	Yes	Tropics, Islands, Cities	
Syntheosciurus brochus	No	NA	NA	NA	
Tamias striatus	Yes	Eastern USA	NA	NA	Also Japan weirdly
Tamiasciurus douglasii	Yes	Southern British Colombia, Canada; Washington State, Oregon, California, USA	NA	NA	
Tamiasciurus fremonti	No	NA	NA	NA	
Tamiasciurus hudsonicus	Yes	USA and Canada	Yes	Cities, Old-growth forests	
Tamiasciurus mearnsi	Yes	Baja California, Mexico	NA	NA	
Tamiops maritimus	Yes	China; Taiwan	Yes	Islands, Cities	Also Jamaica weirdly
Tamiops mcclellandii	Yes	Myanmar; North-eastern India; Yunnan, China; Malaysian Peninsula; Thailand	Yes	Tropics	
Tamiops rodolphii	Yes	Cambodia; Vietnam	Yes	Tropics	
Tamiops swinhoei	Yes	China	NA	NA	Also the Netherlands weirdly
Trogopterus xanthipes	No	NA	NA	NA	
Urocitellus armatus	Yes	Utah, Wyoming, Colorado, USA	NA	NA	
Urocitellus beldingi	Yes	The western USA	NA	NA	
Urocitellus brunneus	Yes	Idaho, Oregon, Washington State, USA	NA	NA	
Urocitellus canus	Yes	Idaho, Oregon, Washington State, USA	NA	NA	
Urocitellus columbianus	Yes	Idaho, Oregon, Washington State, Montana, USA; British Colombia, Canada	NA	NA	
Urocitellus elegans	Yes	Northwestern USA	NA	NA	
Urocitellus mollis	Yes	Idaho, Wyoming, Utah, Nevada, USA	NA	NA	

(continued)

Table 3.3 (continued)

Species	SDM available	Range description	Region of/under high risk participant	Allocated region of/under high risk	Comments
Urocitellus parryii	Yes	Alaska, USA; Yukon, Canada	Yes	Old-growth forests	
Urocitellus richardsonii	Yes	Northern-central North America	NA	NA	
Urocitellus townsendii	Yes	Idaho, Washington State, USA	NA	NA	
Urocitellus undulatus	Yes	Siberia, Russia; Northern Mongolia	NA	NA	
Urocitellus washingtoni	Yes	Washington State, USA	NA	NA	
Xerospermophilus mohavensis	Yes	Arizona, southern California, USA	NA	NA	
Xerospermophilus perotensis	Yes	Mexico; Venezuela	Yes	Tropics, Cities	
Xerospermophilus spilosoma	Yes	New Mexico, Texas, USA; Mexico	NA	NA	
Xerospermophilus tereticaudus	Yes	Northern Mexico	NA	NA	
Xerus rutilus	Yes	Tanzania; Kenya; Ethiopia	Yes	Tropics	

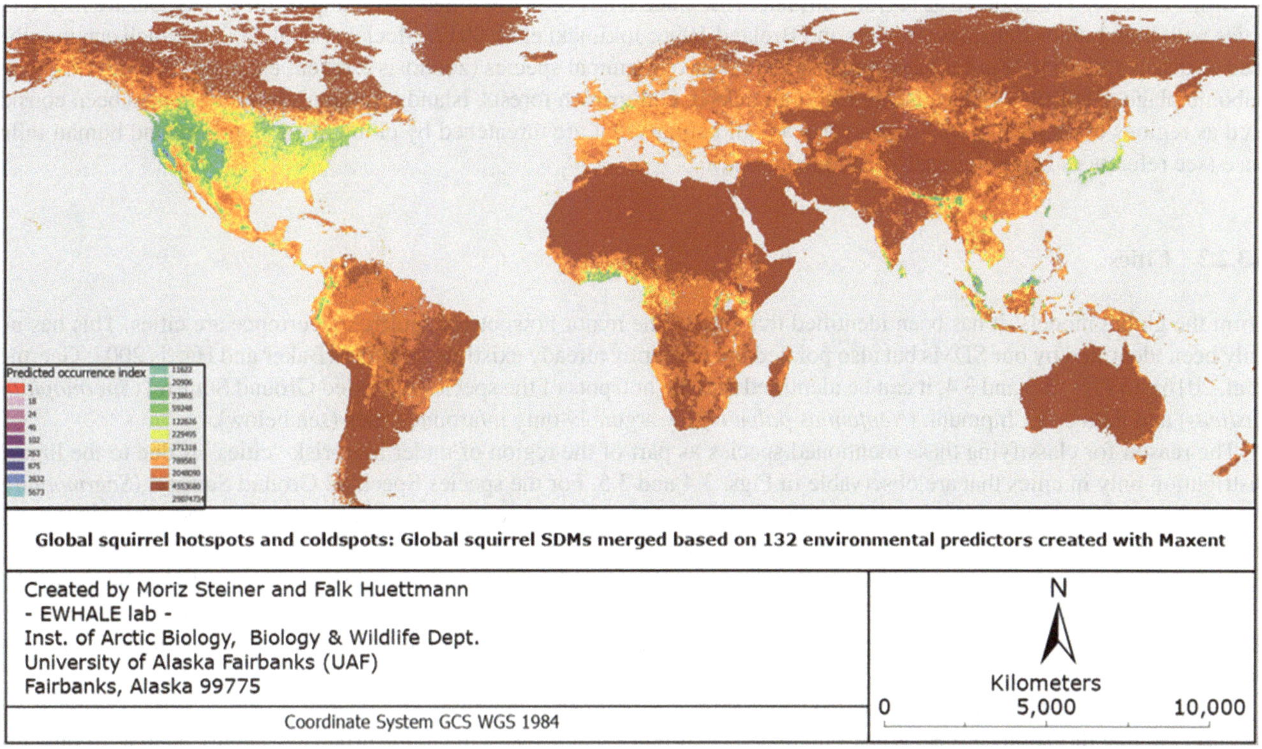

Fig. 3.3 Global squirrel hot–/coldspots: A global overview

3.3.2 Hotspots/Regions of High-Risk Identification of the Global Squirrel Population

In this section, the authors focus on assessing the resulting SDMs, inspecting them further, evaluating and classifying the distribution of each squirrel species, and identifying whether the species at hand occupies one of the spatial hotspots. In case the species at stake inhabits special hotspot niches, these hotspots are identified and broken down into the threats occurring there and how these species can ideally be managed and conserved there. In this study, these hotspots are outlined and the most crucial influences are outlined. An in-depth discussion and a case-unique approach for each region can be found in Chaps. 5, 6, 7, and 8 for cities, tropics, islands, and old-growth forests respectively.

3.3.2.1 Global Squirrel Species Distribution Analysis with a Focus on Occurrence Hotspots

As described in the method section above, the 233 rapid assessment SDMs that have been created have consecutively been assessed and summarized in Table 3.3 below. The global squirrel species distribution analysis with a focus on their occurrence hotspots can be observed there.

This table provides for each squirrel species two main descriptive characteristics, firstly, the primary distribution range, and secondly, it identifies whether this species occurs and has established its habitat hotspot in one or more so-called "regions of/under high risk". These regions of/under high risk have been identified to be threatening the future of the inhabiting squirrel species for their conservation and survival. These "regions of/under high risk" include cities, old-growth forests (for primarily tree squirrels), tropics, and islands. To classify these mentioned habitats as "regions of/under high risk" extensive literature has been performed to identify which are the habitats where squirrels are currently, and in the future, in danger to become extinct. The reasons for each habitat to be 'in danger' is shortly described here and in further detail explained in each dedicated section in Sect. 3.4. Additionally, an in-depth discussion of each region of/under high risk, its dangers, fates, management problems, and suggested solutions is presented in the dedicated chapters.

Cities have been considered as regions of/under risk since it has been identified that many squirrel hotspots are near or in cities with high human densities (Bowers and Breland 1996; Jokimäki et al. 2017; Mccleery et al. 2008), which can possibly lead to disease transmission between humans and invasive mammal species (zoonosis – recent examples: Covid-19, rabies, bubonic plague, and even Lyme disease (through ticks)). Old-growth forests, islands, and the tropics have also been considered as regions of/under high risk since these are all habitats that are threatened by climatic, geologic, or/and human influence (see references below for each mentioned threat).

3.3.2.2 Cities

From the global models, it has been identified that one of the major hotspots of squirrel occurrence are cities. This has not only been identified by our SDMs but also pointed out earlier by already existing literature (Baker and Harris 2007; Gaertner et al. 2016). In Figs. 3.3 and 3.4, it can be identified that the hotspots of the species Speckled Ground Squirrel (*Spermophilus suslicus*) and Palmer's Chipmunk (*Neotamias palmeri*) lay arguably only in/around cities (see below).

The reason for classifying these mentioned species as part of the region of/under high-risk "cities" is due to the limited distribution only in cities that are observable in Figs. 3.4 and 3.5. For the species Speckled Ground Squirrel (*Spermophilus suslicus*), it can be observed that its distribution is seemingly restricted to cities only. These cities are predominately located in Russia (RU), Ukraine (UA), and Kazakhstan (KZ) (see Table 3.3 and Fig. 3.4). In these regions, the cities in which the squirrel is predicted to occur are Kyzylorda (KZ), Atyrau (KZ), Uralsk (KZ), Volgograd (RU), Saratov (RU), Samara (RU), Ulyanovsk (RU), Kazan (RU), Nizhny Novgorod (RU), Kostroma (RU), Kyiv (UA), Dnipro (UA), Lviv (UA). This indicates that this squirrel species is widely spread in the upper-mentioned countries and has specialized its niche in the local urban environment.

The second species distribution presented above (Fig. 3.5), Palmer's Chipmunk (*Neotamias palmeri*), is less spread. It can be observed that this squirrel species has a predicted occurrence that is restricted to the urban and suburban areas of only one city, Las Vegas, Nevada (USA). Since the distribution range and their sizes vary significantly between the two species presented in Figs. 3.4 and 3.5, the threats found for these squirrels also differ. However, not only the threats to the squirrels but also the ones originating from them differ. This will be discussed and followed up in more detail in Chap. 5 (Steiner and Huettmann unpublished).

Following these figures, it seems important to mention that the most significant threats for squirrels, and also originating from squirrels in such habitats are expected in cities. The main reason for squirrels often occurring in cities, their backyards,

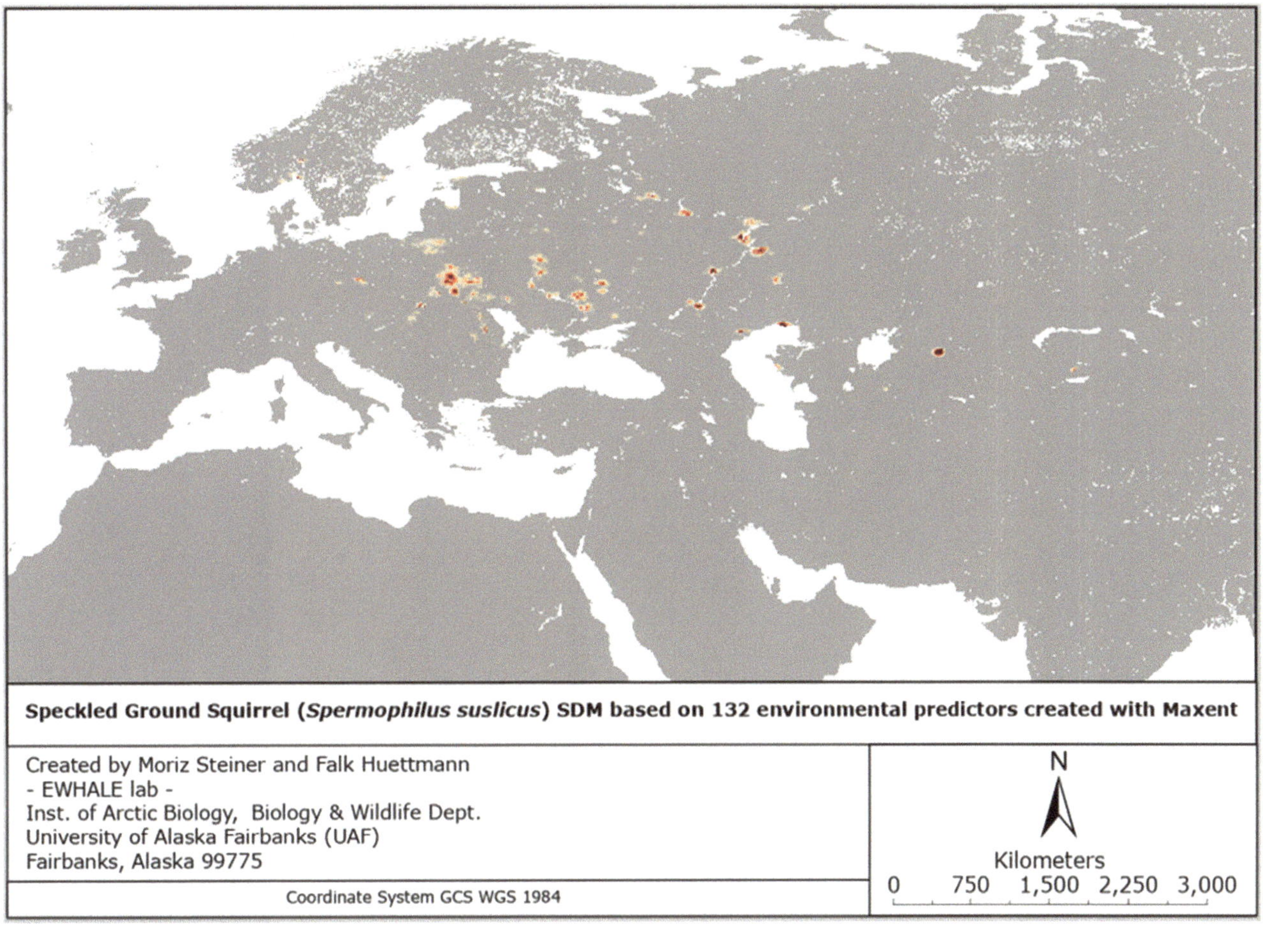

Fig. 3.4 SDM of the squirrel species Speckled Ground Squirrel (*Spermophilus suslicus*)

parks, and green spaces are that in modern days, cities are continuously expanding and enlarging the anthropological influences (Baker and Harris 2007; Eriksen and Nielsen 2013). As a result, one finds that the wilderness habitat of squirrels is shrinking (Baker and Harris 2007). By observing the mentioned threats for many squirrel species (on IUC Red List under the "Threats" section, or other institutions that mention them), one of the most observed threats is habitat loss. One of the major reasons for habitat loss is the expansion of global cities; urbanization. Due to this reduced wilderness, squirrels actually are in the continuous process to expand or disperse their habitats away from the wilderness (due to the lack of space) toward anthropological settings (Krauze-Gryz et al. 2021). Another reason why the squirrels' occurrence in and around cities has highly increased in the last years, is the anthropological feeding of the squirrels (Bonnington et al. 2014; Uchida et al. 2021). Urban areas are sinks but attractive ones, e.g. for food and a warmer climate, likely also the absence of large predators (Bowers and Breland 1996; Rademacher 2019).

With squirrels occurring and inhabiting urban/suburban areas, threats are difficult to exclude. Common threats for squirrels in urban/suburban areas include road kills (Mccleery et al. 2008; Rustiati 2009), poisoning (Mccleery et al. 2008), target practice hunting (Lioy et al. 2019), predation by exotic urban predators, and increased disease transmission risks (Bradley and Altizer 2007; Riley et al. 2014; Van Horne 2008). However, it seems that there are also several benefits for squirrels in cities. Such benefits include highly nutritious food sources (Thompson and Thompson 1980), and fewer natural predation risks (Bowers and Breland 1996; Rademacher 2019). To sustainably conserve squirrels in cities and even integrate them into the anthropological setting of modern cities, a science-based strategic setting and conservation planning are necessary. Some examples of cities that have accomplished this already are Kuopio (Finland) with the Siberian flying squirrel (*Pteromys volans*) (Mäkeläinen et al. 2015). Other cities are also trying to include squirrels more in the city planning e.g. Novosibirsk in Russia which has a squirrel avenue, and a park with many squirrels. Citizen-driven approaches might be relevant here as well. However, there have also been cities/urban settings where such an attempt to manage squirrels has been attempted but failed (e.g. with the Eurasian Red Squirrel (*Sciurus vulgaris*) conservation in North Italy (Lombardia et al. 2019; La Morgia et al. 2017)). Similar projects have also been tried for other wildlife species such as the Ringed Storm Petrel (*Hydrobates hornbyi*) where it actually experienced significant success (Cancino 2015).

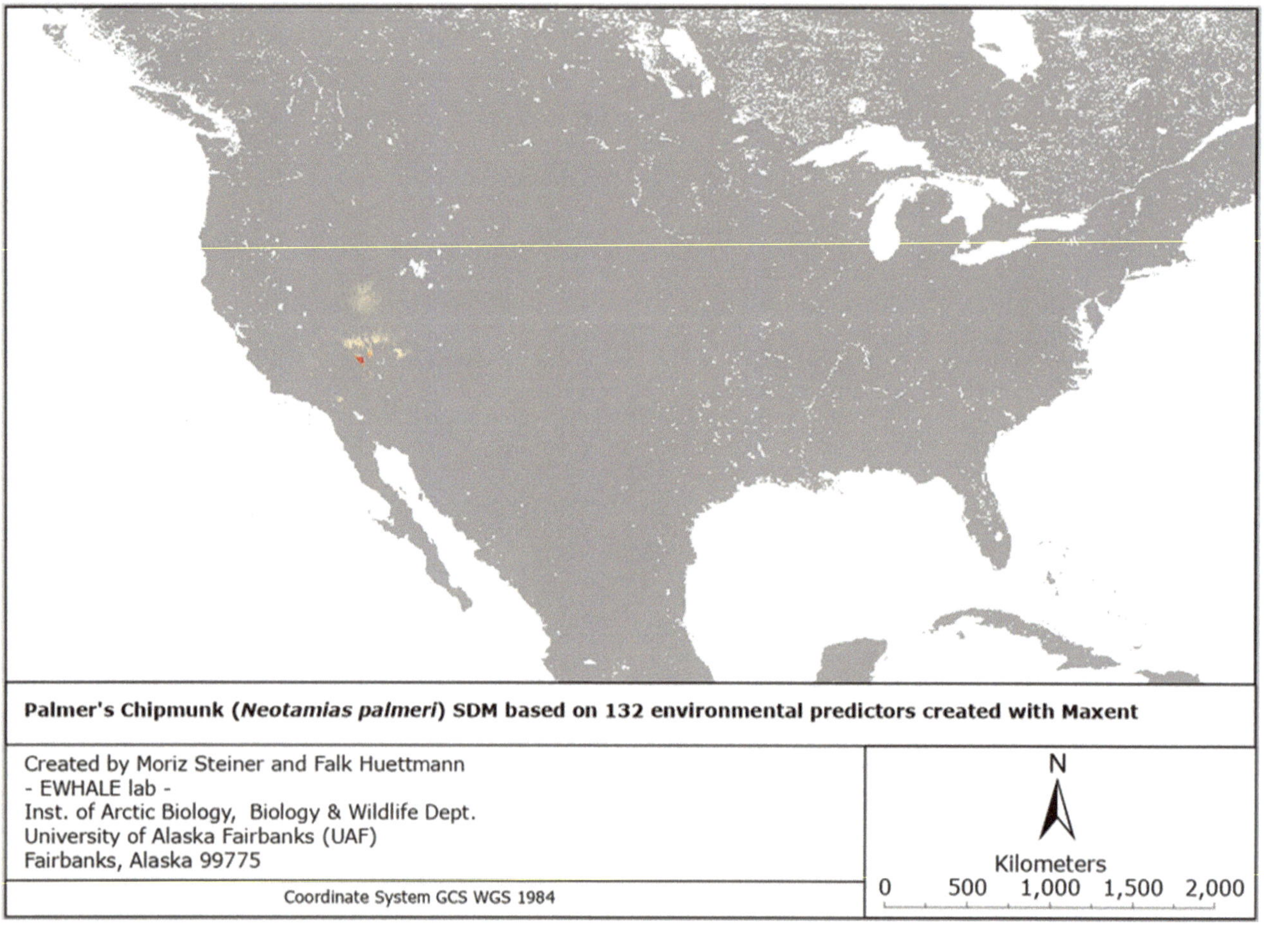

Fig. 3.5 SDM of the squirrel species Palmer's Chipmunk (*Neotamias palmeri*)

As mentioned above in Sect. 3.4.1, there are several diseases also involved with squirrel encounters. These risks are originating from squirrels, putting humans and other animals in disease danger and vice versa is discussed in detail in the dedicated Chap. 5 (Steiner and Huettmann 2021). Another important point to mention here is that the identification of some squirrel species to have their predicted occurrence near or in urban/suburban areas can be faulty. This is because sampling biases of the squirrel presence points in the raw data set cannot be excluded. It is suspected that the data entry for squirrels is higher near more densely populated areas, near road access, as there are simply more surveyors present compared to very isolated and remote locations e.g. forests (old-growth forests or tropics), wilderness, steep slopes, or environments like small islands (Raup 1995). This should in theory not be a problem when SDMs are being created, especially with well-developed ML/AI algorithms, even with Maxent. However, it is something that should be kept in mind (see Huettmann 2018 for coastal and small island data problems), especially when it is operating with a low number of predictors and occurrence points.

3.3.2.3 Old-Growth Forests

For some squirrels (see Table 3.3) it has been identified that their modeled distribution hotspot is located in old-growth forests. It is important to mention that this is only observed for tree squirrels and some flying squirrels, barely any ground squirrels or similar (those might still be located in old-growth type landscapes and ancient wildernesses though). Examples of these latter-mentioned species are Caucasian Mountain ground squirrel (*Spermophilus musicus*), and Olympic marmot (*Marmota olympus*). Figures 3.5 and 3.6 depict the SDMs for the upper-mentioned squirrel species where this old-growth forest prediction can be observed. As a reference for the classification of which squirrel species occur primarily in old-growth forests, a recent study from Hansen et al. (2013) has been utilized. The reason for classifying the old-growth forests as a "region of/under high risk" is because nowadays old-growth forests are heavily threatened. They are threatened by climate change (Dai et al. 2013; Ellis 2013; Spies 2004), human-induced habitat degradation (Ellis 2013; Frank et al. 2009; Tabarelli

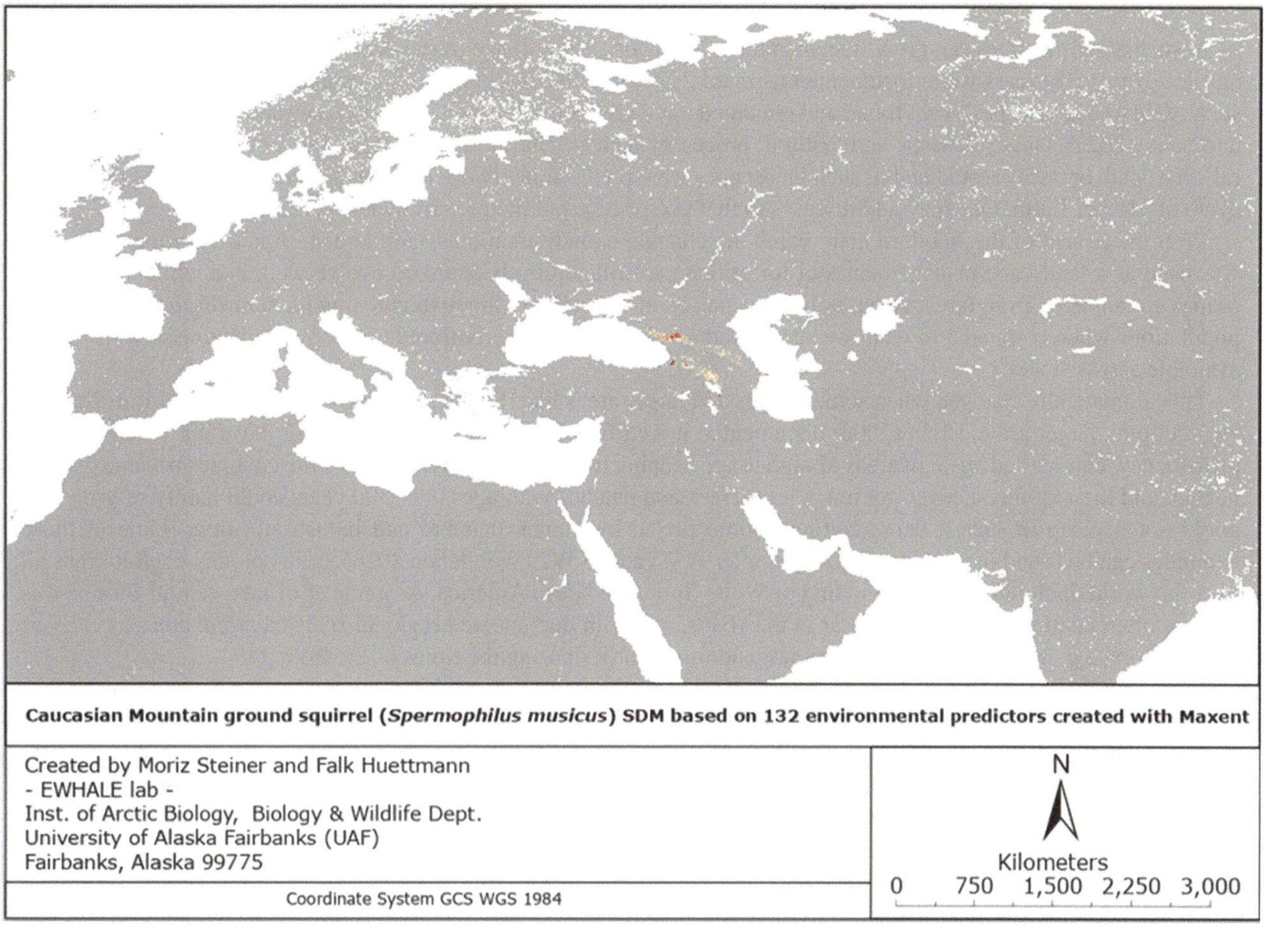

Fig. 3.6 SDM for the old-growth species Caucasian Mountain ground squirrel (*Spermophilus musicus*)

et al. 2004), logging (Shorohova et al. 2011; Tabarelli et al. 2004), invasive species, habitat fragmentation (Ellis 2013), biodiversity loss (White and Lloyd 1995), deforestation (Shorohova et al. 2011; Tabarelli et al. 2004), and more. In order to conserve the squirrels in old-growth forests, the forests have to be systemically conserved to sustainably maintain the long-established balance in these wilderness areas (Textbox 3.7).

Textbox 3.7: Old-Growth Forest Wilderness and Squirrels: The Past, the Anthropocene, and What a Future to Manage and How?

Falk Huettmann Moriz Steiner

Most tree and flying squirrels are tree-living and thus they evolved around trees; they co-evolved overall with such landscapes and the human population within. With initially few human beings on earth, humans interfered assumingly little with the tree-living squirrels of the world directly. For most of the earth's living history, many squirrels used the tree canopy that humans cannot really reach.

Even in the 18th century the squirrels of the world were usually not much affected by humans. The forest cover was primarily part of the global environmental cycle, glaciation phases, fires, and squirrels evolved with it. Species got lost in that process and new species emerged. For the most part of human history, humans had little direct influence on most squirrel species (considering squirrels handled man-made fires well, similar to natural fires). Most humans of that time saw nature, its animals, and plants, as part of themselves, as a wider spiritual framework, that everybody lived under (e.g. Cousteau and Richards 1999 for Papua New Guinea; specific squirrel details are found with indigenous people in North America, etc.).

In such a regime, trees can grow free and tall, and they do. Record-size trees come from such an environment; tree ages of over 9,000 years were documented (Simard 2021; Wohlleben 2016). Many examples exist worldwide where such 'wild tree landscapes' were found and remained there for a long time with vast coverage and canopies (Huettmann and Young 2022). Only stochastic and 'natural' processes such as natural fires and lightning, or earthquakes and 'natural' tree death or decay would end it; such landscapes were predominantly driven by ecological dynamics as part of the cycle of Mother Earth. One may add diseases to that perspective but the pattern widely remains.

With the advent of the 'modern man' came the change, namely man-made fire and slash-and-burn methods. Fire was used as a landscape management tool for millennia. Still, squirrels could co-evolve well, and while some landscapes got removed from forest cover, or became patchy, it was still not much threatening for squirrel species and their populations when compared to what we see nowadays. It was probably buffered on the wider landmass with forests available across scales.

This dramatically changed when some islands and larger areas like the UK, Iceland, or part of Central Europe were clearcut on a larger scale. Elvin (2006) documented it well for China, and Lines (1991) for Australia. Even after re-growth, new cuts started again in areas of high usage keeping trees consistently low and slim, canopies became disconnected, and the vast forest cover got much less. Trees and patches with ages over 300 years could hardly re-grow and could not develop in such a fierce cutting regime fueled by human demand and industrialization. Some of those examples can be found in the repatriation cuts after WWI and WW2 (Wohlleben 2016), or in colonial exploitations all over the world, including the high-grading of White Spruce in North America. As a matter of fact, central Europe was already often cut during the 30-year war in the 1640s, and with documented regional-scale clear-cut areas as early as 11th century, e.g. in Austria for the salt mines, and much earlier during the times of the Romans.

Numerous other vast losses of virgin landscapes can be shown in Europe, Northern Africa, China, and beyond.

A more-global footprint concerning forest loss started earlier with the empires and colonial periods, with the Egyptian, Roman, Greece, and later the Spanish, British, French, Danish, Italian, and German royal courts. It usually resulted in an onslaught of big trees and their landscapes; specifically bad for islands, while steep inaccessible slopes, hill tops and 'taboo' areas survived. The impact on squirrels remains unknown, but the wider loss and transition of old-growth forests can be somewhat tracked globally from the 17th century onwards with science and presented in GIS maps (e.g. Goldewijk and Ramankutty 2009).

As most squirrel diversity is located in the tropics, there remains also the biggest impact, widely unresolved, and still ongoing. It is hardly acknowledged though (Steiner and Huettmann 2021).

In the meantime, after WW2 a more structured and organized approach was implemented for the world's forestry and resource extraction (Huettmann and Young 2022). The era of assumed management and sustainable management started. Still, impacts on squirrels within that 'modern' approach got widely ignored putting the entire operation and 'modernity' in doubt for its sustainability (Steiner and Huettmann 2021).

This problem has not improved until this day, e.g. despite using machine learning and AI computing opportunities in 'the cloud', and it is hardly acknowledged even. Squirrels are not better off yet. And while some squirrel species can live in urbanized environments, many are affected by human landscape changes and urban landscapes on a national or continental scale.

The Anthropocene comes with its own burden, e.g. the marginalizing and failed management scheme trying to achieve sustainability and sustainable development, e.g. via SDGs. Already seen from an old-growth forest wilderness perspective it is an oxymoron. Virtually none of the so-called sustainable forestry, nor its institutions and funders, have dealt with old-growth forests well, or with squirrels of the world, or show even awareness for the environment, global well-being, Mother Earth, or the Universe. They lack the long-term record to convince. Virtually all wilderness areas we are aware of are declining, usually at a record rate and just in the last 50 years, e.g. Borneo, Malaysia, Costa Rica, Patagonia, West Africa, Congo, Papua New Guinea, and the Boreal Forest. As a matter of fact, the approach and institutions, as well as their expertise and business model, are not suited for it whatsoever. If those 'tools' try to cope with the tasks of sustainability, old-growth forest management, wilderness maintenance, and squirrel conservation it achieves the opposite, the clear destruction as it is witnessed in the current global status and its abysmal metrics.

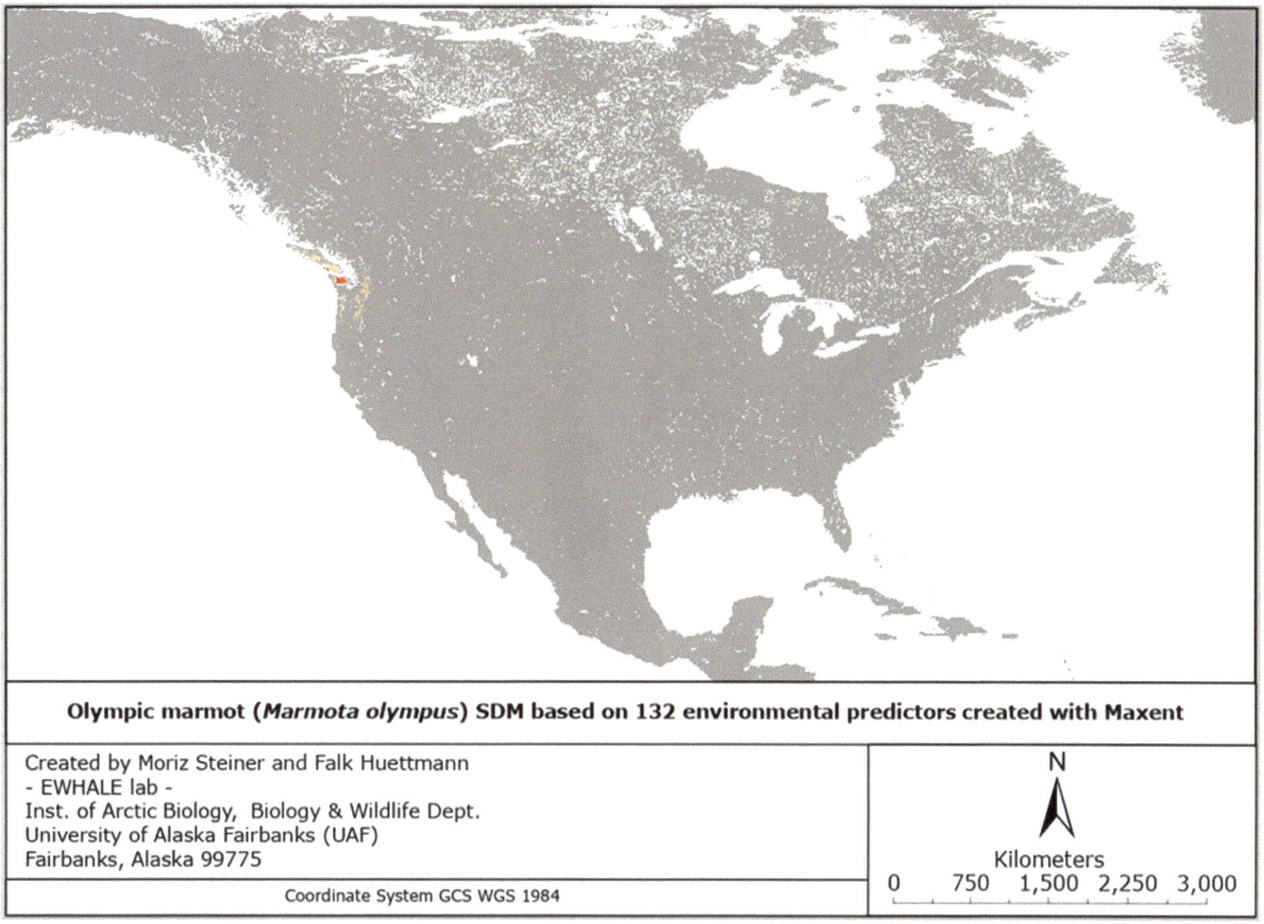

Fig. 3.7 SDM for the old-growth landscape species Olympic marmot (*Marmota olympus*)

The reason for classifying these mentioned species as part of the region of/under high-risk "old-growth forests" is that both of these species' habitat is highly restricted to an environment classified as old-growth forest. This classification is based on a study conducted by Hansen et al. (2013), that assessed the global old-growth forest distribution. The little ground squirrel (*Spermophilus pygmaeu)* is endemic to the virgin forests between the borders of Russia and Georgia with a concerningly small distribution range. This can be observed in Fig. 3.6 above. The habitat of the squirrel species Olympic marmot (*Marmota olympus*) is even more restricted. This species is endemic to a single mid-sized National Park "Olympic National Park" in Washington State, USA. The size of this National Park and with it its protected old-growth forest is just 922,650 acres (3733.8 km^2). The limited distribution range of this species can be observed in Fig. 3.7 above. From these figures presented above, it can be identified that the distribution range of these species is highly restricted to the mentioned old-growth forests, and therefore they have been added to the list of being part of the species inhabiting the "regions of/under high risk". The most severe threats to these squirrels are low genetic diversity (Fitak et al. 2013; Lance et al. 2003), no migration/dispersal opportunities as they seem to be locked to these habitats (Koprowski et al. 2005), and habitat degradation. In the case of the latter-mentioned species (*M. olympus*), habitat degradation is less of a threat as the park that the squirrel is inhabiting is a UNESCO World Heritage site. However, such a label carries little real-world budget and does not guarantee protection (for an opposite view see D'Eramo (2014)), and climate change and biodiversity loss might bear unknown severe problems in the future (Van Horne 2008). An in-depth discussion about the threats mentioned above with applications to more species that are confronted with such issues is presented in Chaps. 5, 6, 7, and 8 (Steiner and Huettmann 2021).

3.3.2.4 Tropics

The tropics are defined as "*the hottest area of the earth, the area on either side of the equator reaching to 23.5 degrees to the north and south*" according to the Cambridge Dictionary (2021b). Within these latitudes, climatic conditions allow biodiversity to flourish. This is not only valid for plants, but also for animals and even squirrels. 92 squirrel species out of the 233

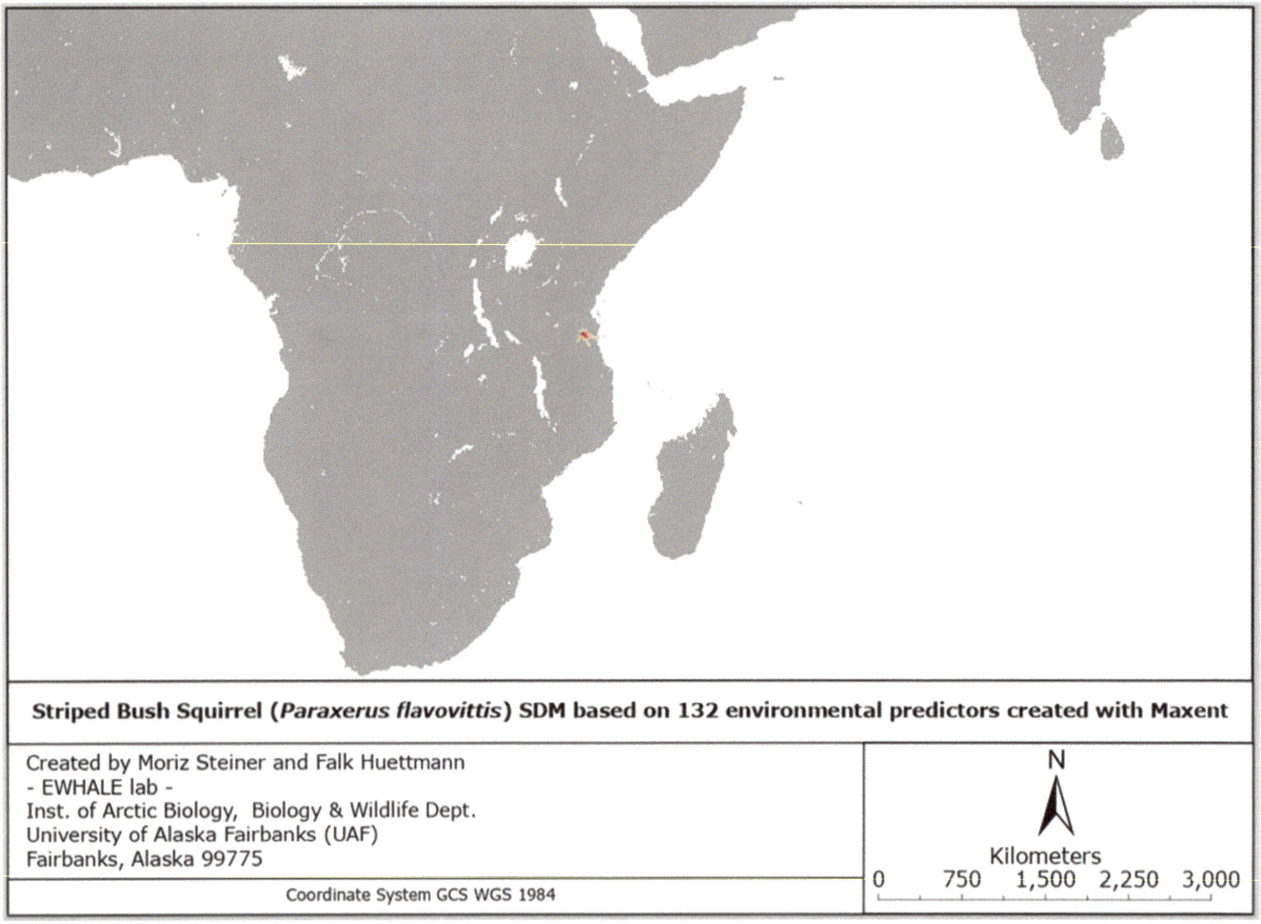

Fig. 3.8 SDM for the tropic species Striped Bush Squirrel (*Paraxerus flavovittis*)

primarily inhabit these latitudes according to our models (See Table 3.3 and the SDMs in the online appendix). The SDMs for the species Striped Bush Squirrel (*Paraxerus flavovittis*) and Niobe Ground Squirrel (*Lariscus Niobe*) indicate a primary occurrence in the tropics for these species. These two SDMs have been presented below in Figs. 3.8 and 3.9.

The reason for classifying these mentioned species as part of the region of/under high-risk "tropics" is that these belong to the 92 identified species that predominately occur within 23.5 degrees north and south of the equator. The first species presented in Fig. 3.8, Striped Bush Squirrel (*Paraxerus flavovittis*), only occurs according to our models in one small and relatively remote region of Tanzania. This highly restricted ecological niche can easily vanish due to human impact on the tropical ecosystem in this regional setting, leading to the extinction of this species. Consecutively, identifying and classifying this species to inhabit one of the regions of/under high risk seems necessary. The second species presented above, Niobe Ground Squirrel (*Lariscus niobe*), is not only considered to inhabit one of the regions of/under high risk but two (tropics and islands). As observed in Fig. 3.9 above, the predicted occurrence for this species is highly restricted to two small areas. One of these two areas is located in West Sumatra (Indonesia), where the occurrence of this squirrel species is confirmed by the GBIF occurrence data. The second identified modeled hotspot is located in Papua New Guinea (PNG) where the occurrence is only predicted, and no data is available from this location. The models possibly indicate a vacant or fitting niche for squirrels, but yet today there are no squirrels living in PNG (Flannery 1990). The risk for a species not to survive increases with the increase of regions of/under high risk this species inhabits. Therefore, inhabiting very restricted niches in regions that are under more threats puts the species itself under multiple risks. Therefore, it is crucial to discuss the major threats for the primary threat to the tropics.

Additionally, it is noteworthy that the major part of the global squirrel population inhabits this threatened region, the "tropics", and therefore it is crucial to assess these habitats for an improved understanding and conservation success. Additionally, due to climate change and other human habitat disruptions, many habitats are heavily threatened in these regions, which reasons to include the tropics in the "regions of/under high risk". Next to these two mentioned threats, the biotic world of the tropics often also suffers from being less studied than other regions in the world and not all the ecosystem

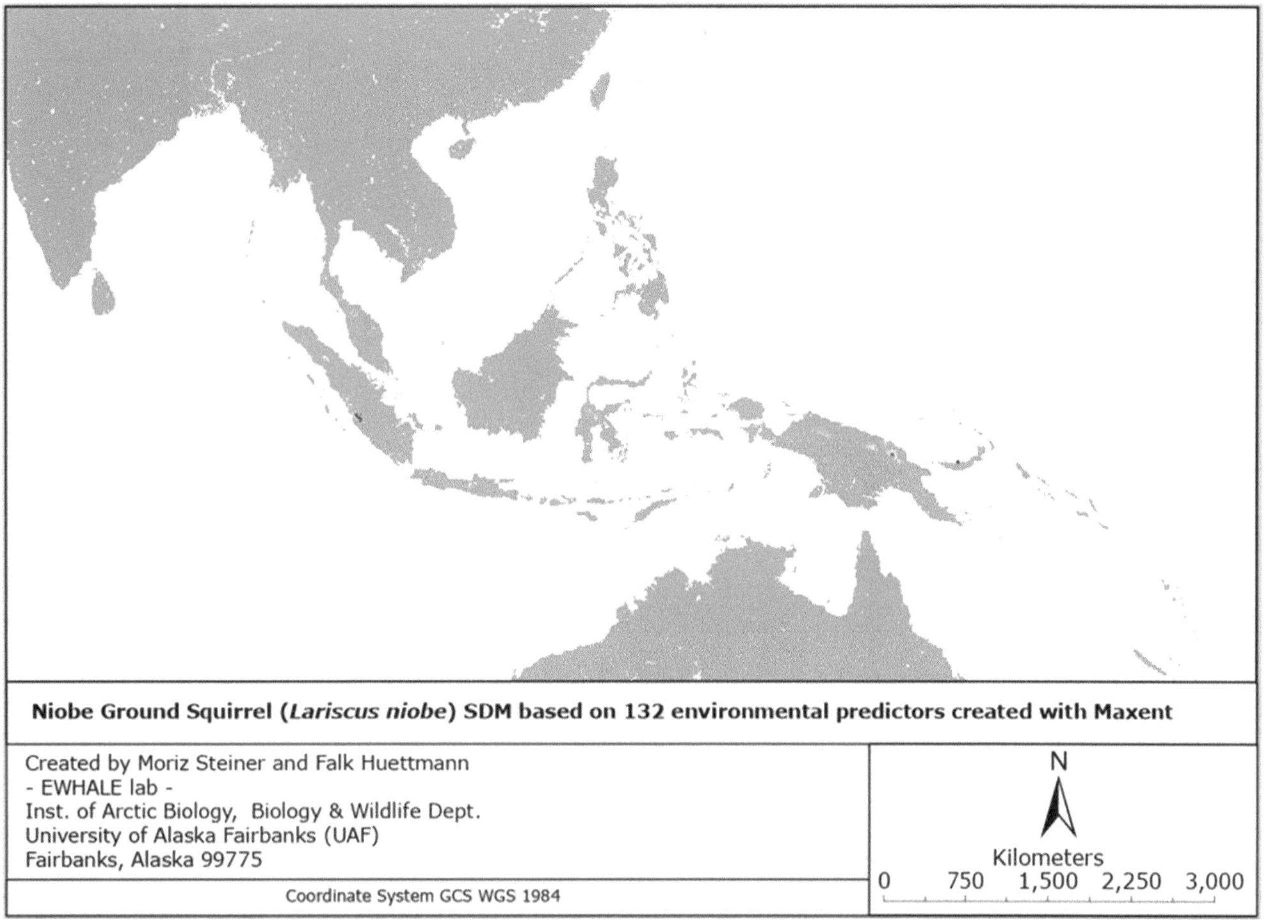

Fig. 3.9 SDM for the tropic species Niobe Ground Squirrel (*Lariscus niobe*)

interactions are yet fully understood (Homeier et al. 2017). The remoteness and inaccessibility of some regions contribute to the fewer research studies executed there (Hickisch et al. 2019). On top of these topographic factors, socio-economic factors play a role here too. Poverty is one of the main ones. Warfare (of relevance for the species *Sciurus anomalus* for instance, with its predicted occurrence hotspot in Turkey, Syria, and Lebanon (see online appendix for the SMD of this species)), is one cause of more marginalized management governance, and since many squirrel species occur in developing countries, this additionally contributes to the lack of conservation management and related research performed in these regions (Hickisch et al. 2019). Due to this lack of research performed in the tropics, the ecosystem interactions are not yet fully understood, and conservation management is not always optimal, and is hardly done with a legal framework. This is, as mentioned before, often due to financial shortcomings since the developing economy of some countries focuses the financial investments on other sectors than small mammal conservation. To improve this entire system and finally the squirrel conservation, a lot needs to happen still. Chap. 6 (Steiner and Huettmann 2021) will discuss these upper-mentioned threats in detail again and present additional solutions and assessment tools to improve sustainable species conservation in the tropics.

3.3.2.5 Islands

Islands are considered habitats that are "*a piece of land entirely surrounded by water*" according to the Cambridge dictionary (2021c). This isolation can be caused by geographical features that cut off either terrestrial mass by surrounding it with water, which is commonly perceived as an island, or terrestrial/aquatic distinctive regions that are land/water locked, respectively. The reason for adding islands to the list of "regions of/under high risk" is because they are heavily threatened by climate change, and more (Courchamp et al. 2014; Leatherman and Beller-Simms 1997; Mimura 1999). These terrestrial water-locked islands continuously suffer from rising sea levels that can be traced back to an increased global temperature rise affecting glaciers, ice masses, and a consecutive increase in the snow and ice-melting rate (see references above and ones

Fig. 3.10 Island indication map; (selected cases)

within). Another reason for islands being part of the habitats classified as "regions of/under high risk" is that many squirrel species are endemic to one or a few terrestrial water-locked islands. Not only island-endemic species but even many others live primarily on islands. The main islands that will be discussed in this section and book have been presented in Fig. 3.10 below.

Figure 3.10 represents the major islands that will be discussed here since most "island species" occur on these and surrounding islands with similar environmental features. Two examples of these "island species" have been presented below in Figs. 3.11 and 3.12, for the species Palawan flying squirrel (*Hylopetes nigripes*), and Black-eared squirrel (*Nannosciurus melanotis*) respectively.

The reason for classifying these mentioned species as part of the region of/under high-risk "islands" is because they primarily occur on islands accordingly to our SDMs and the definition above. The two species presented above (in Figs. 3.11 and 3.12 have been chosen to act as risk example species because both show different threats to species occurring primarily on islands that can severely influence the survival of the species if managed wrongly. The first species, Palawan flying squirrel (*Hylopetes nigripes*), is endemic to only one island, namely Palawan (Philippines). The island has a size of approximately 12,000 km^2 and is located in the western Philippines. Because this species solely occurs on one island worldwide, its risk of extinction is arguable high. This is because if something with a major impact happens on this island (e.g. invasive species/predator introduction, or severe climate events) the species might vanish from this island and planet. Therefore, it seems crucial to investigate further research to assure its survival and conserve this species for future generations.

In this context, the second species mentioned above, Black-eared squirrel (*Nannosciurus melanotis*), has a very distinctive but unfortunately not uncommon, predicted distribution. Its highest predicted occurrence has been identified by the SDM analysis to be near the coasts of Malaysia, Philippines, Brunei, and Indonesia. It can clearly be identified that the predicted occurrence is highly restricted to the coasts of some regions of these mentioned countries. This distinctive distribution might be highly dangerous for the survival of this species since the sea level is predicted to be threatening such niches in the future (Oppenheimer and Hinkel 2019). This is mainly due to the rising sea levels, increased monsoons, tsunamis, typhoons, etc. (see Oppenheimer and Hinkel 2019 and references within). These climate-originated threats (e.g. tornadoes, precipitation intensity, and amounts, or thunderstorms), can become increasingly severe in the future due to human-accelerated climate change (Brooks 2013). Further in-depth analysis of the threats to species endemic to islands and how these should be ideally managed can be found in Chap. 7 (Steiner and Huettmann 2021).

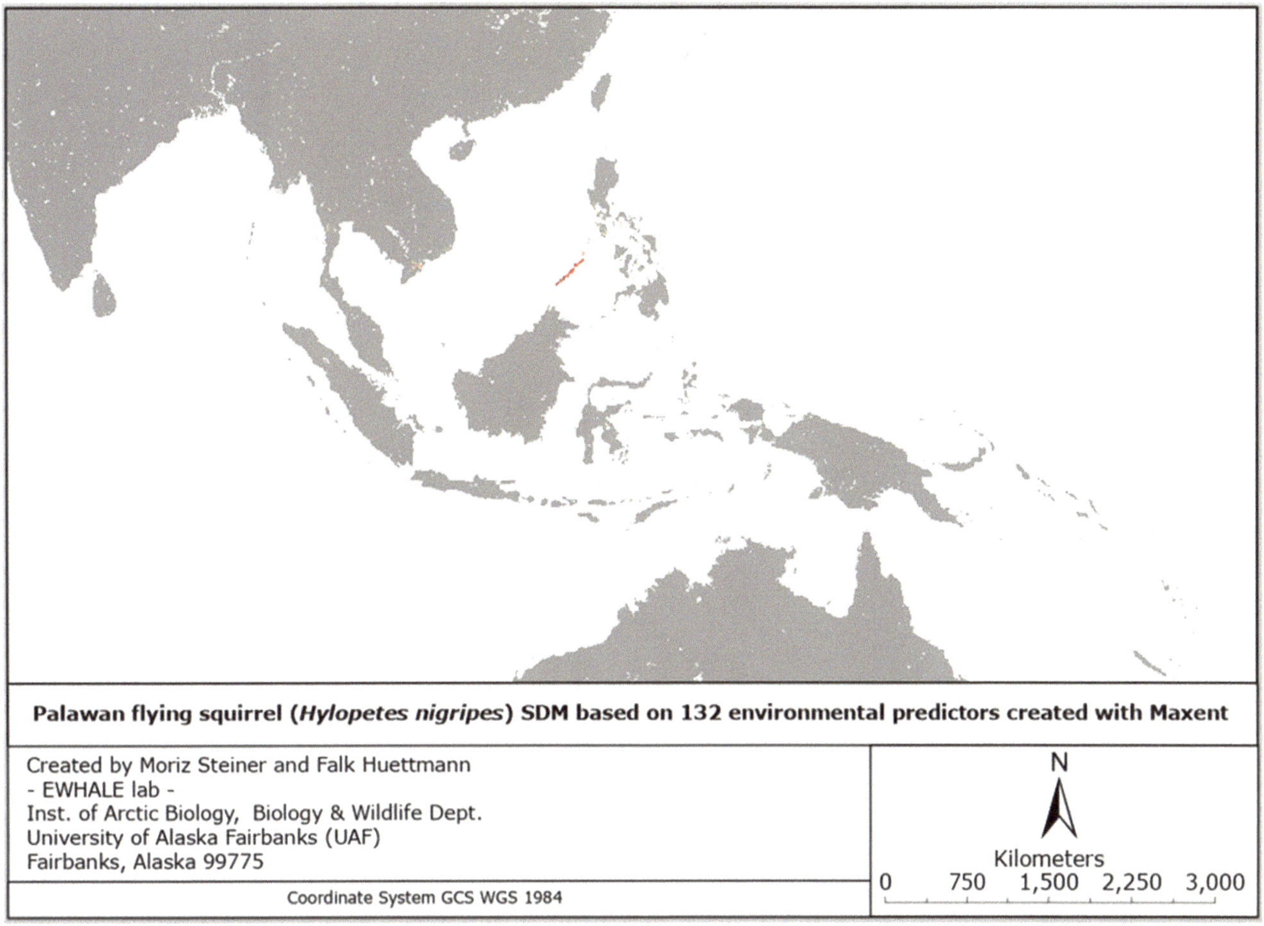

Fig. 3.11 SDM of the squirrel species Palawan flying squirrel (*Hylopetes nigripes*) endemic to the single island Palawan, Philippines

3.4 Discussion

Being 'modern' these days means being digital and using models of the latest technology and with a money-centric extraction economy ignoring essentially all else. This study and the species distribution models (SDMs) created here, have crucial importance for this book on squirrels. This is because digital squirrel approaches are widely missing for most species (but see Robold and Huettmann 2021), and it sets the basics for further in-depth assessments of the identified "regions of/under high risk" in subsequent studies shown in Chaps. 5, 6, 7, and 8 (Steiner and Huettmann 2021). Creating, assessing, and evaluating these SDMs can be one of the most important tasks of a species' conservation. This is because by identifying the current and preferred (realized and fundamental) niches of a squirrel species, the threats of these regions can be assessed and quantified, and help to improve conservation. That way these threats can ideally be eradicated or minimized. This is the reason why the SDMs have been created for these chapters/studies and why one focus of this study was to identify the "regions of/under high risk" for squirrels. Additionally, a workflow model has been presented (Fig. 3.1) that aids to summarize the process of creating SDMs. Afterward, the core content of this chapter is presented, where for each of the regions of/under high risk two example SDMs were presented. For these two example species, their distribution is being discussed, why these have been classified as being in/under "high risk" and some of the major threats have been discussed. Table 3.3 and especially Fig. 3.3 greatly illustrate where in the world the hotspots and coldspots of the squirrel distribution are located. Having such a global overview, aids to create a big-picture idea of what the global squirrel population range looks like and where special attention is necessary.

A large number of species (48 out of 223) have been identified to occur on tropical Asian islands. It is expected that this high number of species is due to speciation caused by geographic isolation (Safran and Nosil 2012). Additionally, the

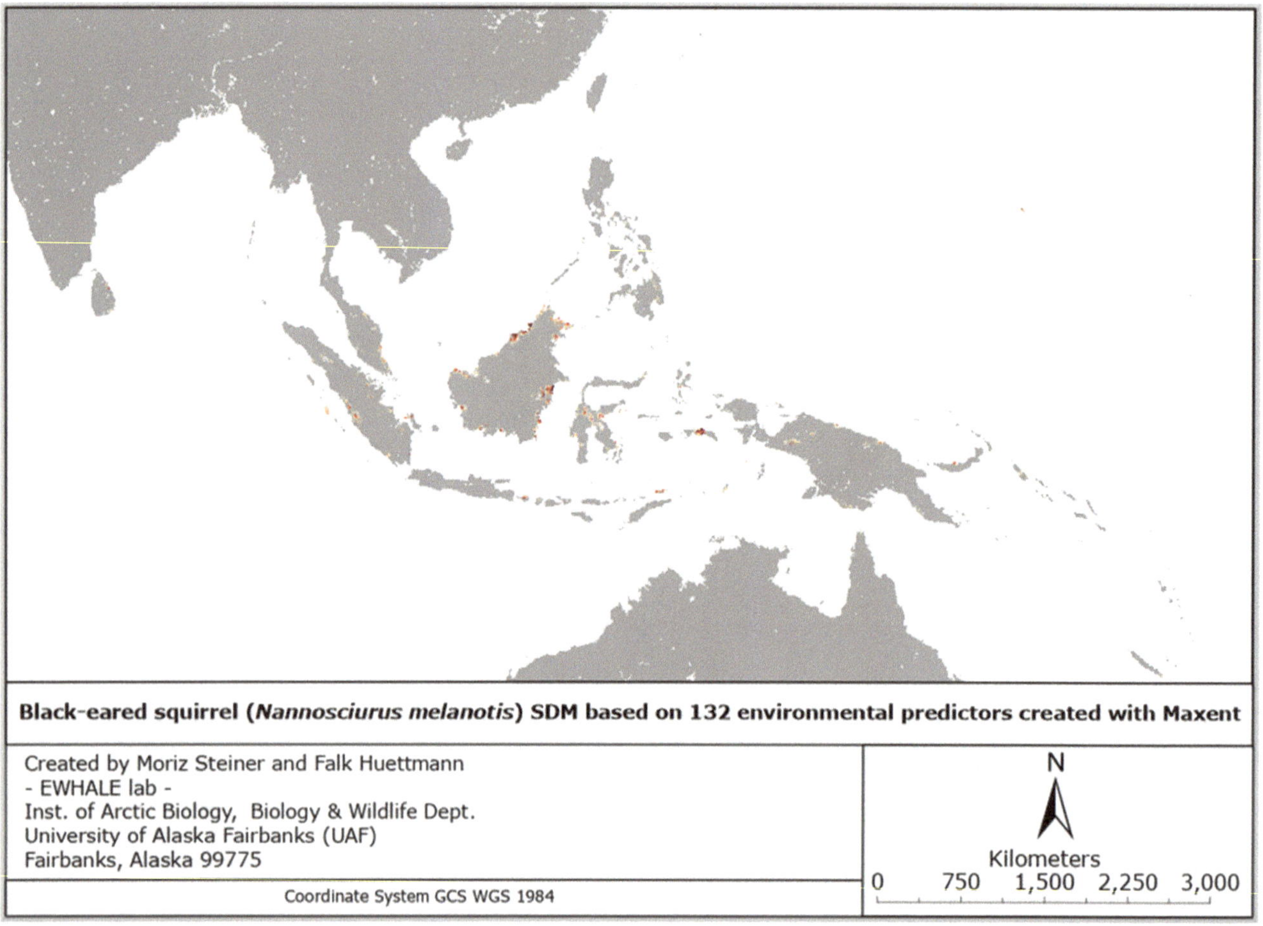

Fig. 3.12 SDM of the squirrel species Black-eared squirrel (*Nannosciurus melanotis*) a species under the climate change threat of sea-level rise

resources and habitats in these latitudes support a high diversity of flora and fauna (Brown 2014), which allows more squirrel species to inhabit the same niche by specializing in their diet and distribution. Due to these high numbers of available niche spaces, different squirrels can also live in the same spatial area, but on different tree heights (Saiful et al. 2001). This high species diversity/richness calls for intensified conservation efforts in these regions.

3.4.1 Digital Data Availability, Accessibility, and Metadata for Squirrels, and Its Influences on 'Modern' Squirrel Conservation

During the data collection for this study, some pitfalls and throwbacks have been detected. This section aims to identify and address these pitfalls and throwbacks and to make this information available. By making these issues publicly available, it calls for a change by the entities from which these issues arise.

One major issue that has been encountered is the non-publicly available and meaningful point data of the distribution of the world's squirrels by the IUCN Red List. The IUCN Red List is commonly known to be the world's most eminent institution for presenting species conservation statuses, trends, and distributing ranges. The data of the IUCN Red List has been repetitively used in this chapter and book, however, there are some issues to be addressed. IUCN Red List presents for each species in their database the assumed/most probable distribution. These presented distribution ranges seem to be unfortunately hand-drawn. This distribution data can be usually downloaded as range maps (polygons – shapefile) or as distribution points (point data). The distribution polygons are apparently based on the distribution point data. However, for unknown reasons, the distribution point data cannot be downloaded for mammals. After getting in touch with IUCN and IUCN Red

List staff members, it has been confirmed to the authors that IUCN actually does NOT provide these distribution point data for mammals, which includes all squirrel species. As described above in Sect. 3.5.1, even though IUCN Red List presents these range maps, they do not present the underlying point data that is supposed to be used to build the polygon range maps. Because these range maps are available for almost all global species, the point data that constitutes these range maps could be highly valuable for many diverse research purposes, including higher-quality SDMs. If the distribution point data would be accessible to the public, it could be used for SDMs and would make the occurrence data set more complete. Instead of presenting 200+ SDMs, around 280 SDMs could be created (not all 307, as these species are not present in the IUCN distribution data). This is especially an issue because the data that has been used for the SDMs have been retrieved from a citizen-science-based institution (GBIF.org). There, the issue is often that there is no data available for rare species (which are in many cases associated with an endangered conservation status). This is the reason for the "over-positivism" of some of the findings presented in this study here and in many public studies. This comes back even stronger in a follow-up study (Steiner and Huettmann 2021 – Chap. 4), where the meta-analysis can only be presented for 223 species instead of 307.

In addition, another aspect should be considered here. If such valuable distribution data would be available, research can be conducted with it, it would be possible to include them in this study and in other conservation assessments which ultimately aim to improve the conservation of the species. However, if this data is not available to the public and also the scientists cannot use it for conservation assessments, and conclusively it cannot be used for efficient and sustainable, science-based conservation planning. Therefore, by connecting this information, it seems like an absurd reality. Because the funding for IUCN Red List is mostly originating to present and execute conservation improvements (see the IUCN Red List sections "Sponsors", and the latest "IUCN Financial Plan"), however, as explained above, this institution does not provide data that is of crucial importance for conservation planning. Because this conservation planning cannot, or only hardly, be executed, the conservation success seems to decrease which consecutively assumingly increases the funding for IUCN Red List as more species are threatened.

By presenting and addressing this information to the public, the authors call for open-access data sharing in the scientific world (as stated in Zuckerberg et al. 2011; Huettmann 2015; examples with Elith et al.), and by the IUCN Red List, with the greater aim to work together to conserve the biodiversity of this planet and the health of nature.

3.4.2 Maxent SDM's Pitfalls, User Skill, and The Consecutive Created Data Selection Bias

As it has been discussed already in the introduction and also in the appendix of this study, in order to create SDMs with Maxent, an intense data clean-up is necessary to fulfill technical requirements, which arguably can create a massive bias and self-fulfilling prophecy. This means that to successfully use distribution data as presence points in Maxent, the raw distribution data must be heavily cleaned up (altered). Here only three columns are allowed in the input CSV file, namely "species", "latitude", and "longitude". In case there are more columns present in the dataset or the columns are named differently, the file is not accepted by Maxent to run SDMs. During this cleaning process, data must be removed, merged, and corrected. By doing so, a data selection bias can easily be created by the executive SDM modeler. Modern software such as Salford models (Breiman et al. 1984) which are primarily working with Machine Learning (ML) can better overcome these strict restrictions. Therefore, such software does not require such drastic clean-up processes which reduce the possibly created data selection bias and lay out a concept Maxent can use as a development example for further improvements. All data are here made available to do so.

3.4.3 GBIF Distribution Bias and How to Handle and Resolve It Well

As described in the introduction of this study, the global GBIF squirrel distribution dataset contains occurrence points ranging from 1 to 65,000 (see Table 3.1). This means, unfortunately, that for many species there are only one, or two distribution points available, and for others tens of thousands. This makes it difficult to compare the SDMs' quality on both ends of the point data count spectrum. In order to produce high-quality SDMs, that can significantly aid in assessing the conservation status and threats of squirrels and generally, species, one must aim to collect sufficient confirmed distribution data for all species. Understandably, this is not always and for every species possible, as it is financially very expensive and requires a high amount of manpower and time. However, if institutions with similar aims (e.g. collecting and presenting distribution data/conservation data) can collaborate and merge their data on a high-quality basis, major leaps forward could have already been made. This has been discussed thoroughly in Sect. 3.5.1. Conclusively, the authors call for a documented open-access approach in science and specifically in conservation ecology in order to successfully conserve the biodiversity of this planet, and also for future generations (see examples with Huettmann and Ickert-Bond 2017).

3.4.4 Further Improvements of Rapid Assessment SDMs and Possibilities to Verify Their Accuracy

SDMs have been in common use for already a few decades and developed since then considerably (e.g. Humphries et al. 2018). Here, we present one rapid assessment method to further improve SDMs and an additional method that allows us to verify the accuracy of created SDMs. Firstly, the method of using additions of landscape metrics to SDM to improve the predictions is being discussed. According to Hasui et al. (2017), SDMs have been heavily criticized for their neglect of landscape metrics. Thus, Hasui et al. (2017) applied landscape metrics in the form of the addition of landscape variables to the predictor set. These landscape variables included climate, topography, vegetation, and soil types. By inspecting the predictor set used for this study (see Table 3.2, and especially Chap. 4 in Steiner and Huettmann 2021), it can be observed that all of these proposed environmental variables have a strong presence in the predictor dataset of this study. As it is also clearly observable in Chap. 4 (Steiner and Huettmann unpublished), these landscape metrics played a crucial role in SDM creation as they heavily contributed to SDM creation. Hasui et al. (2017) additionally mentioned in the findings that the accuracy of SDMs can be significantly increased by using these landscape metrics which suggests the importance of including them in predictor sets for SDMs.

The second method that is being discussed here to further improve SDMs is ground-truth sampling locations for verifying the accuracy of the created SDMs. Milanovich et al. (2012) performed such a study in an urban setting in Ohio, USA. For this study, the authors first performed SDMs for 23 amphibian species and later on conducted a survey in the study area to detect these amphibians. During this survey, the participating authors aimed to verify the predicted distribution of the study species. The outcome of the study indicates that the SDMs were not highly accurate as they were expected, however, it has been still suggested that verifying the SDMs in the field has high benefits verifying their accuracy.

3.4.5 The Benefits of Using a High Number (130+) Environmental Predictors and Presence Points for SDMs

For this study, for the first time, 132 environmental predictors have been compiled and used an overview of these predictors can be observed in Table 3.2 and in the appendix. It is a follow-up from Sriram and Huettmann (unpublished). The reason for compiling this large number of global environmental predictors is that the software that is being used for the SDM creation can thereby identify the specific niche of each occurrence point more accurately. This in turn can affect positively the accuracy of the created SDMs because the original niche can be described in more detail which provides more information for the software to base the predictions on. When using a low number of predictors, it can often be seen that the determining factor of the species distribution is the characteristics these few predictors represent (see e.g. Herkt et al. 2016). Additionally, it is suggested to use a large occurrence point data set too. For this study over 200,000 occurrence points have been utilized to create the SDMs. A large number of occurrence points decreases not only the possible variation of the predictions but also the error margin of possibly wrongly taken occurrence points. In some cases, as seen in this study, only one or two occurrence points were present in the dataset. For these species, the SDMs were mostly not successfully created and had to be discarded. By using a large number of occurrence points such unfortunate and poor occurrences can be avoided.

Additionally, a benefit of such large datasets, both for the occurrence points and also for the predictors is that the data can be used for Machine Learning (ML) studies to further assess, and validate the SDMs and their outcomes.

3.4.6 A Setup for Improved SDMs

By combining these two upper-mentioned improvement points of arguments (in Sects. 3.5.4 and 3.5.5) and using a large and diverse environmental predictor set with large numbers of species occurrence presence points ("BIG DATA") it is assumed to perform the most accurate SDMs possible. Thus, it is suggested to follow these SDM improvement points to create the most precise prediction models.

3.5 Conclusion

Conserving the squirrel species in the world, and especially the threatened ones, is not an easy task, where many highly diverse systems (e.g. tropics, islands, cities, and old-growth forests) must be considered thoroughly. However, by following modern approaches such as using a high number of occurrence points and an extensive dataset of environmental predictors ('Big Data'), the prediction quality and accuracy of the SDMs can be significantly improved. Leaving out data remains harmful. Additionally, with an appropriate research design, by assessing the quality of the predicted training data models (e.g. by ground-truthing the predicted occurrences), and by adding additional landscape metrics to further improve SDM inference, the most modern and accurate outcomes are expected and suggested to be used in the future.

3.5.1 Data Open Access Statement

All the SDMs for the global squirrels that have been created for this study are available in the online appendix with ISO-compliant metadata. This includes all output files, for not only the example species discussed in this chapter, but also for all other squirrel species. Additionally, also all the environmental predictors and the occurrence presence data can be found in an open-access form in the online appendix.

References

Andreadis KM, Schumann GJP, Pavelsky T (2013) A simple global river bankfull width and depth database. Water Resour Res 49(10):7164–7168

Baker PJ, Harris S (2007) Urban mammals: what does the future hold? An analysis of the factors affecting patterns of use of residential gardens in Great Britain. Mammal Rev 37(4):297–315

Bonnington C, Gaston KJ, Evans KL (2014) Relative roles of grey squirrels, supplementary feeding, and habitat in shaping urban bird assemblages. PLoS One 9(10):e109397

Bowers MA, Breland B (1996) Foraging of gray squirrels on an urban-rural gradient: use of the GUD to assess anthropogenic impact. Ecol Appl 6(4):1135–1142

Bradley CA, Altizer S (2007) Urbanization and the ecology of wildlife diseases. Trends Ecol Evol 22(2):95–102

Breiman L (2001) Random forests. Mach Learn 45(1):5–32

Breiman L, Friedman J, Olshen R, Stone C (1984) Classification and regression trees. Vol 37(15). Wadsworth Int. Group, pp 237–251

Brooks HE (2013) Severe thunderstorms and climate change. Atmos Res 123:129–138

Brown JH (2014) Why are there so many species in the tropics?. Journal of biogeography, 41(1), 8–22

Cambridge Dictionary (2021a) Raster. https://dictionary.cambridge.org/dictionary/english/raster. Accessed 15 June 2021

Cambridge Dictionary (2021b) Tropics. https://dictionary.cambridge.org/dictionary/english/tropics. Accessed 15 June 2021

Cambridge Dictionary (2021c) Island. https://dictionary.cambridge.org/dictionary/english/island?q=islands. Accessed 15 June 2021

Cancino L (2015) Ringed storm petrel project: building a link between conservation and cities. https://www.academia.edu/49477266/Ringed_Storm_Petrel_Project_Building_a_Link_Between_Conservation_and_Cities?email_work_card=title. Accessed 31 July 2022

Chantrey J, Dale TD, Read JM, White S, Whitfield F, Jones D, Begon M et al (2014) European red squirrel population dynamics driven by squirrelpox at a gray squirrel invasion interface. Ecol Evol 4(19):3788–3799

Courchamp F, Hoffmann BD, Russell JC, Leclerc C, Bellard C (2014) Climate change, sea-level rise, and conservation: keeping island biodiversity afloat. Trends Ecol Evol 29(3):127–130

Cousteau JM, Richards M (1999) Cousteau's Papua New Guinea journey. Harris Harper Publisher, New York

D'Eramo M (2014) UNESCOCIDE. New Left Review 88. https://newleftreview.org/issues/ii88/articles/marco-d-eramo-unescocide. Accessed 30 July 2022

Dai L, Jia J, Yu D, Lewis BJ, Zhou L, Zhou W, Jiang L et al (2013) Effects of climate change on biomass carbon sequestration in old-growth forest ecosystems on Changbai Mountain in Northeast China. For Ecol Manag 300:106–116

de Abreu-Jr EF, Pavan SE, Tsuchiya MT, Wilson DE, Percequillo AR, Maldonado JE (2020) Museomics of tree squirrels: a dense taxon sampling of mitogenomes reveals hidden diversity, phenotypic convergence, and the need of a taxonomic overhaul. BMC Evol Biol 20(1):1–25

Deneu B, Servajean M, Bonnet P, Botella C, Munoz F, Joly A (2021) Convolutional neural networks improve species distribution modelling by capturing the spatial structure of the environment. PLoS Comput Biol 17(4):e1008856

Elith J, Leathwick JR (2009) Species distribution models: ecological explanation and prediction across space and time. Annu Rev Ecol Evol Syst 40:677–697

Elith JH, Graham CP, Anderson R, Dudík M, Ferrier S, Guisan A, Zimmermann N et al (2006) Novel methods improve prediction of species' distributions from occurrence data. Ecography 29(2):129–151

Ellis CJ (2013) A risk-based model of climate change threat: hazard, exposure, and vulnerability in the ecology of lichen epiphytes. Botany 91(1):1–11

Elvin M (2006) The retreat of the elephants: an environmental history of China. Yale University Press. New Haven, Connecticut, USA

Eriksen TH, Nielsen FS (2013) A history of anthropology. Pluto Press. London, United Kingdom

Fick SE, Hijmans RJ (2017) WorldClim 2: new 1-km spatial resolution climate surfaces for global land areas. Int J Climatol 37(12):4302–4315

Fitak RR, Koprowski JL, Culver M (2013) Severe reduction in genetic variation in a montane isolate: the endangered mount Graham red squirrel (Tamiasciurus hudsonicus grahamensis). Conserv Genet 14(6):1233–1241

Flannery FT (1990) Mammals of New Guinea. The Australian Museum, Sydney

Frank D, Finckh M, Wirth C (2009) Impacts of land use on habitat functions of old-growth forests and their biodiversity. Old-Growth Forests, pp 429–450

Gaertner M, Larson BM, Irlich UM, Holmes PM, Stafford L, van Wilgen BW, Richardson DM (2016) Managing invasive species in cities: a framework from Cape Town, South Africa. Landsc Urban Plan 151:1–9

Goldewijk KK, Ramankutty N (2009) Land use changes during the last 300 years. Land Use, Land Cover Soil Sci: 1:147–168

Goldstein EA, Merrick MJ, Koprowski JL (2017) Functional semelparity drives population dynamics and endangers a peripheral population. Biol Conserv 205:52–59

Grubbs SC (1914) Plague-prevention: report of work in Mobile, Ala., during August and September, 1914. Public Health Rep (1896–1970):2967–2973

Hansen MC, Potapov PV, Moore R, Hancher M, Turubanova SA, Tyukavina A, Townshend J et al (2013) High-resolution global maps of 21st-century forest cover change. Science 342(6160):850–853

Hasui E, Silva VX, Cunha RG, Ramos FN, Ribeiro MC, Sacramento M, Ribeiro BR et al (2017) Additions of landscape metrics improve predictions of occurrence of species distribution models. J For Res 28(5):963–974

Hegel TM, Cushman SA, Evans J, Huettmann F (2010) Current state of the art for statistical modelling of species distributions. In Spatial complexity, informatics, and wildlife conservation. Springer, Tokyo, pp 273–311

Herkt KMB, Barnikel G, Skidmore AK, Fahr J (2016) A high-resolution model of bat diversity and endemism for continental Africa. Ecol Model 320:9–28

Hickisch R, Hodgetts T, Johnson PJ, Sillero-Zubiri C, Tockner K, Macdonald DW (2019) Effects of publication bias on conservation planning. Conserv Biol 33(5):1151–1163

Homeier J, Báez S, Hertel D, Leuschner C (2017) Tropical forest ecosystem responses to increasing nutrient availability. Front Earth Sci 5:27

Huettmann F (2015) On the relevance and moral impediment of digital data management, data sharing, and public open access and open source code in (tropical) research: the Rio convention revisited towards mega science and best professional research practices. In Central American Biodiversity. Springer, New York, pp 391–417

Huettmann F (2018) Advanced Data Mining (Cloning) of Predicted Climate-Scapes and Their Variances Assessed with Machine Learning: An Example from Southern Alaska Shows Topographical Biases and Strong Differences. In Machine Learning for Ecology and Sustainable Natural Resource Management. Springer, Cham, pp 227–241

Huettmann F, Ickert-Bond SM (2017) On open access, data mining and plant conservation in the circumpolar north with an online data example of the herbarium, University of Alaska Museum of the north. Arctic Science 4(4):433–470

Huettmann F, Young BD (2022) The So-called Modern 'Sustainable Forestry' Destroys Wilderness, Old-Growth Forest Landscapes and Ecological Services Worldwide: A Short First-Hand Review and Global Narrative on the Use of 'Growth-and-Yield' as a Destructive and Even Impossible Goal. In: Forest Dynamics and Conservation. Springer, Singapore, pp 53–82

Humphries GR, Huettmann F (2018) Machine Learning and 'The Cloud for Natural Resource Applications: Autonomous Online Robots Driving Sustainable Conservation Management Worldwide?. In Machine Learning for Ecology and Sustainable Natural Resource Management. Springer, Cham, pp 353–377

Humphries GR, Magness DR, Huettmann F (Eds.) (2018) Machine learning for ecology and sustainable natural resource management. Switzerland: Springer, p 3

Jain P, Coogan SC, Subramanian SG, Crowley M, Taylor S, Flannigan MD (2020) A review of machine learning applications in wildfire science and management. Environ Rev 28(4):478–505

Jenkins CN, Pimm SL, Joppa LN (2013) Global patterns of terrestrial vertebrate diversity and conservation. Proc Natl Acad Sci 110(28):E2602–E2610

Jokimäki J, Selonen V, Lehikoinen A, Kaisanlahti-Jokimäki ML (2017) The role of urban habitats in the abundance of red squirrels (Sciurus vulgaris, L.) in Finland. Urban For Urban Green 27:100–108

Jokinen M, Mäkeläinen S, Ovaskainen O (2015) 'Strict', yet ineffective: legal protection of breeding sites and resting places fails with the S iberian flying squirrel. Anim Conserv 18(2):167–175

Jones P, Wint W (2015) Data set produced by Waen Associates for Environmental Research Group Oxford, Limited, funded by the International Research Consortium on Dengue Risk Assessment, Management and Surveillance (IDAMS), European Union's Seventh Framework Programme for research, technological development and demonstration under grant agreement no 281803

Joshua J, Johnsingh AJT (1994) Impact of biotic disturbances on the habitat and population of the endangered grizzled giant squirrel Ratufa macroura in South India. Biol Conserv 68(1):29–34

Koprowski JL, Nandini R (2008) Global hotspots and knowledge gaps for tree and flying squirrels. Current Science, pp 851–856

Koprowski JL, Alanen MI, Lynch AM (2005) Nowhere to run and nowhere to hide: response of endemic Mt. Graham red squirrels to catastrophic forest damage. Biol Conserv 126(4):491–498

Krauze-Gryz D, Gryz J, Brach M (2021) Spatial organization, behaviour and feeding habits of red squirrels: differences between an urban park and an urban forest. Journal of Zoology, 315(1), 69–78

La Morgia V, Paoloni D, Genovesi P (2017) Eradicating the grey squirrel Sciurus carolinensis from urban areas: an innovative decision-making approach based on lessons learnt in Italy. Pest Manag Sci 73(2):354–363

Lance SL, Maldonado JE, Bocetti CI, Pattee OH, Ballou JD, Fleischer RC (2003) Genetic variation in natural and translocated populations of the endangered Delmarva fox squirrel (Sciurus niger cinereus). Conserv Genet 4(6):707–718

Leatherman SP, Beller-Simms N (1997) Sea-level rise and small island states: an overview. J Coast Res:1–16

Lines WJ (1991) Taming the great south land: a history of the conquest of nature in Australia. Univ of California Press

Lioy S, Marsan A, Balduzzi A, Wauters LA, Martinoli A, Bertolino S (2019) The management of the introduced grey squirrel seen through the eyes of the media. Biol Invasions 21(12):3723–3733

Lombardia R, Piemonte R, Liguria R, Bertolino S (2019) Preventing grey squirrel spread in northwest Italy. Invasive Alien Species Colonisation Prevention: your guide to early detection and rapid response, 101

Mäkeläinen S, Schrader M, Hanski IK (2015) Factors explaining the occurrence of the Siberian flying squirrel in urban forest landscape. Urban Ecosyst 18(1):223–238

Matsinos YG, Papadopoulou E (2004) Investigating the viability of squirrel populations; a modeling approach for the island of Lesvos, Greece. Nat Resour Model 17(4):423–444

Mccleery RA, Lopez RR, Silvy NJ, Gallant DL (2008) Fox squirrel survival in urban and rural environments. J Wildl Manag 72(1):133–137

Milanovich JR, Peterman WE, Barrett K, Hopton ME (2012) Do species distribution models predict species richness in urban and natural green spaces? A case study using amphibians. Landsc Urban Plan 107(4):409–418

Mimura N (1999) Vulnerability of island countries in the South Pacific to sea level rise and climate change. Clim Res 12(2–3):137–143

Oppenheimer M, Hinkel J (2019) Sea Level Rise and Implications for Low Lying Islands, Coasts and Communities Supplementary Material. In: IPCC special report on the ocean and cryosphere in a changing climate

Phillips SJ (2005) A brief tutorial on Maxent. AT&T Research 190(4):231–259

Pimm SL, Jenkins CN, Abell R, Brooks TM, Gittleman JL, Joppa LN, Sexton JO et al (2014) The biodiversity of species and their rates of extinction, distribution, and protection. Science 344(6187)

Rademacher E (2019) Responses of urban gray squirrels (Scurius carolinensis) to humans and conspecifics in an area of Boston Common. Arts and Science Writing Program. 150

Raup DM (1995) In: Lawton JH, May RM (eds) Extinction rates, vol 11. Oxford University Press, Oxford

Riley SP, Serieys LE, Moriarty JG (2014) Infectious disease and contaminants in urban wildlife: unseen and often overlooked threats. In Urban Wildlife Conservation. Springer, Boston, pp. 175–215

Robold R, Huettmann F (2021) High-resolution prediction of American red squirrel in interior Alaska: a role model for conservation using open access data, machine learning. GIS and LIDAR, PEERJ. https://peerj.com/articles/11830/

Rustiati EL (2009) Traffics and wildlife: a preliminary study on road-kill. Biospecies 2(2)

Safran RJ, Nosil P (2012) Speciation: the origin of new species. Nature Education Knowledge 3(10):17

Saiful AA, Idris AH, Rashid YN, Tamura N, Hayashi F (2001) Home range size of sympatric squirrel species inhabiting a lowland dipterocarp Forest in Malaysia 1. Biotropica 33(2):346–351

Shorohova E, Kneeshaw D, Kuuluvainen T, Gauthier S (2011) Variability and dynamics of old-growth forests in the circumbolear zone: implications for conservation, restoration and management

Signorile AL, Reuman DC, Lurz PWW, Bertolino S, Carbone C, Wang J (2016) Using DNA profiling to investigate human-mediated translocations of an invasive species. Biol Conserv 195:97–105

Simard S (2021) Finding the mother tree: discovering the wisdom of trees. Knopf Publishers, Chicago

Spies TA (2004) Ecological concepts and diversity of old-growth forests. J For 102(3):14–20

Sriram S, Huettmann F (unpublished) A global model of predicted peregrine falcon (Falco peregrinus) distribution with open source GIS code and 104 open access layers for use by the global public. Earth System Science Data Discussions:1–39. https://doi.org/10.5194/essd-2016-65

Steiner M, Huettmann F (2021) Justification for a taxonomic conservation update of the rodent genus Tamiasciurus: addressing marginalization and mis-prioritization of research efforts and conservation laissez-faire for a sustainability outlook. Eur Zool J 88(1):86–116. https://doi.org/10.1080/24750263.2020.1857852

Tabarelli M, Da Silva JMC, Gascon C (2004) Forest fragmentation, synergisms and the impoverishment of neotropical forests. Biodivers Conserv 13(7):1419–1425

Thompson DC, Thompson PS (1980) Food habits and caching behavior of urban grey squirrels. Can J Zool 58(5):701–710

Uchida K, Yamazaki T, Ohkubo Y, Yanagawa H (2021) Do green park characteristics influence human-wildlife distance in arboreal squirrels? Urban For Urban Green 58:126952

UNEP-WCMC, IUCN (2020) Protected Planet: the World Database on Protected Areas (WDPA) [Online], September 2020, Cambridge, UK: UNEP-WCMC and IUCN. Available at: www.protectedplanet.net

Van Horne B (2008) Conservation of ground squirrels. In: Rodent Societies. University of Chicago Press, pp 463–471

Vanitharani J (2018) Conservation status and guidelines for the maintenance of endangered grizzled giant squirrel Ratufa macroura in Srivilliputhur Wildlife Sanctuary. In: Indian Hotspots. Springer, Singapore, pp 297–307

White DL, Lloyd FT (1995) Defining old-growth: implications for management. In: Proceedings of the eight biennial southern silvicultural research conference. Gen. Tech. Rep. SRS-1. United States Department of Agriculture, Forest Service, Southern Research Station, Asheville, North Carolina, pp. 51–62

Wohlleben P (2016) The hidden lives of trees: how they feel, what they communicate – discoveries from a secret world. Greystone Books

Wright SJ, Zeballos H, Domínguez I, Gallardo MM, Moreno MC, Ibáñez R (2000) Poachers alter mammal abundance, seed dispersal, and seed predation in a Neotropical forest. Conserv Biol 14(1):227–239

Zuckerberg B, Huettmann F, Frair J (2011) Proper data management as a scientific foundation for reliable species distribution modeling. In Predictive species and habitat modeling in landscape ecology. Springer, New York, pp. 45–70

Chapter 4
A First Meta-Analysis Based on Open Access Big Data Mining of Global Predicted Squirrel Distribution Models with Machine Learning for IUCN Conservation Status and Population Trend Policy Assessments

Abstract Machine Learning (ML) is an analysis platform and a set of algorithms as part of a modern approach toward unified data analysis and robust inference of complex data. Here, a ML approach has been applied to 233 global species distribution models of all squirrel species worldwide with available data. This yielded a Meta-analysis based on open-access data ("Big Data" squirrel datasets). By using a ML (Boosted Regression Tree; Treenet SPM 8.3) analysis for two models (IUCN species Conservation Status, and IUCN species Population Trend) we conducted a Meta-Analysis for all these 233 species (out of over 300 species) and 132 environmental predictors. This is the first Meta-Analysis on this topic with these large amounts of data, and we observed that the ML outcome was not similar to such an analysis when created with Maxent (see Chap. 3 in Steiner and Huettmann 2023). Additionally, it showed that the species distribution is not single-predictor dependent and cannot be explained by a parsimonious approach. Rather, a more holistic and wider ecological multi-predictor perspective is suggested to be followed, with a large dataset of high-quality, fine-grain, and interacting environmental predictors. In addition, we confirmed a positive relationship between the IUCN species Conservation Status classes, and the IUCN species Population Trend classifications as it seems the more threatened a species is the less promising is also its population trend (indicating increased conservation needs). Further, the classification of squirrels in the so-called 'data deficient' conservation status classes showed coherence with the absence of data also in the population trend (population trend "unknown" and thus not classified and taken widely off the agenda). This indicates that these categories are used interchangeably for policy, supporting the evidence for the inappropriate marginalization of this species group and its possible unknown and unresolved extinction threats. The so-called "absence of data" should not stop good squirrel conservation actions and should not be used that way politically for inaction.

Keywords Squirrels · Global analysis · Data mining · Machine learning · Stochastic gradient boosting · TreeNet · IUCN · Big data

4.1 Introduction

Despite their popularity in media and with the public, squirrels actually join the long list of thousands (= majority) of species on earth that are still poorly studied and deficient of data (Koprowski 2005). But good efforts, methods, and tools exist that can add some progress to such topics. Squirrels are naturally found almost all over the world, except for Australia & Oceania, the Antarctic continent, and a few harsh regions (Thorington et al. 2012). This group of mammals is hunted, consumed as bushmeat, and consists of over 300 species (approx. 307, see Chap. 1 in Steiner and Huettmann 2023). While we live in the Anthropocene, with a harsh climate change outlook, only some of the squirrel species are even listed currently as being of conservation concern (approx. 100, see Table 1.4 in Chap. 1 in Steiner and Huettmann 2023). Many species (more than 10%) are listed as Data Deficient (35 according to IUCN Red List, see Table 1.4 in Chap. 1 in Steiner and Huettmann 2023), and thus, not correctly classified and left for virtually no action. For even more species, the conservation trend is unknown (115–125 (c. 40%) out of approx. 300, see Table 1.4 in Chap. 1 in Steiner and Huettmann 2023 for details).

For good progress, the discipline of wildlife informatics and related fields are widely on the rise (Cushman and Huettmann 2010; Shahinfar et al. 2020), many of the efforts include the use of species distribution models (SDMs) to fill data gaps (see Elith et al. 2006; Humphries et al. 2018; Zhang et al. 2019). The concept of machine learning (ML) plays a central role in this discussion, and how technology, the world wide web (WWW), and 'The Cloud' can be used for progress in conservation

Supplementary Information The online version contains supplementary material available at https://doi.org/10.1007/978-3-031-23547-4_4.

M. Steiner, F. Huettmann, *Sustainable Squirrel Conservation*, https://doi.org/10.1007/978-3-031-23547-4_4

(Graham et al. 2004; Humphries and Huettmann 2018a). But despite their global abundance and being hunted, squirrels have not been studied well, except for a few over-represented ones (e.g. Eastern gray squirrel (*Sciurus carolinensis*), Eurasian red squirrel (*Sciurus vulgaris*), or Fox squirrel *(Sciurus niger)* (see Chap. 3 in Steiner and Huettmann 2023 for details). Additionally, distribution data is hardly available for other squirrel species than the upper-mentioned ones, which once more represents the global lack of science, SDMs, and machine learning applications for policy progress and better conservation.

While we seem to drown in technically created data and commercial technology, we have fewer means to truly use, summarize, and apply them for good policy, especially with accessibility for the public (open-access). But it is the meta-analysis that can help to achieve that (Lambert et al. 2013; Moher and Olkin 1995; van de Kaa et al. 2007), and often those are done manually or simplistically. Employing machine learning for meta-analysis is still less common but offers promises (Gatta et al. 2020; Jenks et al. 2014). Here we try to carry out a first study that supports the use of meta-analysis by data mining of virtually all publicly available SDMs of the squirrels of the world. Those Rapid Assessment SDMs have been created with a dominant 'quick and dirty' machine learning algorithm (Maxent – vers. 3.4, Elith et al. 2006; Humphries and Huettmann 2018b; Phillips et al. 2006). This is considered 'shallow learning' in the ML/AI world. This has been done for all the Open-Access squirrel species data in the world available at www.GBIF.org. Further, those squirrel SDMs are state-of-the-art, a certain world record, as they are based on 132 Open-Access predictor layers in GIS that we compiled, made available, and then utilized to run 233 SDMs (see details in Chap. 3 in Steiner and Huettmann 2023). Those SDMs have never been available before in one compact set, as we provide here. Subsequently, linking the SDM details for what drives the models, here we try to make sense of the data cube when leading into the latest policy information from IUCN's conservation status and population trend.

This work presents a Big Data world summary for an entire species group – the global squirrels (family Sciuridae) - that has not been well studied overall and where good conservation policy information and subsequent action are widely lacking but potentially promising. Using an Open-Access approach here, we are attempting to present a workflow, a digital pipeline, and discuss whether, and how such work can be improved and extended. Additionally, we assess shortcomings and improvements, especially in the data availability, to promote further data sharing ambitions and their highly beneficial influences on SDMs, ML, AI, and consecutively conservation (*sensu* Zuckerberg et al. 2011; Huettmann 2015).

4.2 Methods

We compiled 132 open-access GIS predictor layers from various public sources, including Sriram and Huettmann (unpublished). Details of that work are presented in Chap. 3 in Steiner and Huettmann (2023).

We then used the squirrel taxonomic list of Steiner and Huettmann (2023) from www.GBIF.org, representing the best publicly available and compiled open-access data for occurrence records of squirrels in the world (yet with possible improvements – see Chap. 3 in Steiner and Huettmann 2023).

Subsequently, we ran Maxent (version 3.4) for each species and obtained the model output, including an .ASC raster (ASCII format, that was afterwards converted into a TIFF raster for illustrative purposes) for the predicted relative index of occurrence (RIO), all model diagnostics, and the top 7 predictors.

4.2.1 Multi Predictor Analysis

The top 7 predictors from the SDM results were linked with the IUCN conservation status and population trend. This allowed us to run a Machine Learning (Boosted Regression Tree; Treenet SPM 8.3) analysis on the subsequent data cube for two models: IUCN Conservation Status, as well as Population Trend, driven by the Top 7-ranked predictors from the SDMs. This equals essentially a multiple regression concept but instead uses unconstrained machine learning analyses by parametric assumptions to find the signal in the data, based on earlier Big Data model predictions and their diagnostics for all squirrels of the world with available data (233 out of 307, c. 76%).

We also visualized the relationship between IUCN Conservation Status and IUCN Population Trend using Treenet, allowing us to analyze a non-linear trend between those two ranks (see Table 4.1).

Table 4.1 Overview of IUCN conservation status and population trend indices

IUCN conservation status	Description of IUCN conservation status category	IUCN population trend	Description of IUCN population trend category
0	Least concern	0	Increasing
1	Near threatened	1	Stable
2	Vulnerable	2	Decreasing
3	Endangered	3	Unknown
4	Critically endangered	4	NA
5	Data deficient		
6	Not evaluated		
7	NA		

Table 4.2 Predictor ranks of theTreeNet model assessing the IUCN conservation status vs Top7 predictors

Predictor	Percent in the TreeNet model
First-ranked	100
Fourth-ranked	95
Fifth-ranked	84
Second-ranked	79
Third-ranked	68

4.2.2 Single Predictor Analysis

This first meta-analysis mainly focused on the interaction between the top 7 predictors and how interlinked these are. To compare this with an approach that ignored the linkage between the predators, a second analysis has been performed. For this second analysis, the top 7 of all SDMs have been identified and entered in a table, unlinked. This aimed to present the ranks and how often individual predictors have been identified by the software Maxent to be essential for the creation of the SDMs and for being among the "top 7 predictors".

Thus, the first analysis focuses on the interaction of the top 7 predictors of all created SDMs for the global squirrel species and the frequency of the same set of combinations of these top 7 predictors. Whereas the second analysis focuses on the identification of the frequency of each predictor, unlinked, to be among the top 7 predictors. This resulted in an overview of the predictors indicated by Maxent to have the overall highest contribution to the global SDMs.

4.3 Results

Our work is based on the world's best-publicly available digital squirrel occurrence data and global habitat data for all extant squirrel species. It further presents the best SDMs for all squirrels in the world.

4.3.1 IUCN Red List Conservation Status Analysis

Further, our findings from the IUCN Red List Conservation Status show that the model can find signals in the link between the response and the top 7 predictor ranks that were unknown before. The rank of predictors is confirmed for the First Predictor, but not afterwards for other ranks (Seventh, Sixth, Fourth, Second, Third, and Fifth), see Table 4.2. Essentially, it shows that the IUCN Conservation Status mostly changes and flips the Maxent ranked predictor ranks of the underlying SDMs illustrating variation and lack of a consistent parsimonious finding across algorithms. While this is fine in the approach of Leo Breiman (2001; inference from predictions) it should be noted for inference from Maxent.

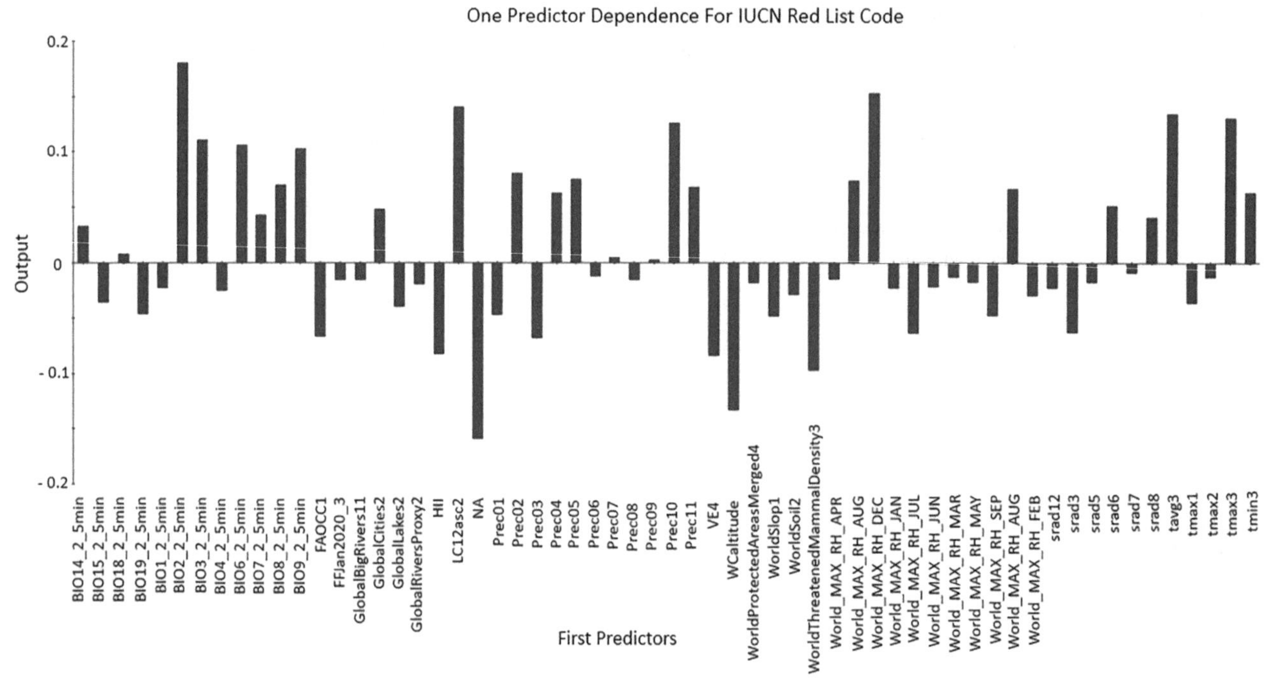

Fig. 4.1 List of Top7 predictors presented as the most important predictor in the model of IUCN conservation Status vs Top7 predictors in SDMs

Table 4.3 Predictor ranks of the TreeNet model assessing the IUCN population trend vs Top7 predictors

Predictor	Percent in the TreeNet model
Secondranked	100
Fourthranked	85
Fifthranked	81
Firstranked	81
Thirdranked	59

It further shows that on a global scale a widely diverse multivariate complex set of factors comes to play (Fig. 4.1), namely climatic and a mix of other categories. The highest peaks we found in the data mining analysis were for the predictors of Bio2 (Mean Diurnal Range (Mean of monthly (max temp - min temp))), LC12 (Global Land Cover), MAX_RH_DEC (Maximum Relative Humidity in December 2020), Tavrg3 (Temperature averages in March 2020), and Precipitation10 (Total precipitation in October 2020). For the most relevant species-genus interaction of squirrels, we identified the genera *Petinomys, Ictidomys* and *Geoscirius* and the species within (model details shown in Appendix 4.2, last slide).

4.3.2 IUCN Red List Population trend Analysis

Our finding from the IUCN Conservation Population Trend (Table 4.3) shows that the model can find signals in the link between the response and the top 7 predictor ranks. The rank of predictors is not in agreement with Table 4.1 and uses its own rank order. The Second Predictor ranks first but is followed by the Sixth, First, Seventh, Fourth, Fifth, and Third; see Table 4.3). These ranks should theoretically, without inference, be in sequence, this is however not the case here. This shows that each model has its own ranks and that it does not agree much with the Maxent model output for ranks. This matters for inference from such algorithms and that predictions should be seen as the main focus (*sensu* Breiman 2001), not the underlying model structure and predictors.

In the IUCN Population Trend analysis we also found that on a global scale a widely diverse multivariate complex set of factors comes to play (Fig. 4.2), especially the predictors Bioclim 3 (Isothermality (BIO2/BIO7) (×100)), Bioclim 6 (Min Temperature of Coldest Month), and WorldSoil (Soil characteristics).

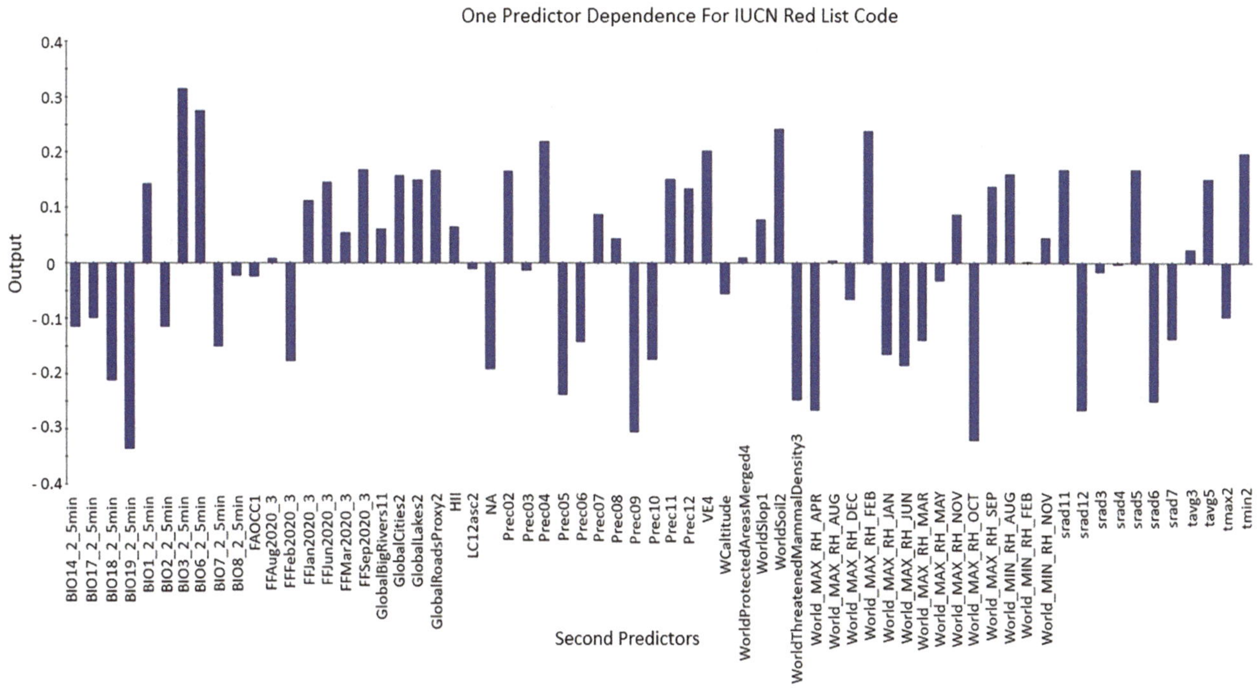

Fig. 4.2 List of Top7 predictors presented as the most important predictor in the model of IUCN population trend vs Top7 predictors in SDMs

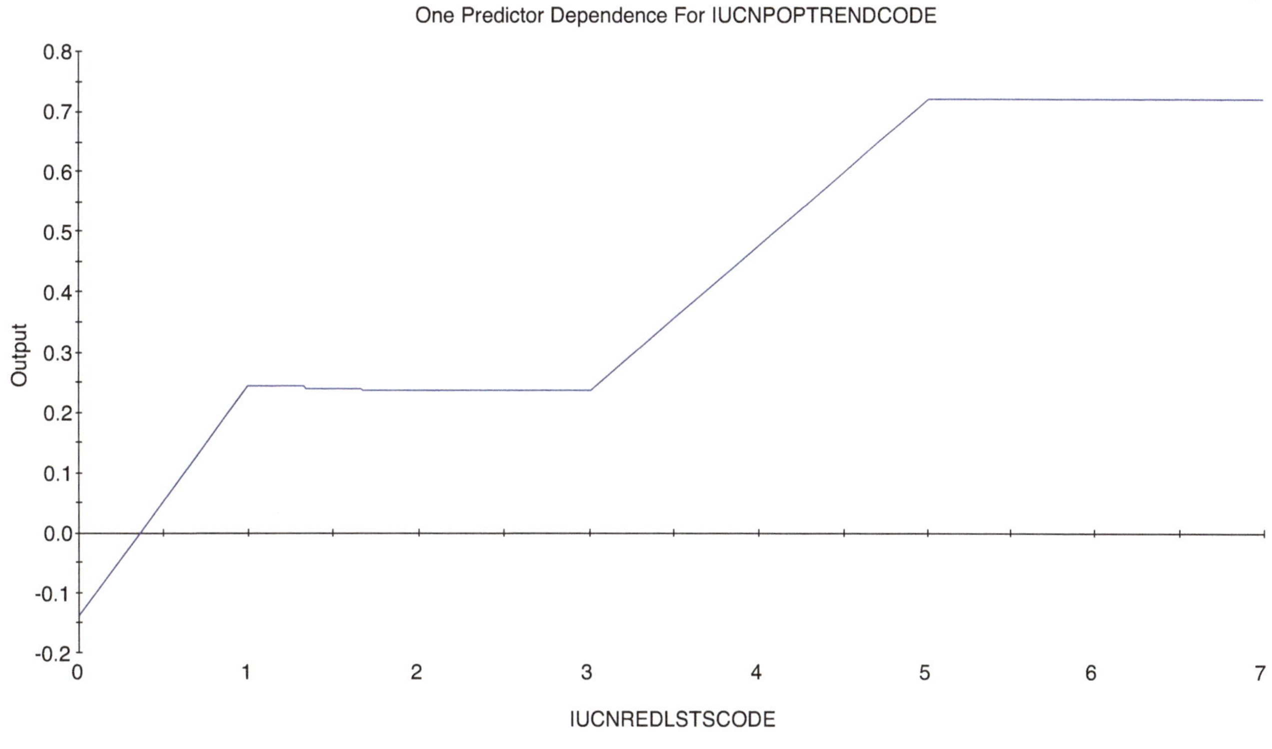

Fig. 4.3 Positive - but not linear - relationship between IUCN population trend and IUCN conservation status, as obtained from Machine Learning (ML)

For the most relevant species-genus interaction of squirrels, we identified the genus *Paraxerus* as well as *Helioscirius* and the species within (model details shown in Appendix 4.3) (Fig. 4.3).

Finally, looking at the relationship on how the IUCN Population Trend relates to the IUCN Conservation Status, we confirm an increasing trend: a higher IUCN Population Trend generally relates to a higher IUCN Conservation Status (higher in this case is not beneficial for the species, see Table 4.1 for details). However, this trend is not linear, and we find a plateau with a weaker link in the middle ranges. This is likely because there are fewer classes for the population trend classes compared to the conservation status classes. But overall it likely indicates that the mid ranges have a relatively constant (not increasing nor decreasing) population trend. Noteworthy here is specifically the data deficient categories (Data deficiency, Unknown, Not evaluated, and NA due to taxonomic mismatches; Table 4.1), showing on a global level (for over 300 species) vast uncertainty in this relationship, and not truly allowing for a meaningful quantitative assessment and inference for any of the categories. This has large policy implications beyond squirrels. Figure 4.3 essentially shows that the low index numbers of the IUCN Red List conservation status classes predominately link with the low index classes of population trend (LC and NT species often have Increasing or Stable population trends), whereas high conservation status classes link well with high population trend classes (EN, CR, and DD species often have a Decreasing and Unknown population trend).

Subsequently, in order to present a ranked table, illustrating the predictors that have been generally contributed most to all SDMs, a single-predictor analysis has been performed. For this analysis, the top 7 predictors of each SDM have been added to a table and ranked. This analysis has been performed to illustrate which predictors have overall been identified to be among the top single predictors – free of interactions-, and which predictors have barely or even never been among the top 7 predictors (Table 4.4).

Table 4.4 Maxent Single-predictor analysis

Variable	Predictor contribution in top 7	Percent total contribution Combined	Percent total involvement of species
WorldSoil2	99	5.696	44.395
FAOCC1	78	4.488	34.978
WorldProtectedAreasMerged4	77	4.430	34.529
WorldThreatenedMammalDensity3	74	4.258	33.184
WCaltitude	55	3.165	24.664
WorldSlope1	49	2.819	21.973
GlobalCities2	48	2.762	21.525
HII1	43	2.474	19.283
Prec09	41	2.359	18.386
BIO19_2_5min	40	2.301	17.937
BIO2_2_5min	38	2.186	17.040
srad6	36	2.071	16.143
World_MAX_RH_OCT	33	1.899	14.798
Prec01	32	1.841	14.350
BIO3_2_5min	32	1.841	14.350
Prec07	30	1.726	13.453
Prec06	30	1.726	13.453
World_MAX_RH_APR	28	1.611	12.556
World_MAX_RH_NOV	25	1.438	11.211
BIO18_2_5min	25	1.438	11.211
BIO14_2_5min	25	1.438	11.211
World_MAX_RH_AUG	23	1.323	10.314
VE4	22	1.266	9.865
tmax2	22	1.266	9.865
World_MAX_RH_MAY	21	1.208	9.417
srad12	21	1.208	9.417
Prec11	21	1.208	9.417
LC12asc2	20	1.151	8.969
Prec12	20	1.151	8.969
srad7	18	1.036	8.072
BIO15_2_5min	18	1.036	8.072
Prec05	18	1.036	8.072
World_MAX_RH_SEP	18	1.036	8.072
World_MAX_RH_DEC	17	0.978	7.623
srad8	17	0.978	7.623
FFJan2021_3	17	0.978	7.623
FFFeb2020_3	17	0.978	7.623

(continued)

Table 4.4 (continued)

Variable	Predictor contribution in top 7	Percent total contribution Combined	Percent total involvement of species
BIO4_2_5min	17	0.978	7.623
World_MAX_RH_JAN	17	0.978	7.623
World_MAX_RH_JUL	16	0.921	7.175
GlobalRiversProxy2	16	0.921	7.175
BIO7_2_5min	16	0.921	7.175
World_MAX_RH_MAR	15	0.863	6.726
World_MAX_RH_JUN	15	0.863	6.726
GlobalRoadsProxy2	15	0.863	6.726
Prec08	15	0.863	6.726
srad4	15	0.863	6.726
Prec04	14	0.806	6.278
Prec10	14	0.806	6.278
srad5	13	0.748	5.830
GlobalLakes2	13	0.748	5.830
GlobalBigRivers11	13	0.748	5.830
Prec03	13	0.748	5.830
World_MAX_RH_FEB	12	0.690	5.381
BIO8_2_5min	11	0.633	4.933
BIO17_2_5min	11	0.633	4.933
srad3	11	0.633	4.933
FFAug2020_3	9	0.518	4.036
tmax1	9	0.518	4.036
FFNov2020_3	9	0.518	4.036
tmax12	9	0.518	4.036
FFSep2020_3	8	0.460	3.587
tmin3	8	0.460	3.587
tmax11	8	0.460	3.587
World_MIN_RH_AUG	7	0.403	3.139
BIO9_2_5min	7	0.403	3.139
World_MIN_RH_NOV	7	0.403	3.139
tmax3	7	0.403	3.139
BIO1_2_5min	7	0.403	3.139
FFMay2020_3	6	0.345	2.691
BIO6_2_5min	6	0.345	2.691
FFJun2020_3	5	0.288	2.242
FFJan2020_3	5	0.288	2.242
FFMar2020_3	5	0.288	2.242
srad11	5	0.288	2.242
tavg1	5	0.288	2.242
srad9	5	0.288	2.242
tmin2	5	0.288	2.242
World_MIN_RH_FEB	5	0.288	2.242
tavg11	4	0.230	1.794
Prec02	4	0.230	1.794
srad1	4	0.230	1.794
World_MIN_RH_APR	3	0.173	1.345
World_MIN_RH_MAR	3	0.173	1.345
tavg2	3	0.173	1.345
tavg3	3	0.173	1.345
World_MIN_RH_MAY	3	0.173	1.345
World_MIN_RH_JUN	3	0.173	1.345
FFOct2020_3	2	0.115	0.897
FFJul2020_3	2	0.115	0.897
tavg10	2	0.115	0.897
tavg5	2	0.115	0.897

(continued)

Table 4.4 (continued)

Variable	Predictor contribution in top 7	Percent total contribution Combined	Percent total involvement of species
tmax10	2	0.115	0.897
tmin12	2	0.115	0.897
tmin6	2	0.115	0.897
tmin1	2	0.115	0.897
World_MIN_RH_DEC	1	0.058	0.448
World_MIN_RH_SEP	1	0.058	0.448
srad2	1	0.058	0.448
World_MIN_RH_JAN	1	0.058	0.448
BIO5_2_5min	1	0.058	0.448
BIO10_2_5min	1	0.058	0.448
BIO11_2_5min	1	0.058	0.448
BIO16_2_5min	1	0.058	0.448
tmax5	1	0.058	0.448
tmax8	1	0.058	0.448
tmin5	1	0.058	0.448
tmin8	1	0.058	0.448
tmin9	1	0.058	0.448
tmax6	1	0.058	0.448
tmin7	1	0.058	0.448
World_MIN_RH_OCT	0	0.000	0.000
World_MIN_RH_JUL	0	0.000	0.000
tmax7	0	0.000	0.000
tmin11	0	0.000	0.000
tmin4	0	0.000	0.000
BIO12_2_5min	0	0.000	0.000
BIO13_2_5min	0	0.000	0.000
GlobalBirdDensity2	0	0.000	0.000
GlobalSnowCoverMonthJan2021_7	0	0.000	0.000
srad10	0	0.000	0.000
tavg12	0	0.000	0.000
tavg4	0	0.000	0.000
tavg6	0	0.000	0.000
tavg7	0	0.000	0.000
tavg8	0	0.000	0.000
tavg9	0	0.000	0.000
tmax4	0	0.000	0.000
tmax9	0	0.000	0.000
tmin10	0	0.000	0.000
WorldMammaldensity4	0	0.000	0.000
WorldRodentDensity3	0	0.000	0.000

On the one hand, it can clearly be observed that the predictors WorldSoil2 (99), FAOCC1 (78), WorldProtectedAreasMerged4 (77), WorldThreatenedMammalDensity3 (74), and WCaltitude (55) are the top 5 predictors overall that have all been identified in over 50 SDMs to be among the top 7 predictors [Explanation: The number in the brackets indicates how often each predictor has been among the top 7 predictors out of 233 runs]. It is noteworthy, that for this analysis, the rank of each predictor in the top 7 sets has not been given any attention, the multi-predictor analysis instead respected the ranks carefully. These five predictors mentioned above present the world's soil types and characteristics, the global climate classes presented from www.FAO.org, the proximity to all global protected areas, the density of the world's threatened mammals, and the altitude respectively.

On the other hand, 21 out of the 132 predictors have never been identified to be among the top 7 predictors. These predictors can be found in Table 4.3 at the very bottom with the predictor contribution count "0". Arguably, on a global scale, these predictors appear to be wider 'background noise' and can be considered to be exchanged/swapped with others that have not been used in this study.

4.4 Discussion and Conclusion

Here we were able to compile and use for the first time the best occurrence, GIS layer, and SDM data set for the squirrels of the world, *sensu* Biotime project (see also Sriram and Huettmann (unpublished) for global analysis, and Huettmann (2021) for a field example). All of this work is presented here for the first time, with Open Access and ISO-compliant metadata documentation which offers itself for improvements. These data were linked with the IUCN conservation status classes and population trends. We were then able to data-mine this unique data set with machine learning methods to find and describe for the first time signals in the data while models are not very strong as such. The focus in Data Mining sits on detecting signals in otherwise complex and 'noisy' data.

We find that the Maxent ranks do not match well our findings from IUCN Conservation Status or Population Trend categories. Instead, our work, to explain IUCN Conservation Status and IUCN Population Trend linkage, shows that a variety of multivariate predictors come to play globally and matter for many individual pixels globally, e.g. for endemic species and many range edge pixels, where climate and some landcover predictors are most prominent. This makes an inherent link between squirrels and how they should be managed and for a conservation priority in the Anthropocene: Climate and its management on land. This is also a topic addressed by the UN with the Sustainable Development Goal (SDG) "Life on Land". It is also further discussed in a future study (Chap. 13 in Steiner and Huettmann 2023), unfortunately it still widely fails in actually improving sustainability with the current approach.

We confirm that IUCN Conservation Status links with IUCN Population Trend, but this link is not linear and carries a large section with data deficiency, not evaluated, or taxonomic mismatches; meaningful and complete assessments for any categories cannot be done that way other than patchwork. This has global implications when using the IUCN approach.

The single-predictor analysis provides great insights into the frequency of each predictor being among the top 7 predictors. The outcome of this analysis was overall coherent with the multi-predictor analysis. However, with the difference that the multi-predictor analysis provides insights into the interactions among the predictors and whether they matter, the single-predictor analysis rather classifies the importance of each predictor. From this single-predictor analysis, it can be identified which ones are the top 20, or even the top 50 predictors used, and which predictors are being used less. This ranking table can be highly useful for further studies such as prioritization assessments, as the predictors that have never – or only once – been identified to be among the top 7 predictors can be left out from the analysis. This can highly increase the time efficiency of running SDMs and can provide space for other predictors that have not been used in this study that might be of high importance. Therefore, by observing Table 4.4 presented above, it can be decided for future studies, which predictors are essential to include in the predictor set for a ML/AI analysis and which ones can be replaced and swapped with other, new predictors (see Huettmann et al. (2018) for swapping and shaving predictors in ML/AI analysis).

Using a global big data open-access approach, we find overall new and synergy insights for the conservation of the world's squirrels. We find that our approach offers a new, relatively simple but powerful way to assess conservation policy for a complex species group. It shows promise and good use for other biodiversity and conservation questions and species. Additionally, it shows that there is no single, or hand full of predictors that dominates the SDMs. This confirms wider ecological and holistic approaches. Instead, we found that many, almost all of the predictors used were of significant importance for different SDMs. From this, it can be concluded that it is suggested to approach such holistically and global studies always with a large and diverse set of environmental predictors instead of only a hand full (examples seen in Sriram and Huettmann (unpublished), Huettmann et al. (2018)). This serves not only for higher diversity purposes but also for a higher quality of the SDMs and better policy. Additionally, with a more diverse and larger set of predictors, the individual niche of each species can be identified and predicted more sophisticatedly, aiming to create state-of-the-art ecological predictions for urgently needed policy improvements in times of a global crisis.

Acknowledgments We are grateful to all data providers for these data, squirrel and GIS ones, in GBIF and online. FH acknowledges the great data work done by S. Sriram; further help and support are greatly acknowledged for D. Steinberg of Salford and by MINITAB SPM Software Suite, H. Berrios, and E. Huettmann and family as well as the incredible Chrome team. This is EWHALE publication # 260.

References

Breiman L (2001) Statistical modeling: The two cultures (with comments and a rejoinder by the author). Statistical science, 16(3): 199–231

Cushman SA, Huettmann F (eds) (2010) Spatial complexity, informatics, and wildlife conservation. Springer, Tokyo, pp 83–108

Elith J, Graham CH, Anderson RP, Dudík M, Ferrier S, Guisan A, Zimmermann NE et al (2006) Novel methods improve prediction of species' distributions from occurrence data. Ecography 29(2):129–151

Gatta R, Depeursinge A, Ratib O, Michielin O, Leimgruber A (2020) Integrating radiomics into holomics for personalised oncology: from algorithms to bedside. Eur Radiol Exp 4(1):1–9

Graham CH, Ferrier S, Huettman F, Moritz C, Peterson AT (2004) New developments in museum-based informatics and applications in biodiversity analysis. Trends Ecol Evol 19(9):497–503

Huettmann F (2015) On the relevance and moral impediment of digital data management, data sharing, and public open access and open source code in (tropical) research: the Rio convention revisited towards mega science and best professional research practices. In: Central American biodiversity. Springer, New York, pp 391–417

Huettmann F (2021) Investigating Matschie's tree kangaroos with 'Modern'Methods: digital workflows, big data project infrastructure, and mandated approaches for a holistic conservation governance. In: Tree Kangaroos. Academic, pp 379–391

Huettmann F, Mi C, Guo Y (2018) 'Batteries' in machine learning: a first experimental assessment of inference for Siberian crane breeding grounds in the Russian high Arctic based on 'Shaving'74 predictors. In: Machine learning for ecology and sustainable natural resource management. Springer, Cham, pp 163–184

Humphries GRW, Huettmann F (2018a) Machine learning and 'the cloud' for natural resource applications: autonomous online robots driving sustainable conservation. Management worldwide? In: Humphries G, Magness DR, Huettmann F (eds) Machine learning for ecology and sustainable natural resource management. Springer, Cham, pp 353–377

Humphries GR, Huettmann F (2018b) Machine learning in wildlife biology: algorithms, data issues and availability, workflows, citizen science, code sharing, metadata and a brief historical perspective. In: Machine learning for ecology and sustainable natural resource management. Springer, Cham, pp 3–26

Humphries GR, Magness DR, Huettmann F (2018) Machine learning for ecology and sustainable natural resource management. Springer, Berlin/Heidelberg

Jenks KE, Songsasen N, Kanchanasaka B, Leimgruber P, Fuller TK (2014) Local people's attitudes and perceptions of dholes (Cuon alpinus) around protected areas in southeastern Thailand. Trop Conserv Sci 7(4):765–780

Koprowski JL (2005) Management and conservation of tree squirrels: the importance of endemism, species richness, and forest condition. In: Gottfried GJ, Gebow BS, Eskew LG, Edminster CB (eds) Connecting mountain islands and desert seas: biodiversity and management of the Madrean Archipelago II. Proceedings of RMRS-P-36, Forest service, rocky mountain research station: 245–250. US Department of Agriculture, Fort Collins, p 36

Lambert JC, Ibrahim-Verbaas CA, Harold D, Naj AC, Sims R, Bellenguez C, Nalls MA et al (2013) Meta-analysis of 74,046 individuals identifies 11 new susceptibility loci for Alzheimer's disease. Nat Genet 45(12):1452–1458

Moher D, Olkin I (1995) Meta-analysis of randomized controlled trials: a concern for standards. JAMA 274(24):1962–1964

Phillips SJ, Anderson RP, Schapire RE (2006) Maximum entropy modeling of species geographic distributions. Ecol Model 190(3–4):231–259

Shahinfar S, Meek P, Falzon G (2020) "How many images do I need?" Understanding how sample size per class affects deep learning model performance metrics for balanced designs in autonomous wildlife monitoring. Eco Inform 57:101085

Sriram S, Huettmann F (unpublished) A global model of predicted peregrine falcon (Falco peregrinus) distribution with open source GIS code and 104 open access layers for use by the global public. Earth System Science Data Discussions:1–39. https://doi.org/10.5194/essd-2016-65

Thorington RW Jr, Koprowski JL, Steele MA, Whatton JF (2012) Squirrels of the world. JHU Press

van de Kaa G, De Vries HJ, van Heck E, van den Ende J (2007) The emergence of standards: a meta-analysis. In: 2007 40th annual Hawaii international conference on system sciences (HICSS'07). IEEE, p 173a

Zhang L, Huettmann F, Liu S, Sun P, Yu Z, Zhang X, Mi C (2019) Classification and regression with random forests as a standard method for presence-only data SDMs: a future conservation example using China tree species. Eco Inform 52:46–56

Zuckerberg B, Huettmann F, Frair J (2011) Proper data management as a scientific foundation for reliable species distribution modeling. In: Predictive species and habitat modeling in landscape ecology. Springer, New York, pp 45–70

Chapter 5
Squirrels in Cities: Meeting the Anthropological Conservation *Conundrum* of the World's Squirrels

Abstract As occurrences and even entire populations of squirrels in cities, and especially around them, become increasingly more frequent, addressing this from a conservation aspect is not trivial. With urbanization on the rise, it cannot be forgotten and left out in any serious elaboration of the world's squirrels' conservation and wilderness. Here we aim to identify how squirrels are managed in some megacities and their parks (e.g. New York City NYC Central Park and several others in Helsinki (Finland), Seattle, and Vancouver (Canada)), and even zoos. Additionally, we focus on anthropological aspects of the conservation attempts such as citizen science and "bird feeders". Even though it appears only as an indirect, unintended action, we found that it greatly influences the squirrel's presence in urban areas. Also, it is discussed how squirrels follow human activity, with data obtained from citizen science-based online archives such as "www.feederwatch.org" on a continental scale (for North America). Similar "urban" food sources for the squirrels are included here, such as trash bins, and public water sinks. However, besides sources, also the sinks and threats for squirrels to live in urban areas are important (e.g. being readily killed by cars on the streets, urban diseases, exotic predators, urban pollution/contamination, and more). To demonstrate this, a literature review has been performed for some specialized urban-environment inhibiting squirrel species. For those species, Species Distribution Models (SDMs) and Species Distribution Forecasts (SDFs) for the year 2100 have been created to visualize their current urban distribution trends and how it is predicted to change by 2100 (using three different Global Climate Models as scenarios). This approach aims for a model-based assessment for a better science-based outlook for squirrels. In addition, as the cities are part of the "regions of/under high risk", we focus on the threats to humans originating from squirrel disease transmissions (zoonosis), when interactions are left unevaluated, as supported by another extended literature review. Last but not least, suggestions are made on how to perform sustainable conservation actions in and around cities, to create a safe environment for both parties (humans and squirrels). This includes suggestions such as a possible reallocation of high squirrel densities out of the cities to decrease disease contamination risks, and to seek greater conservation success (e.g. limiting the isolation of populations through extensions of human civilizations).

Keywords Squirrels · Anthropocene · Management · Citizen science · Sources and sinks · Companionship · Disease transmission (zoonosis wild animals to humans) · Species distribution models (SDMs) · Species distribution forecasts (SDFs) · Maxent · TreeNet

5.1 Introduction

Squirrels are among the few mammals that can be observed in both nature and the urbanized Anthropocene. Squirrels are ecological members of many city parks, green spaces, zoos, and even gardens in global urban and suburban areas. The percentage of urbanization is dramatically on the rise, with more than 80% of the world population living in urban regions covered by concrete, asphalt, and buildings (Chrysanthou et al. 2014; Esbah et al. 2009; Luo and Lau 2018; Osseiran and Chriscaden 2016; an estimate for impervious surfaces with remote sensing affecting albedo and water run-off, etc.). Some metropolis city parks are even known for their high occurrence of squirrels (New York City Central Park, Vancouver Park, etc. (Gonzales 2005; Kheraj 2012; Munshi-South and Nagy 2014; Slaughter 2019)). Having these small mammals in urban areas and in the human contact zone, threats and dangers are originating for both the animal and the human side, e.g. zoonotic diseases, reservoirs, and those should not be ignored, but currently are.

Supplementary Information The online version contains supplementary material available at https://doi.org/10.1007/978-3-031-23547-4_5.

M. Steiner, F. Huettmann, *Sustainable Squirrel Conservation*, https://doi.org/10.1007/978-3-031-23547-4_5

During the last few centuries, and specifically during the last decades, virtually all global cities have greatly increased in size (Bradley and Altizer 2007; Brockerhoff 1999; Duranton and Turner 2012; see data in the Appendix), as well as their population sizes (Alonso 1973; Bradley and Altizer 2007; Buhaug and Urdal 2013; Kuang et al. 2016), and thereby also their ecological footprints (Ahmed et al. 2020a; Ahmed et al. 2020b; Hubacek et al. 2009; Nathaniel et al. 2019). In order to quantify and illustrate such changes, Fig. 5.1 below illustrates the population gain of the hundred largest cities on the planet from 1950 to 2020. Apart from the human population hotspots (depicted using the 100 largest cities in the world), Fig. 5.1 also includes the global squirrel hotspots (in red), which aim provide a better understanding of how and where human and squirrel population hotspots meet.

With this greatly increased human population size, the demands for infrastructure and housing drastically increase (Linn 1982; Madlener and Sunak 2011; Wang et al. 2019). In order to fulfill these increased demands for housing, cities must grow, either by increasing the density of urban areas, in height, density, or by enlarging the cities' boundaries (Bennett 2000; Ichimura 2003; Linn 1982). This enlargement or expansion happens all on a finite space (Daly 1977; Dietz and O'Neill 2013; Farley 2014), negatively affecting the surrounding landscape, especially wilderness areas (Apostolopoulou and Adams 2019; Bennett 2000). Slums, ghettos, and favelas are on the rise too, and so is squatting (see Davis 2004). Therefore, the surrounding landscape and wilderness areas are compromised, shrinking in size further due to this chaotic virtually unplanned, and highly unsustainable urban expansion. This shrinkage negatively influences the correlated species living within those habitats and often even forced them to inhabit these new suburban areas (Arcari et al. 2020; Lourenço et al. 2017). The remaining greenspaces get 'packed' with species.

However, not only the increased human population and the concluding increased housing demand are forcing species, including squirrels, to live in the Anthropocene, but also increased urban animal feeding is an issue. In urban and suburban areas animal feeding is very popular, it is a part of missing wildlife and outdoor experiences and a wilderness loss

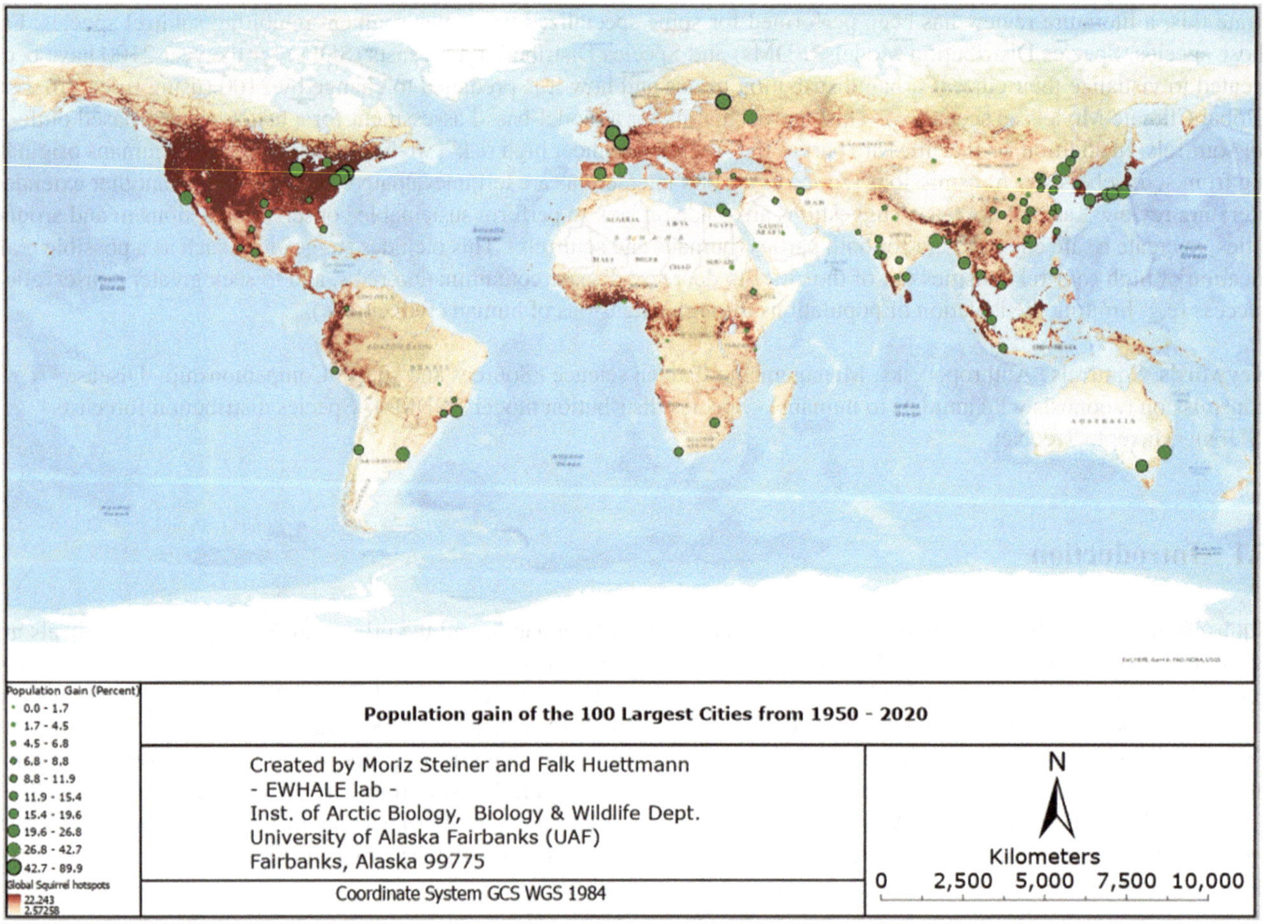

Fig. 5.1 Population gain over the last 70 years of the 100 largest cities on earth (sized green dots) plotted over global squirrel hotpots (red index) The population size data for 1950 and 2020 has been obtained from International Institute for Environment and Development (2020), and the geographical coordinates of the cities (for mapping purposes) have been obtained from SimpleMaps (2021). The data and files of these maps can be found in Appendix 5.2.

compensation activity. This usually results in an unnatural increase in the abundance and occurrence of many wild species in the Anthropocene, including squirrels. The carrying capacity of their ecological niche gets artificially expanded. In order to visualize such an increase in animal feeding sites, Fig. 5.2 have been created to show a generic cultural trend while land gets increasingly urbanized (Fig. 5.3).

Figure 5.2a represents the bird feeder stations in North America from the year 1990, and Fig. 5.2b represents the same data, for the year 2020. By comparing the two figures, a clear increase in the number of feeding stations can be observed within just the last 30 years (from 1990 to 2020). This indicates an increase in food provisioning in urban and suburban landscapes which does not only lure birds in cities but also other animals, often squirrels (see a squirrel feeding station in Fig. 5.3a and bird feeding station in Fig. 5.3b) (Adler 2014; Corrigan 2019).

From this, it can safely be concluded that the situation is going in a direction that is not beneficial for both, humans, and squirrels, while wilderness is lost. This is because the trend shows that human abundance and density increased over time which can be assumed to be continued for the upcoming decades (Bradley and Altizer 2007; Brockerhoff 1999; Duranton and Turner 2012). In addition to this human increase in cities, humans also force and lure squirrels into the cities and thereby, increasing the number of squirrels in urban areas. While cute, squirrels are often also considered predators (Bradley and Marzluff 2003; Callahan 1993; Steiner and Huettmann 2021). As it has been recorded in the past and highlighted recently, zoonosis (e.g. Covid-19, rabies, and bubonic plague) can present major threats to human societies and the global ecosystem overall (Estrada-Peña et al. 2014; Wolfe et al. 2005). Another very recent example is ectoparasites (Franchini et al. 2021). With an increase of both, humans and squirrels in cities, the probability of possible zoonosis transmissions increases likewise. This topic will be discussed in broader detail in Section 5.3.4. However, to provide an overview of how impactful and widespread global zoonosis diseases are, Fig. 5.4 have been created.

Figures 5.4 represent the global zoonosis distribution density for both mammals (a) and rodents (b). These maps aim to illustrate the current situation and severity of the problem which has been caused by humans and their often *laissez-faire* approach towards the living world (unsustainable construction of suburban areas with minimal respect to the wilderness areas and biodiversity hotspots which seem to be "in the way" (Apostolopoulou and Adams 2019; Bennett 2000)). Apart from the transmissible zoonosis diseases depicted in the Figures above, also the population gain of the 100 largest cities of the world has been included to better link the zoonotic hotspots to the human population hotspots.

Another disputed topic concerns the situation of squirrels in zoos. To better understand this, Textbox 5.1. has been created.

Textbox 5.1: Squirrels in Zoos and the Captive Breeding of Squirrels

Do zoos offer heroic solutions for future squirrels, the new *status quo* habitat, a model of the past, or the inevitable last zombie refuge on this planet?

At first look, captive breeding appears as a good choice for saving species. Entire nations bought into such concepts, with China being one of the recognized world leaders. Zoos offer themselves for such efforts, on a national and global level. Breeding programs are often globally networked allowing for the best options. Further, zoos are heavily frequented by children, their parents, and animal lovers, and are often seen as tourist attractions, fundraisers, and serve as a national pride of the capital (e.g. Beijing Zoo, San Diego Zoo, Bronx Zoo, Berlin Zoological Garden/ Berliner Zoos, Singapore Zoo, etc.). However, it faces the question that remains still unanswered after centuries of running zoos: what really is the role of zoos while wilderness – as the only resource reservoir for animals - becomes gradually extinct? No zoo can harbor the world's wilderness.

Zoos typically offer access to animals that are otherwise difficult to experience first-hand. They are often the homes of special, very rare, or/ and charismatic animal species which are hard to detect in nature, inaccessible to most people, or heavily reduced in abundance. Apart from these charismatic and (as generally interpreted) beautiful animal species, zoos also house some few remaining individuals of almost extinct animal species (a long list of these exists such as Arabian Oryx (*Oryx leucoryx*), California Condor (*Gymnogyps californianus*), Corroboree Frog (*Pseudophryne spp.*), Bongo (*Tragelaphus eurycerus*), Regent Honeyeater (*Anthochaera phrygia*), Golden Lion Tamarin (*Leontopithecus rosalia*), Bellinger River Snapping Turtle (*Myuchelys georgesi*), Amur Leopard (*Panthera pardus orientalis*), etc.). This "last step" of some species' existence – the living zombies - is even represented by a specifically assigned IUCN conservation status so-called "Extinct in the wild". It can be found for several species (e.g. Banded Allotoca (*Allotoca goslinei*), Franklin Tree (*Franklinia alatamaha*), Yellow Fatu (*Abutilon pitcairnense*) (see www.iucnredlist.org)). It gives an impression that the species is still well and under management to be dealt with, whereas it is not, and the genetic pool has been dramatically wiped out. The species that end up in the zoos in a fenced box of a certain size, representing their "last step" on earth having their natural habitat and natural population commonly perceived as being destroyed or vanished from this planet. Arguably, this is all caused and done by humans, specifically during the last four decades.

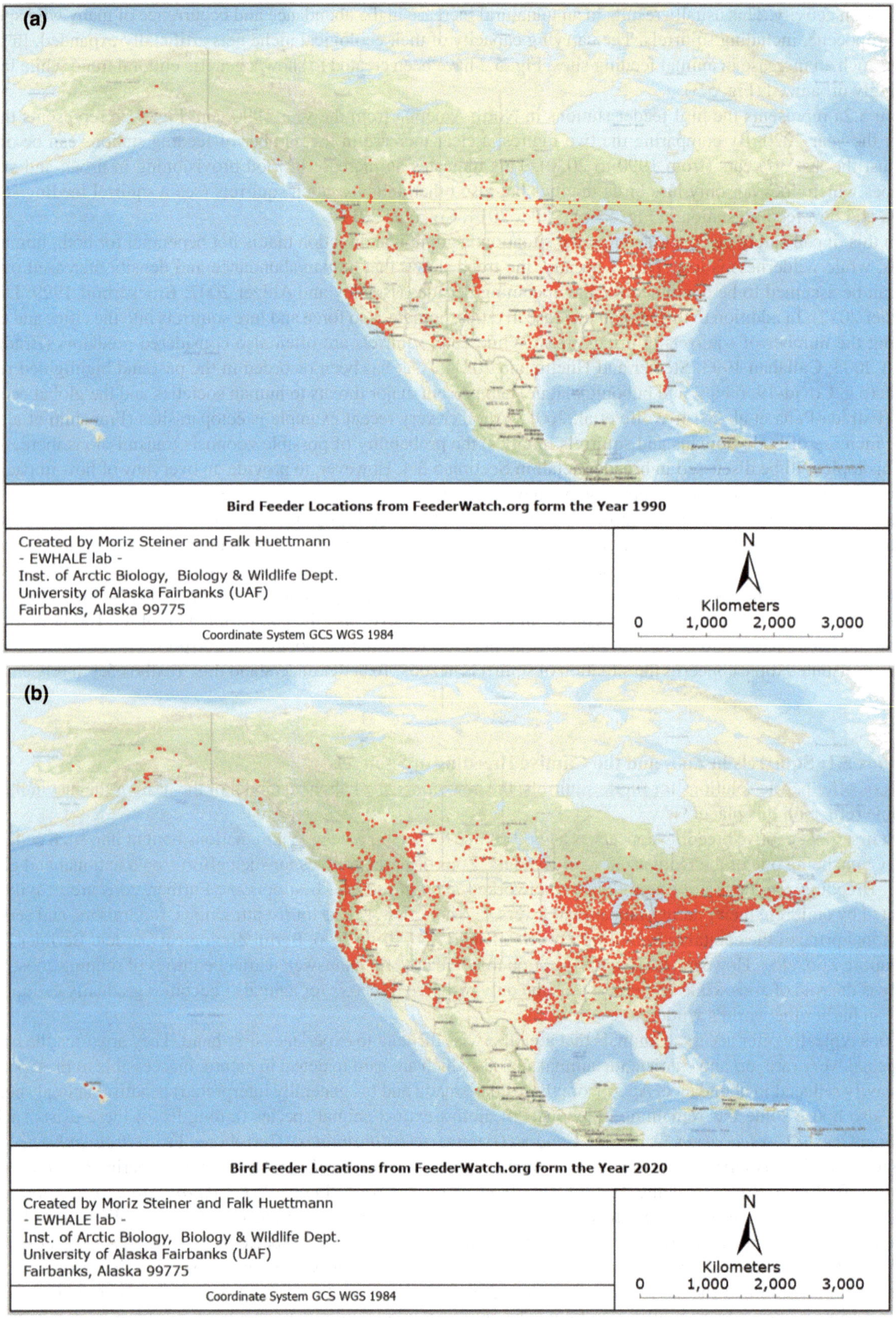

Fig. 5.2 Bird feeder stations distribution throughout North America for the year (**a**) 1990, (**b**) 2020. (Public Data source: Cornell Lab (https://feederwatch.org/))

Fig. 5.3 Squirrel feeding on a squirrel feeder (**a**) and bird feeder (**b**) in an urban setting. (Source: Pixelbay/Unsplash)

In our study focusing on squirrels, the seemingly obvious follow-up question is: Are zoos also going to be this "final step", a living graveyard, for many of the endangered squirrel species? Will the world's human population be able to see squirrels in the future only in zoos, if at all? And should one invest in zoos as the last arch for squirrels and beyond, while wilderness protection efforts are written off and de-funded?

In contrast, wilderness actually remains the world's heritage, legacy, and resource of any species we know of. Wilderness has no alternative. Without wilderness, mankind cannot live. Any loss of wilderness means a loss of life quality. We believe that species should not really be kept in captivity as long as their natural habitat is intact and allows the species to thrive. This should be goal number #1 of any conservation management and policy; it is to be prioritized. Reality shows otherwise though. Once the natural environment has either been widely destroyed, compromised, or vanished from this planet, an end stage is reached. If there is will and a good option, indeed the species shall be recovered from the field. Such zoo species are now essentially living in a guarded museum, with fences and locks, and present essentially the 'living zombies' created by the Anthropocene. This should be fully acknowledged, and also that this is a sustainability sink, that is promoted at the cost of other finite resources and species. Once these species have been recovered, if ever, science-based housing systems should be installed in place to guarantee the highest survival rate possible.

Furthermore, experts should be consulted to develop conservation plans with the final aim to increase the population numbers and restore the original environment to the extent to be able to reintroduce these species into the wild, as per the carrying capacity available in the future. The absurdity of those concepts can easily be seen already with Great Panda (where many individuals are worse off in protected areas than outside) and with the epic polar bear 'Kurt' (which died in Berlin, far away from his real habitat: Arctic sea ice), while Germany remains a key polluter for man-made climate change melting Arctic summer ice and this destroying a basis for 'Kurt' to live anyways. Kurt's death has not changed that reality.

Retrieving wild animals from their natural environment should only be carried out in the worst-case scenario when all other conservation measures failed or have been initiated too late in the conservation process. Unfortunately, this can be observed many times in ecological conservation where authorities fail to act on time to truly conserve habitats and the species endemic to them. Not being proactive and risk-aware clearly puts doubt on the entire undertaking of species recovery and zoos.

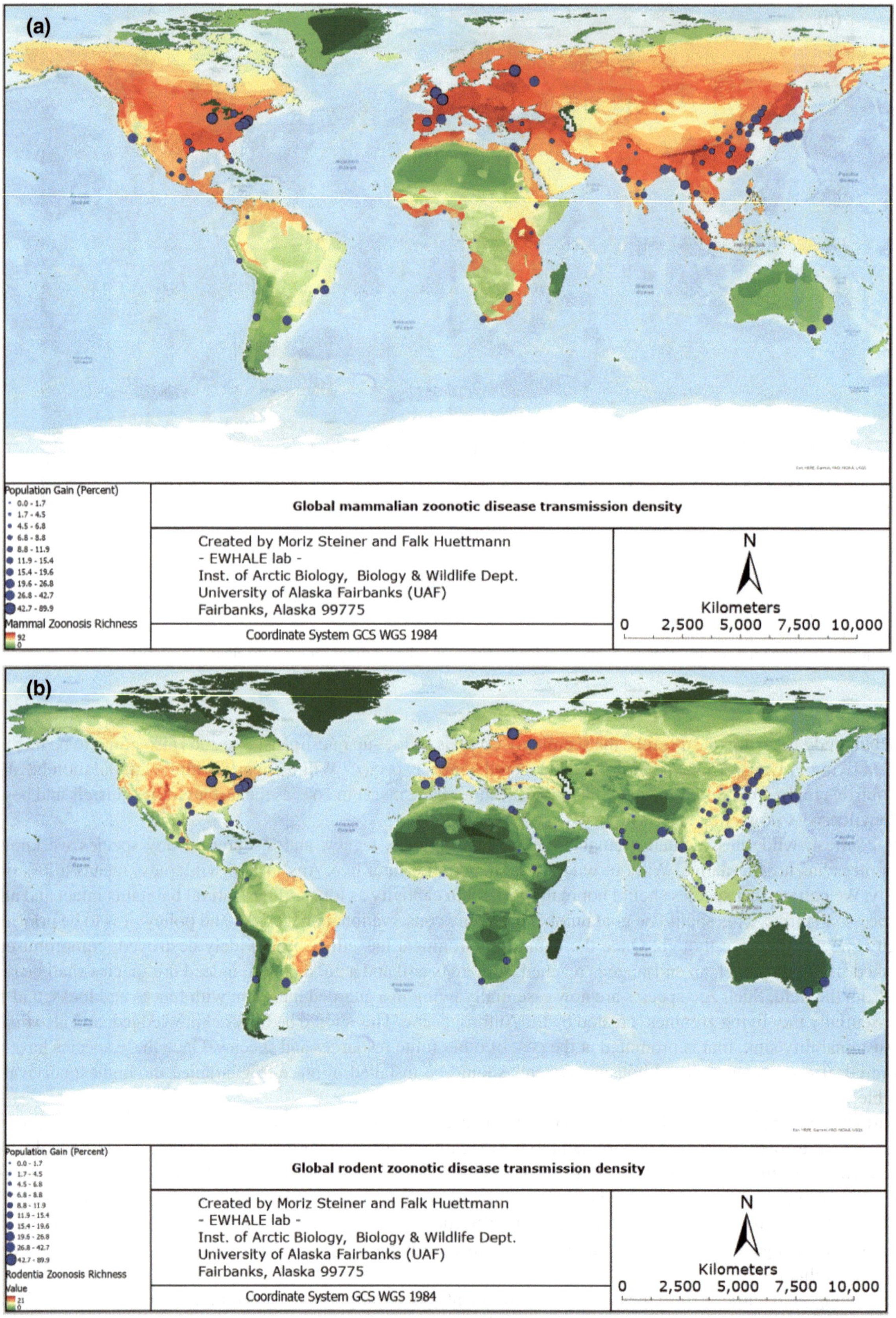

Fig. 5.4 Density of global transmissible zoonosis diseases density of (**a**) mammals, (**b**) rodents. (This data has been obtained in GIS format from the authors of Han et al. (2016))

Therefore, for any further conservation work, we suggest interacting as early as possible right when a conservation concern originates. Proactive actions remain the policy of the day, as mandated by the U.N. Thereby, it can help to avoid "being too late". If one simply proceeds in this manner, cases of this upper-mentioned worst-case scenario can be widely avoided. Other than these rare species that outside of zoos are no longer able to survive, we recommend not housing many species in zoos as a proactive and last-resort effort. It is very expensive, takes resources from ex-situ projects, and remains a sustainability sink.

Zoos are described as places where colonialism, constraint, abuse, and suffering on such mindsets can be found (MacKinnon 1977). Zoos clearly run such schemes with a business model for profit; many zoos are private or semi-private enterprises. Zoos need life supplies to sustain their animal populations. They are a place where the public can observe wildlife and have a personal, albeit highly biased, interaction with "wildlife". This can enforce the protection instinct and generate funding for the conservation of the species housed in these zoos. The same is valid for zoo entrance tickets/ fees, and certainly for the all-abundant souvenir shops (where often plastic toys "*Made in China*" style are sold for profit by a vendor and private contractor). At a minimum, these financial resources shall be used effectively to cover the zoo's expenses and fund conservation initiatives for the zoo's species in the wild and their rewilding projects, if possible and necessary. But those details are to be assessed for administrative efficiency and progress.

However, considering that just the last 40 years of the Anthropocene have probably been the most damaging for the world's wilderness and wildlife (Bengston and Dockry 2014; Dawson 2016; Johnson et al. 2017; etc.), the classic zoo argument is beyond dubious. Zoos simply do NOT halt wilderness loss, nor do they make wildlife conservation much better or provide more habitat than before for most ecosystems and their animals. Wildlife does not live well in boxed-up fence zones with signs on the outside. It can easily be seen in the rainforests of this world or the coral reefs, seagrass beds, mangroves, the cold climate zones, or the atmosphere. Albeit exciting like a circus perhaps, zoos remain an endpoint for animals of the wild and do not really stop it. Zoos are a sustainability sink, a last step before havoc rules the species, and humans eventually.

In regards to the current situation of squirrels in zoos, it widely follows the upper-mentioned trends. Big, colorful, and charismatic species are mostly held in zoos. This is likely because these species are most appreciated by visitors, as such species are often the medium of attraction in zoos; the money-makers. These zoo-prominent species include Prevost's squirrel or Asian tri-colored squirrel (*Callosciurus prevostii*) (Smithsonian's National Zoo & Conservation Biology Institute (2022)), and Javan Black giant squirrels (*Ratufa bicolor bicolor*) (Zoo Chat 2011), Sri Lankan or grizzled giant squirrels (*Ratufa macroura*), Finlayson's squirrels (*Callosciurus finlaysoni*), Grey-bellied squirrels (*Callosciurus caniceps*), Pallas's or red-bellied squirrels (*Callosciurus erythraeus*), Plaintain or oriental squirrels (e.g. *Callosciurus notatus*), Perny's long-nosed ground squirrels (*Dremomys pernyi*), Northern or five-striped palm squirrels (*Funambulus pennantii*), Himalayan striped squirrels (*Tamiops mcclellandii*), and Swinhoe's striped squirrels (*Tamiops swinhoei*). These are among the species that are observed the most often in zoos. A detailed list of which zoos contain which species can be found on the following website https://www.zootierliste.de/en/. Overall, the trend remains the same as with many other animals in zoos: data deficient species are marginalized, colorful and charismatic species are displayed, and little contribution to science-based conservation projects can be observed overall while wilderness is lost in parallel; A classic case can be found with the absurd expansion of palm oil plantations on the costs of virgin pristine rainforest and its species.

In literature, squirrels in zoos are also often connected with disease issues and their spreading (see Allendorf et al. 2021; Beest et al. 2017; Cadar et al. 2021; Nelder et al. 2009; Tappe et al. 2019; etc.). Zoos actually seem to be a source of diseases on squirrels and on the ground where transmissions occur frequently. This highlights the currently mismanaged concept of ecologically-conserving zoos, and it calls for a change if species shall really benefit from zoos. Here we point to this problem and acknowledge it, and provide a call for agreement, improvement, and progress.

So far, we have discussed the threats originating from squirrels affecting humans and how 'modern' humans accelerate and enforce this process. However, as the modern world develops, mammals and especially small mammals can and should also have their space in urban and suburban areas. In order to make that happen, it is of crucial importance to consider those species and sustainably plan and manage squirrels in parks, green spaces, and even the gardens of cities. In order to provide an overview of the current situation and numbers, here we created a general overview (Fig. 5.5), which illustrates 41 global metropolis cities and their proportion of green spaces in relation to the entire city. We additionally included the global squirrel hotspots (in red) to better link the these green spaces with actual squirrel hotspots (areas in need of green spaces). We can observe that the percentage of green spaces in North America is very low compared to its needs for the extreme squirrel

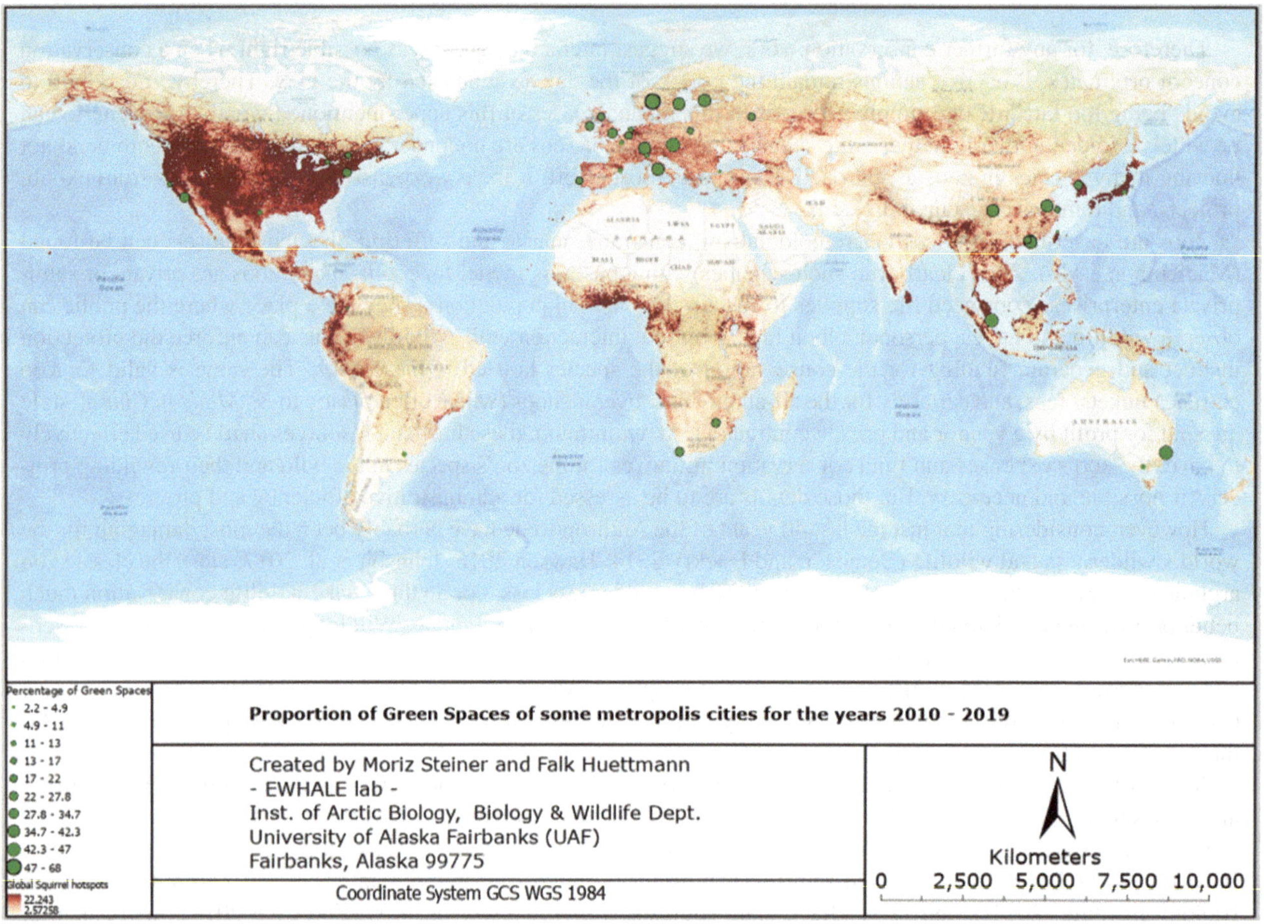

Fig. 5.5 Proportion of green spaces compared to the entire city. (Data source: World Cities Culture Forum (2021) (for the green space proportions), and SimpleMaps (2021) again for geographical coordinates of the included cities (see reproduced in Appendices 5.1 and 5.2 respectively) overlaid with global squirrel hotspots (red index)

hotspots there. In Europe, we can observe that the percentage of green spaces are larger and more coherent with the squirrels hotspots there. In Asia, we can observe a sort of in-between situation of North America and Europe.

Additionally, in this study we discuss some major threats originating from both humans and squirrels, affecting each other. This aims to present the status quo and initiate a wider discussion on these topics. It is meant to put these kinds of topics on the agendas of the modern city and greenspace planning, and management, as well as species and conservation management (see urban wildlife management chapter in Fryxell et al. 2014). To gain some base knowledge about the squirrels' distribution and occurrence in cities, a literature review has been performed. Unfortunately, the number of literature records is very limited for non-western and 'non-prominent' species (unlike for Eurasian Red Squirrel (*Sciurus vulgaris*), Eastern Gray Squirrel (*Sciurus carolinensis*), Eastern Fox Squirrel (*Sciurus niger*), and North American Red Squirrel (*Tamiasciurus hudsonicus*)). Therefore, only some specific squirrel species, which have been identified to inhabit primarily urban areas, will be analyzed in this study and discussed thoroughly. All of this aims to make society, and therefore also scientists and policymakers, more aware of the current issues. It tries to name problems and shortcomings which urge to be discussed more broadly and addressed properly. Additionally, some global discussion points and problems for and with squirrels in cities are being presented and discussed. Lastly, some conservation suggestions are being included with which conservation and its planning in urban and suburban areas can be improved. It must be noted that such an assessment and analysis can be applied not only to squirrels but also to many other species for a wider and more holistic assessment. Therefore, we aim that the assessment methods presented here can be utilized as a template for other species with urban interferences.

5.2 Methods

5.2.1 Literature Review of Squirrel Occurrence in Cities

A first literature review has been performed in order to assess the literature records of the squirrels' distribution and occurrence in the cities of this planet. This literature review has been performed using primarily Google and Google Scholar entries to scout for literature records. To a small extent also personal work and online libraries have been used to complete the literature reference list. This literature review focuses on the five urban environment-inhabiting squirrel species selected for this study (see next paragraph). The literature review has been limited to these five species as they have been identified to depend quite heavily on the urban environment as their 'natural' habitat. These five species can also bee seen as reference species for any other urban-living mammal species, where te findings from this literature review widely apply.

5.2.2 Species Distribution Models (SDMs) and Species Distribution Forecasts (SDFs) for some Urban Environment-Inhabiting Squirrel Species

Study species

In Chap. 3 (Steiner and Huettmann 2023), it has been identified for each global squirrel species whether they occur in one, or more region(s) of/under high risk. The results of this identification have been summarized in Table 3.3 in Chap. 3. Based on this table and the SDMs created there, six species have been chosen to be analyzed here in greater detail due to their predominant distribution in urban areas. These six species are Prevost's Squirrel (*Callosciurus prevostii*) (predominately occurring in Mainland Malaysia, Borneo, and Sumatra (Indonesia)), Southern Flying Squirrel (*Glaucomys volans*) (predominately occurring in Eastern North America, Mexico, Guatemala, and Honduras), Three-striped Ground Squirrel (*Lariscus insignis*) (predominately occurring in Mainland Malaysia, Borneo, Sumatra & Java (Indonesia)), Merriam's Chipmunk (*Neotamias merriami*) (predominately occurring in California (USA)), Palmer's Chipmunk (*Neotamias palmeri*) (predominately occurring near Las Vegas (Nevada – USA)), and Speckled Ground Squirrel (*Spermophilus suslicus*) (predominately occurring in Ukraine, Belarus, and Eastern Russia). The distribution information for these species has been retrieved from www.iucnredlist.org. The last species of this short list (*S. suslicus*) is only marginally discussed in this study; however, all the models have likewise been performed for it and have been presented for the overall completion in Appendix 5.3. An overview of those species, their life phylogeny, reported distribution, etc. can be found in Table 5.1.

Modeling process

For these six species, SDMs and SDFs have been created. This means that for each species five different models have been performed. The first model (a) has already been created in Chap. 3 and is presented with the additional feature of polygon boundaries, which will be discussed further in this section. The second model (b) represents the current (the year 2000) distribution of the species at stake based on 7 BioClim, and one elevation predictor (source: Worldclim (2021)). The third model (c) represents the future (the year 2100) distribution based on the same predictors as used for Figures b, using the MIROC Global Climate Model (GCM) as a future climate scenario (approximate decrease of 1 °C – see details on MIROC GCM in Chap. 12, and Tatebe et al. (2019)). The fourth model (d) represents the future (the year 2100) distribution based on the same predictors as used for Figures b, and c using the MRI GCM (an approximate increase of 2 °C – see Chap. 12, and Yukimoto et al. (2019)). The fifth model (e) represents the future (the year 2100) distribution based on the same predictors as used for Figures b, c, and d using the IPSL GCM (an approximate increase of 3 °C – see Chap. 12, and Boucher et al. (2020)).

Polygon boundaries of the high-occurrence hotspot identification

In order to be able to compare these newly created SDMs and SDFs with the originally created SDM (using 132 predictors), polygon boundaries have been created. This means that the original SDM raster file has been polygonized (= converted from a raster into a polygon) in QGIS. For this vector/feature file, the symbology has been adjusted, which resulted in the bright-green outline of the original core distribution. This newly created file has been overlaid over the SDMs and SDFs which results in the presented illustrations.

Modeling algorithm and software

These SDMs and SDFs have been modeled primarily with the software Maxent (Maxent 2022). The results are presented in Section 5.3.2 below. To include another, more holistic approach and modeling prediction for squirrels in the distribution forecasting, the SDFs have also been created with another software (TreeNet 2022). These models have been presented in

Table 5.1 Life history and distribution overview of 5 urban environment-inhabiting squirrel species in the focus of this study

Common species name	Scientific species name	Description	Countries and regions of endemic distribution	IUCN Conservation status class	IUCN population trend	Most crucial threats	References
Prevost's squirrel	*Callosciurus prevostii*	Size HB: 237–241 mm, T: 234 mm, mass: 353–403 g, colorations: Black-dark chest, red-ish belly. 17 subspecies. Exceedingly variable species regarding colorations, habitat, and consumed feed.	Peninsula Thailand, Malaysia, Sumatra (Indonesia), Borneo	Least concern	Decreasing	Very small wild habitat left due to agricultural deforestation, and this species is severely hunted for pets.	Cassola 2016a; Thorington et al., 2012 page 148–149; Wilson and Reeder 2005 page 779
Southern flying squirrel	*Glaucomys volans*	Size HB: 131–134 mm, T: 100–103 mm, mass: 53–70 g, colorations: Grey-ish, white. 11 subspecies. Primarily nocturnal and feeding on nuts, seeds, fruit, and fungi. Flying squirrel prefer oak-hickory forests, and occasionally mixed forests consisting of deciduous and coniferous tree species.	Eastern USA, Southeast Canada, Mexico to Honduras	Least concern	Stable	Local and small-scale habitat and cavity-bearing loss.	Cassola 2016b; Thorington et al. 2012 page 95–97; Wilson and Reeder 2005 page 768
Three-striped ground squirrel	*Lariscus insignis*	Size HB: 182–194 mm, T: 100–109 mm, mass: 175-182 g, colorations: Dark brown with three black stripes, ventrally white-pale. 5 subspecies. Diurnal predominately found on the ground and on/around fallen trees.	Malaysia, Sumatra and Java (Indonesia), Borneo	Least concern	Decreasing	None have been identified	Tizard 2016; Thorington et al. 2012 page 169–170; Wilson and Reeder 2005 page 783
Merriam's chipmunk	*Neotamias merriami*	Size HB: 131–135 mm, T: 106–116 mm, mass: 68–71 g, coloration: Brown-white-greyish. 3 subspecies. Diurnal and "*usually occupies shrubby chaparral vegetation often associated with pine (Pinus) and oak (Quercus)*".	California (USA), very northern Baja California (Mexico)	Least concern	Stable	None have been identified	Álvarez-Castañeda et al. 2016; Thorington et al. 2012 page 324–325; Wilson and Reeder 2005 page 815
Palmer's chipmunk	*Neotamias palmeri*	Size HB: 125–125 mm, T: 80–98 mm, mass: 52–60 g, coloration: Brown-grey. No subspecies. Diurnal living in high elevations (2100–3600 m asl) with a granivorous diet (including coniferous seeds and fruits).	Spring Mountains near Las Vegas (Nevada, USA)	Endangered	Decreasing	Residential and commercial development, human habitat intrusion and disturbance, natural system modification, invasive and other problematic species (genes & diseases)	Lowrey 2016; Thorington et al. 2012 page 330–331; Wilson and Reeder 2005 page 815

Chap. 12 (Steiner and Huettmann 2023, see details there). For this chapter, these figures have been slightly adjusted by adding urban focus regions. These regions are highlighted with a surrounding blue rectangle that highlights the major changes in the distribution range of all global squirrels. In addition, these focus regions are of high interest as they contain several metropolis cities on this planet, and thus, distribution changes in high-human density urban areas can be assessed with the trend of the global squirrel population.

5.2.3 *Identification of the Risks for Squirrels Living in Urban Environments (as Part of the Regions of/under High Risk)*

In Chap. 3, cities have been identified as regions of high risk. In this study, we performed a second literature review on all major risks for squirrels originating from cities. As this is the only "region of high risk" which is currently not particularly "under" risk, as cities continuously enlarge and grow (in size and human density (Bradley and Altizer 2007; Hummel 2020)), here are no threats to cities but threats originating from cities reviewed and presented from the literature and then extrapolated for a wider application.

5.3 Results

5.3.1 *Literature Review of the Global Squirrel Occurrence in Cities*

Due to the lack of research papers published on the distribution of the 5 species which are being discussed in this study, the main literature resources that have been used for this literature review were the IUCN Red List, Thorington et al. (2012), and Wilson and Reeder (2005). The findings of the literature review have been summarized in Table 5.1.

The overview in Table 5.1 and thus, also the literature review, focused primarily on the recorded distribution of the squirrels. For completion, also general information on the life history, conservation status, population trend, and primary threats to the species, have been included.

5.3.2 *SDMs and SDFs of Five Urban Environment-Inhabiting Squirrel Species*

We were able to create for each of the five species (*Callosciurus prevostii, Glaucomys volans, Lariscus insignis, Neotamias merriami,* and *Neotamias palmeri*), SDMs and SDFs (Figs. 5.6, 5.7, 5.8, 5.9, 5.10).

According to Fig. 5.6a, the species Prevost's Squirrel (*Callosciurus prevostii*) can primarily be found in Southeast Asia and Sri Lanka. Figure 5.6b shows a distribution in the same regions, however with a much wider hotspot distribution. This wider hotspot distribution is coherent with all the SDFs of this species (Fig. 5.6c, d, e). The greatest differences between these SDFs can be observed in Central Borneo and Sumatra. This species is considered Least Concerned with a decreasing population trend according to the IUCN Red List and Cassola (2016a).

According to Fig. 5.7a, the Southern Flying Squirrel (*Glaucomys volans*) can primarily be found in the eastern part of the USA and Ontario (Canada). This distribution remains the same also for the other four figures (Figs. 5.7b, c, d, e). However, the distribution boundaries seem less defined. Additionally, a small hotspot can be observed in the South Dakotan Black Hills National Forest (in proximity to several indigenous people reserves). According to IUCN Red List and Cassola (2016b), the species is considered of Least Concern with a stable population trend.

Figure 5.8a indicates that the Three-striped Ground Squirrel (*Lariscus insignis*) occurs primarily in Java and Sumatra (Indonesia), Northwestern Borneo, Peninsular Malaysia, and Sri Lanka. The trend of the distribution observed for the other Figs (5.8b, c, d, e) is overall similar, with a wider hotspot distribution. The main hotspots seem to be Malaysia, Borneo (and small islands surrounding it), Sri Lanka, the Southwestern coast of India, broader Coastal New Guinea, Indonesia, and the Philippines. This species is considered Least concerned with a decreasing population trend according to the IUCN Red List and Tizard (2016).

Figure 5.9a illustrates that the modeled primary distribution of Merriam's Chipmunk (*Neotamias merriami*) lies in California (USA). The other models (Figs. 5.9b, c, d, e), which have been created with 7 BioClim predictors and one

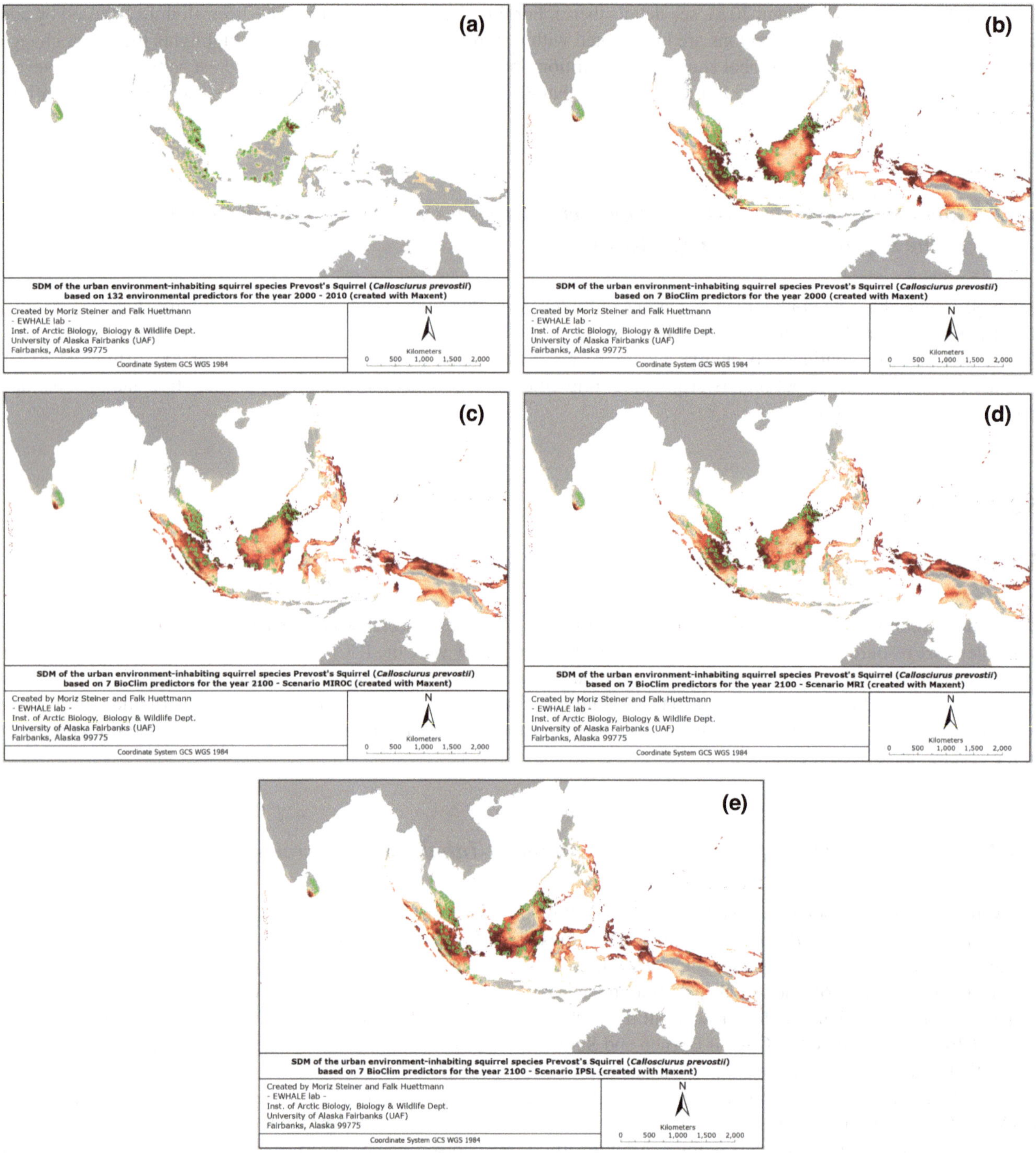

Fig. 5.6 SDM of the urban environment-inhabiting squirrel species Prevost's Squirrel (*Callosciurus prevostii*) based on (**a**) 132 environmental predictors for the year 2000–2010 (**b**) 7 BioClim predictors for the year 2000 (**c**) 7 BioClim predictors for the year 2100 – Scenario MIROC – (**d**) 7 BioClim predictors for the year 2100 – Scenario MRI – (**e**) 7 BioClim predictors for the year 2100 – Scenario IPSL

elevation predictor instead of 132 environmental predictors, show a similar distribution with an additional hotspot in central Mexico. From these SDFs, the last figure (Fig. 5.9e) with the IPSL GCM shows the smallest increase in the distribution range. This species is considered as Least Concerned with a stable population trend according to the IUCN Red List and Álvarez-Castañeda et al. (2016).

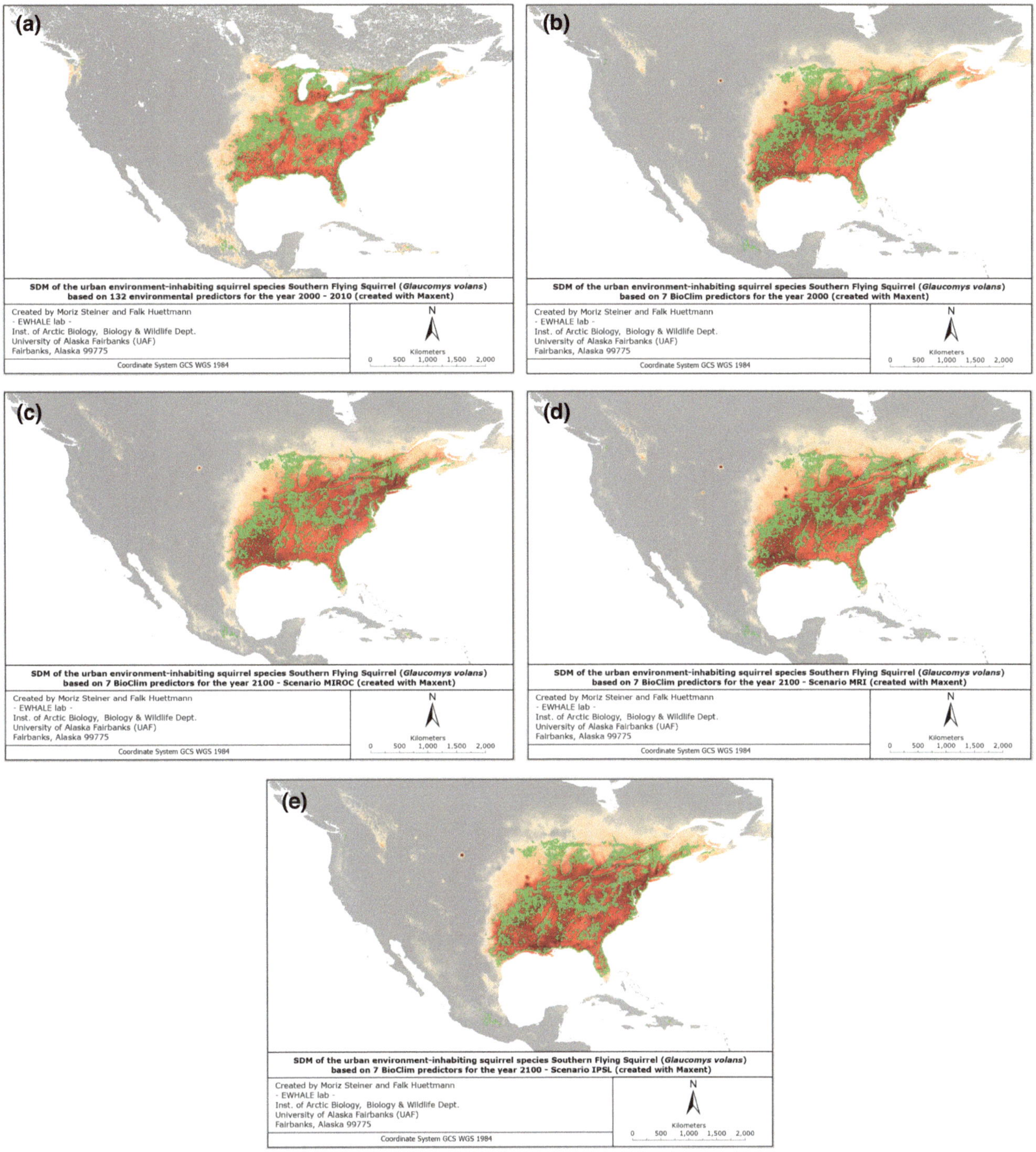

Fig. 5.7 SDM of the urban environment-inhabiting squirrel species Southern Flying Squirrel (*Glaucomys volans*) based on (**a**) 132 environmental predictors for the year 2000–2010 (**b**) 7 BioClim predictors for the year 2000 (**c**) 7 BioClim predictors for the year 2100 – Scenario MIROC – (**d**) 7 BioClim predictors for the year 2100 – Scenario MRI – (**e**) 7 BioClim predictors for the year 2100 – Scenario IPSL

According to Fig. 5.10a, Palmer's Chipmunk (*Neotamias palmeri*) solely occurs in the suburban areas of Las Vegas (Nevada, USA). The other figures (Figs. 5.10b, c, d, e) illustrate a much larger distribution, ranging from Nevada to Oregon, Idaho, Utah, Wyoming, Colorado, and Arizona (USA). All the latter figures show the same distribution ranges, however, with different intensities (MIROC with the highest, and MRI & IPSL with lower hotspot intensity). According to the IUCN Red List and Lowrey (2016), this species is considered Endangered with a decreasing population trend.

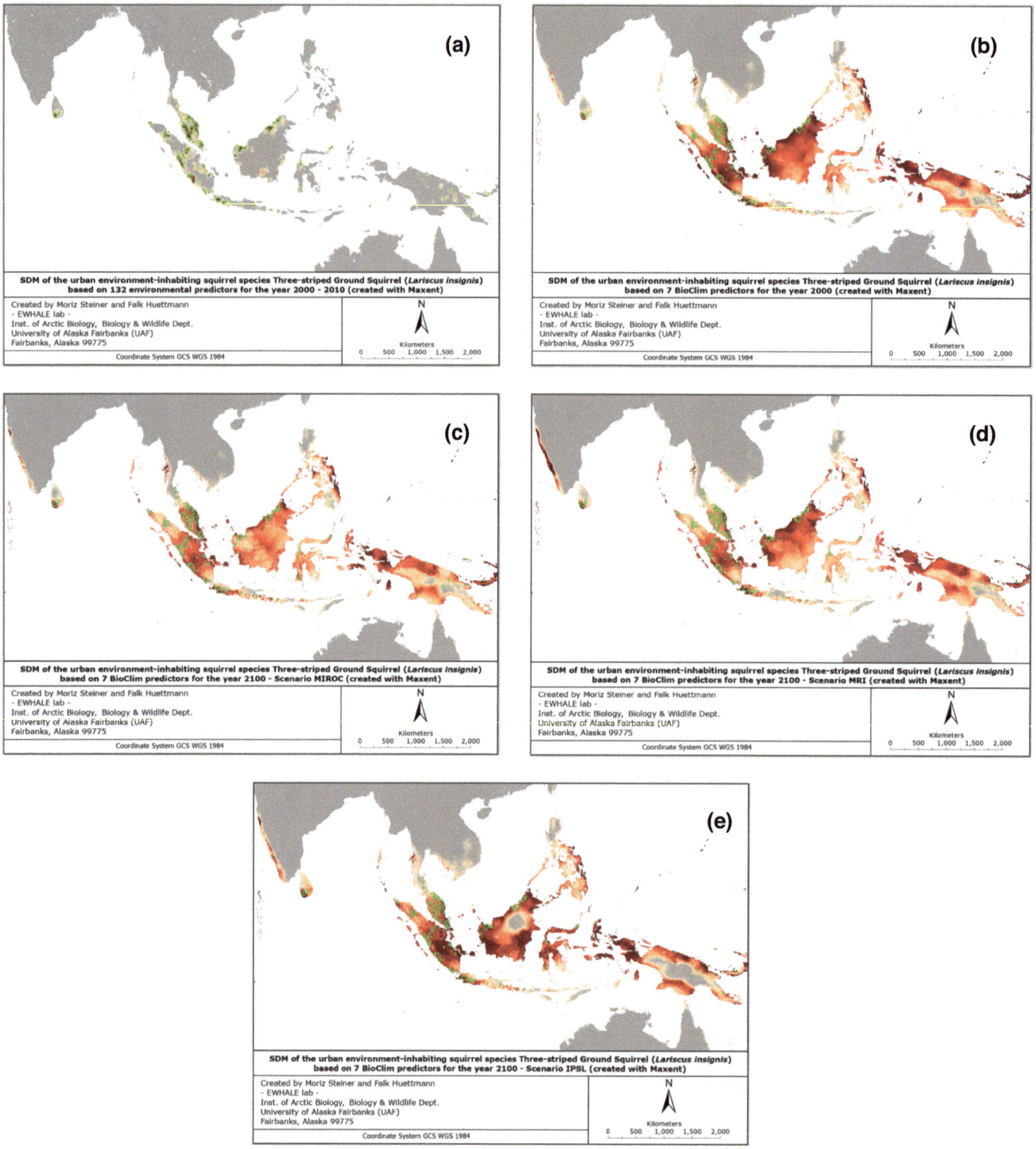

Fig. 5.8 SDM of the urban environment-inhabiting squirrel species Three-striped Ground Squirrel (*Lariscus insignis*) based on (**a**) 132 environmental predictors for the year 2000–2010 (**b**) 7 BioClim predictors for the year 2000 (**c**) 7 BioClim predictors for the year 2100 – Scenario MIROC – (**d**) 7 BioClim predictors for the year 2100 – Scenario MRI – (**e**) 7 BioClim predictors for the year 2100 – Scenario IPSL

5.3.3 *Global Species Distribution Forecast for all Squirrel Species Using the Modeling Software TreeNet*

Compared to the models created with Maxent (which are species-specific), the SDFs created with TreeNet provide a much broader, holistic approach to all global squirrel species.

Figure 5.11a illustrates the global squirrel distribution created with TreeNet based on 132 environmental predictors. The other figures (Figs. 5.11b, c, d) have also been created with TreeNet but with 7 BioClim and one elevation layer. Therefore,

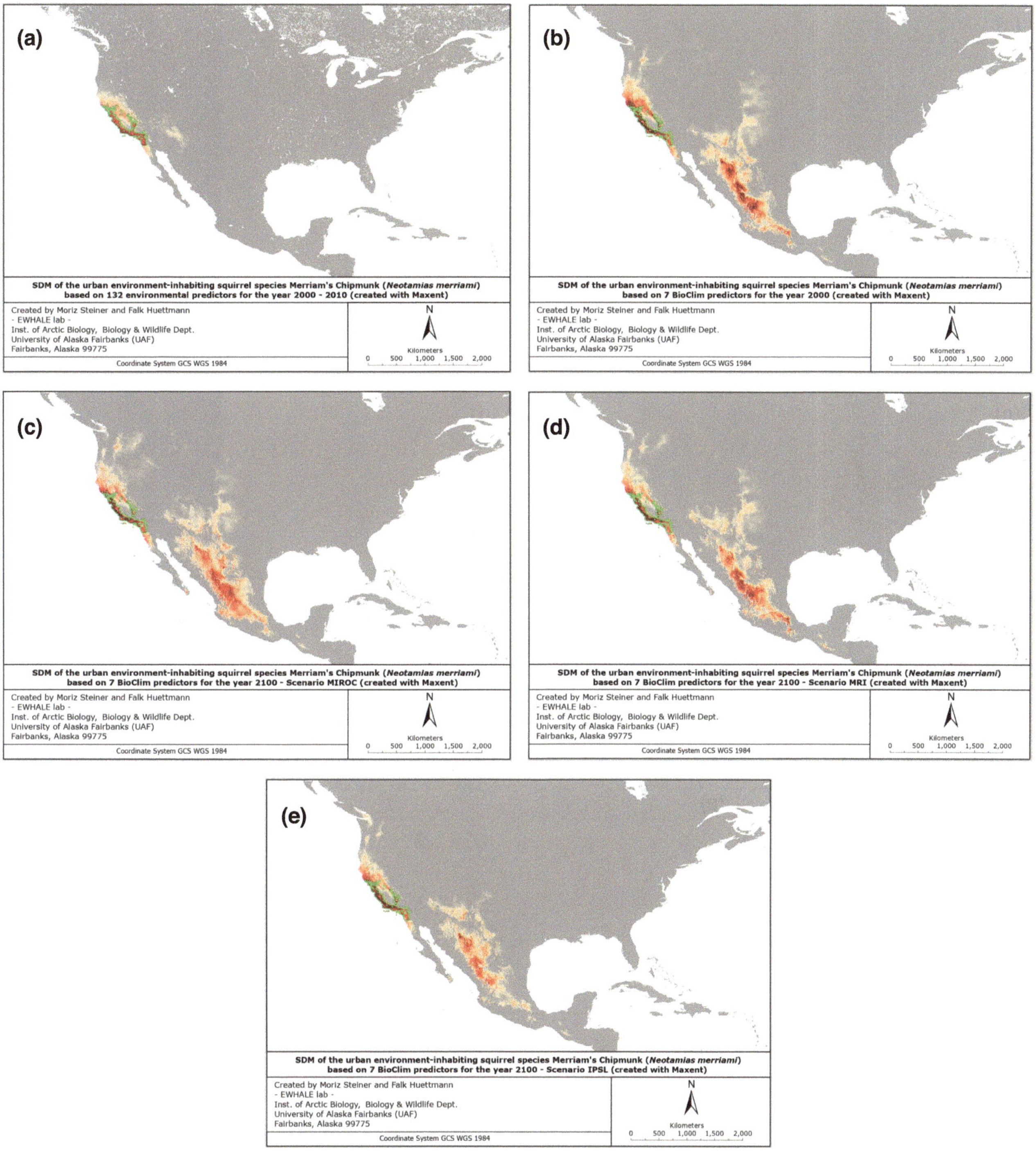

Fig. 5.9 SDM of the urban environment-inhabiting squirrel species Merriam's Chipmunk (*Neotamias merriami*) based on (**a**) 132 environmental predictors for the year 2000–2010 (**b**) 7 BioClim predictors for the year 2000 (**c**) 7 BioClim predictors for the year 2100 – Scenario MIROC – (**d**) 7 BioClim predictors for the year 2100 – Scenario MRI – (**e**) 7 BioClim predictors for the year 2100 – Scenario IPSL

a clear decrease hotspot intensity can be observed when one compares Fig. 5.11a with Figs. 5.11b, c, d. In Fig. 5.11b (GCM scenario MIROC), the focus region has been identified as Northern East America. This is because when compared with the other TreeNet figures, and specifically the base figure (Fig. 5.11a), this region loses significant ranges for squirrels, suggesting a clear shift southward. With this shift, major populations appear to move southwards, and thereby increasingly densify in the human-population-rich regions (Lower North American East coast). Figure 5.11c (GCM scenario MRI), the indicated focus region is Asia, more specifically China. This is again because many of the world's largest cities can be found in China,

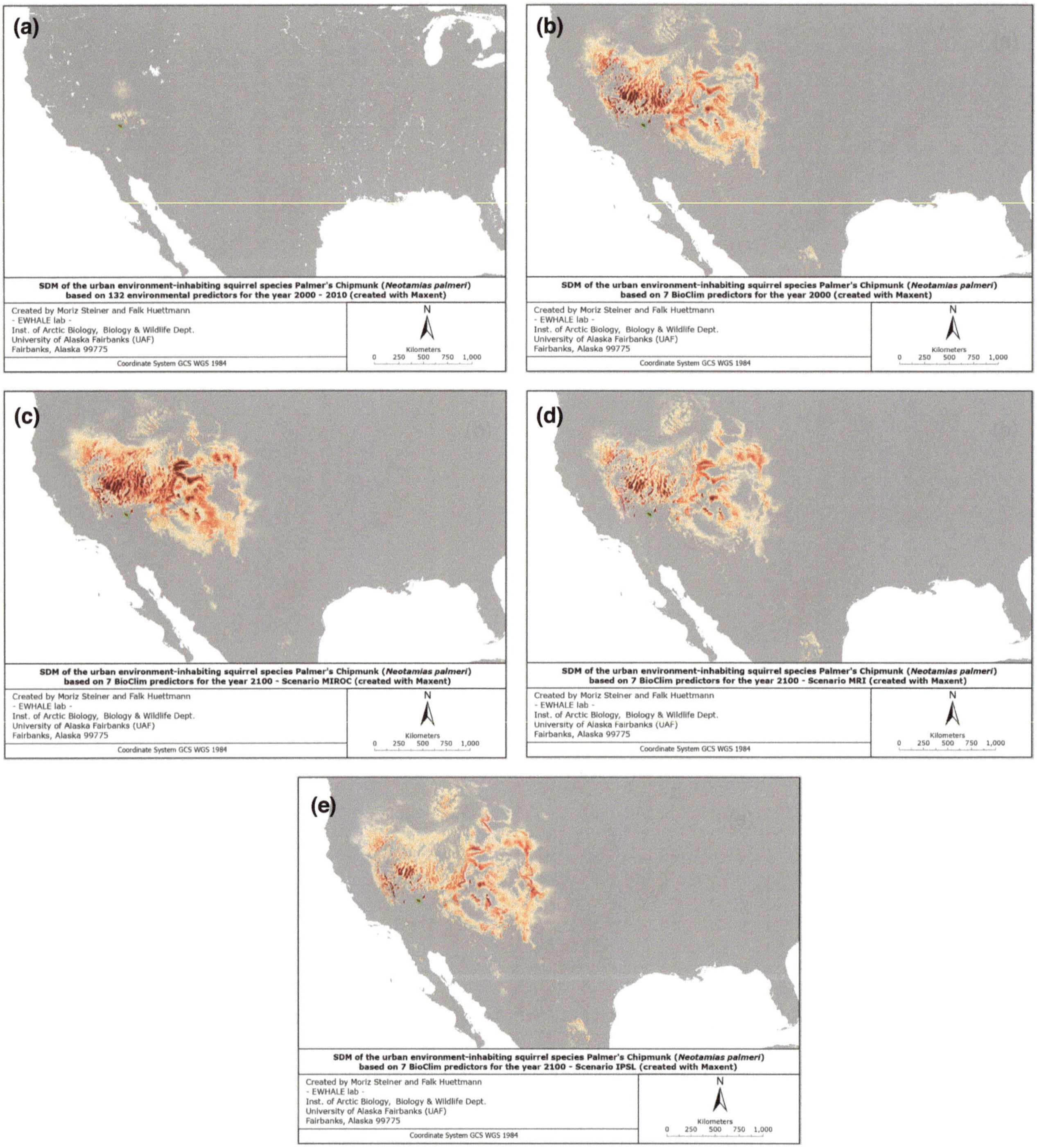

Fig. 5.10 SDM of the urban environment-inhabiting squirrel species Palmer's Chipmunk (*Neotamias palmeri*) based on (**a**) 132 environmental predictors for the year 2000–2010 (**b**) 7 BioClim predictors for the year 2000 (**c**) 7 BioClim predictors for the year 2100 – Scenario MIROC – (**d**) 7 BioClim predictors for the year 2100 – Scenario MRI – (**e**) 7 BioClim predictors for the year 2100 – Scenario IPSL

next to pretty high squirrel biodiversity. This indicates that when business-as-usual is continued further, the squirrel biodiversity in China is predicted to decline quite significantly. Here it must be noted that the TreeNet model only incorporates climate predictors, whereas highly increased urbanization, habitat fragmentation, land-use changes, etc. are not included in this predictor set yet. Therefore, it can be expected that the real-life declines are much larger in magnitude compared to these rather minimum-loss predictions presented here. Lastly, Fig. 5.11d indicates two focus regions. These regions can be

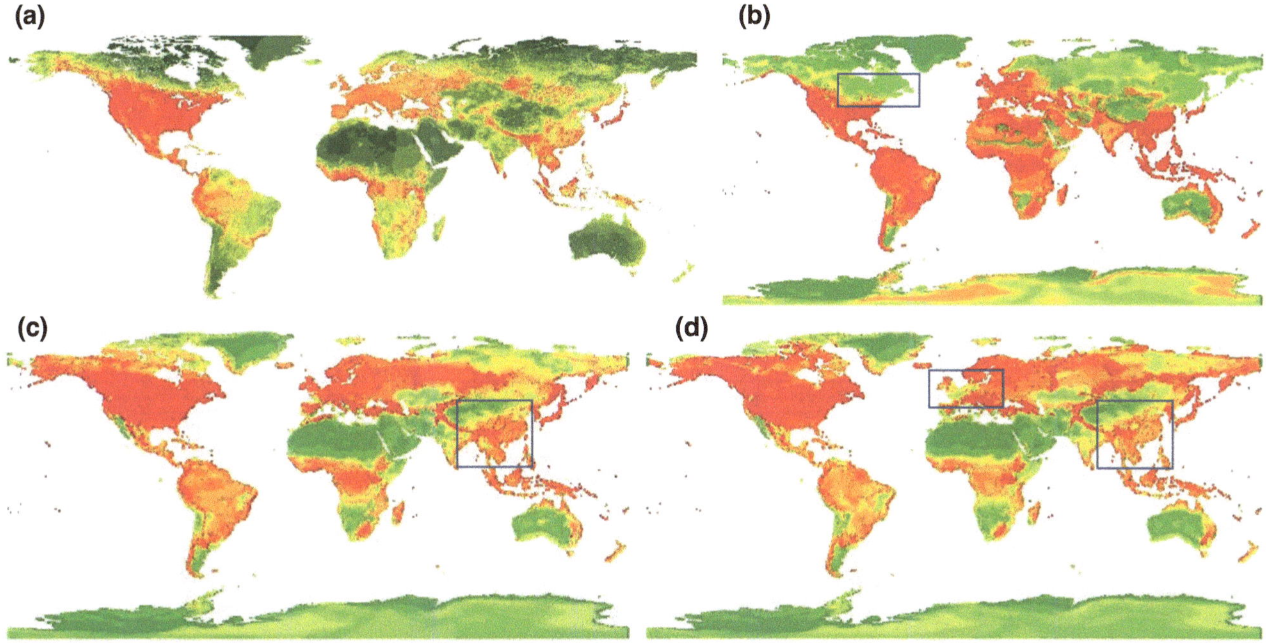

Fig. 5.11 (**a**) SDM of all global squirrel species based on 132 environmental predictors for the year 2000–2010 created with TreeNet, (**b–d**) Species distribution forecast using TreeNet for the Global Climate Models scenario (**b**) MIROC, (**c**) MRI, (**d**) IPSL

observed in Western Europe and Asia (China) again. For these regions, the same applies as to Fig. 5.11c. The difference here is that this last figure (IPSL GCM scenario) assumes a global increase of temperature of approximately 3 degrees Celsius, compared to 2 degrees Celsius in the MRI GCM scenario. This one-degree increase in the global temperature would lead to a predicted drastic decrease in the squirrel population in Western Europe in addition to the one in China. However, unfortunately, current climate policies and actions lead to an increase in temperature of 1.5 °C by 2034 and an increase of 2.7–3.1 °C by 2100 (see Climate Action Tracker (2022)). This means that the Paris Agreement is not being met and policies for betterment have globally failed (Roberts et al. 2021; Van Renssen 2018; Yang et al. 2020). The consequences are well known and it is suggested that the last scenario seems the most likely one to occur.

5.3.4 Risk Identification of the Major Threats to/by Squirrels in Urban Environments

The literature review yielded great insights into the major threats/risks originating from cities to squirrels and vice versa. These major risks consist of the following components.

City expansion

During the last few decades/centuries, cities have been continuously expanding and enlarging the anthropological influences (Baker and Harris 2007; Chen et al. 2018; Eriksen and Nielsen 2013). This is true for the city itself but also for the supply areas of cities, which tend to be global (Bruckner et al. 2015; Eakin et al. 2014; Green et al. 2019; Liu et al. 2018; Lenschow et al. 2016, and see concepts of telecoupling). As a result of this, the forested wilderness habitat of squirrels surrounding urban and suburban areas is shrinking (Baker and Harris 2007). Due to this reduced wilderness, squirrels have to continuously expand or disperse their habitats away from the wilderness (due to the lack of space) toward anthropological settings. In order to provide a squirrel-specific example of city expansion, and road development, Fig. 5.12 have been created, compiled, and presented below.

Figures 5.12 aim to demonstrate the growth and construction development of a very modernized region of the world over the last 100 years. It can be identified by observing these figures, that the endemic and endangered squirrel species – Palmer's chipmunk (*Neotamias palmeri*) – is restricted and captured in its distribution and dispersion by a major city and road construction (Lowrey 2016; McKelvey et al. 2013). This type of species isolation is new. It represents isolation due to fast urbanization (Ng et al. 2011; Piano et al. 2020). It should be a key element to include endemic and conservationally

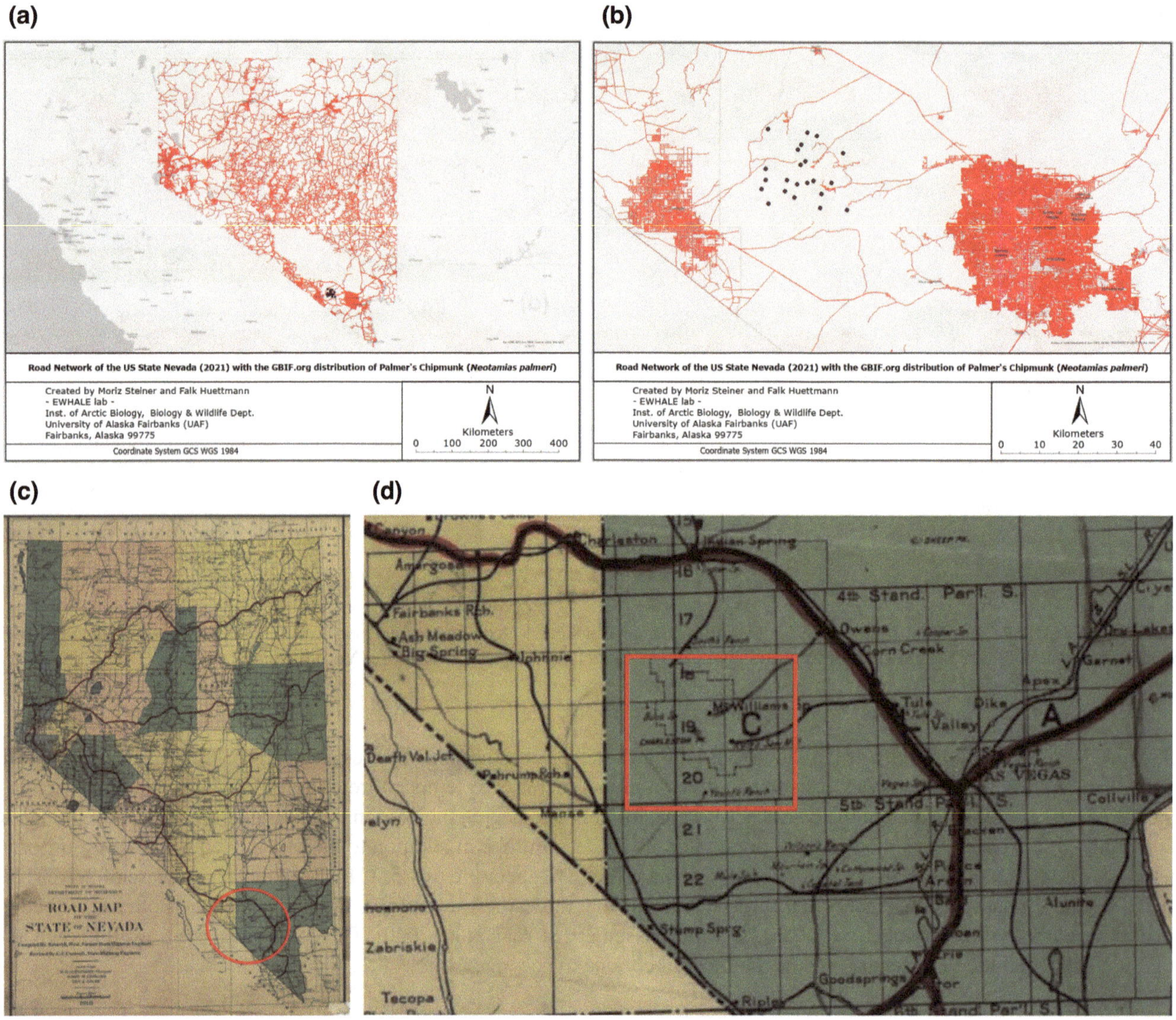

Fig. 5.12 Road network of the US state Nevada with the GBIF.org distribution of Palmer's chipmunk (*Neotamias palmeri*) for (**a**) 2021 (**b**) 2021 zoomed-in version for Las Vegas and *N. palmeri* distribution (**c**) 1919 (**d**) 1919 zoomed-in version for Las Vegas and *N. palmeri* territory Roadmap source for 1919: Nevada Department of Transportation (Nevada Dot 2021a). Roadmap source for 2021: Nevada Department of Transportation (Nevada Dot 2021b)

threatened species in city expansion and construction planning. The example presented in Fig. 5.12, should city planners and construction corporations encourage to include evaluations and consultancy of ecologists/conservationists before and during the construction planning of cities.

Anthropological feeding of squirrels

This is a major issue as it lures continuously more squirrels into urban areas, putting them at risk to cities (Bonnington et al. 2014; Uchida et al. 2021). In addition to this attractant issue, often squirrels (and other urban-environment inhabiting species) are fed with a type of feed that is unnatural for them and results in metabolism problems (Corcoran et al. 2013; Orams 2002; Schwarzkopf 1984). In the introduction section, the North American bird feeder locations have already been presented to illustrate the wide range of these bird feeder locations (Fig. 5.2). Comparing it with the occurrence of squirrels in North America, it can be confidently stated that in most regions with a high density of bird feeders, the squirrel density is likewise high. Figure 5.3b illustrates a squirrel feeding on such a bird-feeding location.

Road kills

While exact data are missing, road kills of squirrels have widely been recorded and have happened for a long time (e.g. see Francis et al. 2015; Desmond 2013; Mccleery et al. 2008; Rustiati 2009; Smith-Patten and Patten 2008), but are barely

Fig. 5.13 Urban and sub-urban environment in Colorado: what is left of nature and squirrels? Picture source: FH

Fig. 5.14 American Red Squirrel roadkill picture from FH/ HB, Alaska USA

kept track of, e.g. a requirement for valid conservation management of game species. In some locations, it has even been recorded to have squirrel road kills almost every 100 feet (30.5 m) (Farmer's Almanac (2021)). The authors can add their observations in a 'wilderness boreal forest area' of interior Alaska where very high (not quantified) roadkill numbers can equally be found daily (FH and MS in pers). This represents an incredible toll on squirrel populations vastly unaccounted for (Fig. 5.14).

With an increased number of squirrels in the Anthropocene and an increased number of roads, habitats become more fragmented, and roadkill numbers increase.

Poisoning

According to Mccleery et al. (2008), the poisoning of squirrels in cities is a major concern and often even widely advertised by pest-fighting companies (BPCA 2021; Do-it-Yourself Pest Control 2021; Pest UK 2021; Terminix 2021; etc.). Poisoning leads in most cases to the death of the animal, they certainly suffer. Squirrels are often seen as 'pests' in cities and are therefore in many cities under the threat of being poisoned (Key 1990; Khan and Siddiqi 1980; Linsdale 1931; Wood and Phillipson 1977). This impact is also vastly unaccounted for.

Disease (zoonosis) transmissions

Zoonosis transmissions can be of major concern to the human population (Bradley and Altizer 2007; Riley et al. 2014; Van Horne 2008). Recent examples such as Covid-19, rabies, and bubonic plague showed that such topics are of the highest relevance for the modern and growing society. The more squirrels are occurring in cities with a high human population and density, the higher the likelihood that zoonotic diseases are transmitted from squirrels to humans. Figures 5.4 in the introduction section illustrate the zoonotic disease density of a) mammals and b) rodents on a global scale. It can be clearly observed that the major density hotspots for mammals lay in Europe, the Middle East, Central Asia, the Southern East coast of Africa, and the Southeast USA. For rodents (Fig. 5.4b), the hotspots can be observed in Central and Eastern Europe, Central Asia, Continental USA, and the northern and southern coastal regions of Brazil. Figures 5.4 have additionally the maps from Fig. 5.1 overlaid, illustrating the population gain of the 100 largest cities globally. By presenting it this way, it is possible to identify the world's 100 largest cities and how these are affected by zoonotic diseases. It appears that a very high number of these 100 largest cities are located in zoonotic hotspot regions. This enforces the statement above, the higher the number and density of people in a city, and the more squirrels occur there, the higher the chance of zoonotic disease transmission.

Habitat destruction around urbanized areas for urban and commercial constructions

In order to satisfy the increased housing demands in and around cities, new homes must be constructed. This requires continuously more land for construction and impervious surfaces, which supposedly can only be obtained by increasing the city boundaries and changing the land use of the surrounding landscape (natural areas) (Apostolopoulou and Adams 2019). A typical example can be found in Beijing (China), where now the sixth city ring and expansion was planned, virtually without any greenspaces (Huang et al. 2015; Tang et al. 2016; Zhao et al. 2010). Due to this increased wildlife habitat destruction, wildlife heavily suffers (Bennett 2000). Another example of this is Owens Lake and its history (Wikipedia 2021). There, the entire lake has been drained and all the local ecosystem destroyed in order to support the development of one city (Las Vegas, Nevada, USA). This has been done in a desert for 'entertainment', catering to one of the largest industrialized economies in the world and earth's history. As this is a major threat for many species, the IUCN Red List acknowledges it by creating a specific threat category for this, entitled "Residential and commercial development", which ranks first in the threats on www.iucnredlist.org. This habitat destruction leads to additional dangerous situations for squirrels and species in general e.g. habitat fragmentation (Chapman et al. 2012; DeMarco et al. 2021; Thomas et al. 2018), population isolations (Lourenço et al. 2017), increased frequency of road kills (see literature references in roadkill section), etc.

Urban and industrial pollution

Environmental air, light, noise pollution, etc. are all components of urban pollution, affecting squirrels and their close relatives (Bradley and Altizer 2007; Chow et al. 2021; Francis et al. 2015; Tangley 1986). As mentioned in these literature references, urban pollution can cause stress (health stress, reproduction stress, feed stress, survival stress, etc). Conclusively, it is of no good for the squirrels being (trapped) in cities in the absence of other options. Urban pollution also includes the production and contamination of Greenhouse gasses (GHG) e.g. CO_2 (Herzog 2009; Olivier et al. 2017). To provide a global overview of the CO_2 emissions of the 100 largest cities in the world aligned with the global squirrel hotspots, Fig. 5.15 has been created.

Due to the absence of the CO_2 emissions data for several African metropolis cities in the dataset, it is possible that some cities in Africa do not appear correctly in Fig. 5.15.

Decreased fertility of squirrels

Many external factors can lead to decreased fertility and reproductive disadvantages in mammals. Stress is an important factor, as well as the quality of surrounding habitats (Merrick et al. 2016). Commonly species thrive and reproduce at natural rates only in intact habitats (except for very flexible species) (Jansen and Yoshimura 1998; Peterson et al. 2007; Yang and Kim 2016), and cities are generally not considered to be very intact.

Squirrels as target practice (mainly for kids/teenagers)

Squirrels are often seen as targets for shooting practices (Arluke 2002; Carpenter 2015; Crowley et al. 2018; Flynn 2002; Hampton et al. 2020; MacRae 2010; Wright 1995; etc.), this is especially true for North America and kids/teenagers with

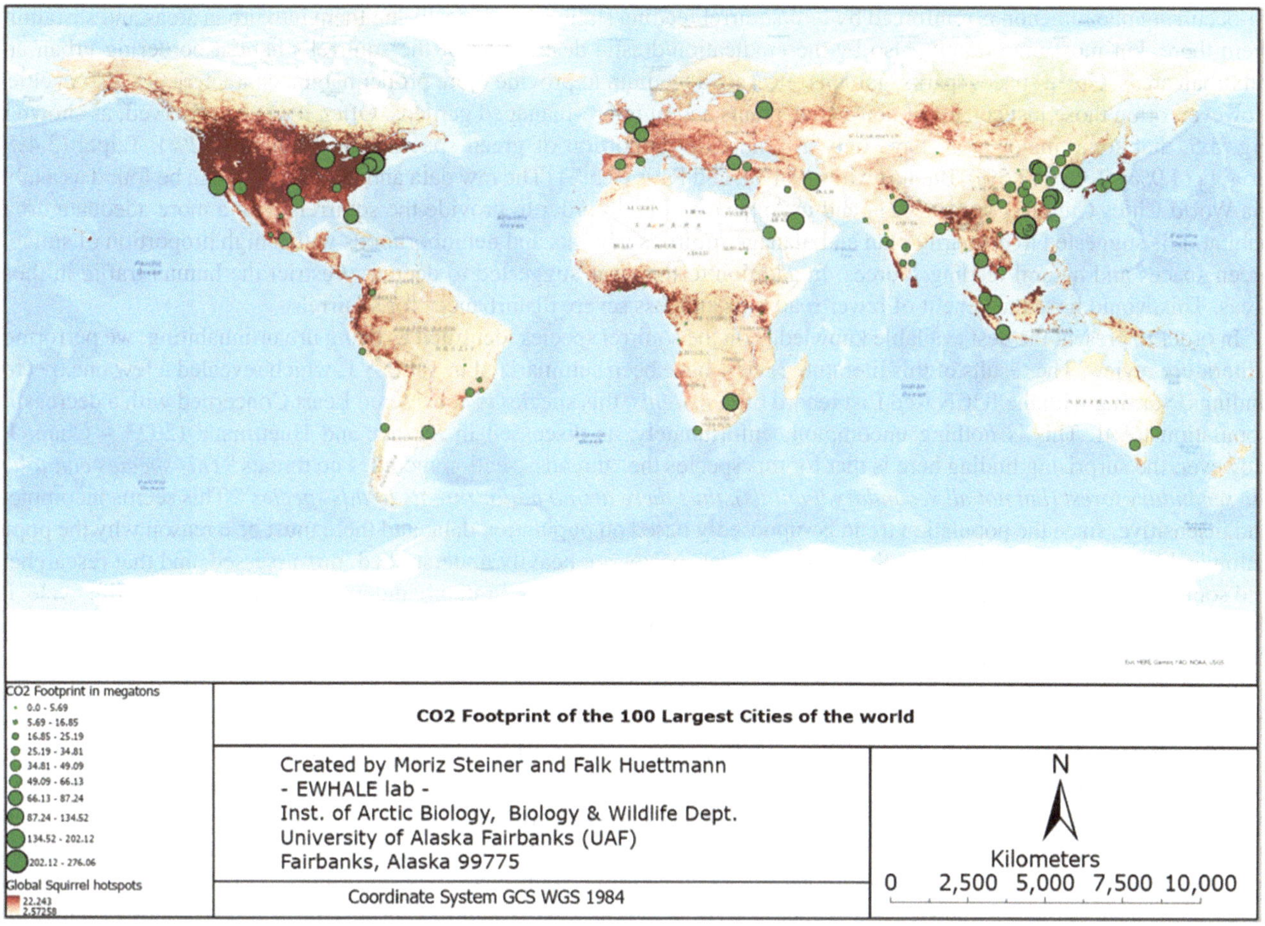

Fig. 5.15 CO_2 emission of the 100 largest cities of the world (seized green dots). The data source of the CO_2 emissions of the 100 largest cities has been obtained from Moran et al. (2018), and the geographical location of the 100 largest cities has been obtained from SimpleMaps (2021). The data and files of this map can be found in Appendix 5.2

rifles (Miller and Hemenway 2004; Sanders 2011). In addition to plenty of research studies documenting these unregulated activities but no quantification of squirrel populations, and often the results of such activities are posted and shared online (e.g. see hunting Facebook groups entitled "Extreme Squirrel Hunting", etc.).

Unregulated harvest regimes and no limits and no science or valid data

Squirrels are widely unmanaged and harvest limits are for most squirrels missing (see Steiner and Huettmann 2021 and Fish and Wildlife Department of the U.S. states). Squirrels are often quite intensively hunted/trapped, however, without any hunting regulations and law enforcement. Additionally, responsible entities are widely ignoring the public interest on such topics and an open-access approach seems to be a foreign concept for most entities (see Steiner and Huettmann, 2023 – Chaps. 11 and 13 for details). This can lead to severe population declines of sub-species or even species, happening without anyone noticing. Therefore, in this current unsustainable and disrespectful situation, increased efforts must flow into the creation of hunting regulations, its legal enforcement, and the awareness of the current situation, and general efforts for squirrels must increase. Ideally, the mindset of the current hunters should be formed more sustainably too, moving away from a playful massacring one towards science-based and environmentally-aware management, for a biodiversity-rich and sustainable future.

5.4 Discussion/Conclusion

Observing squirrels in cities is no rare phenomenon anymore. The fact that they can be observed more frequently in cities is strongly correlated with the continuous increase of the global squirrel populations in cities (Benson 2013; Merrick et al. 2016). In addition to this, the human population also continuously increases in number and density all around the world. This co-occurring phenomenon is reinforced by bird/squirrel feeding stations in cities, luring them into urban areas and sustaining them there, but most importantly, also by the eradication/drastic destruction of the squirrel's habitat bordering urban and suburban areas. Green spaces (parks, gardens, etc.) are a medium to provide more proper habitat characteristics also in cities. However, often those just consist of non-native plants and of micro-managed gardens. Often it can be observed, as shown in Fig. 5.5, that some major cities have only a minuscule proportion of green spaces (e.g. Istanbul (2.2%), Taipei (3.4%), Bogota (4.9%), Tokyo (7.5%), Buenos Aires (9.4%), and Paris (9.5%) (The raw data and the overview can be found as a table via World Cities Culture Forum (2021) and in Appendix 5.1). In order to provide the squirrels with a more adequate urban habitat, it is suggested to primarily plan and manage squirrels in parks and neighborhoods with a high proportion of suitable green spaces and natural feeding sources. In addition to this, it is suggested to decrease/restrict the human traffic in those areas. This would have the benefit of fewer road kills and less severe disturbances for squirrels.

In order to present the best available knowledge on the squirrel species identified as being urban-inhabiting, we performed a literature review. The results of this literature review have been summarized in Table 5.1, which revealed a few unexpected findings. Starting with the IUCN Red List record of *L. insignis*, this species is considered Least Concerned with a decreasing population trend. This is nothing uncommon, unfortunately, as discussed in Steiner and Huettmann (2023 – Chap. 1). However, the surprising finding here is that for this species the "threads" section includes no threats "*This species can tolerate secondary forest (but not all secondary habitats), thus there are no major threats to this species*". This seems incomplete and insensitive, since the population trend is supposedly based on population data, and there must be a reason why the population is declining. This shows again that many squirrel species are heavily understudied, mis-assessed, and that researchers and society are most of the time walking in the dark when it comes to the in-depth understanding of a species' life cycle, its threats, population trend, the population size of mature individuals, and behavior (see Chap. 10 in Steiner and Huettmann 2023). The squirrel species conservation suffers accordingly and everybody loses. Additionally, during the literature review, only very few scientific records have been found to be publicly available on the world wide web that discusses and studies the squirrel's distribution, their life cycle, etc. The main contributors were the IUCN Red List and its corresponding species-specific citations, Wilson and Reeder (2005), and Thorington et al. (2012). This lacks diversity and peer-reviewed quality checks. We argue that the world of the literature is not only narrowly represented by those three anglophone studies (which seem to reference themselves for most species), but it is also very restricted to squirrels occurring in western regions. The bulk of squirrel species are found in tropical regions but are not represented whatsoever (see Textbox 5.2). Therefore, research efforts and the number of publications are not equal for all squirrel species, unfortunately. This is commonly found in research, and it biases our knowledge and approach (Boakes et al. 2010; Raup 1995; Thévenin et al. 2020; Titley et al. 2017).

Textbox 5.2: Incorrect, Dubious, Confusing, Misleading, and Fluctuating Species Representation in the Public Record and Ecological Sciences

Species occur within their ecological niche and can be found in their special ranges. Humans can access, intrude, and observe those niches; the species' habitat. Due to the topography of this world and its human distribution, some habitats are easier to access than others. The same is valid for different ethnicities and their cultural distribution on this planet. As a consequence of this, certain people have access to certain places and habitats endemic to these places which are inaccessible to other people.

This pattern can also be reflected in the occurrence record counts of global species. An example of this can be found in the global occurrence point dataset of squirrels provided in the citizen science database repository www.GBIF.org. In 2020, this dataset contained over 227,000 squirrel occurrence entries. Over 170,000 (approx. 75%) out of these 227,000+ records come from only 5–6 species (out of over 300 squirrel species, mostly located in the tropics). All these 5–6 species occur in western countries (the USA and Europe). The remaining approx. 25% of the records were logged for the remaining 98% of the species (see details on numbers in Steiner and Huettmann 2023 – Table 3.1, Chap. 3). These numbers already display a certain and ongoing bias in the squirrel dataset. Either these 5–6 species have extraordinary population numbers compared to the remaining approx. 300 species or the bias lies in the recording of the occurrences. Following ecological principles and logical thinking processes, this recording bias can likely be assigned to an incorrect species representation of the population numbers in the dataset and how it gets collected. Reality remains widely hidden. Compared to the real-world population numbers, the represented record number for these 5–6 hyper-present species are likely overrepresented and the remaining approx. 300 species are most likely underrepresented. So what do we really know about squirrels?

The reason for this inequality in record representation is likely that many more recording citizens, that are reporting to GBIF.org, are located in these western countries compared to all the other parts of the world where the reporting of species is severely underrepresented, hardly popular. This has previously also been mentioned by other authors e.g.

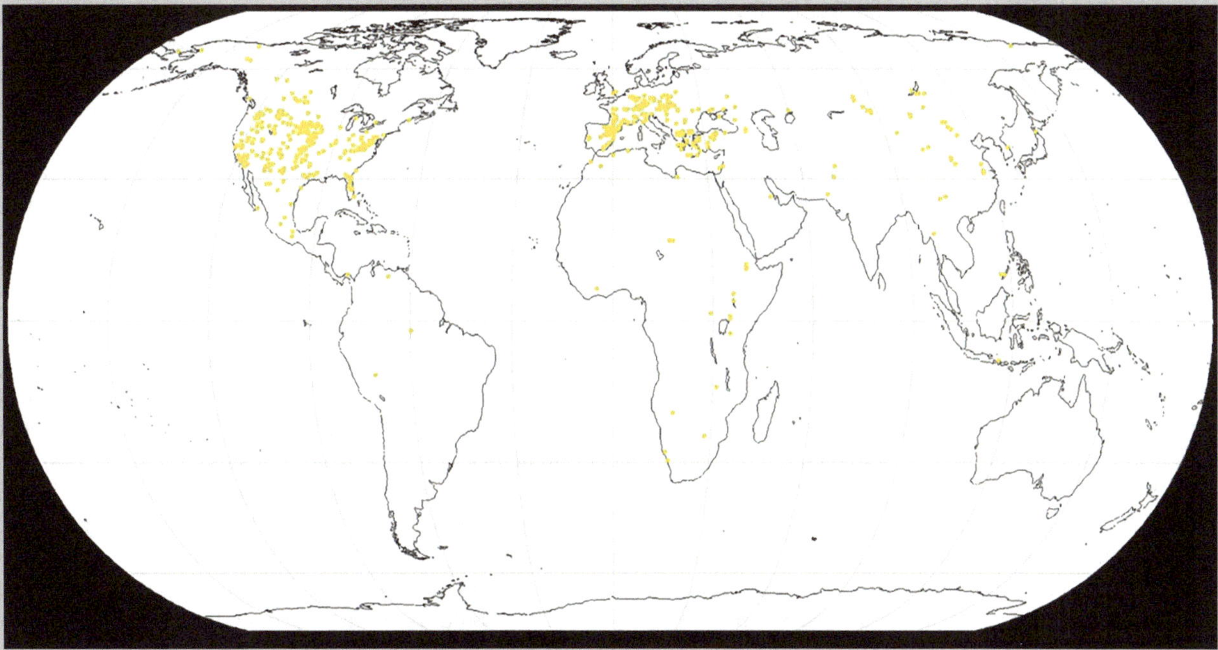

Textbox Fig. 5.1 Squirrel Museum specimen (Source: www.fossilworks.org)

Raup (1995) Note: For many other species, certainly for microbes but which make the bulk of the world's biodiversity, this is much worse.

Therefore, since some habitats are more difficult to access than others and are accessed by people with different scopes, the species numbers recorded within those habitats can be significantly different than for easily accessible habitats. On the one hand, a great example of this is these upper-mentioned over-presented species from the USA and Europe, where access to nature and most squirrel habitats are fairly easy and subsequently recorded numbers are extraordinarily high. A widely developed road network in an urbanized landscape with leisure time and 'hunger for nature' makes it possible. On the other hand, species inhabiting tropical habitats on poorly populated islands with a low national science effort have very low to zero occurrence records for their endemic squirrel species. Species from wilderness areas are underrepresented in the public record, as shown here for squirrels.

Conclusively, we warrant any scientist and citizen that is handling species occurrence data and similar datasets not to solely rely on occurrence counts without correcting model-predictions as estimation of population numbers but consider location, habitat access, national science interest, global species importance, conservation status, etc. Modeling will help to overcome such problems. Additionally, we suggest investing increased research and financial efforts into the research of the species in less-represented species and rarely observed habitats. This is especially important for the species that are not even been reported in GBIF.org, indicating that no citizen has ever recorded the species at stake and is widely marginalized.

A very similar bias as presented above can be observed for museum specimens and where they have been found. This can be observed by observing the map from fossilworks.org below (Textbox Fig. 5.1). Each dot represents a museum specimen of a squirrel and where it has been found. Here again, a clear Western-dominated bias/ trend can be observed.

This case is very obvious for squirrels. However, another problem relates to the actual identification of the species. The assumption is that the public record is more or less error-free, e.g. for species names and locations. But that is hardly the case. In addition, the actual species names are changing and fluctuating, e.g. due to genetic work and quick but frequent revisions in the taxonomic authority (see Steiner and Huettmann 2023 – Chap. 1 for details). The actual 'authority' for taxonomic names, as it relates to citizens or researchers and in GBIF remains also very dubious. For instance, ITIS.gov is stating that they are the mandate and authoritative taxonomy resource, but it is not widely used, and hardly available for all species globally. In the meantime, national standards differ, so they do in a DNA database (which is often used as the taxonomic research authority, but then not used in ITIS or by nations, agencies, and court cases). Needless to state that some genetic data are private, owned by corporations, or not shared, nor geo-referenced and thus lack any scrutiny.

Where does that really leave us?

Another great bias was observed by us in the process of performing this literature review since most literature records only include the following species: Eurasian Red Squirrel (*Sciurus vulgaris*), Eastern Gray Squirrel (*Sciurus carolinensis*), Eastern Fox Squirrel (*Sciurus niger*), and North American Red Squirrel (*Tamiasciurus hudsonicus*). For those above-mentioned species, there are plenty of studies published, on many topics but very few real-world policy issues are followed and the basic ecology or limit of resources is widely not acknowledged even. These four species seem almost 'overstudied' compared to all the other approx. 300 squirrel species in the world which are widely ignored.

Apart from these strong signals in our findings, for all the species included in the literature review, the recorded distribution aligned with the modeled distribution of these species (see Steiner and Huettmann 2023 – Chap. 3, and Sub-figures (a) from Figs. 5.6, 5.7, 5.8, 5.9 and 5.10).

To create a forecast for those distribution ranges for the year 2100, we performed Species Distribution Forecasts (SDFs) for all the here included urban environment inhabiting squirrel species. These SDFs aim to forecast the distribution of these individual squirrel species for the year 2100, based on three different Global Climate Models (GCMs – MIROC, MRI, and IPSL). By observing the results of these SDFs, it can be concluded that for these species-specific models no major changes can be observed. For most species, the distribution range expands into similar regions around the original distribution with varying hotspot shrinkages/ changes. This is presumably however not what occurs in the real future. This is because for each species also a current distribution is modeled based on the same predictors as for the 2100 SDFs, but then for 2000. These models (Sub-figures (b) from Figs. 5.6, 5.7, 5.8, 5.9, 5.10) illustrate the current distribution of the 5 squirrels using 7 BioClim predictors and one elevation predictor. The SDFs are only minimally different from the current distribution. This can be due to very few changes that are happening to the squirrel, and its distribution range, but also due to the rather weak SDFs. These SDFs are likely considered rather weak because they consist of only 7-8 predictors, which is a small fraction compared to the SDMs of the Sub-figures (a) from Figs. 5.6, 5.7, 5.8, 5.9, 5.10. This could mean that these models create a rather weak forecast thus far. To improve this, we strongly encourage any data providers, data developers, and similar, to investigate further resources and time to develop future data sets, especially for 2100. If more such datasets, relevant for squirrels are available, including future road maps, biodiversity richness, forest fire occurrences, land cover, land use, etc. a more precise species distribution forecast modeling would be possible. With more accurate species distribution forecasts, conservation planning, and in this case also city planning, can be conducted better, more efficiently, and with higher reliability. Next to investing more resources into the development of future datasets, it is also important to look at the current threats to squirrels in urban areas. In this study, the major urban environmental threats to squirrels have been researched from literature, presented, and consecutively discussed. These topics include the expansion of urban areas, feedings of squirrels by humans, road kills, poisoning, disease transmissions (zoonosis), and habitat destruction around urban and suburban areas for urban and commercial constructions. Thus far, we fall short of a good global culture to live and interact with squirrels. Once these aspects and risks are also included in future city planning and species management, e.g. as part of a sustainability plan and certification, the future outlook for squirrels can become brighter, especially in the Anthropocene running currently a crisis mode.

References

Adler B (2014) Outwitting squirrels: 101 cunning stratagems to reduce dramatically the egregious misappropriation of seed from your birdfeeder. Chicago Review Press, Chicago

Ahmed Z, Asghar MM, Malik MN, Nawaz K (2020a) Moving towards a sustainable environment: the dynamic linkage between natural resources, human capital, urbanization, economic growth, and ecological footprint in China. Resour Policy 67:101677

Ahmed Z, Zafar MW, Ali S (2020b) Linking urbanization, human capital, and the ecological footprint in G7 countries: an empirical analysis. Sustain Cities Soc 55:102064

Allendorf V, Rubbenstroth D, Schlottau K, Hoffmann D, Frank C, Amler S, Homeier-Bachmann T et al (2021) Assessing the occurrence of the novel zoonotic variegated squirrel bornavirus 1 in captive squirrels in Germany—a prevalence study. Zoonoses Public Health 68(2):110–120

Alonso W (1973) Urban zero population growth. Daedalus, pp 191–206

Álvarez-Castañeda ST, Castro-Arellano I, Lacher T (2016) Neotamias merriami. The IUCN Red List of Threatened Species 2016: e.T21358A22269203. https://doi.org/10.2305/IUCN.UK.2016-2.RLTS.T21358A22269203.en. Accessed 10 Oct 2021

Apostolopoulou E, Adams WM (2019) Cutting nature to fit: urbanization, neoliberalism and biodiversity offsetting in England. Geoforum 98:214–225

Arcari P, Probyn-Rapsey F, Singer H (2020) Where species don't meet: invisibilized animals, urban nature and city limits. Environ Plan E: Nat Space 4(3):2514848620939870

Arluke A (2002) Animal abuse as dirty play. Symb Interact 25(4):405–430

Baker PJ, Harris S (2007) Urban mammals: what does the future hold? An analysis of the factors affecting patterns of use of residential gardens in Great Britain. Mammal Rev 37(4):297–315

Beest JT, Cushing A, McClean M, Hsu W, Bildfell R (2017) Disease surveillance of California ground squirrels (Spermophilus beecheyi) in a drive-through zoo in Oregon, USA. J Wildl Dis 53(3):667–670
Bengston D, Dockry MJ (2014) Forest futures in the Anthropocene: can trees and humans survive together? The Futurist 48(4):34–39
Bennett AJ (2000) Environmental consequences of increasing production: some current perspectives. Agric Ecosyst Environ 82(1–3):89–95
Benson E (2013) The urbanization of the eastern gray squirrel in the United States. J Am Hist 100(3):691–710
Boakes EH, McGowan PJ, Fuller RA, Chang-qing D, Clark NE, O'Connor K, Mace GM (2010) Distorted views of biodiversity: spatial and temporal bias in species occurrence data. PLoS Biol 8(6):e1000385
Bonnington C, Gaston KJ, Evans KL (2014) Relative roles of grey squirrels, supplementary feeding, and habitat in shaping urban bird assemblages. PLoS One 9(10):e109397
Boucher O, Servonnat J, Albright AL, Aumont O, Balkanski Y, Bastrikov V, Vuichard N et al (2020) Presentation and evaluation of the IPSL-CM6A-LR climate model. J Adv Model Earth Syst 12(7):e2019MS002010
BPCA (2021) Pest advice for controlling Grey Squirrels. https://bpca.org.uk/a-z-of-pest-advice/squirrel-control-how-to-get-rid-of-squirrels-bpca-a-z-of-pests-/188983. Accessed 10 Oct 2021
Bradley CA, Altizer S (2007) Urbanization and the ecology of wildlife diseases. Trends Ecol Evol 22(2):95–102
Bradley JE, Marzluff JM (2003) Rodents as nest predators: influences on predatory behavior and consequences to nesting birds. Auk 120(4):1180–1187
Brockerhoff M (1999) Urban growth in developing countries: a review of projections and predictions. Popul Dev Rev 25(4):757–778
Bruckner M, Fischer G, Tramberend S, Giljum S (2015) Measuring telecouplings in the global land system: a review and comparative evaluation of land footprint accounting methods. Ecol Econ 114:11–21
Buhaug H, Urdal H (2013) An urbanization bomb? Population growth and social disorder in cities. Glob Environ Chang 23(1):1–10
Cadar D, Allendorf V, Schulze V, Ulrich RG, Schlottau K, Ebinger A, Tappe D et al (2021) Introduction and spread of variegated squirrel bornavirus 1 (VSBV-1) between exotic squirrels and spill-over infections to humans in Germany. Emerg Microbes Infect 10(1):602–611
Callahan JR (1993) Squirrels as predators. Great Basin Nat:137–144
Carpenter T (2015) Small-game hunting. ABDO
Cassola F (2016a) Callosciurus prevostii. The IUCN Red List of Threatened Species 2016: e.T3603A22253650. https://doi.org/10.2305/IUCN.UK.2016-2.RLTS.T3603A22253650.en. Accessed 10 Oct 2021
Cassola F (2016b) Glaucomys volans (errata version published in 2017). The IUCN Red List of Threatened Species 2016: e.T9240A115091392. https://doi.org/10.2305/IUCN.UK.2016-3.RLTS.T9240A22257175.en. Accessed 10 Oct 2021
Chapman T, Rymer T, Pillay N (2012) Behavioural correlates of urbanisation in the cape ground squirrel Xerus inauris. Naturwissenschaften 99(11):893–902
Chen J, John R, Sun G, Fan P, Henebry GM, Fernández-Giménez ME, Qi J et al (2018) Prospects for the sustainability of social-ecological systems (SES) on the Mongolian plateau: five critical issues. Environ Res Lett 13(12):123004
Chow PKY, Uchida K, von Bayern AM, Koizumi I (2021) Characteristics of urban environments and novel problem-solving performance in Eurasian red squirrels. Proc R Soc B 288(1947):20202832
Chrysanthou A, Van der Schrier GEJM, Van Den Besselaar EJM, Klein Tank AMG, Brandsma T (2014) The effects of urbanization on the rise of the European temperature since 1960. Geophys Res Lett 41(21):7716–7722
Climate Action Tracker (2022) Addressing global warming. https://climateactiontracker.org/global/temperatures/. Accessed 5 Aug 2022
Corcoran MJ, Wetherbee BM, Shivji MS, Potenski MD, Chapman DD, Harvey GM (2013) Supplemental feeding for ecotourism reverses diel activity and alters movement patterns and spatial distribution of the southern stingray, Dasyatis americana. PLoS One 8(3):e59235
Corrigan DH (2019) Nuts about squirrels: the rodents that conquered popular culture. McFarland, North Carolina
Crowley SL, Hinchliffe S, McDonald RA (2018) Killing squirrels: exploring motivations and practices of lethal wildlife management. Environ Plan E: Nat Space 1(1–2):120–143
Daly HE (1977) The steady-state economy. In: W. H. Freeman (ed) illustrated edition (January 1, 1973)
Davis M (2004) Planet of slums. New Left Rev 26:5
Dawson A (2016) Extinction: a radical history. Or Books, La Vergne
DeMarco C, Cooper DS, Torres E, Muchlinski A, Aguilar A (2021) Effects of urbanization on population genetic structure of western gray squirrels. Conserv Genet 22(1):67–81
Desmond J (2013) Requiem for roadkill: death and denial on America's roads. In: Environmental Anthropology. Routledge, Abingdon, pp 60–72
Dietz R, O'Neill D (2013) Enough is enough: building a sustainable economy in a world of finite resources. Routledge
Do-it-Yourself Pest Control (2021) How To Get Rid Of Squirrels. https://www.doyourownpestcontrol.com/squirrelcontrol.htm. Accessed 10 Oct 2021
Duranton G, Turner MA (2012) Urban growth and transportation. Rev Econ Stud 79(4):1407–1440
Eakin H, DeFries R, Kerr S, Lambin EF, Liu J, Marcotullio PJ, Zimmerer K, et al. (2014) Significance of telecoupling for exploration of land-use change. In: Rethinking global land use in an urban era. MIT Press, Cambridge, pp. 141–161
Eriksen TH, Nielsen FS (2013) A history of anthropology. Pluto Press, London
Esbah H, Cook EA, Ewan J (2009) Effects of increasing urbanization on the ecological integrity of open space preserves. Environ Manag 43(5):846–862
Estrada-Peña A, Ostfeld RS, Peterson AT, Poulin R, de la Fuente J (2014) Effects of environmental change on zoonotic disease risk: an ecological primer. Trends Parasitol 30(4):205–214
Farley J (2014) Steady state economics. In: Degrowth. Routledge, pp 77–80
Farmer's Almanac (2021) What's Going On With All The Dead Squirrels? https://www.farmersalmanac.com/dead-squirrels-32602. Accessed 10 Oct 2021
Flynn C (2002) Hunting and illegal violence against humans and other animals: exploring the relationship. Soc Anim 10(2):137–154
Franchini M, Mazza G, Mori E (2021) First assessment of ectoparasite prevalence in Apennine populations of Eurasian red squirrel: does habitat fragmentation affect parasite presence? Ethol Ecol Evol:1–17

Francis MJ, Spooner PG, Matthews A (2015) The influence of urban encroachment on squirrel gliders (Petaurus norfolcensis): effects of road density, light and noise pollution. Wildl Res 42(4):324–333

Fryxell JM, Sinclair AR, Caughley G (2014) Wildlife ecology, conservation, and management. John Wiley & Sons, Chichester

Gonzales EK (2005) The distribution and habitat selection of introduced Eastern Grey Squirrels, Sciurus carolinensis, in British Columbia. Can Field-Nat 119(3):343–350

Green JM, Croft SA, Durán AP, Balmford AP, Burgess ND, Fick S, West CD et al (2019) Linking global drivers of agricultural trade to on-the-ground impacts on biodiversity. Proc Natl Acad Sci 116(46):23202–23208

Hampton JO, DeNicola AJ, Forsyth DM (2020) Assessment of Lead-free. 22 LR bullets for shooting European rabbits. Wildl Soc Bull 44(4):760–765

Han BA, Kramer AM, Drake JM (2016) Global patterns of zoonotic disease in mammals. Trends Parasitol 32(7):565–577

Herzog T (2009) World greenhouse gas emissions in 2005. World Res Inst 7:2009

Huang D, Liu Z, Zhao X (2015) Monocentric or polycentric? The urban spatial structure of employment in Beijing. Sustainability 7(9):11632–11656

Hubacek K, Guan D, Barrett J, Wiedmann T (2009) Environmental implications of urbanization and lifestyle change in China: ecological and water footprints. J Clean Prod 17(14):1241–1248

Hummel D (2020) The effects of population and housing density in urban areas on income in the United States. Local Econ 35(1):27–47

Ichimura M (2003) Urbanization, urban environment and land use: challenges and opportunities. In: Asia-Pacific Forum for Environment and Development, Expert Meeting, Vol. 23. pp 1–14

International Institute for Environment and Development (2020) The world's 100 largest cities from 1800 to 2020, and beyond. https://www.iied.org/worlds-100-largest-cities-1800-2020-beyond. Accessed 10 Oct 2021

Jansen VA, Yoshimura J (1998) Populations can persist in an environment consisting of sink habitats only. Proc Natl Acad Sci 95(7):3696–3698

Johnson CN, Balmford A, Brook BW, Buettel JC, Galetti M, Guangchun L, Wilmshurst JM (2017) Biodiversity losses and conservation responses in the Anthropocene. Science 356(6335):270–275

Key G (1990) Control of the African striped ground squirrel, Xerus erythropus, in Kenya. In: Proceedings of the Vertebrate Pest Conference, Vol. 14, No. 14

Khan Z, Siddiqi JI (1980) Poison-and bait-shyness of northern palm squirrel, Funambulas pennanti pennanti WROUGHTON. Appl Entomol Zool 15(2):175–177

Kheraj S (2012) Demonstration wildlife: negotiating the animal landscape of Vancouver's Stanley Park, 1888–1996. Environ Hist 18(4):497–527

Kuang W, Liu J, Dong J, Chi W, Zhang C (2016) The rapid and massive urban and industrial land expansions in China between 1990 and 2010: a CLUD-based analysis of their trajectories, patterns, and drivers. Landsc Urban Plan 145:21–33

Lenschow A, Newig J, Challies E (2016) Globalization's limits to the environmental state? Integrating telecoupling into global environmental governance. Environ Polit 25(1):136–159

Linn JF (1982) The costs of urbanization in developing countries. Econ Dev Cult Chang 30(3):625–648

Linsdale JM (1931) Facts concerning the use of thallium in California to poison rodents: its destructiveness to game birds, song birds and other valuable wild life. Condor 33(3):92–106

Liu J, Dou Y, Batistella M, Challies E, Connor T, Friis C, Sun J et al (2018) Spillover systems in a telecoupled Anthropocene: typology, methods, and governance for global sustainability. Curr Opin Environ Sustain 33:58–69

Lourenço A, Álvarez D, Wang IJ, Velo-Antón G (2017) Trapped within the city: integrating demography, time since isolation and population-specific traits to assess the genetic effects of urbanization. Mol Ecol 26(6):1498–1514

Lowrey C (2016) Neotamias palmeri. The IUCN Red List of Threatened Species 2016: e.T21355A22267875. https://doi.org/10.2305/IUCN.UK.2016-3.RLTS.T21355A22267875.en. Accessed 10 Oct 2021

Luo M, Lau NC (2018) Increasing heat stress in urban areas of eastern China: acceleration by urbanization. Geophys Res Lett 45(23):13–060

MacKinnon B (1977) "Animal liberation: a new ethics for our treatment of animals," by Peter Singer. Mod Sch 54(4)

MacRae S (2010) Small-game Hunting: rabbits, Squirrels, and Other Small Animals. The Rosen Publishing Group, Inc.

Madlener R, Sunak Y (2011) Impacts of urbanization on urban structures and energy demand: what can we learn for urban energy planning and urbanization management? Sustain Cities Soc 1(1):45–53

Maxent (2022) Maxent software for modeling species niches and distributions.https://biodiversityinformatics.amnh.org/open_source/maxent/. Accessed 5 Aug 2022

Mccleery RA, Lopez RR, Silvy NJ, Gallant DL (2008) Fox squirrel survival in urban and rural environments. J Wildl Manag 72(1):133–137

McKelvey KS, Ramirez JE, Pilgrim KL, Cushman SA, Schwartz MK (2013) Genetic sampling of Palmer's chipmunks in the Spring Mountains, Nevada. West N Am Nat 73(2):198–210

Merrick MJ, Evans KL, Bertolino S (2016) Urban grey squirrel ecology, associated impacts and management challenges. In: The grey squirrel: ecology management of an invasive species in Europe, European Squirrel Initiative, Warwickshire, pp 57–77

Miller M, Hemenway D (2004) Unsupervised firearm handling by California adolescents. Inj Prev 10(3):163–168

Moran D, Kanemoto K, Jiborn M, Wood R, Többen J, Seto KC (2018) Carbon footprints of 13 000 cities. Environ Res Lett 13(6):064041

Munshi-South J, Nagy C (2014) Urban park characteristics, genetic variation, and historical demography of white-footed mouse (Peromyscus leucopus) populations in New York City. PeerJ 2:e310

Nathaniel S, Nwodo O, Adediran A, Sharma G, Shah M, Adeleye N (2019) Ecological footprint, urbanization, and energy consumption in South Africa: including the excluded. Environ Sci Pollut Res 26(26):27168–27179

National Zoo & Conservation Biology Institute (2022) Prevost's squirrel. https://nationalzoo.si.edu/animals/prevosts-squirrel. Accessed 5 Aug 2022

Nelder MP, Reeves WK, Adler PH, Wozniak A, Wills W (2009) Ectoparasites and associated pathogens of free-roaming and captive animals in zoos of South Carolina. Vector-Borne Zoonotic Dis 9(5):469–477

Nevada Dot (2021a) Historical state highway maps. https://www.dot.nv.gov/travel-info/maps/historical-maps. Accessed 10 Oct 2021

Nevada Dot (2021b) Statewide routes. https://geohub-ndot.hub.arcgis.com/search?groupIds=9641867973f84b97bb1225c826a3274d. Accessed 10 Oct 2021

Ng CN, Xie YJ, Yu XJ (2011) Measuring the spatio-temporal variation of habitat isolation due to rapid urbanization: a case study of the Shenzhen River cross-boundary catchment, China. Landsc Urban Plan 103(1):44–54

Olivier JG, Schure KM, Peters JAHW (2017) Trends in global CO2 and total greenhouse gas emissions. PBL Netherlands Environmental Assessment Agency. The Hague, p 5

Orams MB (2002) Feeding wildlife as a tourism attraction: a review of issues and impacts. Tour Manag 23(3):281–293
Osseiran N, Chriscaden K (2016) Air pollution levels rising in many of the world's poorest cities. WHO, Geneva
Pest UK (2021) Get rid of squirrel problems | squirrels removal specialists. https://www.pestuk.com/pest/get-rid-of-squirrel-problems-squirrels-removal-specialists/. Accessed 10 Oct 2021
Peterson DL, Vecsei P, Jennings CA (2007) Ecology and biology of the lake sturgeon: a synthesis of current knowledge of a threatened north American Acipenseridae. Rev Fish Biol Fish 17(1):59–76
Piano E, Giuliano D, Isaia M (2020) Islands in cities: urbanization and fragmentation drive taxonomic and functional variation in ground arthropods. Basic Appl Ecol 43:86–98
Raup DM (1995) Extinction rates (Vol. 11). J. H. Lawton, & R. M. May (Eds.). Oxford: Oxford University Press
Riley SP, Serieys LE, Moriarty JG (2014) Infectious disease and contaminants in urban wildlife: unseen and often overlooked threats. In: Urban Wildlife Conservation. Springer, Boston, pp 175–215
Roberts JT, Weikmans R, Robinson SA, Ciplet D, Khan M, Falzon D (2021) Rebooting a failed promise of climate finance. Nat Clim Chang 11(3):180–182
Rustiati EL (2009) Traffics and wildlife: a preliminary study on road-kill. Biospecies 2
Sanders JM (2011) Coming of age: how adolescent boys construct masculinities via substance use, juvenile delinquency, and recreation. J Ethn Subst Abus 10(1):48–70
Schwarzkopf SK (1984) Feeding of golden-mantled ground squirrels by park visitors at Crater Lake National Park
SimpleMaps (2021) World cities database. https://simplemaps.com/data/world-cities. Accessed 10 Oct 2021
Slaughter NG IV (2019) Central Park Squirrel Census Maps. Abstracts of the ICA 1:NA–NA
Smith-Patten BD, Patten MA (2008) Diversity, seasonality, and context of mammalian roadkills in the southern Great Plains. Environ Manag 41(6):844–852
Steiner M, Huettmann F (2021) Justification for a taxonomic conservation update of the rodent genus Tamiasciurus: addressing marginalization and mis-prioritization of research efforts and conservation laissez-faire for a sustainability outlook. The European Zoological Journal, 88(1), 86–116
Tang Y, Chen A, Zhao S (2016) Carbon storage and sequestration of urban street trees in Beijing. China Front Ecol Evol 4:53
Tangley L (1986) The urban ecologist. Bioscience 36(2):68–71
Tappe D, Frank C, Homeier-Bachmann T, Wilking H, Allendorf V, Schlottau K, Schmidt-Chanasit J (2019) Analysis of exotic squirrel trade and detection of human infections with variegated squirrel bornavirus 1, Germany, 2005 to 2018. Eur Secur 24(8):1800483
Tatebe H, Ogura T, Nitta T, Komuro Y, Ogochi K, Takemura T, Kimoto M et al (2019) Description and basic evaluation of simulated mean state, internal variability, and climate sensitivity in MIROC6. Geosci Model Dev 12(7):2727–2765
Terminix (2021) How to Get rid of squirrels?. https://www.terminix.com/pest-control/squirrels/removal/. Accessed 10 Oct 2021
Thévenin C, Morin A, Kerbiriou C, Sarrazin F, Robert A (2020) Heterogeneity in the allocation of reintroduction efforts among terrestrial mammals in Europe. Biol Conserv 241:108346
Thomas LS, Teich E, Dausmann KH, Reher S, Turner J (2018) Degree of urbanisation affects Eurasian red squirrel activity patterns. Hystrix 29(2):175–180
Thorington RW Jr, Koprowski JL, Steele MA, Whatton JF (2012) Squirrels of the world. JHU Press, Baltimore
Titley MA, Snaddon JL, Turner EC (2017) Scientific research on animal biodiversity is systematically biased towards vertebrates and temperate regions. PLoS One 12(12):e0189577
Tizard RJ (2016) Lariscus insignis. The IUCN Red List of Threatened Species 2016: e.T11305A22242321. https://doi.org/10.2305/IUCN.UK.2016-2.RLTS.T11305A22242321.en. Accessed 10 Oct 2021
Treenet (2022) Saldford Predictive Modeler. https://www.minitab.com/en-us/products/spm/. Accessed 5 Aug 2022
Uchida K, Yamazaki T, Ohkubo Y, Yanagawa H (2021) Do green park characteristics influence human-wildlife distance in arboreal squirrels? Urban Urban Green 58:126952
Van Horne B (2008) Conservation of ground squirrels. In: Rodent Societies. University of Chicago Press, Chicago, pp 463–471
Van Renssen S (2018) The inconvenient truth of failed climate policies. Nat Clim Chang 8(5):355–358
Wang Z, Ahmed Z, Zhang B, Wang B (2019) The nexus between urbanization, road infrastructure, and transport energy demand: empirical evidence from Pakistan. Environ Sci Pollut Res 26(34):34884–34895
Wikipedia (2021) Owens lake. https://en.wikipedia.org/wiki/Owens_Lake. Accessed 10 Oct 2021
Wilson DE, Reeder DM (Eds.) (2005) Mammal species of the world: a taxonomic and geographic reference vol. 1. JHU Press
Wolfe ND, Daszak P, Kilpatrick AM, Burke DS (2005, 1822) Bushmeat hunting, deforestation, and prediction of zoonotic disease. Emerg Infect Dis 11(12)
Wood DA, Phillipson J (1977) The utilisation of poison hoppers designed for grey squirrel (Sciurus carolinensis Gmelin) control. Biol Conserv 11(2):119–127
World Cities Culture Forum (2021) Public Green spaces. http://www.worldcitiescultureforum.com/data/of-public-green-space-parks-and-gardens. Accessed 10 Oct 2021
Worldclim (2021) Future climate data. https://www.worldclim.org/ data/cmip6/cmip6climate.html. Accessed 10 Oct 2021
Wright JD (1995) Ten essential observations on guns in America. Society 32(3):63–68
Yang YY, Kim JG (2016) The optimal balance between sexual and asexual reproduction in variable environments: a systematic review. J Ecol Environ 40(1):1–18
Yang P, Cui C, Li L, Chen W, Shi Y, Mi Z, Guan D (2020) Carbon emissions in countries that failed to ratify the intended nationally determined contributions: a case study of Kyrgyzstan. J Environ Manag 255:109892
Yukimoto S, Kawai H, Koshiro T, Oshima N, Yoshida K, Urakawa S, Ishii M, et al. (2019) The Meteorological Research Institute Earth System Model version 2.0, MRI-ESM 2. 0: description and basic evaluation of the physical component. J Meteorol Soc Japan. Ser. II
Zhao P, Lü B, de Roo G (2010) Urban expansion and transportation: the impact of urban form on commuting patterns on the city fringe of Beijing. Environ Plan A 42(10):2467–2486
Zoo Chat (2011) Squirrels in captivity. https://www.zoochat.com/community/threads/squirrels-in-captivity.470398/. Accessed 5 Aug 2022

Chapter 6
Squirrels in the Tropics: A Specific Synthesis of their Fate, Stress, Declines, and Extinctions

Abstract The squirrel conservation assessments in this book focus thus far mainly on the northern hemisphere; much of our research has been performed there. However, to have more complete global and holistic coverage and where much biodiversity actually sits, this study is centered around the population, habitat declines, and the urgent conservation need for squirrels in tropical regions. The included species here are all endemic to the tropics and occur predominately in Middle America/Northern South America, or Central Africa. As it has been previously identified and predicted in Chap. 3 (Steiner and Huettmann 2023), squirrels' occurrence in tropical regions is particularly high. The tropics are a squirrel hotspot indeed (e.g. Amazonia, Central American low elevation rainforests, West Africa, and Southeast Asia). To show this, a literature analysis has been performed on the literature records for the squirrel's distribution in the tropics. To contribute to this literature knowledge, state-of-the-art Species Distribution Models (SDMs) and Species Distribution Forecasts (SDFs) for the year 2100 have been created. With these future forecasts, we aim to present possible future climate-based scenarios and assess how these affect the squirrel's distribution. However, to approach it more systematically, additional threats, shortcomings, and problems for the tropics are also being discussed, supported by the literature. This aims to create a more holistic big-picture overview of the situation and its impacting factors for a future outlook. These impacting factors affect the lack of research data available for the tropics and wildlife conservation (especially for small mammals). We show that besides existing poverty, neo-colonialism, and development aid also lack of governmental stability has bad impacts on the habitat development and on the combined influence on squirrel conservation in such regions. Finally, suggestions are presented (e.g. methods to bridge such problems – telecoupling), to improve the current situation, and work towards long-term sustainable conservation of small mammals – and especially squirrels – with a focus on the tropics.

Keywords Squirrels · Tropics · Lack of governmental and development aid · Neocolonialism · Lack of research data availability

6.1 Introduction

Tropics were typically defined as "*Parts of the world that lies between two lines of latitude, the Tropic of Cancer, 23½° north of the equator, and the Tropic of Capricorn, 23½° south of the equator*" (Collins dictionary 2021; Feeley and Stroud 2018). However, in modern science often more than just the variable of latitude plays a role in the definition of this biome. According to Feeley and Stroud (2018), variables such as "*All areas of the Earth's surface that have a net positive energy balance*", "*All areas of the Earth's land surface where mean annual temperatures show little or no change with latitude*", "*All areas of the Earth's surface where temperatures do not go below freezing in a typical year*", *All areas of the Earth's surface where the mean monthly temperature is never less than 18 degrees C.*", etc. play a major role and therefore change the original definition – and the boundaries – of 'the tropics' (Feeley and Stroud 2018).

For this study, and to simplify the identification process of the species endemic to the tropics, a pre-defined area has been identified that we considered here as "the tropics". This area has been highlighted in Fig. 6.1.

Additionally, the tropical region is famous for its occurrence of high biodiversity (Raven et al. 2020). This can also be said for squirrels, as according to the SDMs and Table 3.3 in Chap. 3 (Steiner and Huettmann 2023) where 91 out of 233 squirrels primarily inhabit the tropics. Because for 76 squirrel species, there is not enough data available to create the SDMs, it is

Supplementary Information The online version contains supplementary material available at https://doi.org/10.1007/978-3-031-23547-4_6.

M. Steiner, F. Huettmann, *Sustainable Squirrel Conservation*, https://doi.org/10.1007/978-3-031-23547-4_6

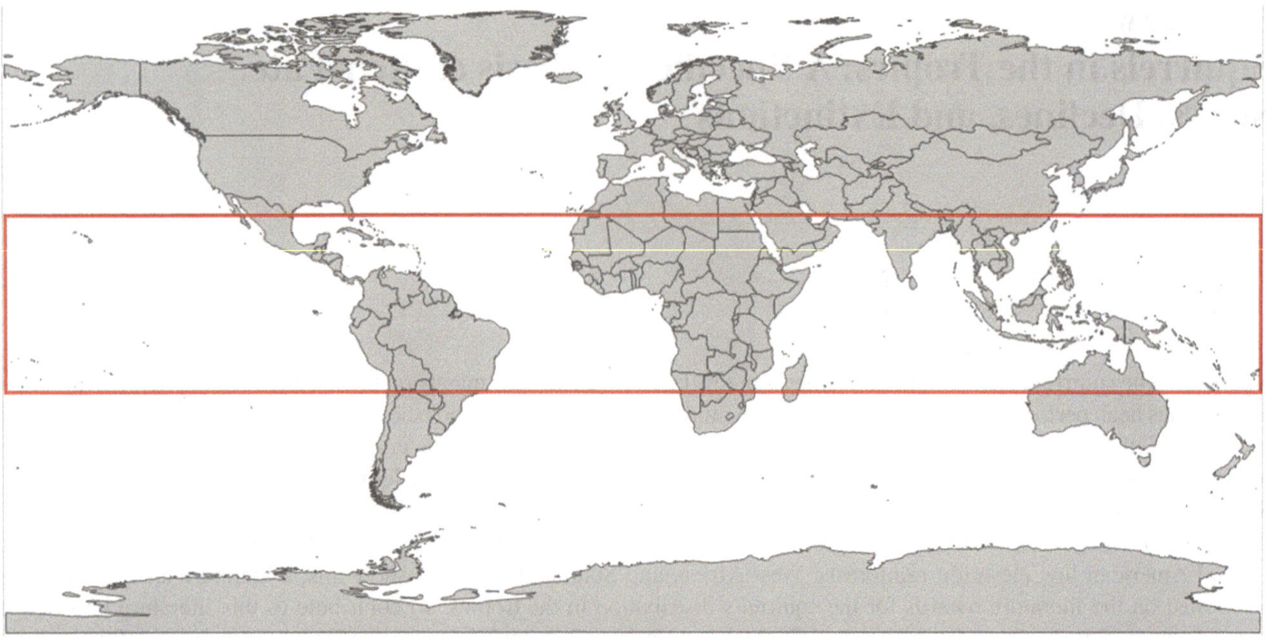

Fig. 6.1 Global map with the identification of the study area (~ the tropics)

expected that the actual number is even (significantly) higher. Due to the paucity of data and range maps, the exact validity of that statement remains unclear. From Table 3.3 in Chap. 3 (Steiner and Huettmann 2023), it can also be observed that "tropics" refers to the allocated "region of/under high risk" that occurs the most often in the performed analyses. Therefore, from that, it can already be concluded that the tropics are the prime biome that is to be considered under high risk, and that concerns the most squirrel species out of the four regions of/under high risk.

The reasons why the 'tropics' have been considered "under high risk" and, therefore, also the species within, will be further discussed in this study. This investigation is designed to begin with a literature review of the squirrel population primarily occurring in the tropics, followed by the presentation of additional SDMs that have been created in Chap. 3 (Steiner and Huettmann 2023). To also assess and provide a future outlook for those species, Species Distribution Forecasts (SDFs) have been created that will support this perspective since future scenarios can be presented that illustrate good support for the fact that the tropics are correctly included in the "regions of/under high risk".

For this study, five species have been selected that will be examined closely, especially in Sects. 6.1.2, 6.2.2, and 6.3.2, (see life history summary and endemic distribution summary in Table 6.1). These species are Layard's Palm Squirrel (*Funambulus layardi*), Ruwenzori Sun Squirrel (*Heliosciurus ruwenzorii*), Western Dwarf Squirrel (*Microsciurus mimulus*), Northern Amazon Red Squirrel (*Sciurus igniventris*), and Richmond's Squirrel (*Sciurus richmondi*) (see Taxonomic Species Numbers (TSN) in Table 6.1). These species are all endemic to the tropics and occur predominately in Middle America/ Northern South America, or Central Africa. Only one species out of these five occurs in Southeast Asia which is also part of the tropics. The reason for this is that in Chap. 7 (Steiner and Huettmann 2023), additional four species are analyzed similarly to here and they occur all in this squirrel hotspot (Southeast Asian tropics). Thus, to include species and regions more widely distributed around the world, here we focus on Central Africa and Middle/Northern South America. Such species, occur in the tropics and on islands, often even also in old-growth forests there, and inhabit therefore more than just one region of/ under high risk. Such species require special attention for conservation as it is expected that they are under a higher risk of extinction than a species that inhabits none, or one region of/under high risk (their vulnerability is multifarious).

6.1.1 Literature Review of Tropical Endemic Squirrel Species

The tropics are known to be the biodiversity-richest biome on this planet (Raven et al. 2020). This is also valid for squirrels. In Chap. 3 (Steiner and Huettmann 2023), more specifically in Table 3.3, the biome and also region of/under high risk that has been observed to be hosting the most squirrel species is represented by the tropics. However, these identifications are purely based on the SDMs created in Chap. 3. In order to support and back up the identification of this area, a brief literature review has been performed. This literature review focuses on the five species that have been mentioned above, and the species Niobe Ground Squirrel (*Lariscus niobe*), and Striped Bush Squirrel (*Paraxerus flavovittis*) which have shortly been discussed in Chap. 3 but are included here in the discussion and literature review.

6.1.2 Species Distribution Models (SDMs) for Tropics-Endemic Squirrel Species

In this study, an in-depth analysis of the modeled distribution of a hand full of tropical-endemic squirrel species is being performed. This aims to set a representative workflow and role model for a conservation species analysis and assessment which can later be repeated for almost any other vertebrate species on this planet.

6.1.3 Global Squirrel Distribution Forecast for 2100 Using Three Different Global Climate Models

For this study, five species have been selected from this global pool of 233 species, for an in-depth study on their distribution, threats to their conservation, and the fate of the future. In addition to those five newly selected species, the two species already selected in Chap. 3 will also be included in parts of this study. For these 7 species, further Species Distribution Forecasts (SDFs) have been performed for each species with the software Maxent using a generic ML algorithm, which will be analyzed and discussed here (see Maxent obtainable for the global public via https://biodiversityinformatics.amnh.org/open_source/maxent/). The 5 species that are newly presented here, will be discussed thoroughly, whereas the species that have been already presented in Chap. 3 will only be discussed marginally. For Niobe Ground Squirrel (*Lariscus niobe*), and Striped Bush Squirrel (*Paraxerus flavovittis*) the same analysis has been performed as for the other five species, these are however only presented in Appendix 6.1. These SDFs aim to forecast the species' distribution for the year 2100.

In order to validate our statements and models in this study, a global overview is being provided using the software TreeNet (see https://www.minitab.com/en-us/products/spm/ using Boosted Regression Tree (see details in Friedman 2001)), in addition to the SDFs created with Maxent. This means, that a global trend of all squirrels, with a focus on the tropics, is being presented with a forecast for 2100. This additional SDF presentation aims to involve another modeling source other than Maxent, to increase the accuracy of our findings, and validate them.

6.1.4 "Regions of/Under High Risk" Threats Identification for the Tropics

Chap. 3 (Steiner and Huettmann 2023) showed already that the tropics belong to the identified "regions of/under high risk", due to risks/shortcomings caused socially, politically, economically, and here most importantly, environmentally. To break these risks down into their primary threats, the tropics are known to be under the major threats of timber logging, logging for farming interests (beef cattle, soya, palm oil, pet food, etc.), mining, hydroelectricity, biodiversity loss, and climate change (droughts, flooding, warming influencing germination capability of species, warming causing too rapid habitat shifts which do not allow species to follow the ideal conditions, etc.). These risks will be discussed in this study while considering also which ones of these threats are affecting squirrels, their distribution range, and possibly even species richness, and abundance. These threat identifications aim to partly explain the changes in species range distribution presented in the SDMs and SDFs using Maxent and TreeNet (for the seven squirrels in detail, and all global squirrels in a general context). It is noteworthy that the changes in the SDFs only depict climate change-derived changes, whereas in the real world, as seen above, many more threats influence the species (and squirrels) on the ground. Therefore, the SDFs presented and discussed in this study are most likely an underestimation of the real changes that the species (the squirrels), will have to face in the future.

6.1.5 Problems in the Tropics: Research Gaps/Shortcomings, Society Details, Warfare, Neo-Colonialism, etc.

For a more meaningful discussion, additional risks and shortcomings of the nations, their societies, and the environment of the tropics are being discussed. Such shortcomings include research gaps, financial shortcomings, access difficulties on the ground, warfare, neocolonialism, difficult politics, micro country politics issues, economical difficulties, etc. These discussion points will be introduced in form of a literature review on the individual topics with options to further learn about these topics by checking out the included literature references. The reason why we introduce these common problems of the tropics is that they often rank higher on the priority list and agenda of governments and management agencies than the ecological conservation of small marginalized mammals (e.g. squirrels) (see Zuidema et al. (2013) for research shortcomings, Sachs (2001) for financial shortcomings, Alamgir et al. (2017) and Bawa et al. (2004) for access difficulties, Falk (1973) for warfare, Boshoff (2009) and Coghe (2014) for Neocolonialism, Coghe (2014) for difficult politics, Cooper and Shaw (2009) for micro country politics issues, and Sachs (2001) for economical difficulties; and more references in Sect. 6.3.5).

6.2 Methods

6.2.1 Literature Reviewing Methods and Focus Points

A literature review has been performed to gain a basic understanding of the species of interest's distribution.

For this study, the three sources IUCN Red List, Thorington et al. (2012), and Wilson and Reeder (2005) were of major help to locate literature records and distribution data for the species of interest. GBIF.org has been a major supporter of the open-access occurrence points of many global squirrel species. Unfortunately, like for many other squirrel species, the number of studies performed on the distribution of the species of interest for this study is very limiting and mostly absent.

6.2.2 Species Distribution Modeling and Forecasting Methods for Tropics-Endemic Squirrel Species Using Maxent

The SDMs have been obtained and used from the Appendix of Chap. 3 (Steiner and Huettmann 2023). In order to make them fit into this chapter, the symbology has been changed in open source QGIS and ArcGIS to make the figures more appealing and illustrative.

The SDFs have been specifically created for this study. Here SDMs have been created using 7 BioClim predictors and one elevation predictor. These predictors have all been obtained from www.worldclim.org, and these SDMs have been created using the software Maxent (https://biodiversityinformatics.amnh.org/open_source/maxent/) for each of the species Layard's Palm Squirrel (*Funambulus layardi*), Ruwenzori Sun Squirrel (*Heliosciurus ruwenzorii*), Western Dwarf Squirrel (*Microsciurus mimulus*), Northern Amazon Red Squirrel (*Sciurus igniventris*), Richmond's Squirrel (*Sciurus richmondi*), Niobe Ground Squirrel (*Lariscus niobe*), and Striped Bush Squirrel (*Paraxerus flavovittis*). The first SDM of each species in the figure presentation below (a) represents their current distribution based on 132 environmental predictors. The second figure (b) represents for each species their moodeled distribution for the year 2000 based on the 7 BioClim layers (BIO1, BIO7, BIO10, BIO11, BIO12, BIO16, BIO17) and one elevation layer. This SDM acts as a baseline of the "current" distribution using 7(+1) predictors. The next three SDMs can actually be called SDFs as they aim to forecast the species' distribution for the year 2100. These SDFs are based on the same 7(+1) predictors as the SDMs before, however, here the difference is that each model predicts the distribution for a different Global Climate Model (GCM). The first SDF always represents a possible cooling scenario compared to the "current" climate. This cooling scenario is represented by the "MIROC scenario"(see details on this GCM in Chap. 12 (Steiner and Huettmann 2023)). The second SDF represents the "business-as-usual" scenario entitled "MRI scenario". This MRI scenario is based on the average warming of the global temperature by approx. 2 °C (see details on this GCM in Chap. 12 (Steiner and Huettmann 2023)). Lastly, the highly probable scenario of a high magnitude warming (IPSL scenario) has been used to forecast each species' distribution. This IPSL scenario predicts global warming of approx. 3 °C (see details on this GCM in Chap. 12 (Steiner and Huettmann 2023)).

In order to further improve the interpretation of these SDMs and SDFs, a polygon boundary outline has been added. This outline symbolizes the hotspot distribution range from the SDMs, created with 132 predictors. Therefore, this outline aims to keep in mind the most accurate modeled distribution range for each species by looking at the different SDFs. These polygon boundaries have been created in QGIS, using the "Polygonize" tool for the SDM hotspots created with 132 predictors. This newly created vector file (Shapefile (.shp)) has then been overlaid over the SDMs and SDFs of interest. Using adjusted symbology settings aids to visualize the polygon outline better.

6.2.3 Global SDM Forecasting Methods for Tropics-Endemic Squirrel Species Using TreeNet

In Chap. 12 (Steiner and Huettmann 2023), FH created SDFs for four different squirrel groups (All global squirrel species merged, the ten most endangered squirrels merged, the genus *Geosciurus*, and the genera *Heliosciurus* and *Paraxerus* merged. In order to provide a tropics-focused perspective, in this study, a closer look has been taken at the global squirrel population and how their distribution range is expected to change by the year 2100 based on the same climatic predictors as presented in Sect. 6.2.2. For this study, MS has prepared the figures in order to make them more comparable with each other and the tropics region within. Out of the four squirrel groups, the one where all global squirrels are included has been chosen to be presented here as it follows our holistic approach in this book.

6.2.4 Methods to Identify the Threats for a Region of/Under High Risk (Tropics)

By analyzing the results of Chap. 3 in Steiner and Huettmann (2023), and especially Table 3.3, it was possible to identify which squirrels primarily occupy the tropics and/or are endemic to them. All these squirrels which are then considered to be endemic to the tropics or primarily occur there are assumed to be affected by the threats which are leading this biome/habitat to be considered "under threat/high risk". Consecutively, these risks have been compiled, presented, and supported by a literature review.

6.2.5 Identification Methods of the Shortcomings in Tropical Regions Regarding Research Shortcomings, Society, Warfare, Neo-colonialism, etc.

Simply by watching the news, reading global newspapers, or following the internet, one will find plenty of headers with negative news on climate change, timber logging in the Amazonas, etc. However, next to these globally covered up-to-date topics, other problems and shortcomings are also present in modern times in the tropical regions but that is barely covered by the western news and their information system. Such less-well discussed topics include consistent research gaps and biases, societal problems, warfare, neo-colonialism, health shortcomings, the absence of access to clean water and basic hygiene resources, etc. Also, this section is supported by a literature review on the mentioned topics. This mainly aims to provide more context for the changes in the squirrel population over time and to initiate a discussion on how such topics can also cause/lead to changes in the species distribution range and possibly even richness and abundance.

6.3 Results

6.3.1 Results of the Literature Review on the Squirrel Distribution in the Tropics

Squirrels are a widely recognized but poorly understood animal group (Family). Until this day, research on squirrels focused mostly on species of western interest, namely one European species; Eurasian Red Squirrel (*Sciurus vulgaris*), and a couple of other dominant species in the United States; Eastern Grey Squirrel (*Sciurus carolinensis*), and Eastern Fox Squirrel (*Sciurus niger*). This is reflected in the funding, budgeting, and subsequent publication record. In Alaska for instance, the last tree squirrel abundance/distribution estimation work was done in the 1970s (Kranowski 1969; Smith 1968). Besides these 'overstudied' squirrels, very few research projects are being initiated and carried out including the squirrels and their distribution. Therefore, for this literature review, three wider sources have been used; the IUCN Red List, Thorington et al. (2012), and Wilson and Reeder (2005).

Data and information are few, also for endemic species. But for instance, for Layard's Palm Squirrel (*Funambulus layardi*) it can be confirmed that the species is endemic to Southern Central Sri Lanka with small patches in the mountains of Southern India (Goonatilake 2016; Thorington et al. 2012, p. 161; Wilson and Reeder 2005, p. 781). This species is considered Vulnerable with a decreasing population trend according to the IUCN Red List and Goonatilake (2016). Further on, for Ruwenzori Sun Squirrel (*Heliosciurus ruwenzorii*) it can be safely assumed that the species is endemic to Rwanda, Burundi, far Eastern DRC, and Western Uganda (Cassola 2016; Thorington et al. 2012, p. 230). This species is considered of Least Concern with an unknown population trend according to the IUCN Red List and Cassola (2016). Next, the Western Dwarf Squirrel (*Microsciurus mimulus*) has been described as being endemic to Panama, Western Colombia, and Northwestern Ecuador (Reid 2016; Thorington et al. 2012, pp. 34–35; Wilson and Reeder 2005, p. 758). According to the IUCN Red List and Reid (2016), the species is considered as Least Concerned with a stable population trend. Additionally, for the species Northern Amazon Red Squirrel (*Sciurus igniventris*) it can be confirmed that the species inhabits the tropical regions of Colombia, Venezuela, Ecuador, Peru, and Brazil (Amori et al. 2016; Thorington et al. 2012, pp. 57–58; Wilson and Reeder 2005, p. 762). According to the IUCN Red List and Amori et al. (2016), the species is considered of Least Concern with an unknown population trend. Lastly, for Richmond's Squirrel (*Sciurus richmondi*) it can be stated that this species is endemic to Nicaragua (Koprowski and Roth 2018; Thorington et al. 2012, p 67; Wilson and Reeder 2005, p. 763). According to the IUCN Red List and Koprowski and Roth (2018), this species is considered as Near Threatened with an unknown population trend. These findings have been summarized in Table 6.1.

Table 6.1 contains the results of the literature review but also general information about the individual species discussed here. In addition to those general pieces of information, the most crucial threats for each species have been described which will be discussed further on.

6.3.2 *Results of the SDMs and SDFs of Some Tropics-Endemic Squirrel Species and Their Analysis*

In Chap. 3 (Steiner and Huettmann 2023), 233 SDMs of the current distribution have been presented for the global squirrel population. Unfortunately, for the sake of space and time, all of these cannot be presented and discussed individually. However, to make the best use of the information obtained, here we used 7 of those SDMs and created for them distribution forecasts for the year 2100. This has been accomplished with the software Maxent in this section (Sect. 6.3.2), and with TreeNet in the upcoming section (Sect. 6.3.3). However, first, a global overview of the distribution points which have been used for these 7 SDMs and SDFs is presented (Fig. 6.2). These distribution points that can be observed in Fig. 6.2 have been obtained from www.GBIF.org (https://cran.r-project.org/web/packages/rgbif/index.html – download DOI: https://doi.org/10.15468/dl.665b59) and have been used here for the species distribution modeling.

The individual SDMs and SDFs are presented in Figs. 6.3, 6.4, 6.5, 6.6 and 6.7.

Figures 6.3 represent the modeled distribution for Layard's Palm Squirrel (*Funambulus layardi*). According to Fig. 6.3a, the species is endemic to Sri Lanka and the Fiji Islands. Figure 6.3b shows a broad distribution pattern, covering most of the Southeast Asian Islands (Malaysia, Indonesia, New Guinea, Borneo, Philippines, etc.). The same can be said for Figs. 6.3c, d, depicting hotspot regions in Western India, Sri Lanka, Borneo, and New Guinea. The same is valid for for Fig. 6.3e without the high hotspot region in Western India. This species is considered Vulnerable with a decreasing population trend according to the IUCN Red List and Goonatilake (2016).

Figures 6.4 represent the modeled distribution for the squirrel species Ruwenzori Sun Squirrel (*Heliosciurus ruwenzorii*). Figure 6.4a illustrates the current modeled distribution of this species based on 132 environmental predictors with a small distribution range around the lakes Lake Kivu, and Lake Edward in Rwanda, Burundi, and Western DRC. The other figures present all a similar distribution just with a wider spread. There, the regions around Lake Victoria, especially in Uganda seem to be of high interest for this squirrel species. For all future scenarios a slight increase in their distribution range can be seen, where Fig. 6.4d seems to show the largest increase in the distribution range of this species. This species is considered of Least Concern with an unknown population trend according to the IUCN Red List and Cassola (2016).

Figures 6.5 represent the modeled distribution for the squirrel species Western Dwarf Squirrel (*Microsciurus mimulus*). This species seems to occur predominately in the Northern Andes. The distribution range is more restricted with the use of 132 predictors (Fig. 6.5a) compared to one created with the use of 7 BioClim and one elevation predictor (Figs. 6.5b–e). The distribution hotspots for Figs. 6.5b, c only differ minimally. However, for the warming climate scenarios (Figs. 6.5d, e) the original distribution hotspots seem to be maintained, with an extension of the range to the more Southern Andes. According to the IUCN Red List and Reid (2016), the species is considered of Least Concerned with a stable population trend.

Table 6.1 Summary of the life history, endemic distribution, and conservation outlook for the tropical-endemic squirrel species in the focus of this chapter

Common species name	Scientific species name	TSN	Description	Countries of endemic distribution	IUCN Conservation status class	IUCN population trend	Most crucial threats	References
Layard's Palm Squirrel	*Funambulus layardi*	632351	Size HB: 144–154 mm, T: 178 mm, Mass: 158–168 g, Colorations: orange yellow – chestnut/russet. No subspecies	Southern Central Sri Lanka, Southern India	Vulnerable	Decreasing	Habitat degradation due to large wood plantations, selective logging and forest fires. "*Habitat loss has been estimated at 20–50% over the last 10 years and is predicted at 30% over the next 10 years (Molur et al.* 2005)"	Goonatilake (2016), Molur et al. (2005), Thorington et al. (2012, p. 161) and Wilson and Reeder (2005, p. 781)
Ruwenzori Sun Squirrel	*Heliosciurus ruwenzorii*	632370	Size HB: 209–227 mm, T: 249–274 mm, Mass: 291–305 g, Colorations: medium grey – pure white, yellow brown – grey. 4 subspecies. Diurnal and arboreal, living near the ground in lower vegetations	Rwanda, Burundi, far Eastern DRC, and Western Uganda	Least concern	Unknown	Primary forests deforestation for agricultural land, extraction of forest resources (e.g. fuelwood), fire, and mining for coltan	Cassola (2016) and Thorington et al. (2012, p. 230)
Western Dwarf Squirrel	*Microsciurus mimulus*	632389	Size HB: 135–148 mm, T: 94–116 mm, Mass: 120 g, Colorations: Grizzled brown with yellow – orange. 3 subspecies. Diurnal living in the lower forest canopy and near the ground	Panama, Western Colombia, and Northwestern Ecuador	Least concern	Stable	Deforestation	Emmons and Feer (1997), Reid (2016), Thorington et al. (2012, pp. 34–35) and Wilson and Reeder (2005, p. 758)
Northern Amazon Red Squirrel	*Sciurus igniventris*	632429	Size HB: 240–95 mm, T: 240–305 mm, Mass: 500–900 g, Coloration: dark chestnut red – rusty orange/yellow with black. 2 subspecies. Diurnal and specialized on large tree seeds with extremely hard endocarps	Colombia, Venezuela, Ecuador, Peru, and Brazil	Least concern	Unknown	Habitat reduction, fragmentation, pet trade, and hunting	Amori et al. (2016), Emmons (1984), Thorington et al. (2012, pp. 57–58) and Wilson and Reeder (2005, p. 762)
Richmond's Squirrel	*Sciurus richmondi*	632434	Size HB: 160–218 mm, T: 130–184 mm, Mass: 235–268 g, Coloration: dark brown with some yellow – orange. No subspecies. Diurnal living in the lower forest parts, solitary	Nicaragua	Near threatened	Unknown	Hunting, trapping, habitat deforestation, and fragmentation	Koprowski and Roth (2018), Thorington et al. (2012, p. 67) and Wilson and Reeder (2005, p. 763)

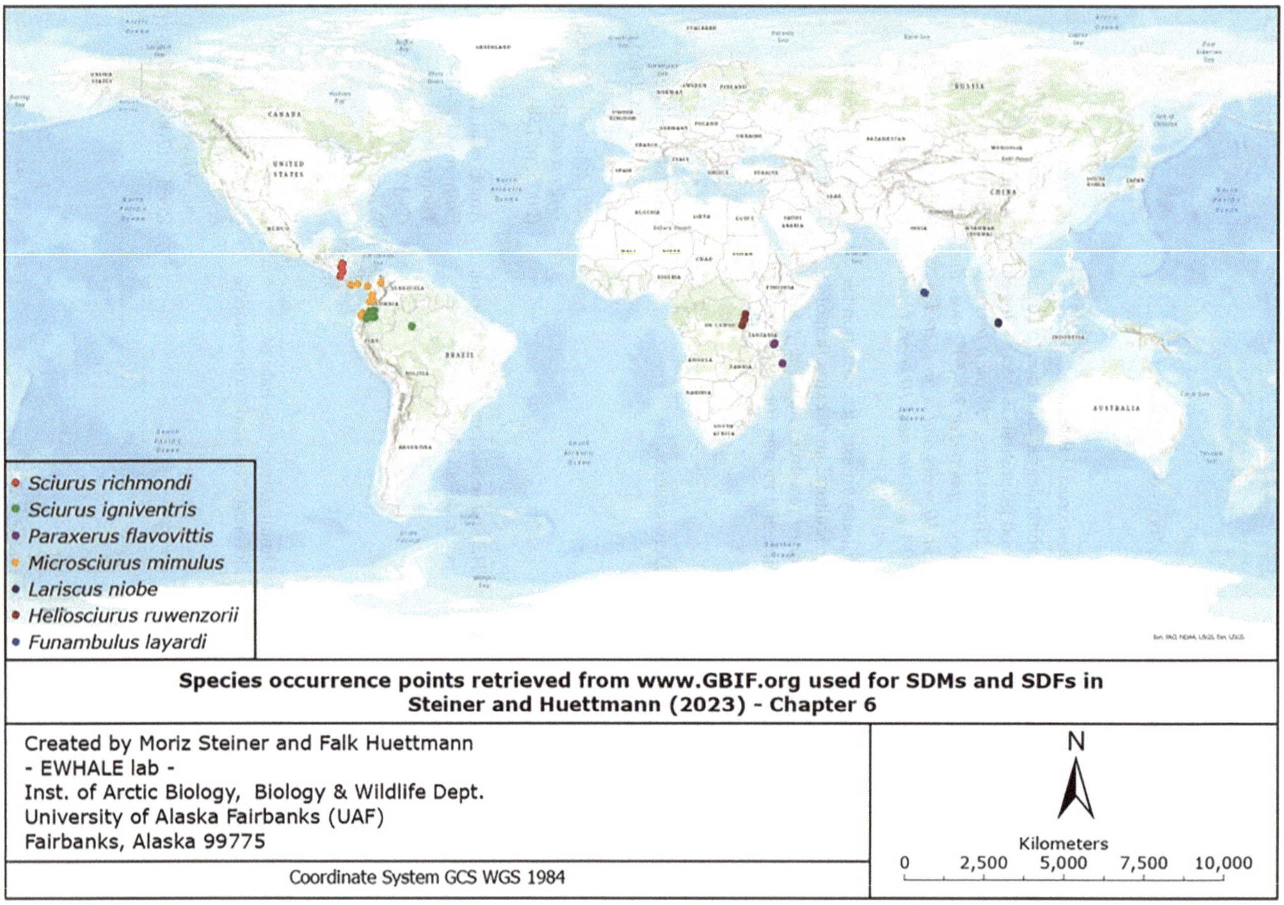

Fig. 6.2 Global distribution overview of the squirrel species using GBIF.org data for Layard's Palm Squirrel (*Funambulus layardi*), Ruwenzori Sun Squirrel (*Heliosciurus ruwenzorii*), Western Dwarf Squirrel (*Microsciurus mimulus*), Northern Amazon Red Squirrel (*Sciurus igniventris*), Richmond's Squirrel (*Sciurus richmondi*), Niobe Ground Squirrel (*Lariscus niobe*), and Striped Bush Squirrel (*Paraxerus flavovittis*)

Figures 6.6 represent the modeled distribution for the Northern Amazon Red Squirrel (*Sciurus igniventris*). This species seems to primarily occur in Southern Colombia, Western Ecuador, Northern Peru, and the Brazilian State of the Amazonas. This species can be truly considered a tropical squirrel species. The distribution does not differ much among the 4 SDFs created with 7(+1) predictors, however, comparing it with the SDM created with 132 predictors, a few distinctive differences can be noted. The 7(+1) predictor SDFs seem to have two major distribution hotspots. One on the Western coast of Colombia, and the other one in the Amazonas Rainforest. The 132 predictor SDM seems to lack the hotspot on the West coast of Colombia and only depicts the Amazonas Rainforest distribution hotspot. According to the IUCN Red List and Amori et al. (2016), the species is considered of Least Concern with an unknown population trend.

Figures 6.7 represent the modeled distribution for Richmond's Squirrel (*Sciurus richmondi*). According to Fig. 6.7a, this species is endemic to the low-elevation Atlantic rainforest in Nicaragua. Figure 6.7b shows major hotspots in Nicaragua, Costa Rica, Panama, and the Brazilian States of Pará, and Amapá. A similar distribution range can be observed in Fig. 6.7c, where, however, the hotspots in the previously mentioned nations decline and new one arises in the Brazilian state of Amazonas. Furthermore, the GCM MRI seems to depict a highly increased distribution range (Fig. 6.7d). Here, the hotpots seem to occur in most of Northern coastal areas of South America and the Brazilian States of Pará, Amapá, and Amazonas. Lastly, for Fig. 6.7e the Northern and Western coast of South America seems to be the most interesting for this squirrel species according to the IPSL climate scenario.

The figures presented in this section are a first attempt to present detailed SDMs and SDFs for these squirrel species. No science-based distribution models/maps are yet available for the science world and these species (mainly due to lack of research conducted). Therefore, these first ones and the estimations within, are still a bit fuzzy. However, as this work is open-access and publicly available with documentation, with further research and studies focusing on these non-western species, great improvements can be achieved (see for tropical forest research Bitariho et al. 2020).

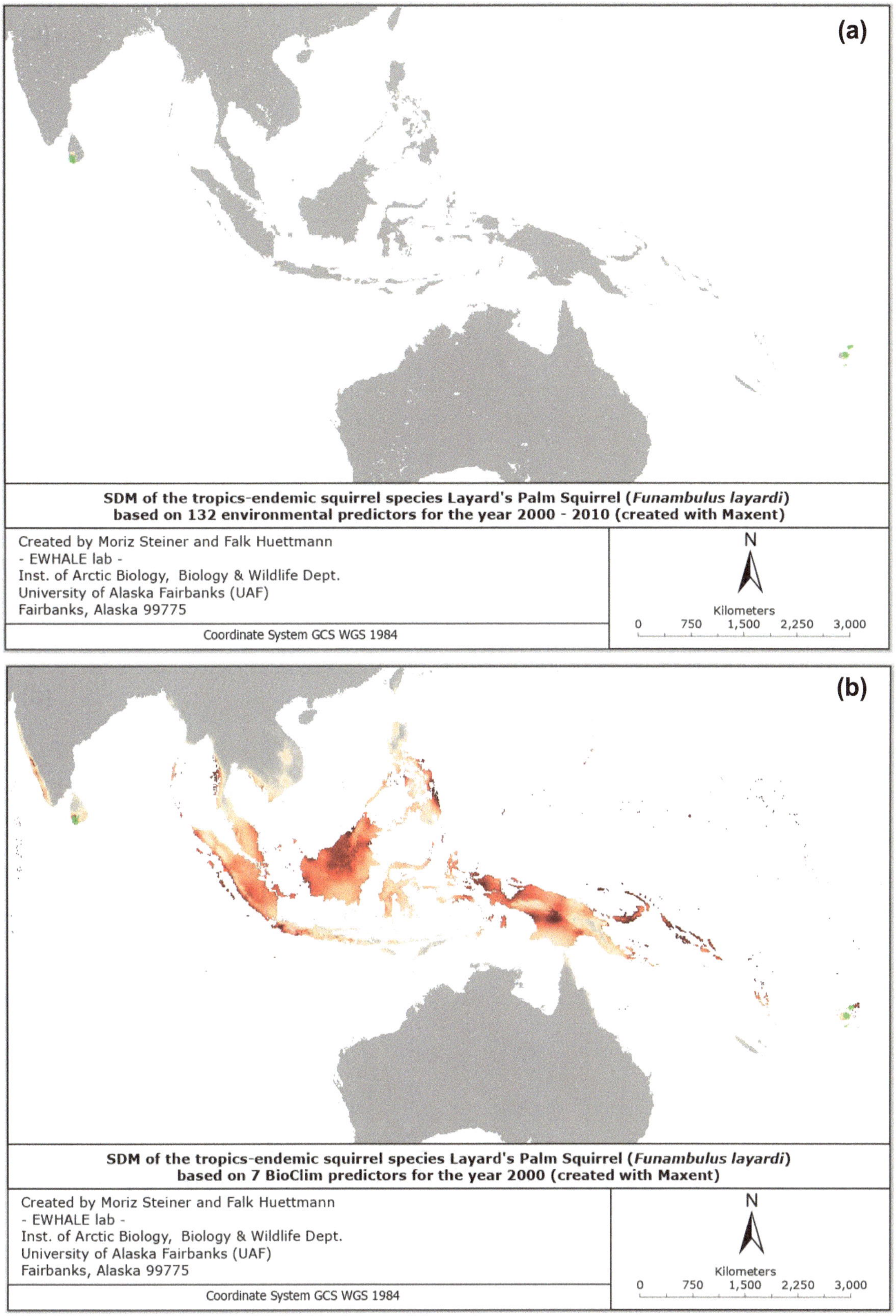

Fig. 6.3 SDM of the tropics-endemic squirrel species Layard's Palm Squirrel (*Funambulus layardi*) based on (**a**) 132 environmental predictors for the year 2000–2010 (**b**) 7 BioClim predictors for the year 2000 (**c**) 7 BioClim predictors for the year 2100 – Scenario MIROC – (**d**) 7 BioClim predictors for the year 2100 – Scenario MRI – (**e**) 7 BioClim predictors for the year 2100 – Scenario IPSL

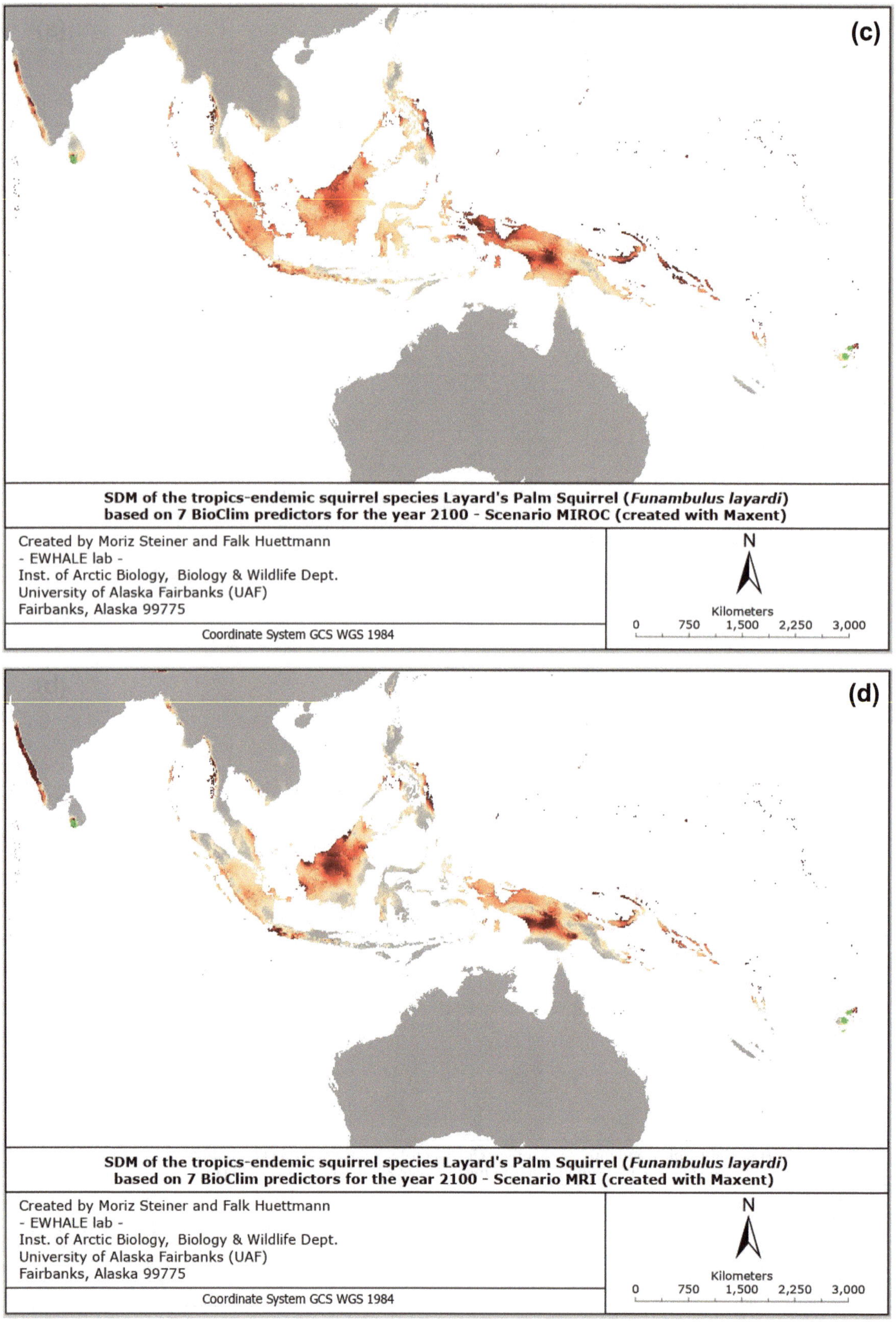

Fig. 6.3 (continued)

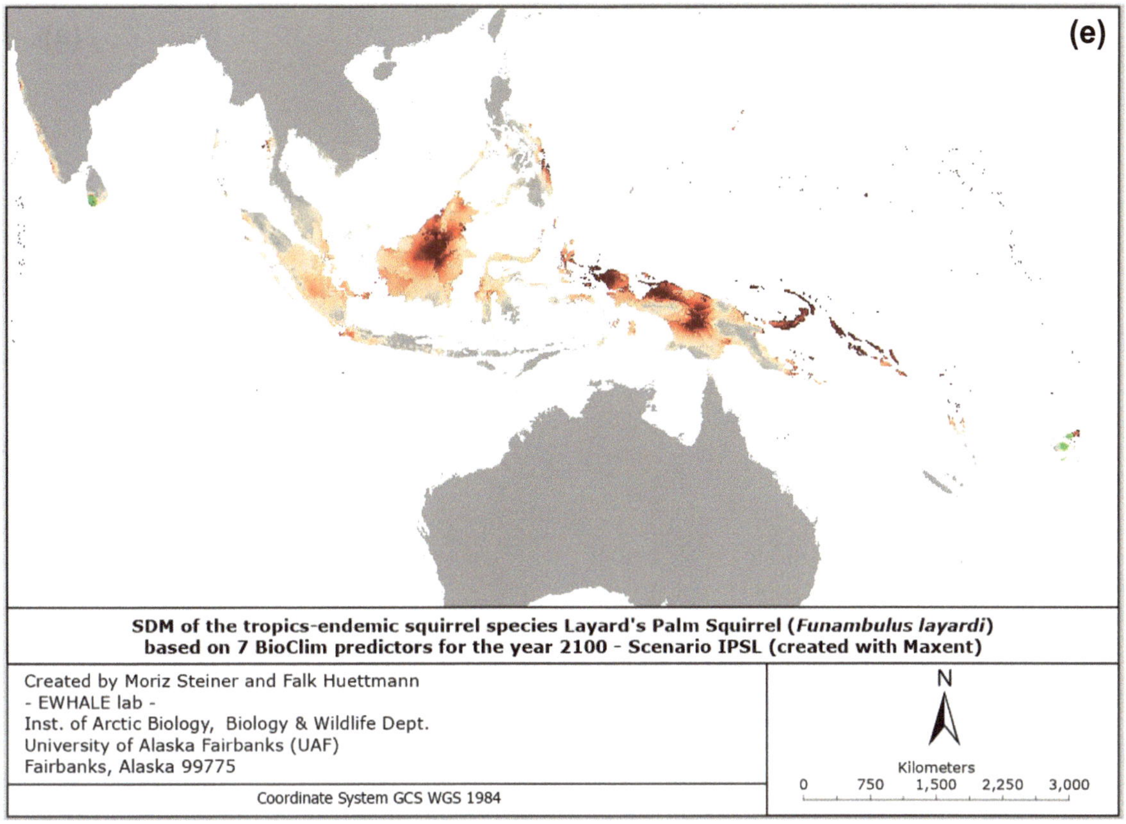

Fig. 6.3 (continued)

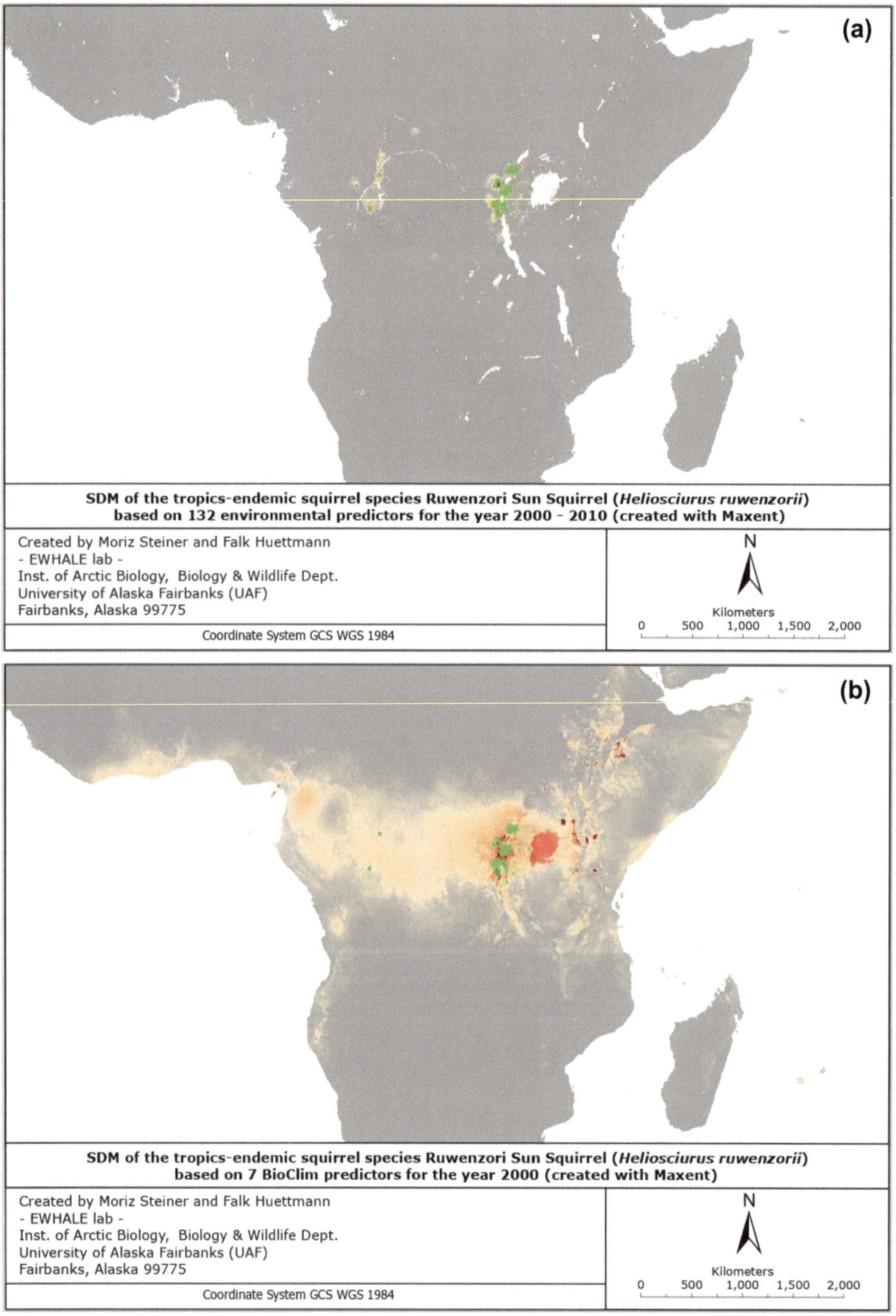

Fig. 6.4 SDM of the tropics-endemic squirrel species Ruwenzori Sun Squirrel (*Heliosciurus ruwenzorii*) based on (**a**) 132 environmental predictors for the year 2000–2010 (**b**) 7 BioClim predictors for the year 2000 (**c**) 7 BioClim predictors for the year 2100 – Scenario MIROC – (**d**) 7 BioClim predictors for the year 2100 – Scenario MRI – (**e**) 7 BioClim predictors for the year 2100 – Scenario IPSL

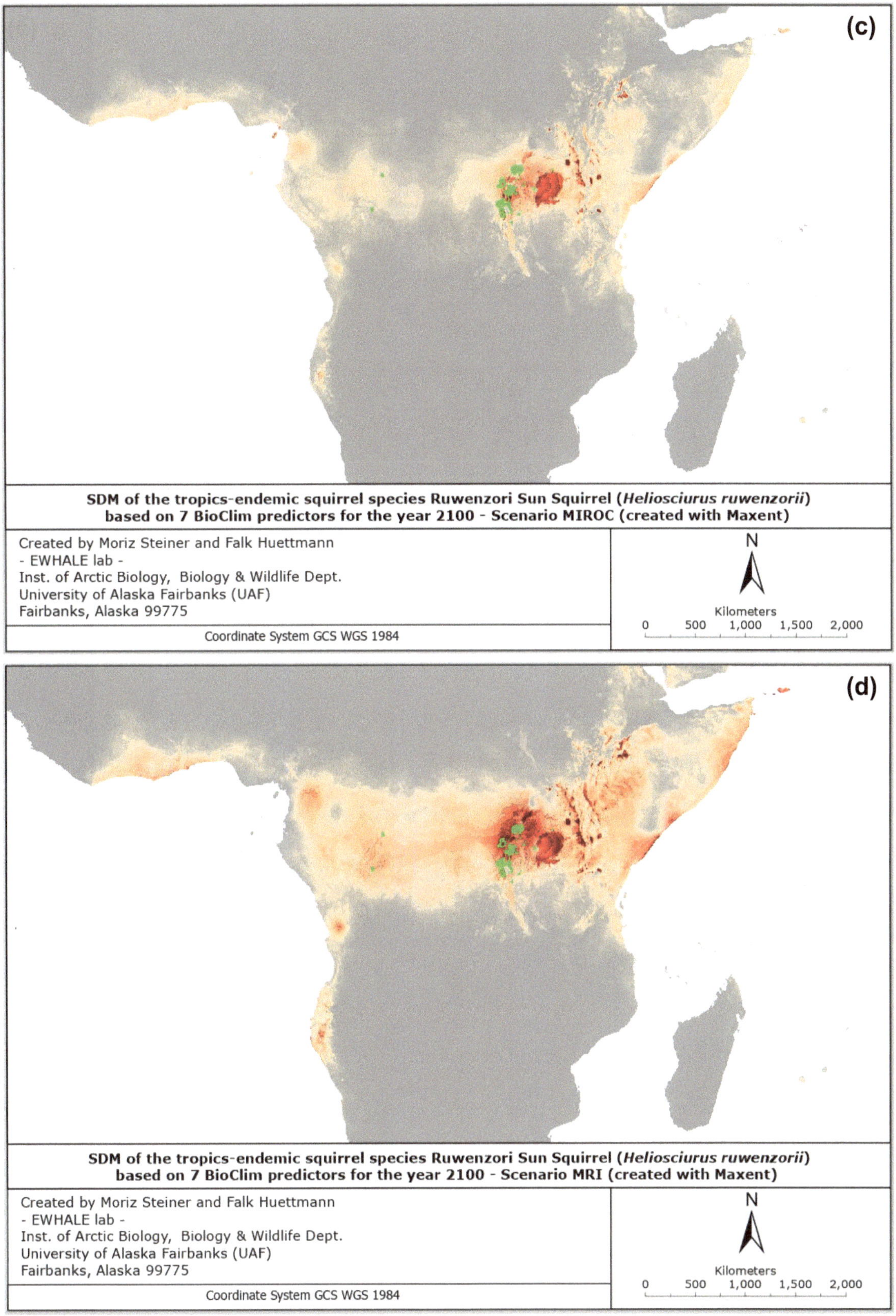

Fig. 6.4 (continued)

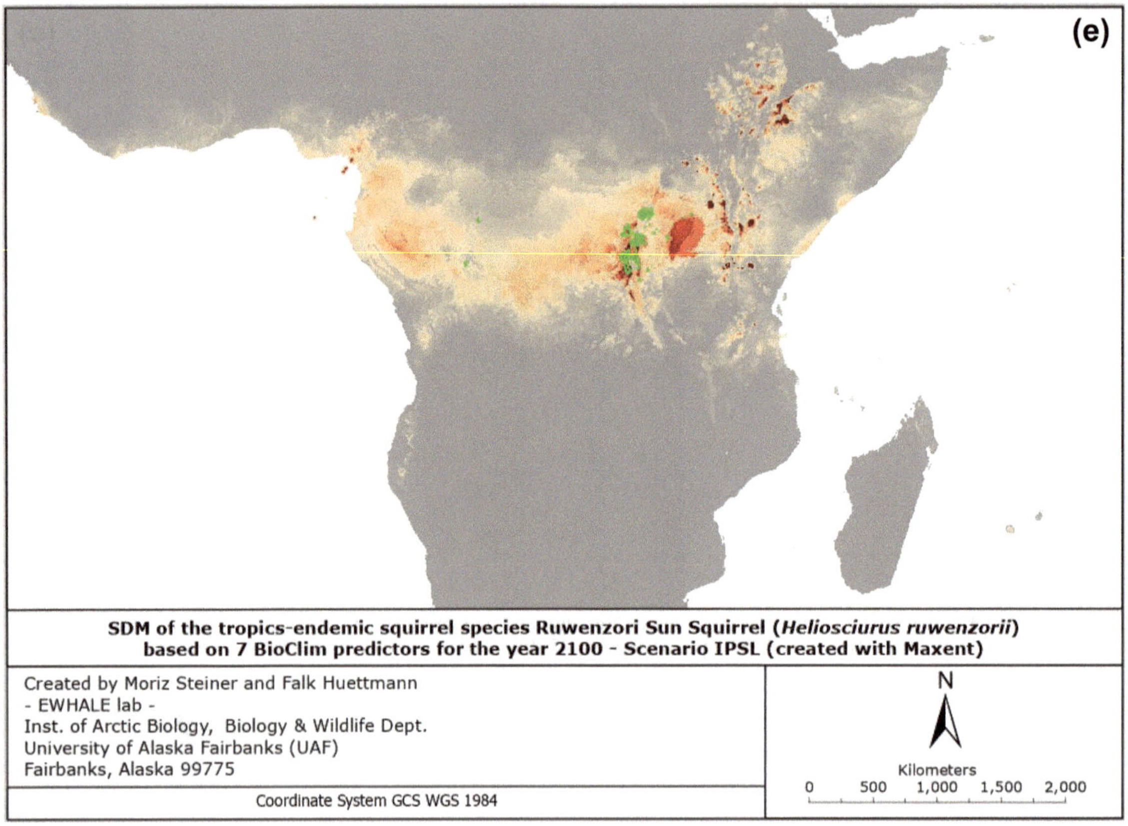

Fig. 6.4 (continued)

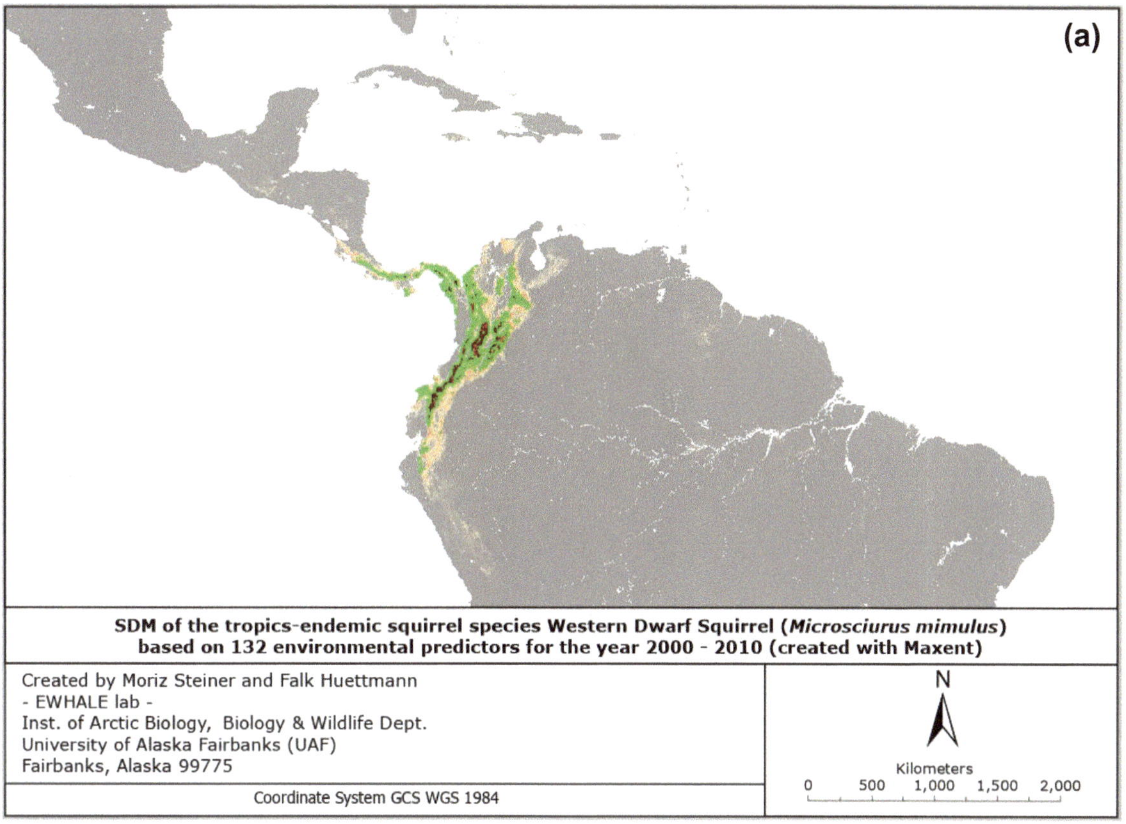

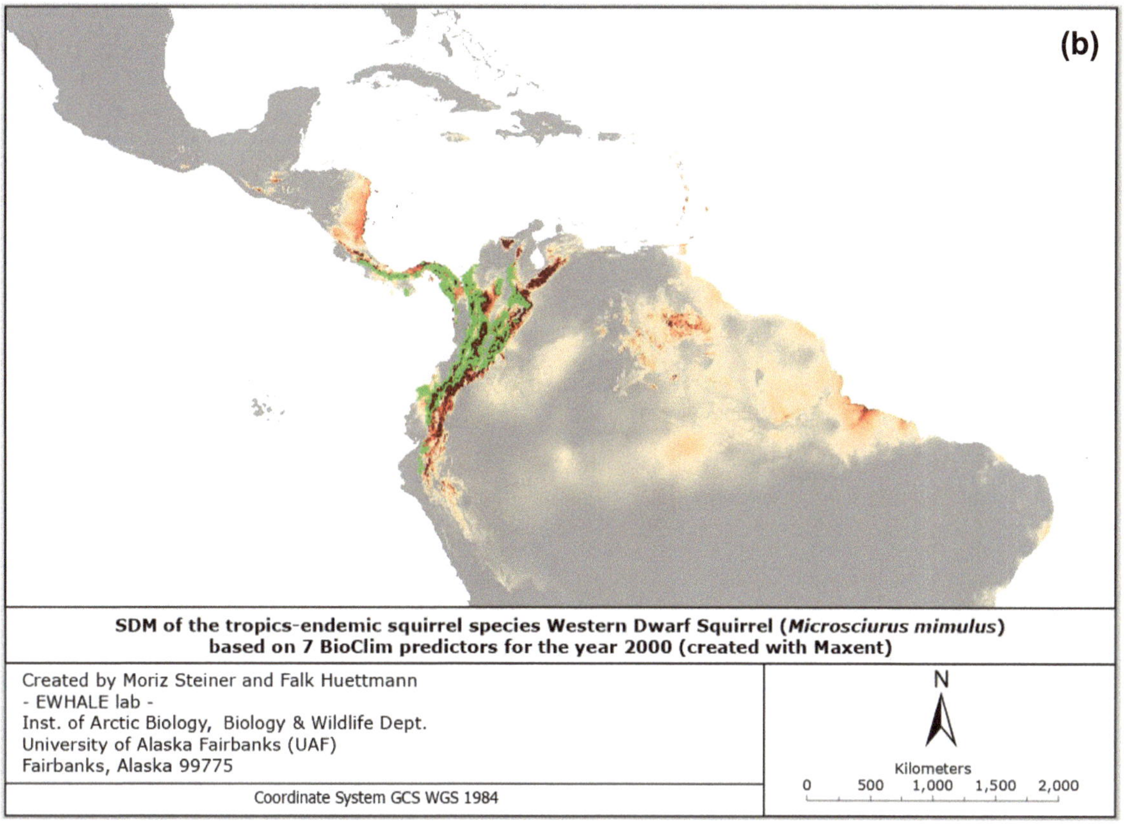

Fig. 6.5 SDM of the tropics-endemic squirrel species Western Dwarf Squirrel (*Microsciurus mimulus*) based on (**a**) 132 environmental predictors for the year 2000–2010 (**b**) 7 BioClim predictors for the year 2000 (**c**) 7 BioClim predictors for the year 2100 – Scenario MIROC – (**d**) 7 BioClim predictors for the year 2100 – Scenario MRI – (**e**) 7 BioClim predictors for the year 2100 – Scenario IPSL

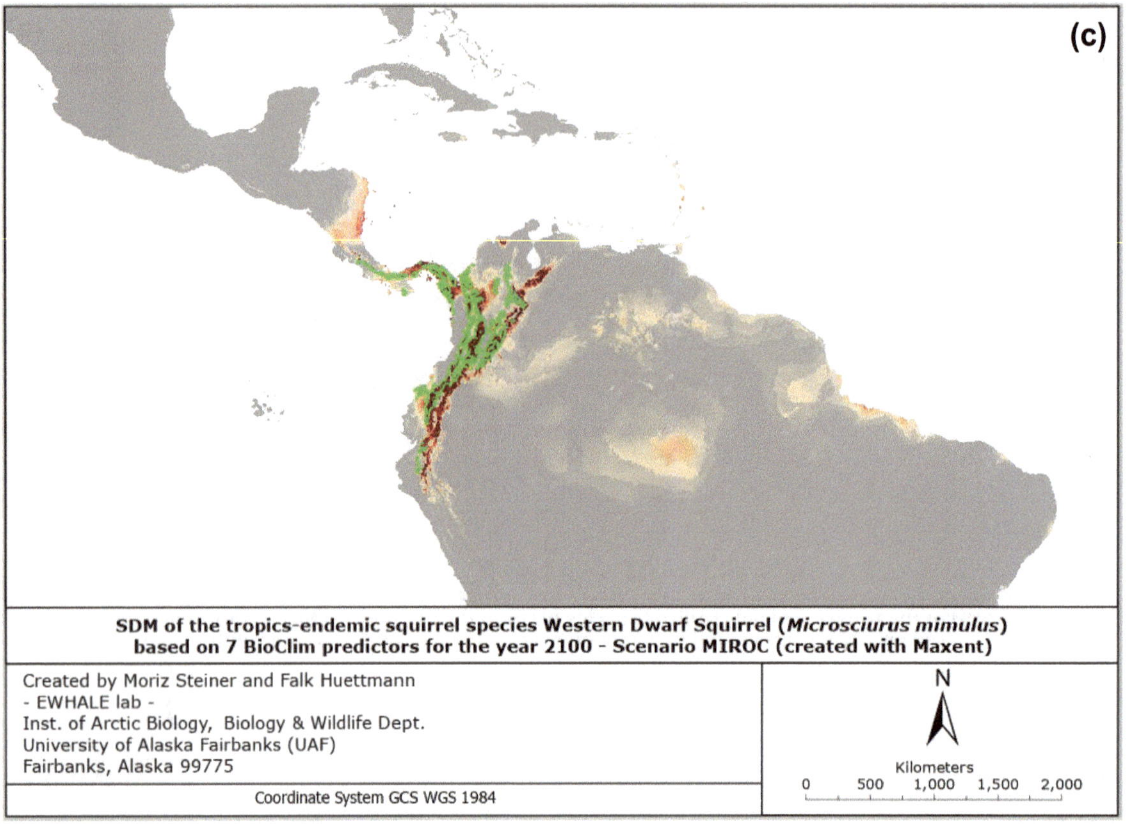

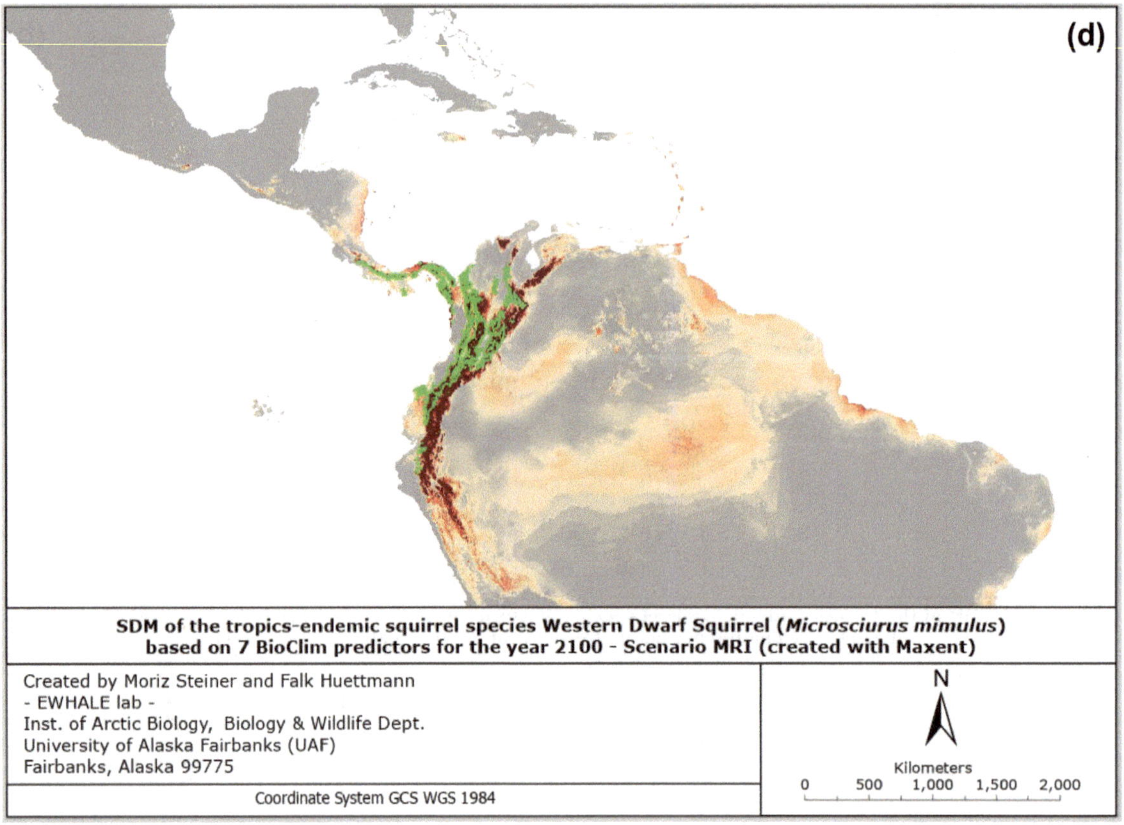

Fig. 6.5 (continued)

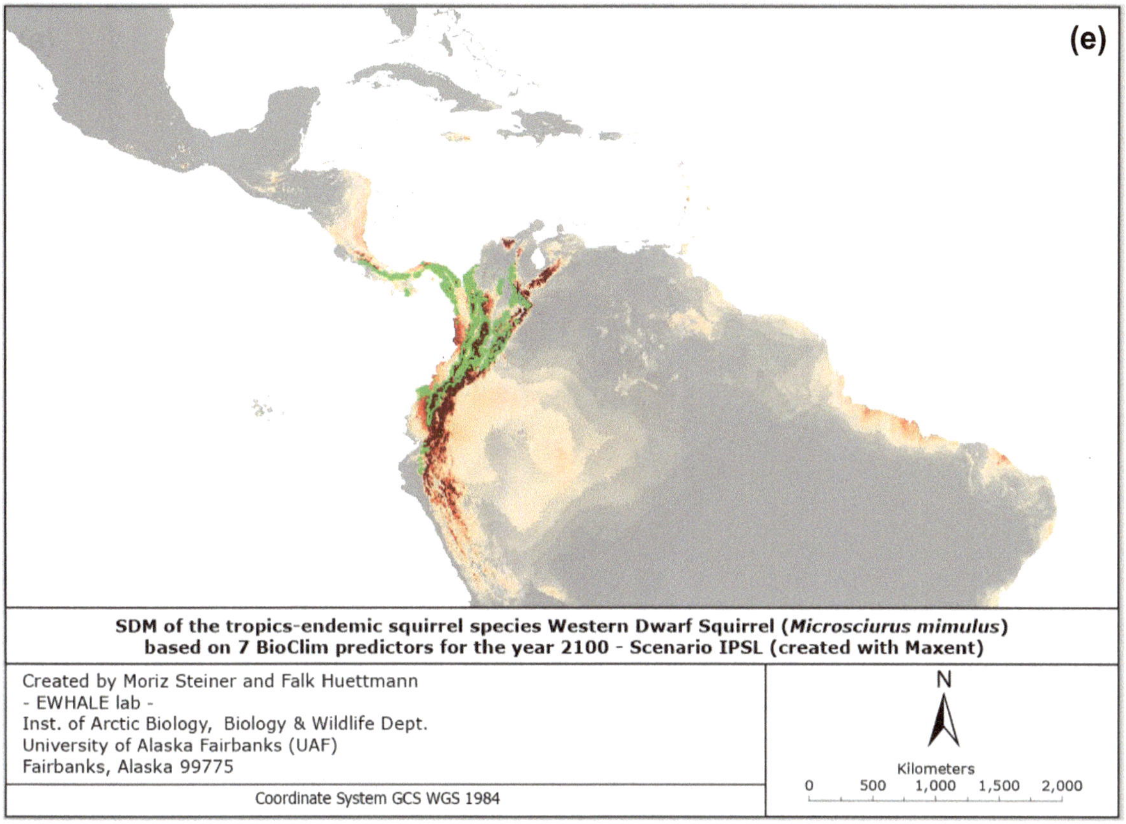

Fig. 6.5 (continued)

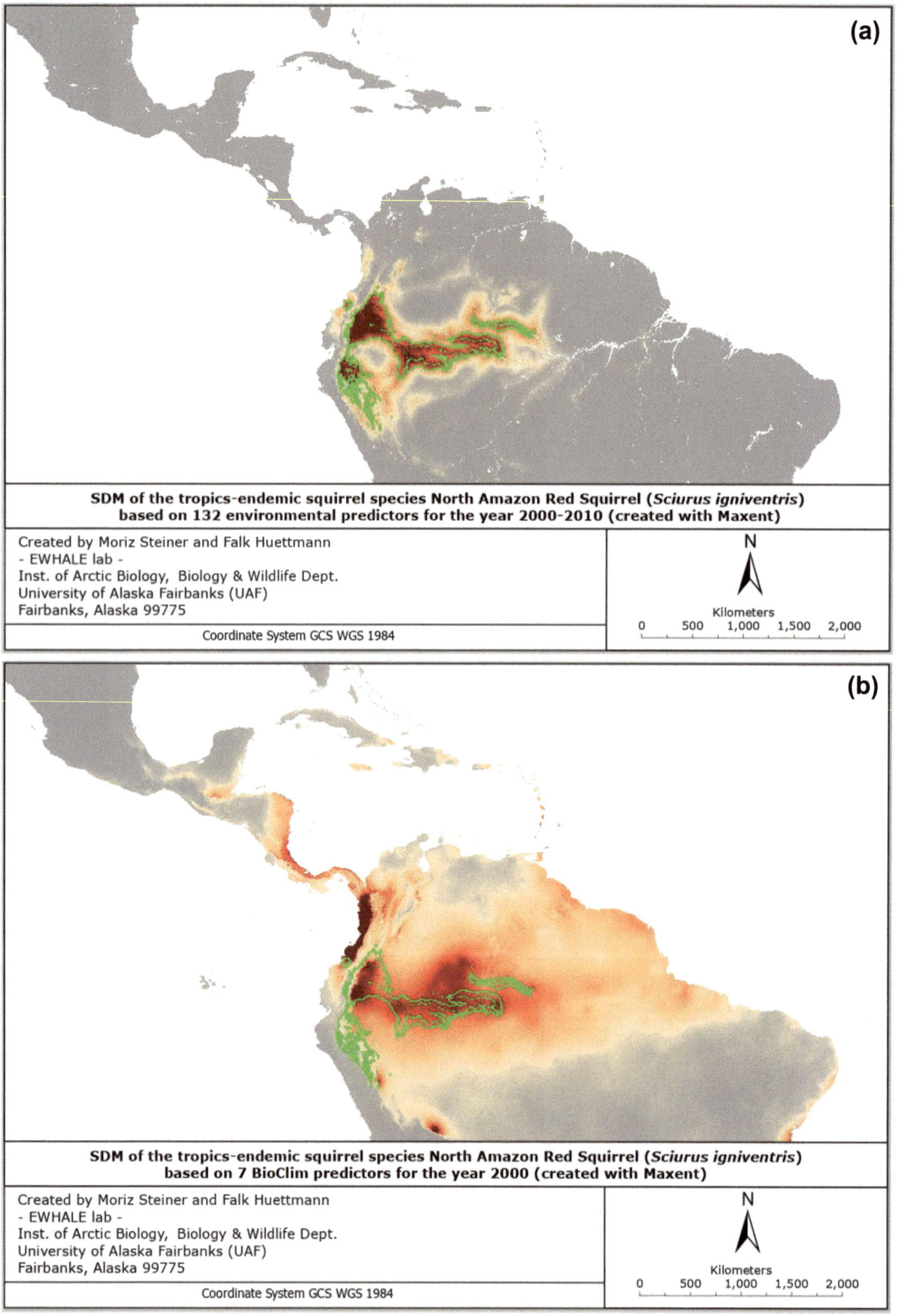

Fig. 6.6 SDM of the tropics-endemic squirrel species Northern Amazon Red Squirrel (*Sciurus igniventris*) based on (**a**) 132 environmental predictors for the year 2000–2010 (**b**) 7 BioClim predictors for the year 2000 (**c**) 7 BioClim predictors for the year 2100 – Scenario MIROC – (**d**) 7 BioClim predictors for the year 2100 – Scenario MRI – (**e**) 7 BioClim predictors for the year 2100 – Scenario IPSL

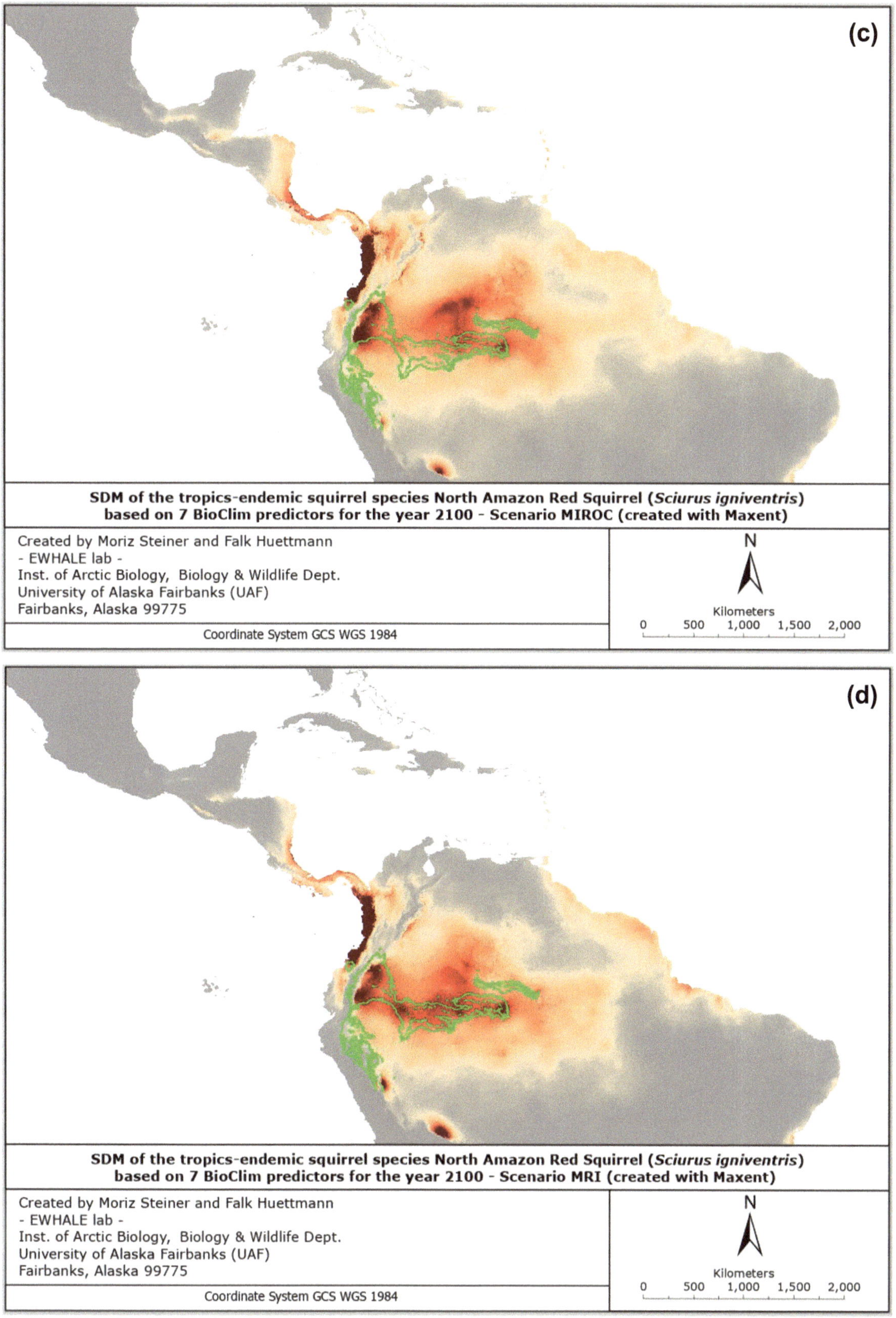

Fig. 6.6 (continued)

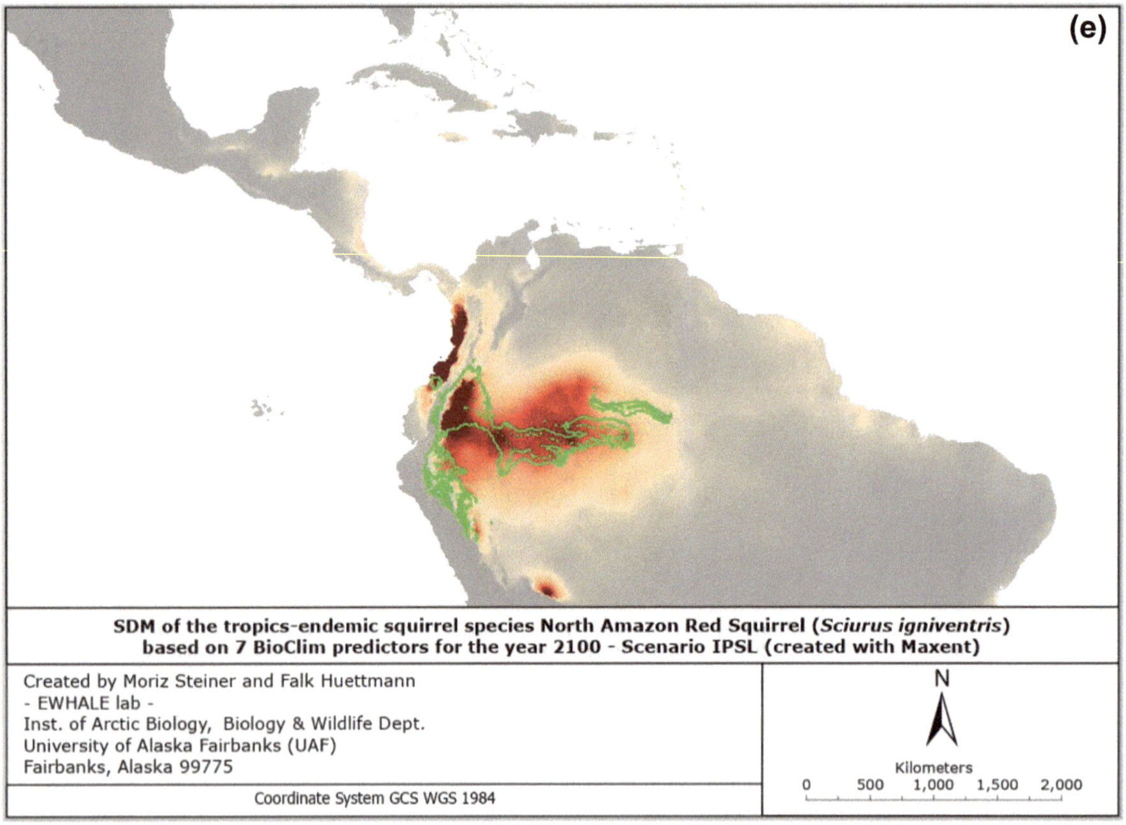

Fig. 6.6 (continued)

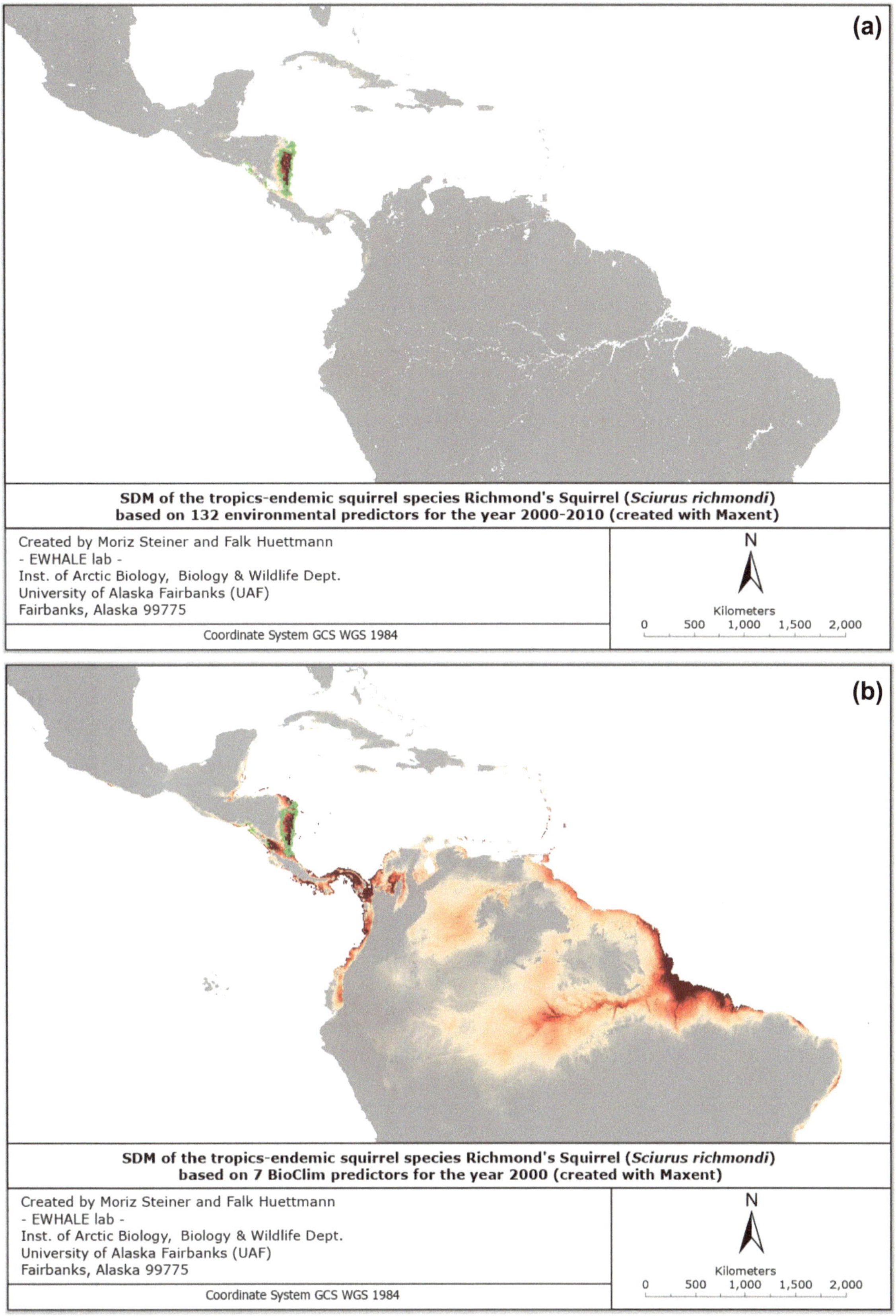

Fig. 6.7 SDM of the tropics-endemic squirrel species Richmond's Squirrel (*Sciurus richmondi*) based on (**a**) 132 environmental predictors for the year 2000–2010 (**b**) 7 BioClim predictors for the year 2000 (**c**) 7 BioClim predictors for the year 2100 – Scenario MIROC – (**d**) 7 BioClim predictors for the year 2100 – Scenario MRI – (**e**) 7 BioClim predictors for the year 2100 – Scenario IPSL

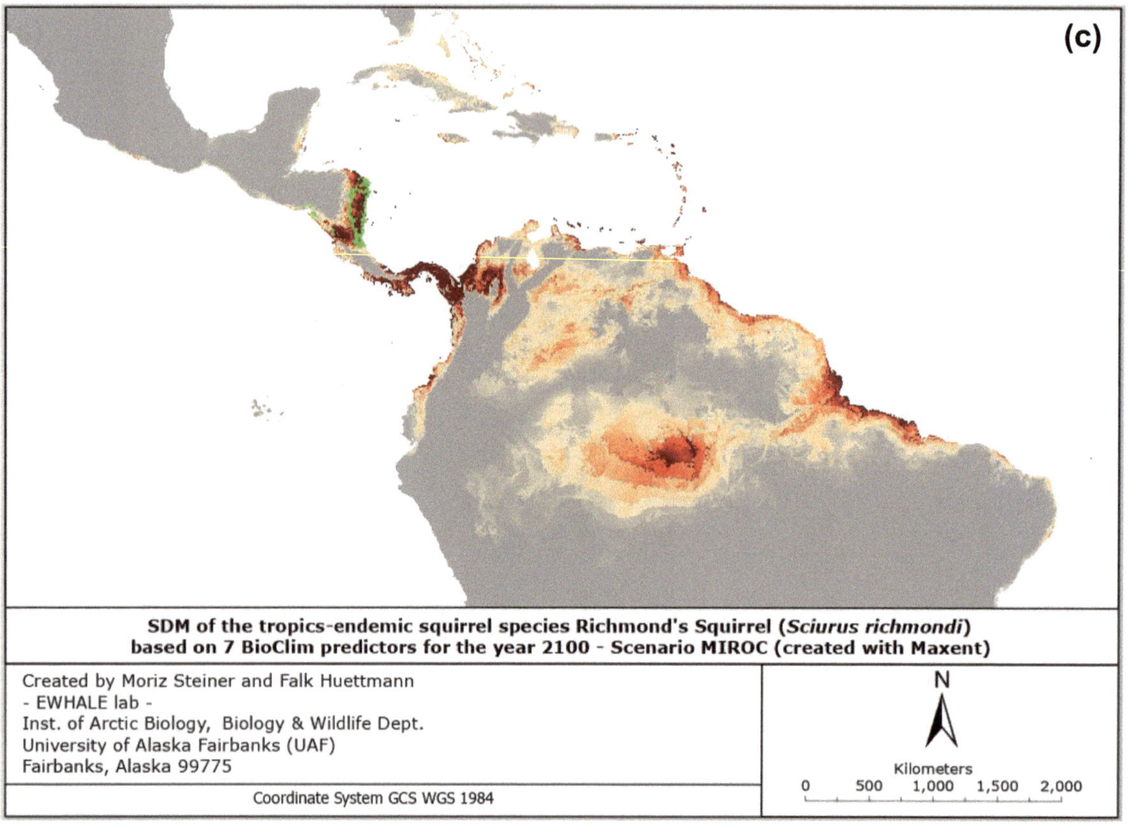

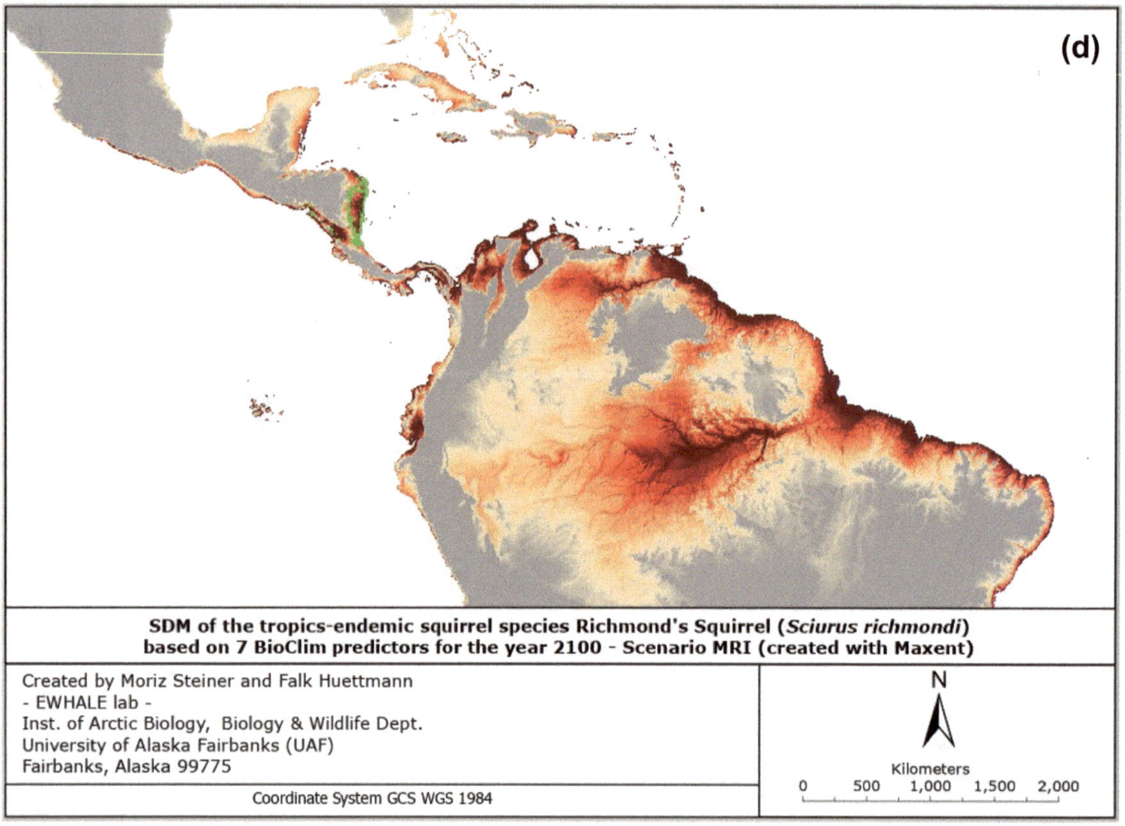

Fig. 6.7 (continued)

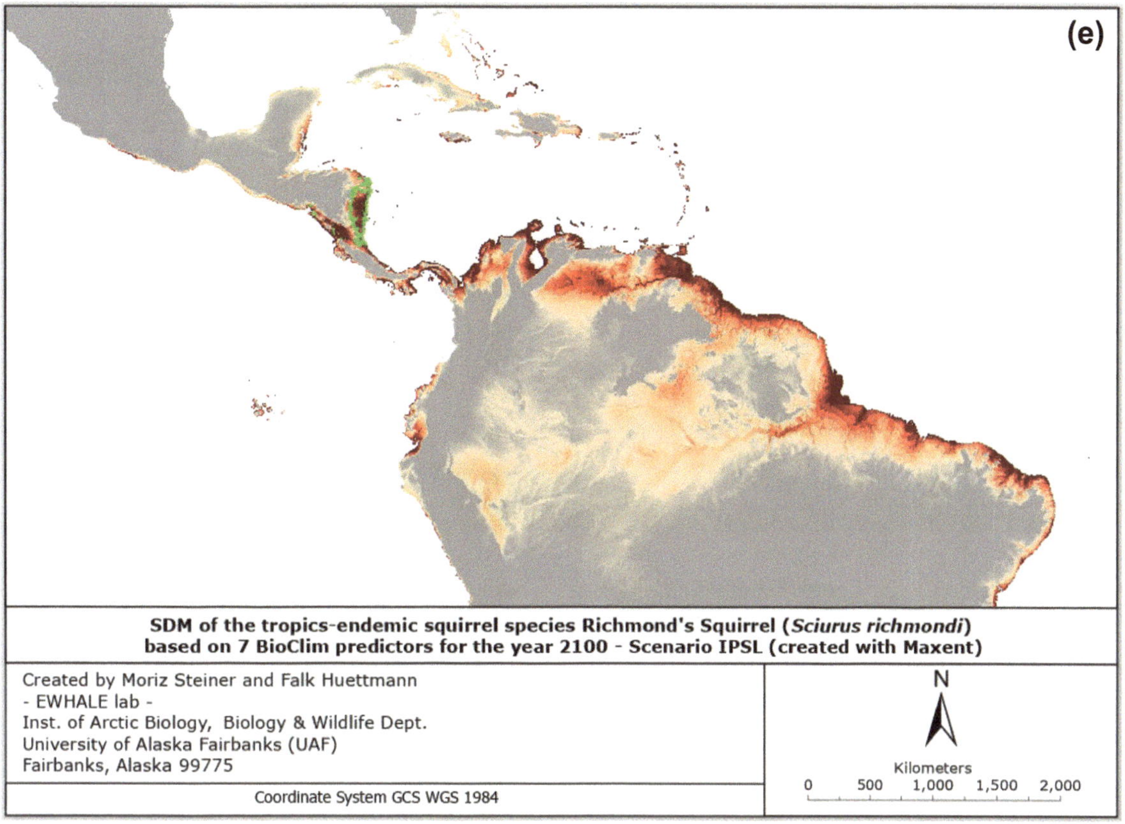

Fig. 6.7 (continued)

6.3.3 *Results of the Global Distribution Forecast (GDF) for All Squirrel Species Occurring in the Tropics*

So far, a species-specific approach has been used to present the current distribution and the resulting forecasts for 2100. Here, a general, more holistic approach is being applied. The global trend of all squirrel species is being observed and future trends are presented. Similarly, as in the previous section (Sect. 6.3.2), the three GCM scenarios for 2100 are MIROC6 (= Cooling scenario, abbreviated from here onwards as MIROC (Tatebe et al. 2019)), MRI-ESM2-0 (= Business-as-usual scenario, abbreviated from here onwards as MRI (Yukimoto et al. 2019)), and IPSL-CM6A-LR (= High warming scenario, abbreviated from here onwards as IPSL (Boucher et al. 2020)). For this analysis and the modeling, the software TreeNet has been used, with which Figs. 6.8a–d have been created. The reason for including another software other than Maxent in this context is to verify and validate the findings from the models created with Maxent.

Here, we focus specifically on the region within the red rectangle shown in Figs. 6.8 since this is considered by most entities as 'the tropics' (Feeley and Stroud 2018). One can observe a clear shift in the major distribution ranges within this rectangle. As Fig. 6.8a represents the current situation, this is considered the baseline. From that distribution, an overall increase in the distribution range can be observed for the MIROC scenario, whereas for the other two scenarios (= warming scenarios MRI, and IPSL) a clear decline can be observed. A further in-depth discussion can be found in the discussion Sect. 6.4.

6.3.4 *Estimated Future Threats to the Tropics and Its Global Squirrel Population*

The tropics have been identified to be part of the regions of/under high risk. This is due to several risks that endanger this biome and the species living in it. To gain a better understanding of these risks, the overall situation, and their severity, in this section the known risks are broken down and presented individually. It must be noted that these topics are only a short introduction to the overall situation and its problems. Many more threats are influencing real-life situations, especially on a country and global scale. These short introductions are supported with references for wider and deeper reads.

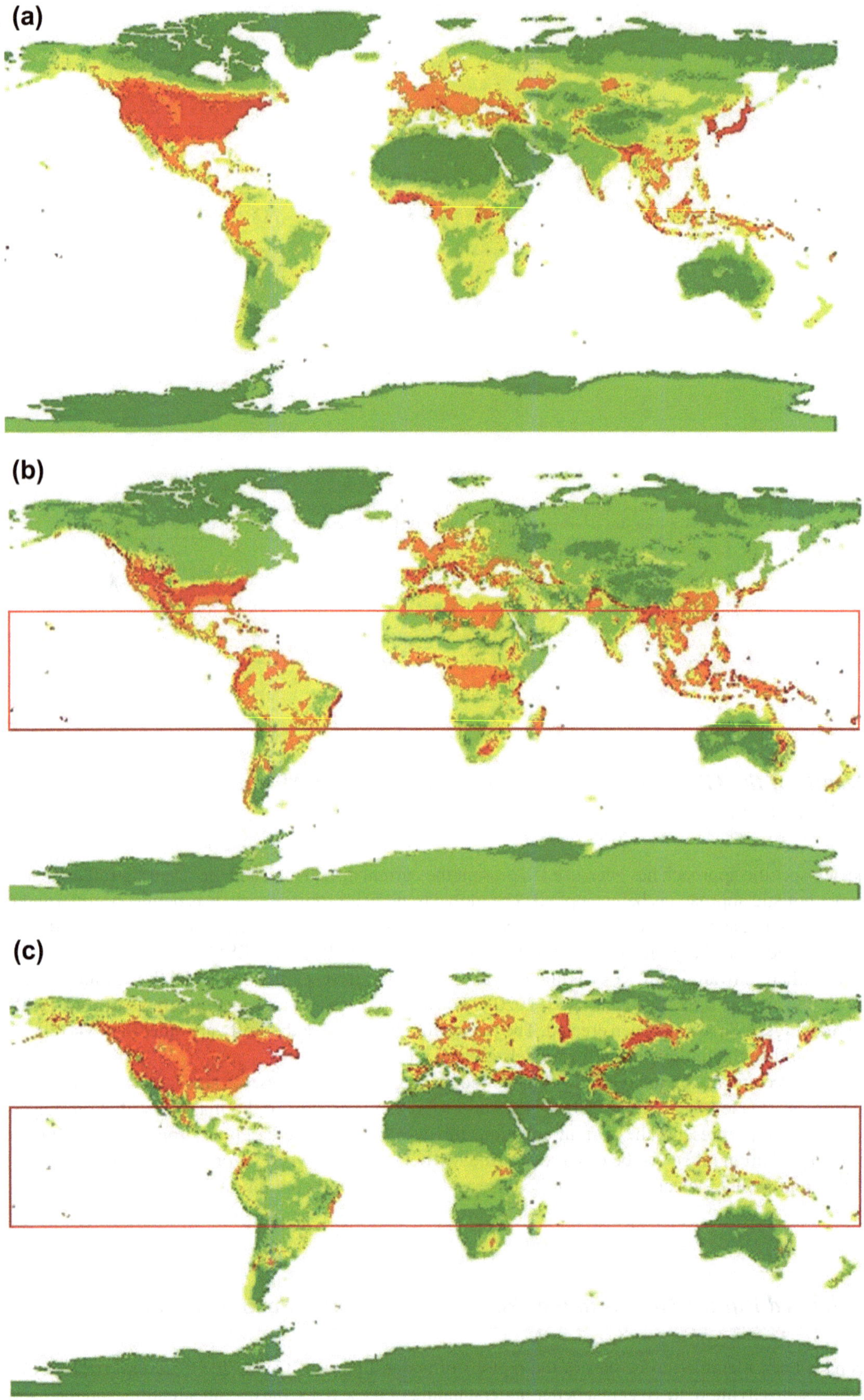

Fig. 6.8 Modeled distribution of all global squirrel species based on the (**a**) Current climate scenario, (**b**) MIROC cooling scenario, (**c**) MRI business-as-usual scenario, and (**d**) IPSL hot scenario

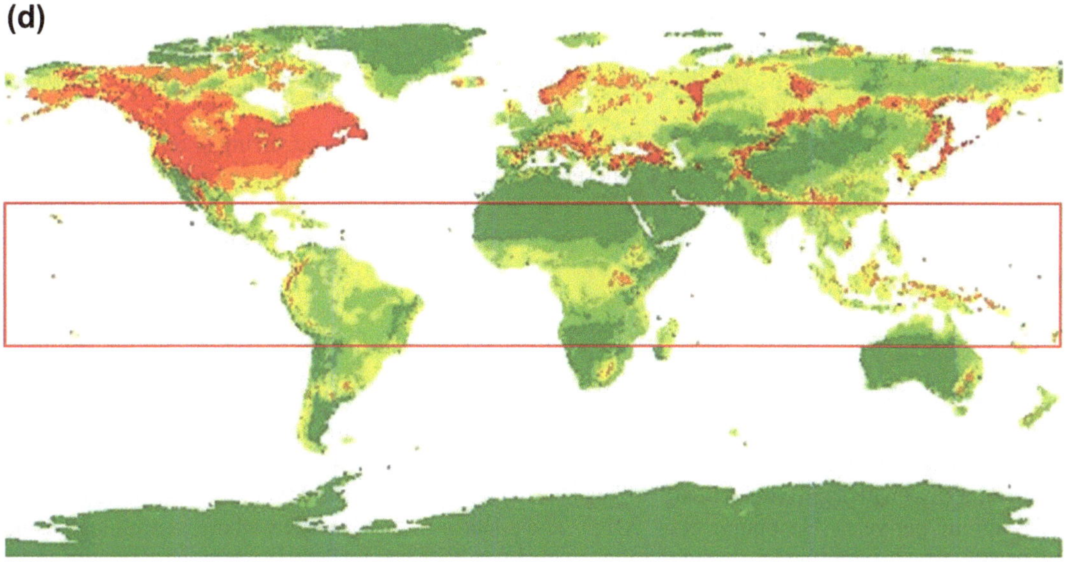

Fig. 6.8 (continued)

6.3.4.1 Timber Logging

This is a major and globally known issue (see Industrial Logging journal (Industrial Logging 2021)). To quantify this activity in the forests of the tropics, according to Weisse and Goldman (2020), the planet lost a football pitch-sized area of primary rainforest every 6 seconds in 2019. Further quantifications indicate an almost 30 million hectares loss of forest cover in the tropics in 2016 alone (Weisse and Goldman 2017), which equals 6.1 million hectares of primary tropical forest (Seymour 2020), and is approximately as large as Latvia or West Virginia (USA). And the recovery of this lost wilderness forest is very complex and difficult to achieve (Katovai et al. 2021). This loss of primary tropical forest is not only impacting the animal and plant species living within it but also ecological services and is associated with 1.6 billion+ people (Chao 2012). This deforestation is also to a large extent illegally practiced (Glastra 2014), and even occurs within (tropical) national parks, e.g. Yen et al. (2005).

6.3.4.2 Logging for Farming Interests (Soya, Palm Oil, Beef Cattle, Animal Food, etc.)

Next to timber logging for timber use and trade, large forest parts are also cut for agricultural use (FAO 2016; Gibbs et al. 2010; Rainforest Alliance 2020). This will virtually remove any canopy habitat for squirrels. According to Curtis et al. (2018), the shift from tropical rainforest to agricultural land accounts for 24% of the overall deforestation and forest loss. FAO (2016) lists more extreme numbers, namely "*There was a net forest loss of 7 million hectares per year in tropical countries in 2000–2010 and a net gain in agricultural land of 6 million hectares per year*". FAO (2016) also mentions "*In tropical and subtropical countries, large-scale commercial agriculture and subsistence agriculture accounted for 73 percent of the deforestation, with significant regional variations. For example, commercial agriculture accounted for almost 70 percent of deforestation in Latin America but for only one-third in Africa, where small-scale agriculture is a more significant driver.*"

According to Kaimowitz et al. (2004), Boucher et al. (2011, Chapter 5), and Rudel et al. (2009), the most dominant new land use for which the tropical forest (Amazon rainforest) is "legally" deforested for the agricultural sector is beef cattle production with over 85%. This trend seems to continuously increase, and factors such as increased meat demand, economic growth, and the linkage between these factors create the high meat demand worldwide (Haley 2001; Milford et al. 2019; Ortega et al. 2009).

According to Boucher et al. (2011, Chapter 4), soybean production in the Amazonas grew by more than 1,000% from 1990 to 2010. This equaled a production of almost 70 million tons of soybean, which required an approximate area of 25 million hectares. Against common belief, soybean is primarily consumed by animals (e.g. beef, chicken) as feed, instead of by humans (Boucher et al. 2011, Chapter 4; and see concepts of food-feed competition). This enormous soy production is heavily influenced by so-called "agrarian colonization" which is led by western countries and China in the last few decades (Nepstad et al. 2006; Steward 2007). According to Nepstad et al. (2006), such interventions have significant roles as *economic "teleconnections*", which have led to significant deforestation events in tropical regions such as the Amazonas. Economic teleconnections are a sub-group of telecoupling, which is defined by Kapsar et al. (2019) as "*Telecoupling is a strategy that comprehensively analyzes both the socioeconomic and environmental impacts of distant, coupled human and natural system's*".

According to Boucher et al. (2011, Chapter 6), the largest palm oil production is occurring in Southeast Asia with approximately 9 million hectares of area harvested in 2008. These palm oil productions and human presence have major impacts on the local tropical wilderness and endemic species (Pardo et al. 2021).

All these logging actions result in low ecosystem integrity with long-term negative consequences on the local environment and species within (Grantham et al. 2020).

6.3.4.3 Mining

FAO (2016) contains global land-use division figures, where the reasons for tropical forest deforestations are illustrated comparatively as figures. These figures indicate overall deforestation caused by mining was approximately 52%, peaking in America (South America) with approximately 70%. This peak of America equals 30.000 km^2/year of deforested area (equal to the size of Belgium, per year) (Hosonuma et al. 2012). The country in the Americas where the mining business peaks is Brazil with over 80% of deforestation attributed to mining (De Sy et al. 2015; FAO 2016). Resource mining and exploitation have been executed in Southern America since the very early days of its colonization. The ancient civilizations of the Aztecs, Maya, Inka, etc. already made great use of their soil resources (Lee 2019; Pillsbury et al. 2017). Later on, Latin America was colonized for its precious metal resources (e.g. gold and silver), predominately by Spanish colonizers (Dean 2010). The Spanish Empire obtained enormous wealth from these resource-rich colonizations in Latin America (Baten 2016). This newly obtained wealth pushed the Spanish Kingdom to continuously nourish the colonization, with which they however penalized their home country, as investments in innovation and education decreased. This led to a decrease in economic strength of Spain in Europe compared to its neighboring countries (Baten 2016). In conclusion, the history of the tropics' intense exploitation is relatively long, and it has ever since been negatively influencing the local human population and environment including local species. One of the parties/stakeholders was always at disadvantage, either the local people, local biodiversity, or the environment, where the actions were predominantly economically driven (Brown 1994; Koh and Wilcove 2008; Meng et al. 2019). Industrial approaches to international markets made it specifically worse for squirrels.

6.3.4.4 Hydroelectricity

Hydroelectricity is electricity created using water. This is according to the International Energy Agency (IEA) the globally most utilized renewable energy source, contributing 16% to the total energy production in 2018. It is widely called 'green energy', see World Commission on Hydro Dams, etc. (International Rivers 2022). The construction of hydroelectricity plants in tropical virgin environments requires space, access, electricity, roads, powerlines, worker camps, and a working force that all requires the deforestation of the tropical forest (Barbier 1991; Regmi and Huettmann 2020). However, in addition to deforestation and holding reservoirs, the dams that are used to create hydroelectricity disturb the natural flow of the wilderness water streams that are of such crucial importance for the forest and its inhabiting species (Dudgeon 2000; Huettmann et al. 2020). This area loss and disturbance impacts the species that are dependent on it and can even lead to mating disturbance, life-cycle incompletion, stresses, and death (Johansson et al. 2006). This topic of biodiversity loss is further addressed in the upcoming paragraph.

6.3.4.5 Endemic Biodiversity Loss

Endemic biodiversity loss can be seen as the result of many of the threats mentioned above and other factors such as poor conservation management (Rosser and Mainka 2002), poor politics (Ehrlich and Ehrlich 1981), poor species marginalization (Gambourg et al. 2012; Steiner and Huettmann 2021), etc. Chap. 7 (in detail Sect. 7.4.5) in Steiner and Huettmann (2023), elaborates on this issue in more detail and provides additional literature references.

6.3.4.6 Climate Change

Climate change is a very broad and overarching topic that entails many components (Hardy 2003; Şekercioğlu et al. 2012; etc.). A few of those components are droughts, flooding, wildfires, warming influencing the germination capability of some plant species, warming causing too rapid habitat shifts which do not allow species to follow the ideal conditions, etc. One reason why climate change is so threatening is it highly impacts the squirrels' habitats, also in the tropics (Imbach et al. 2018). Without their species-specific habitat and ecological niche, the species cannot strive nor survive and either have to adapt to a different habitat or face extinction (Isaac 2009; Simmonds and Isaac 2007; Thuiller et al. 2011; Thuiller et al. 2006). Thus, it requires a thorough analysis of the species' habitat and science-based proper management and conservation actions. In order to do so though, the human-driven global phenomena must be understood and included in the decision-making of policies and management plans mitigating the core problems.

Droughts and floods have been greatly discussed for squirrels in Chap. 7 (in detail Sect. 7.4.5) in Steiner and Huettmann (2023). Wildfires account according to Curtis et al. (2018) for 23% of the global forest loss. According to Juárez-Orozco et al. (2017), the forest fire frequency is strongly related to the role of fire in agricultural and ecosystem traits. These tropical forest fires are also believed to be increasing in size and frequency in the last few years (Cochrane 2003). Cochrane (2003) also mentions that millions of hectares of tropical forests are annually burned due to forest fires in the tropics. 5 million hectares are annually solely burned in the Brazilian state Amazonas and many millions more in neighboring Brazilian states and countries (Cochrane 2003). The issue, however, does not remain just in South America. Also in tropical Asia major forest fires can be/are being observed (Cochrane 2003; Nurdiati et al. 2022). Cochrane (2003) additionally mentions that with these forest fires the planet does not only lose trees and other plant species but also plenty of animals and even thousands of humans fall annually victim to the emerging smoke. Such emerging smoke is usually an event observable originating from major fires, smaller fires produce often less smoke and are less destructive. An aspect that is often overlooked with climate change and an increased global temperature is the influence of the climate on the germination capability of some plant species, e.g. of relevance for squirrels as food items and habitat. This is because some plant species require a certain temperature in order to start germinating or similar processes, including appropriate maturation (Walck et al. 2011). Without the necessary temperatures, these species can hardly grow, proliferate, and survive as species (Chhetri and Rawal 2017). The mentioned literature references (Chhetri and Rawal 2017; Walck et al. 2011) provide a list of species that are already now affected by this threat. Lastly, the topic that is to be discussed here is the issue when due to climate change the species' preferred ecological niche changes, but these cannot follow that change. This incapability of not being able to follow the optimal niche can be caused by terrestrial limitations (altitude, or water-locked regions – see Chap. 7 for details), habitat limitations, habitat fragmentation (e.g. due to timber logging or urbanization), or the high speed of the change. Plants can hardly move several hundreds of km in 80 years, but some models suggest that much greater distances must be covered to remain in a favorable habitat. Such far distribution range shifts are not only difficult to perform for plant species, but also for animal species (e.g. squirrels, see Chap. 12) and will stress them, and/or can lead to local disappearances, changes, or even extinctions.

6.3.5 Results of the Tropics' Shortcomings, and Possible Ways to Overcome Them

Here we focused on providing an overall big-picture overview of the shortcomings in the tropical regions. These shortcomings include topics such as Research gaps, Societal problems (wealth distribution, gender emancipation, etc.), Warfare, Neocolonialism, Health shortcomings, Political corruption, etc. Similarly as in the previous section, here just a short introduction to the most important topics is given. Further in-depth information on these topics can be obtained by reading into the provided literature references.

To begin with, research gaps and shortcomings are directly and indirectly heavily affecting the tropics (Meijaard and Sheil 2007). Directly, in the sense of an overall decreased number of publications compared to some regions in the temperate zone (e.g. Europe and the USA) (Cayuela et al. 2018). Indirectly, in the sense that there is drastically less funding available (Bhadouria et al. 2019), difficult accessibility (Huerlimann et al. 2020), and difficulties with the interaction/communication and collaboration with the local communities and scientists (Molyneux 1997). Next to the research shortcoming, general societal problems have also been recorded. These societal shortcomings include the large injustice of wealth distribution (Birdsall et al. 2006; UN 2020), underdeveloped gender emancipation (Jayachandran 2015), etc. An additional shortcoming in the tropical regions is warfare. Warfare is unfortunately still ongoing in many tropical countries due to territorial conflicts, mining, logging conflicts, and many other reasons (McNeely 2003; Price 2020). In addition to Warfare, Neocolonialism and its effects can still be observed (Boshoff 2009; Laine 2009; Sheridan 1963). Besides those mentioned issues, access to proper health care, clean and drinkable water, diseases, and basic hygiene are also still considered shortcoming factors in tropical regions (Black 2016; Boisson et al. 2021; Molyneux 1997; Molyneux et al. 2020; Schroth et al. 2000; Utzinger and Keiser

2006; Viboud et al. 2006, etc.). Lastly, Political and general corruption can be observed in many tropical countries (Transparency International 2019). Among the 20 most corrupt countries, 14 are tropical countries (Transparency International 2019). Next to these mentioned shortcomings, many more can be found once one pays attention and looks closely. All of these mentioned factors influence the local society, politics, conservation management, and finally also species conservation and their survival. All of these mentioned reasons we introduced here are common problems of the tropics and because they often rank higher on the priority list and agenda of governments and management agencies than ecological conservation of small marginalized mammals (e.g. squirrels), they are often marginalized and not considered in the big picture.

6.4 Discussion and Conclusion

Here a literature review section has been included to review the already existing knowledge on the distribution of the squirrels studied here. This literature review aimed to provide a baseline with which the created models can be compared. Threats to the tropical environment, its species, society, politics, and environment are also being discussed. This final threat discussion aims to create a big-picture overview of the current situation and to understand which factors are behind the current species management.

The results of the literature review and the creation of the current and future SDMs can be compared with each other and show similar findings. Overall, the SDMs that have been created in Chap. 3 and used here are very coherent with the literature sources. The only main difference that has been found was for the species Layard's Palm Squirrel (*Funambulus layardi*). This species occurs according to literature (see sources in Sect. 6.3.1) only in Sri Lanka and in Southeast India. However, according to Fig. 6.3a, this species occurs primarily in Sri Lanka and the Fiji Islands. This far-distant and isolated distribution might be caused by the fact that these two environments are highly similar, however, despite the high environmental similarity, it is suspected that the Fiji Islands are not inhabited by the squirrel species *F. layardi*. Such "unlikely" modeled distributions can be avoided by including space-specific predictors for the SDM creation (lat-long coordinate predictors). This has not been done for this study as the main aim was to model the generic distribution of squirrels based on environmental factors/characteristics instead of geographical ones. Other than that, the SDMs are highly coherent with the literature records.

Besides the good coherence with the literature resources, it is also essential to discuss here the coherence between the SDMs and the SDFs. The SDFs have been created with the SDMs as the baseline. Overall the SDMs and SDFs are likewise highly coherent with each other (considering the use of a different number of predictors). However, between all the SDMs and SDFs, one major discrepancy can be observed, the hotspot difference between Figs. 6.6a and 6.6b–e. In Fig. 6.6a, the core distribution can be observed in the Western Amazonas, whereas in the other figures, the core distribution can be observed predominately on the western coast of Colombia and distributed to a smaller extent in the Amazonas. This hotspot on the western coast of Colombia is absent in Fig. 6.6a. This major hotspot difference is probably caused by the difference in the number of predictors used for the SDMs. For the SDM from Fig. 6.6a, 132 environmental predictors have been used, for Figs. 6.6b–e only 7(+1) predictors have been used. By using only 7 climate predictors and one elevation predictor, the species' niche cannot be described and modeled as accurate as when 132 environmental predictors are being used, describing not only the climate but also vegetation cover, proximity layers (to cities, roads, rivers, lakes, protected areas, etc.), forest fires, snow cover, threatened mammal density, etc. This difference in modeling accuracy and strength due to the different predictor numbers can also be observed for the SDFs of the species *H. ruwenzorii* and *S. richmondi*. There, the MRI SDFs (Figs. 6.4d and 6.7d) show a drastic increase in the distribution range of the species. These SDFs also only include climate variables in the forecasted models, other variables such as land cover are not included (mainly since they are not available for 2100), which can easily flip the board from an increase to a decreasing trend. However, not only the climate and the environment determine the distribution range of the squirrel family. Sections 6.3.4 and 6.3.5 discuss many of those "threats and shortcomings" in the tropics. These discussion points are mainly aimed at the tropical habitat itself and the human population living in those regions. However, these factors also secondarily influence the species management and conservation of squirrels. For example, all the timber logging for timber trade, land-use shift to agriculture, and mining require the deforestation of the virgin and pristine habitat of the tropical forests. With the deforestation of those habitats, the habitats of many squirrel species shrink or vanish, and it puts many squirrel species under stress, possibly even causing local extinctions. Therefore, in order to decrease the negative outlook for the squirrels' future and their habitats, it is suggested to work against corrupt politics and policies (e.g. see Tacconi and Williams (2020)), unfair wealth and funding distribution (also in research), excessive unsustainable habitat destruction, and against global warming practices, habits, and businesses (Soulsbury and White 2015). In addition to those briefly discussed topics, the species are also influenced by other, very modern, and local problems. An example of this is the species Ruwenzori Sun Squirrel (*Heliosciurus ruwenzorii*) with its distribution around Lake Victoria (especially in Uganda). Because in modern days, the primary forest and landscape there have been heavily modified by humans in the last few decades. How the habitat was a few decades ago is incomparable with the current situation. The area is now highly populated by people, including refugee areas, and is heavily used by floristic agriculture businesses for the European flower market (Mlaponi 2011; The East African 2012). With all these areas modified and populated by people, there is little to no space or resources left for squirrels and their habitats, considering the finity of our planet earth (Dietz and O'Neill 2013). To provide space for squirrels and all other species, humanity should seek to live closer to nature and in harmony with mother earth, instead of destroying the very place humans and all other species live on. A less marginalized and capitalism-driven management and a more holistic science-based sustainable conservation management is already a good step towards deeper harmony with mother earth.

References

Alamgir M, Campbell MJ, Sloan S, Goosem M, Clements GR, Mahmoud MI, Laurance WF (2017) Economic, socio-political and environmental risks of road development in the tropics. Curr Biol 27(20):R1130–R1140

Amori G, Koprowski J, Roth L (2016) Sciurus igniventris (errata version published in 2017). The IUCN Red List of threatened species 2016: e.T20013A115154863. https://doi.org/10.2305/IUCN.UK.2016-3.RLTS.T20013A22246972.en

Barbier E (1991) Tropical deforestation. In: Blueprint 2: greening the world economy. Earthscan Publications, London, pp 138–165

Baten J (ed) (2016) A history of the global economy. Cambridge University Press

Bawa KS, Kress WJ, Nadkarni NM, Lele S (2004) Beyond paradise–meeting the challenges in tropical biology in the 21st century. Biotropica 36(4):437–446

Bhadouria R, Tripathi S, Srivastava P, Singh P (eds) (2019) Handbook of research on the conservation and restoration of tropical dry forests. IGI Global

Birdsall N, Vanzetti D, de Cordoba SF (2006) The world is not flat: inequality and injustice in our global economy. Helsinki: World Institute for Development Economics Research

Bitariho R, Blundo C, Carilla J, Grau R, Malizia A, Malizia L, Bradford M et al (2020) Taking the pulse of earth's tropical forests using networks of highly distributed plots

Black M (2016) The atlas of water: mapping the world's most critical resource. University of California Press

Boisson S, Wohlgemuth L, Yajima A, Peralta G, Obiageli N, Matendechero S, Velleman Y (2021) Building on a decade of progress in water, sanitation and hygiene to control, eliminate and eradicate neglected tropical diseases. Trans R Soc Trop Med Hyg 115(2):185–187

Boshoff N (2009) Neo-colonialism and research collaboration in Central Africa. Scientometrics 81(2):413–434

Boucher D, Elias P, Lininger K, May-Tobin C, Roquemore S, Saxon E (2011) The root of the problem: what's driving tropical deforestation today?. The root of the problem: what's driving tropical deforestation today?

Boucher O, Servonnat J, Albright AL, Aumont O, Balkanski Y, Bastrikov V, Vuichard N et al (2020) Presentation and evaluation of the IPSL-CM6A-LR climate model. J Adv Model Earth Syst 12(7):e2019MS002010

Brown DW (1994) The causes of tropical deforestation: the economic and statistical analysis of factors giving rise to the loss of the tropical forests. UBC Press

Cassola F (2016) Heliosciurus ruwenzorii. The IUCN Red List of threatened species 2016: e.T9834A22260596. https://doi.org/10.2305/IUCN.UK.2016-2.RLTS.T9834A22260596.en. Accessed 17 Nov 2021

Cayuela L, Granzow-de la Cerda Í, Méndez M (2018) The state of European research in tropical biology. Biotropica 50(2):202–207

Chao S (2012) Forest peoples: numbers across the world. Forest Peoples Programme, Moreton-in-Marsh, p 27

Chhetri SB, Rawal DS (2017) Germination phenological response identifies flora risk to climate change. Climate 5(3):73

Cochrane MA (2003) Fire science for rainforests. Nature 421(6926):913–919

Coghe S (2014) Population politics in the tropics: demography, health and colonial rule in Portuguese Angola, 1890s–1940s. Doctoral dissertation

Cooper A, Shaw T (Eds.) (2009) The diplomacies of small states: Between vulnerability and resilience. Springer

Curtis PG, Slay CM, Harris NL, Tyukavina A, Hansen MC (2018) Classifying drivers of global forest loss. Science 361(6407):1108–1111

De Sy V, Herold M, Achard F, Beuchle R, Clevers JGPW, Lindquist E, Verchot L (2015) Land use patterns and related carbon losses following deforestation in South America. Environ Res Lett 10(12):124004

Dean C (2010) The after-life of Inka rulers: Andean death before and after Spanish colonization. Death and afterlife in the early modern Hispanic world. John Beusterien and Constance Cortez. Hispanic Issues On Line 7. 27–54. http://hispanicissues.umn.edu/DeathandAfterlife.html

Dietz R, O'Neill D (2013) Enough is enough: building a sustainable economy in a world of finite resources. Routledge

Dudgeon D (2000) The ecology of tropical Asian rivers and streams in relation to biodiversity conservation. Annu Rev Ecol Syst 31(1):239–263

Ehrlich P, Ehrlich A (1981) The politics of extinction. Bull At Sci 37(5):26–30

Emmons LH (1984) Geographic variation in densities and diversities of non-flying mammals in Amazonia. Biotropica 16:210–222

Emmons LH, Feer F (1997) Neotropical rainforest mammals, a field guide. The University of Chicago Press, Chicago/London

Falk RA (1973) Environmental warfare and ecocide – facts, appraisal, and proposals. Bull Peace Propos 4(1):80–96

FAO (2016) State of the world's forests 2016. Forests and agriculture, land-use challenges and opportunities. Forests and agriculture: land-use challenges and opportunities. Rome. http://www.fao.org/3/i5588e/i5588e.pdf

Feeley KJ, Stroud JT (2018) Where on earth are the "tropics"? Front Biogeogr 10(1–2)

Friedman J (2001) TreeNet™. MART – Salford Systems

Gambourg C, Palmer C, Sandoe P (2012) Ethics of wildlife management and conservation: what should we try to protect? Nat Educ Knowl 3(7)

Gibbs HK, Ruesch AS, Achard F, Clayton MK, Holmgren P, Ramankutty N, Foley JA (2010) Tropical forests were the primary sources of new agricultural land in the 1980s and 1990s. Proc Natl Acad Sci 107(38):16732–16737

Glastra R (ed) (2014) Cut and run: illegal logging and timber trade in the tropics. IDRC

Goonatilake de A WLDPTS (2016) Funambulus layardi. The IUCN Red List of threatened species 2016: e.T8700A22260033. https://doi.org/10.2305/IUCN.UK.2016-3.RLTS.T8700A22260033.en. Accessed 17 Nov 2021

Grantham HS, Duncan A, Evans TD, Jones KR, Beyer HL, Schuster R, et al., Watson JEM (2020) Anthropogenic modification of forests means only 40% of remaining forests have high ecosystem integrity. Nature communications 11(1):5978

Haley MM (2001) Changing consumer demand for meat: the US example, 1970–2000. Chang Struct Glob Food Consum Trade 1(1):42–48

Hardy JT (2003) Climate change: causes, effects, and solutions. Wiley

Hosonuma N, Herold M, De Sy V, De Fries RS, Brockhaus M, Verchot L, Romijn E et al (2012) An assessment of deforestation and forest degradation drivers in developing countries. Environ Res Lett 7(4):044009

Huerlimann R, Cooper MK, Edmunds RC, Villacorta-Rath C, Le Port A, Robson HLA, Jerry DR et al (2020) Enhancing tropical conservation and ecology research with aquatic environmental DNA methods: an introduction for non-environmental DNA specialists. Anim Conserv 23(6):632–645

Huettmann F, Shrestha M, Puri G, Regmi GR (2020) Hydrodams again: a closer look at expensive gharials, lost dolphins, dissected fish migration, and destructive energy for non-sustainable mining and societies abroad fueling a non-democratic governance for the designed 'homo electricus'. In: Hindu Kush-Himalaya watersheds downhill: landscape ecology and conservation perspectives. Springer, Cham, pp 649–661

Imbach P, Chou SC, Lyra A, Rodrigues D, Rodriguez D, Latinovic D, Georgiou S et al (2018) Future climate change scenarios in Central America at high spatial resolution. PLoS One 13(4):e0193570

Industrial Logging (2021) Industrial logging. World rainforest movement. https://www.wrm.org.uy/bulletins/issue-53. Accessed 22 Aug 2022

International Rivers (2022) The world commission on dams. https://archive.internationalrivers.org/campaigns/the-world-commission-on-dams. Accessed 22 Aug 2022

Isaac JL (2009) Effects of climate change on life history: implications for extinction risk in mammals. Endanger Species Res 7(2):115–123

Jayachandran S (2015) The roots of gender inequality in developing countries. Economics 7(1):63–88

Johansson B, Sellberg B, Gruber L (2006) Dams under debate. Swedish Research Council Formas

Juárez-Orozco SM, Siebe C, Fernández y Fernández D (2017) Causes and effects of forest fires in tropical rainforests: a bibliometric approach. Trop Conserv Sci 10:1940082917737207

Kaimowitz D, Mertens B, Wunder S, Pacheco P (2004) Hamburger connection fuels Amazon destruction. Center for International Forest Research, Bangor, pp 1–10

Kapsar KE, Hovis CL, Bicudo da Silva RF, Buchholtz EK, Carlson AK, Dou Y, Liu J et al (2019) Telecoupling research: the first five years. Sustainability 11(4):1033

Katovai E, Katovai DD, Campbell M, Laurance SG, Edwards W, Laurance WF (2021) Structural recovery of logged forests in the Solomon Islands: implications for conservation and management. Trop Conserv Sci 14:19400829211028125

Koh LP, Wilcove DS (2008) Is oil palm agriculture really destroying tropical biodiversity? Conserv Lett 1(2):60–64

Koprowski J, Roth L (2018) Sciurus richmondi. The IUCN Red List of threatened species 2018: e.T20020A22246819. https://doi.org/10.2305/IUCN.UK.2018-2.RLTS.T20020A22246819.en. Accessed 17 Nov 2021

Kranowski PV (1969) Aspects of red squirrel (Tamiasciurus hudsonicus) population ecology in interior Alaska. University of Alaska Fairbanks

Laine J (2009) The neo-colonialism of green gold. Interciencia 34(6):378–378

Lee JR (2019) Environmental conflict and cooperation: premise, purpose, persuasion, and promise. Routledge

McNeely JA (2003) Biodiversity, war, and tropical forests. J Sustain For 16(3–4):1–20

Meijaard E, Sheil D (2007) Is wildlife research useful for wildlife conservation in the tropics? A review for Borneo with global implications. Biodivers Conserv 16(11):3053–3065

Meng HH, Zhou SS, Li L, Tan YH, Li JW, Li J (2019) Conflict between biodiversity conservation and economic growth: insight into rare plants in tropical China. Biodivers Conserv 28(2):523–537

Milford AB, Le Mouël C, Bodirsky BL, Rolinski S (2019) Drivers of meat consumption. Appetite 141:104313

Mlaponi E (2011) Spatial and temporal distribution of Nile Perch (Lates niloticus Linnaeus, 1758) in relation to oxygen in Lake Victoria Tanzanian side. The Netherlands Organisation for International Cooperation in Higher Education (NUFFIC), World Bank under the Lake Victoria Environmental Management Project Phase I & European Union Under the Lake Victoria Fisheries Research Project, (LVFRP). https://doi.org/10.13140/RG.2.2.25550.08001

Molur S, Srinivasulu C, Srinivasulu B, Walker S, Nameer PO, Ravikumar L (2005) Status of south Asian non-volant small mammals: conservation assessment and management plan (CAMP) workshop report. Zoo Outreach Organization/CBSG-South Asia, Coimbatore. 618pp

Molyneux ME (1997) Tropical trials: the problems of research in the tropics. Ann Trop Med Parasitol 91(7):841–844

Molyneux DH, Aboe A, Isiyaku S, Bush S (2020) COVID-19 and neglected tropical diseases in Africa: impacts, interactions, consequences. Int Health 12(5):367–372

Nepstad DC, Stickler CM, Almeida OT (2006) Globalization of the Amazon soy and beef industries: opportunities for conservation. Conserv Biol 20(6):1595–1603

Nurdiati S, Sopaheluwakan A, Septiawan P, Ardhana MR (2022) Joint Spatio-Temporal Analysis of Various Wildfire and Drought Indicators in Indonesia. Atmosphere, 13(10):1591

Ortega DL, Wang HH, Eales JS (2009) Meat demand in China. China Agric Econ Rev 1:410–419

Pardo LE, Edwards W, Campbell MJ, Gómez-Valencia B, Clements GR, Laurance WF (2021) Effects of oil palm and human presence on activity patterns of terrestrial mammals in the Colombian Llanos. Mamm Biol 101(6):775–789

Pillsbury J, Potts T, Richter KN (2017) Golden kingdoms: luxury arts in the ancient Americas. Getty Publications

Price SV (2020) War and tropical forests: conservation in areas of armed conflict: conservation in areas of armed conflict

Rainforest Alliance (2020) Preventing agriculture driven deforestation and conversion of natural ecosystems. https://apo.org.au/sites/default/files/resource-files/2020-12/apo-nid310594.pdf

Raven PH, Gereau RE, Phillipson PB, Chatelain C, Jenkins CN, Ulloa CU (2020) The distribution of biodiversity richness in the tropics. Science. Advances 6(37):eabc6228

Regmi GR, Huettmann F (eds) (2020) Hindu Kush-Himalaya watersheds downhill: landscape ecology and conservation perspectives. Springer

Reid F (2016) Microsciurus mimulus. The IUCN Red List of threatened species 2016: e.T13411A22254996. https://doi.org/10.2305/IUCN.UK.2016-2.RLTS.T13411A22254996.en. Accessed 17 Nov 2021
Rosser AM, Mainka SA (2002) Overexploitation and species extinctions. Conserv Biol 16(3):584–586
Rudel TK, Schneider L, Uriarte M, Turner BL, DeFries R, Lawrence D, Grau R et al (2009) Agricultural intensification and changes in cultivated areas, 1970–2005. Proc Natl Acad Sci 106(49):20675–20680
Sachs JD (2001) Tropical underdevelopment. National Bureau of Economic Research. https://www.nber.org/papers/w8119. Accessed 22 Aug 2022
Schroth G, Krauss U, Gasparotto LJAD, Aguilar JD, Vohland K (2000) Pests and diseases in agroforestry systems of the humid tropics. Agrofor Syst 50(3):199–241
Şekercioğlu ÇH, Primack RB, Wormworth J (2012) The effects of climate change on tropical birds. Biol Conserv 148(1):1–18
Seymour F (2020) Is destruction the inevitable fate of our forests? World Resource Institute. https://www.wri.org/insights/destruction-inevitable-fate-our-forests
Sheridan RB (1963) Temperate and tropical: aspects of European penetration into tropical regions. Caribb Stud 3(2):3–21
Simmonds MP, Isaac SJ (2007) The impacts of climate change on marine mammals: early signs of significant problems. Oryx 41(1):19–26
Smith MC (1968) Red squirrel responses to spruce cone failure in interior Alaska. J Wildl Manag 32:305–317
Soulsbury CD, White PC (2015) Human–wildlife interactions in urban areas: a review of conflicts, benefits and opportunities. Wildl Res 42(7):541–553
Steiner M, Huettmann F (2021) Justification for a taxonomic conservation update of the rodent genus Tamiasciurus: addressing marginalization and mis-prioritization of research efforts and conservation laissez-faire for a sustainability outlook. Eur Zool J 88(1):86–116
Steward C (2007) From colonization to "environmental soy": a case study of environmental and socio-economic valuation in the Amazon soy frontier. Agric Hum Values 24(1):107–122
Tacconi L, Williams DA (2020) Corruption and anti-corruption in environmental and resource management. Annu Rev Environ Resour 45(1):305–329
Tatebe H, Ogura T, Nitta T, Komuro Y, Ogochi K, Takemura T, Kimoto M et al (2019) Description and basic evaluation of simulated mean state, internal variability, and climate sensitivity in MIROC6. Geosci Model Dev 12(7):2727–2765
The East African (2012) Uganda projects rise in flower earnings. https://www.theeastafrican.co.ke/tea/business/uganda-projects-rise-in-flower-earnings%2D%2D1313782. Accessed 17 Nov 2021
Thorington RW Jr, Koprowski JL, Steele MA, Whatton JF (2012) Squirrels of the world. JHU Press
Thuiller W, Broennimann O, Hughes G, Alkemade JRM, Midgley GF, Corsi F (2006) Vulnerability of African mammals to anthropogenic climate change under conservative land transformation assumptions. Glob Chang Biol 12(3):424–440
Thuiller W, Lavergne S, Roquet C, Boulangeat I, Lafourcade B, Araujo MB (2011) Consequences of climate change on the tree of life in Europe. Nature 470(7335):531–534
Transparency International (2019) The world's 20 most corrupt countries. Yahoo!finance. https://finance.yahoo.com/news/most-corrupt-countries-130825742.html. Accessed 17 Nov 2021
UN (2020) United Nations Department of Economic and Social Affairs. World social report 2020 inequality in a rapidly changing world
Utzinger J, Keiser J (2006) Urbanization and tropical health – then and now. Ann Trop Med Parasitol 100(5–6):517–533
Viboud C, Alonso WJ, Simonsen L (2006) Influenza in tropical regions. PloS Med 3(4):e89
Walck JL, Hidayati SN, Dixon KW, Thompson KEN, Poschlod P (2011) Climate change and plant regeneration from seed. Glob Chang Biol 17(6):2145–2161
Weisse M, Goldman ED (2017) Global tree cover loss rose 51 percent in 2016. World Resource Institute. https://www.wri.org/insights/global-tree-cover-loss-rose-51-percent-2016. Accessed 17 Nov 2021
Weisse M, Goldman ED (2020) We lost a football pitch of primary rainforest every 6 seconds in 2019. World Resource Institute. https://www.wri.org/insights/we-lost-football-pitch-primary-rainforest-every-6-seconds-2019. Accessed 17 Nov 2021
Wilson DE, Reeder DM (eds) (2005) Mammal species of the world: a taxonomic and geographic reference, vol 1. JHU Press, pp 754–818
Yen P, Ziegler S, Huettmann F, Onyeahialam AI (2005) Change detection of forest and habitat resources from 1973 to 2001 in Bach Ma National Park, Vietnam, using remote sensing imagery. Int For Rev 7(1):1–8
Yukimoto S, Kawai H, Koshiro T, Oshima N, Yoshida K, Urakawa S, Ishii M et al (2019) The meteorological research institute earth system model version 2.0, MRI-ESM2. 0: description and basic evaluation of the physical component. J Meteorol Soc Jpn Ser II 97:931–965
Zuidema PA, Baker PJ, Groenendijk P, Schippers P, van der Sleen P, Vlam M, Sterck F (2013) Tropical forests and global change: filling knowledge gaps. Trends Plant Sci 18(8):413–419

Chapter 7
Squirrels on Islands: The Effect of a 'laissez-faire' Approach from Governments and Their Responsible Entities on the Marginalization and Extinction in Extremely Restricted Habitats

Abstract Squirrels live on land, and like the forest habitats, but land is often fragmented and surrounded by water. Many squirrel species are endemic to very small to mid-sized islands (e.g. in Indonesia[47], Malaysia & Malaysian Peninsula[30], Brunei[19], Philippines[13], Japan[3], Sri Lanka[2], Vancouver Island[1] - exponential numbers indicate the number of endemic squirrel species on the corresponding island/island group). These islands are often isolated and extremely scattered and present an active conservation threat for several dozens of squirrel species. It is a specific textbook example of biogeography. Additionally, many squirrels are highly restricted to these habitats, as they cannot expand/shift their habitat because it is surrounded by water. This can have severe influences on the future of these animals, and squirrels, inhabiting such areas with the rising sea level and other climatic changes (described in Chap. 12). Such highly restricted habitats are therefore an enormous threat for endangered squirrel species and are considered in conservation as "high-risk regions". This situation was never prioritized but calls for increased conservation efforts to avoid species extinction caused by the '*laissez-faire*' approach from governments and their responsible entities for their entrusted entities they have a mandate for (as seen in Chaps. 11 & 13). Here we present and discuss some evidence and suggest improvements (using well-working examples from other biotic species) for a wider future conservation success. In this study, we also take a closer look at the global SDMs previously created for species endemic to island habitats, and additionally, perform SDFs for those species, to forecast their fate and population ranges for the year 2100.

Keywords Squirrels · Extinction threats on islands · Lack of governmental species interest · Solutions for endangered habitats · Climate change

7.1 Introduction

Islands are usually defined as a piece of land surrounded by water. The official definition for an island according to the Cambridge dictionary is "*a piece of land completely surrounded by water*" (Cambridge Dictionary, July 2021).

Following this definition, there are approximately over 340,000 islands on the planet (Sayre et al. 2019). This study will mainly focus on approximately 10 representative big islands to outline our message. These islands are Borneo, Sumatra, Sri Lanka, Madagascar, Japan, Taiwan, New Guinea, Philippines (entire Island group), Vancouver Island (Canada), and Sulawesi.

The reason for focusing on these islands is that it has been identified that many squirrel species are endemic to either one or more of those islands. In this study, the main threats to these squirrels are being discussed, with a special focus on the influences on the squirrels' conservation. We take the evidence from literature and then provide a modeling assessment.

One might wonder why one of the biggest "islands" is not primarily included, Australia (in the wider sense one may consider here New Zealand also; another big island - Greenland - is mostly covered by glaciers and snow and not included in this discussion; Papua New Guinea PNG is devoted a chapter of its own in Huettmann in press). Naturally, Australia does not host species belonging to the squirrel family Sciuridae, however, due to human invasion two species have been introduced, and partly already extirpated from Australia (Di Febbraro et al. 2016; Peacock 2009). In Chap. 12 (Steiner and Huettmann 2023) it has been identified several times that future prediction models predict well suitable habitats for squirrels in Australia. This habitat, and thus, the ecological niche might however not be vacant. Australia hosts several sister taxa to squirrels (e.g. Squirrel Gliders (*Petaurus norfolcensis*), or Quokka (*Setonix brachyurus*) which is also known as Australian Chipmunk, or Tree Kangaroos (*Dendrolagus*)) (Brearley et al. 2012; De Tores et al. 2007; Martin 2005). These mentioned sister taxa are likely to occupy the squirrel's ecological niche, leaving little space for squirrels. Also, squirrels have widely been eradicated

Supplementary Information The online version contains supplementary material available at https://doi.org/10.1007/978-3-031-23547-4_7.

M. Steiner, F. Huettmann, *Sustainable Squirrel Conservation*, https://doi.org/10.1007/978-3-031-23547-4_7

by humans again, after realizing that they can heavily predate on the local rare wildlife (Peacock 2009). Similar approaches are heavily executed in the neighboring country New Zealand with the motto "Predator free 2050", where all predators to local endemic species (e.g. songbirds, kiwi, parrots, etc.) are being eradicated. These species that are being eradicated are mainly small mammals like rats (rodents), stoats, and possums (see governmental website: Department of Conservation (2022)).

While discussing islands, their habitats, and species (squirrels), a concept called "Insular biogeography" or "island biogeography" is worth mentioning (Brown 1978). This theory states that factors like the degree of isolation (distance to the nearest neighbor, and the mainland – the more distant of the mainland the more difficult it becomes for species to reach the islands and terrestrial species richness thus declines -), duration/length of isolation (time – the longer the isolation period the more species have been evolved into new ones -), size of islands (larger islands usually facilitate more species) are heavily determining the species richness and diversity on islands (Brown 1978). Additional factors such as human influence, the location relative to ocean currents and dust flow, and its climate also influence islands biodiversity (Brown 1978; Chadwick et al. 1999).

The main threats that are being discussed in this study are habitat loss, genetic diversity loss, sea-level rise, extreme climate events, and global climate change. But more could be added.

Besides the generic literature, the species distribution models (SDMs) from Chap. 3 (Steiner and Huettmann 2023), and the global climate model (GCM) outcomes from Chap. 12 (Steiner and Huettmann 2023) are being further analyzed with a keen eye on the global islands. There, the SDMs are presented comparatively, presenting the current modeled distribution based on 132 environmental - and 7 BioClim predictors and the future (the year 2100) modeled distribution based on the same 7 BioClim predictors for the three climate scenarios (MIROC, MRI, and IPSL). Consecutively, the future fate of squirrels on islands is being discussed based on the Global Distribution Forecasts using global climate models for the year 2100 for a more generic statement.

7.2 Methods

7.2.1 Literature Review on the Distribution of Island-Endemic Squirrel Species

To present here the current state of the literature on the squirrels' distribution with a focus on islands and climate change-derived threats, a literature review has been performed and will be presented in a summarized form. This literature summary aids in understanding the base knowledge on the distribution of island-endemic squirrels and the connected climate-change-derived threats.

7.2.2 Analysis of the Species Distribution Models (SDMs)

Two-hundred-thirty-three global SDMs have been created and presented in Chap. 3 (Steiner and Huettmann 2023). Here, these SDMs have been further analyzed to identify the species that are endemic to islands and are therefore influenced by its threats, poor governance, conservation marginalization, and climate. Thus, to include more species specifically in this discussion, an additional 5 SDMs that have been created for this study, which are being presented and discussed here. For each species, 5 figures are presented (a–e). These represent the following situations respectively; SDM based on 132 environmental predictors for the year 2000–2010 (a), SDM based on 7 BioClim predictors for the year 2000 (b), SDF based on 7 BioClim predictors for the year 2100 (MIROC climate scenario) (c), SDF based on 7 BioClim predictors for the year 2100 (MRI climate scenario) (d), SDF based on 7 BioClim predictors for the year 2100 (IPSL climate scenario) (e). This representation allows to optimally compare the SDMs created with many (132) predictors and with few (7 climate +1 altitude) predictors for the current time, and then for the year 2100 using 3 different climate scenarios (temperature cooling (MIROC), business-as-usual warming (MRI), and the most likely happening scenario with a more extreme warming scenario (IPSL).

7.2.3 Global Climate Model (GCM) Zoom-In Into Squirrel Species-Rich Islands

In addition to the SDMs, in Chap. 12 (Steiner and Huettmann 2023) global climate models (GCMs) have been presented to forecast the distribution of the global squirrel population based on possible climate scenarios for the year 2100. In detail, 3 different climate scenarios have been presented which act as climate forecasts in combination with the modeled squirrel distribution. Four different groups of squirrels have been utilized for the models in Chap. 12. Only two of these groups (All global squirrels, and the 10 most endangered ones) are being discussed here. This is because the other two groups (genera *Geosciurus,* and *Heliosciurus* & *Paraxerus*) mainly occur in regions of the world where there are no or very few islands (Central Africa). Chapter 12 presents several figures that present the results on a global scale. Here, we present a specific zoom-in into some major and globally (for squirrels) interesting island selections. Then, we compare how different software and climate scenarios predict/forecast the future distribution of squirrels.

7.2.4 Further Man-Caused Climate Aspects Affecting Squirrels Today and in 2100

For a holistic approach, further man-caused climate aspects which are affecting - and predicted to affect - squirrels today and in 2100 have been analyzed. This has been done by identifying the core habitats of island-endemic squirrel species and consecutively compiling a list of threats and dangers based on literature research.

7.3 Results

7.3.1 Literature Review on the Distribution of Island-Endemic Squirrel Species

For the species Kinabalu Squirrel (*Callosciurus baluensis*) it can be confirmed that its distribution is restricted to habitats >300 m above sea level in Sabah and Sarawak, Malaysia, which includes the Mount Kinabalu (Borneo) (Meijaard 2016; Thorington et al. 2012, p. 135; Wilson and Reeder 2005, p. 775). For the next species, Lowland Long-nosed Squirrel (*Hyosciurus ileile*) it can be confirmed that its distribution is restricted to mountainous habitats in Sulawesi (Indonesia) (Chiozza 2016; Thorington et al. 2012, p. 168; Wilson and Reeder 2005, p. 783). Further, for the species Palawan flying Squirrel (*Hylopetes nigripes*), it can be confirmed that its distribution is restricted to Palawan and Bancalan Isis (Philippines) (Kennerley and Ong 2019; Thorington et al. 2012, pp. 100–101; Wilson and Reeder 2005, p. 769). Next, for the species Niobe Ground Squirrel (*Lariscus niobe*) it can be confirmed that its distribution is restricted to the mountains of western Sumatra, and eastern Java (Indonesia) (Gerrie et al. 2016; Thorington et al. 2012, p. 171; Wilson and Reeder 2005, p. 783). Furthermore, for the species Vancouver Island Marmot (*Marmota vancouverensis*) it can be confirmed that its distribution is restricted to Vancouver Island (Canada) (Roach 2017; Thorington et al. 2012, p. 289). For the species Black-eared Squirrel (*Nannosciurus melanotis*), it can be confirmed that its distribution is restricted to the forests of Borneo, Sumatra, Java, and adjacent small islands (Francis et al. 2016; Thorington et al. 2012, p. 174; Wilson and Reeder 2005, p. 784). And lastly, for the species Bornean Mountain Ground Squirrel (*Sundasciurus everetti*), it can be confirmed that its distribution is restricted to the mountains of Sabah and Sarawak (Malaysia), and Kalimantan (Indonesia) (Borneo) (Hawkins et al. 2016a; Thorington et al. 2012, p. 153; Tizard 2016).

It must be noted that this short literature review does not include all island-endemic squirrels species, rather, it should act as an assessment example and template for other species of interest. Additionally, it must be noted that for the squirrels' distribution literature review, mostly only three sources were able to be used, IUCN Red List (www.iucnredlist.org), Thorington et al. (2012), and Wilson and Reeder (2005), because no other reliable sources have been found for these species. This again underlines the marginalization of so many squirrel species and how few studies have been performed on the squirrels' life history, distribution, etc. occurring in non-western countries.

7.3.2 Analysis of the Species Distribution Models (SDMs)

In Chap. 3 (Steiner & Huettmann 2023), 233 species distribution models (SDMs) for all global squirrel species have been performed and presented. From these 233, for 92 squirrel species, it has been identified that their modeled primary distribution consists of island habitats. Many of these 92 squirrels are endemic to only one or a couple of islands. Two examples for this have already been presented as figures in Chap. 3 (Figs. 3.11 and 3.12) for the species Palawan flying Squirrel (*Hylopetes nigripes*) which is endemic to the single island of Palawan, Philippines, and the species Black-eared Squirrel (*Nannosciurus melanotis*), a species which seems to encounter major threats due to climate change, especially sea-level rise. In order to include some more squirrel species and present their distribution models, here additional 5 models will be presented. These models belong to the species Kinabalu Squirrel (*Callosciurus baluensis*), Lowland Long-nosed Squirrel (*Hyosciurus ileile*), Niobe Ground Squirrel (*Lariscus niobe*), Vancouver Island Marmot (*Marmota vancouverensis*), and Bornean Mountain Ground Squirrel (*Sundasciurus everetti*).

To provide an overview of these squirrels' distribution, a global overview has been created (see Fig. 7.1).

The individual SDMs for each species can be found in Figs. 7.2, 7.3, 7.4, 7.5 and 7.6 and Appendices 7.1 and 7.2.

Figures 7.2 represent the modeled distribution for the squirrel species Kinabalu Squirrel (*Callosciurus baluensis*). This squirrel species is endemic to the island Borneo, more specifically to the region of Sabah (Malaysia). This species is considered Least Concerned (LC) with a decreasing population trend according to IUCN Red List and Meijaard (2016).

Figures 7.3 represent the modeled distribution for the squirrel species Lowland Long-nosed Squirrel (*Hyosciurus ileile*). This squirrel species is endemic to the islands Sulawesi (Indonesia), and New Guinea. This species is considered Vulnerable (VU) with a decreasing population trend according to IUCN Red List and Chiozza (2016).

Figures 7.4 represent the modeled distribution for the squirrel species Niobe Ground Squirrel (*Lariscus niobe*). This squirrel species is endemic to the islands of Sumatra, mores specifically West Sumatra, and New Britain (Papua New Guinea), and Madang Province (Papua New Guinea). This species is considered as Data deficient (DD) with an Unknown population trend according to IUCN Red List and Gerrie et al. (2016).

Figures 7.5 represent the modeled distribution for the squirrel species Vancouver Island Marmot (*Marmota vancouverensis*). This squirrel species is endemic to Vancouver Island (British Columbia, Canada). This species is considered Critically endangered (CR) with a decreasing population trend according to IUCN Red List and Roach (2017). For this species only approximately 90 individuals are left living in nature.

Figures 7.6 represent the modeled distribution for the squirrel species Bornean Mountain Ground Squirrel (*Sundasciurus everetti*). This squirrel species is endemic to the coastal regions of northern Borneo (Sabah), and Balikpapan (Indonesia). This species is considered Least Concerned (LC) with a stable population trend according to IUCN Red List and Tizard (2016).

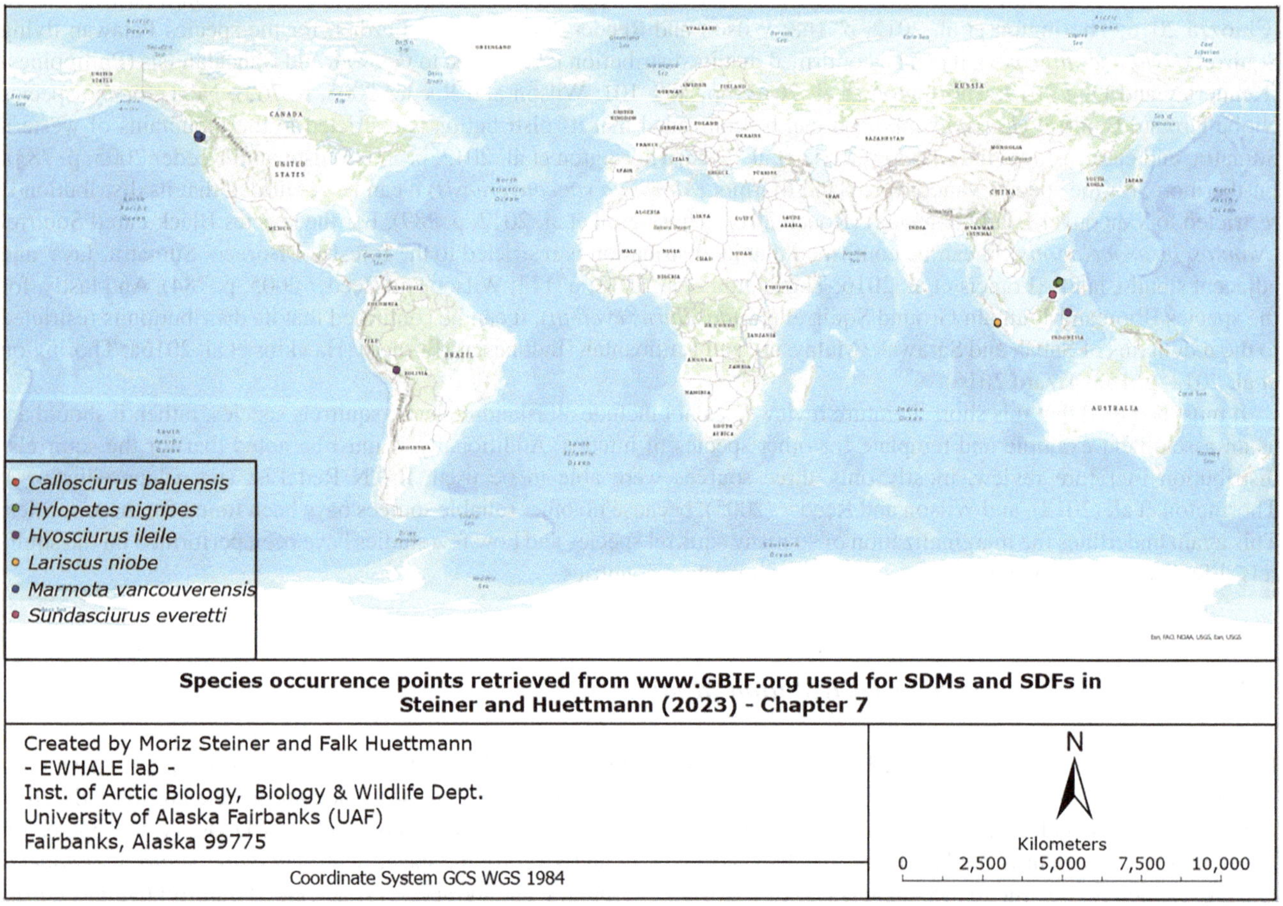

Fig. 7.1 Squirrel distribution of the species *Callosciurus baluensis, Hyosciurus ileile, Hylopetes nigripes, Lariscus niobe, Marmota vancouverensis, Nannosciurus melanotis*, and *Sundasciurus everetti*

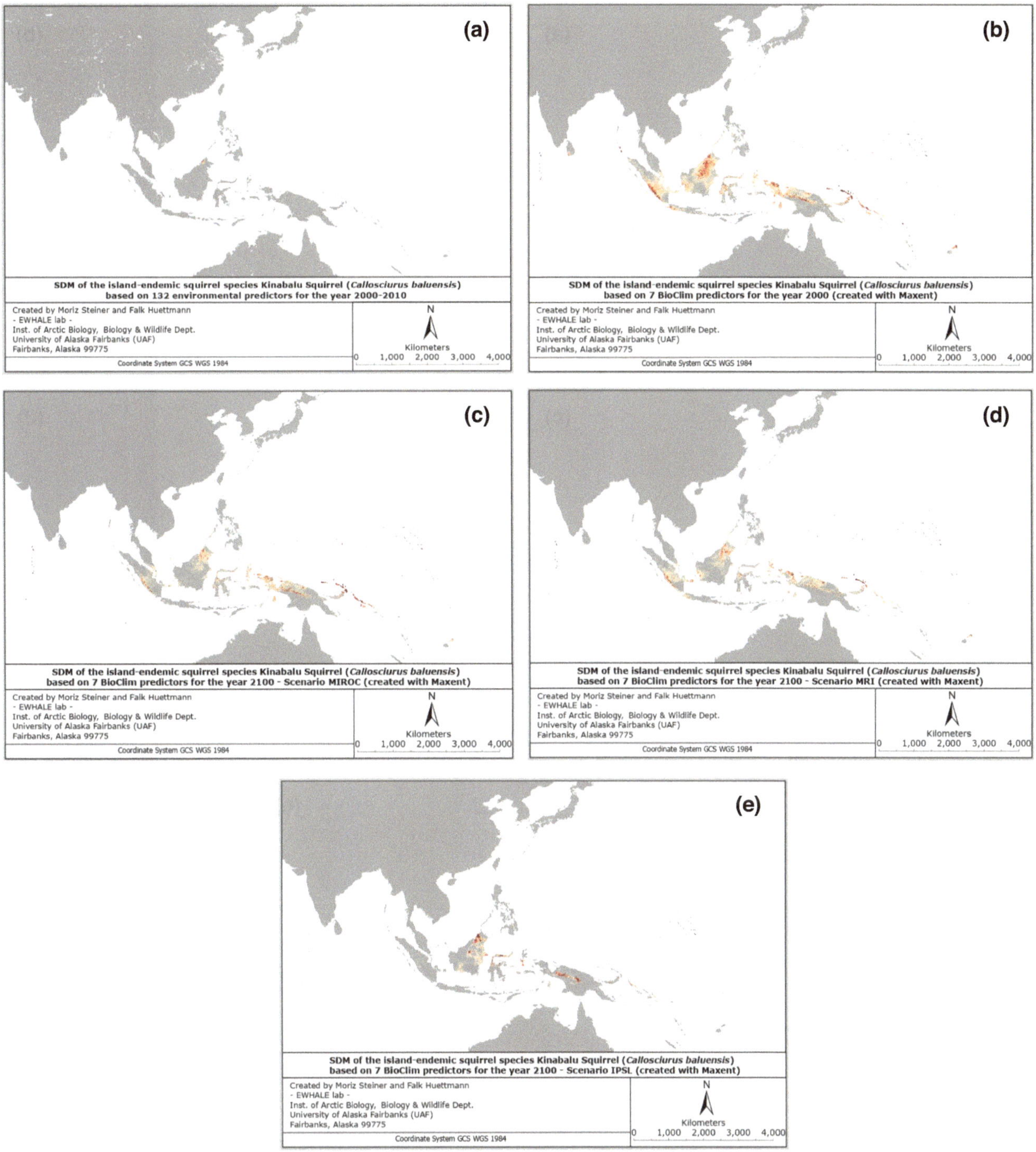

Fig. 7.2 SDM of the island-endemic squirrel species Kinabalu Squirrel (*Callosciurus baluensis*) based on (**a**) 132 environmental predictors for the year 2000–2010, (**b**) 7 BioClim predictors for the year 2000, (**c**) 7 BioClim predictors for the year 2100 – Scenario MIROC –, (**d**) 7 BioClim predictors for the year 2100 – Scenario MRI, (**e**) 7 BioClim predictors for the year 2100 – Scenario IPSL

7.3.3 Squirrel Distribution Forecast (SDF) Zoom-In to Squirrel Species-Rich Islands for 2100

In Chap. 12 (Steiner and Huettmann 2023), several Machine Learning (ML) analyses have been performed to forecast the distribution of the global squirrel population for the year 2100. The figures in Chap. 12 present for each group of squirrels the created model on a global scale. These rapid-assessment methods have here been chosen because the aim was to present a holistic work style and data. This has the use to include all global squirrel species and present the data to make it publicly

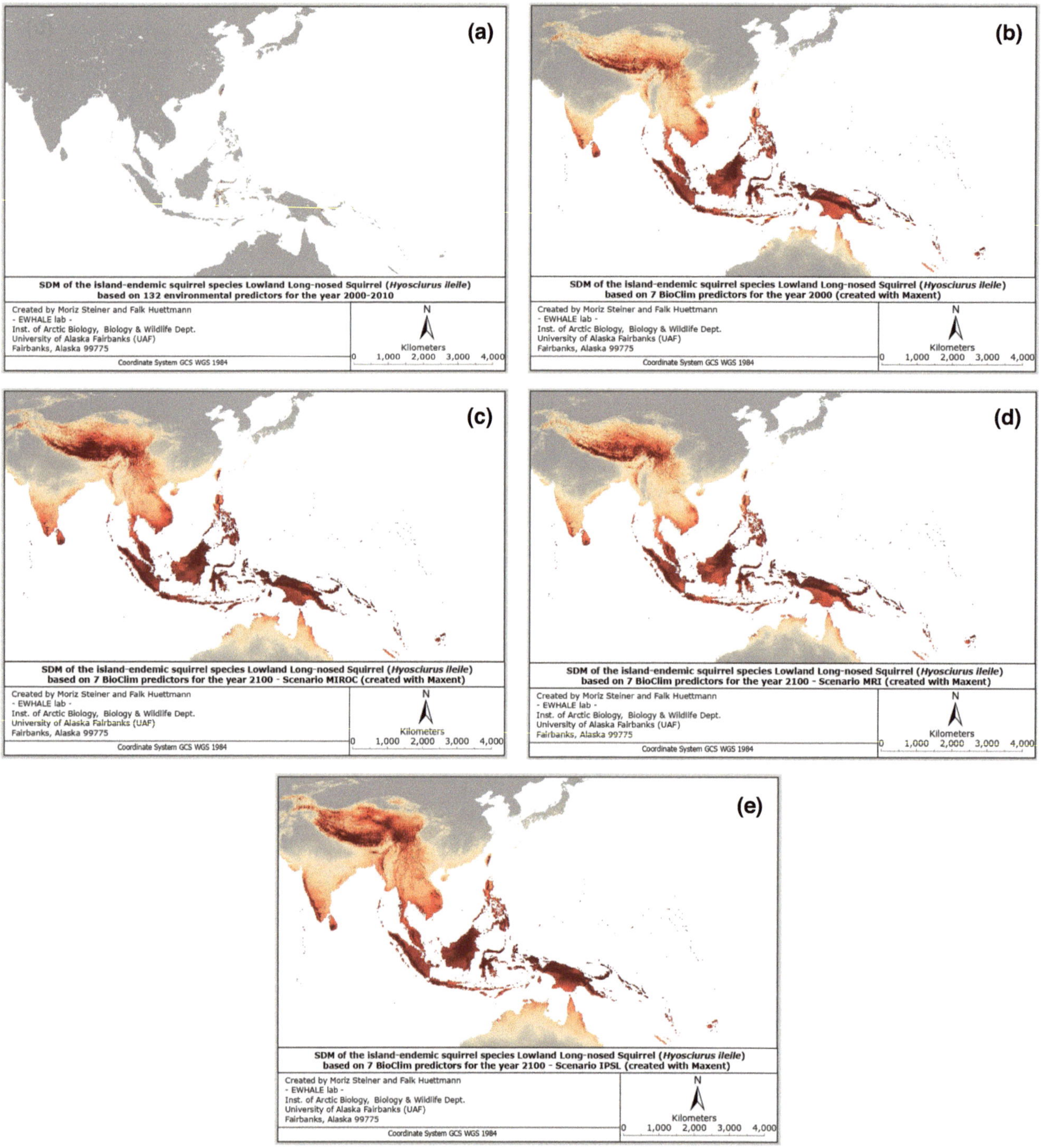

Fig. 7.3 SDM of the island-endemic squirrel species Lowland Long-nosed Squirrel (*Hyosciurus ileile*) based on (**a**) 132 environmental predictors for the year 2000–2010, (**b**) 7 BioClim predictors for the year 2000, (**c**) 7 BioClim predictors for the year 2100 – Scenario MIROC, (**d**) 7 BioClim predictors for the year 2100 – Scenario MRI, (**e**) 7 BioClim predictors for the year 2100 – Scenario IPSL

available and thereby bring the squirrel family further into the discussion of conservation needs and the lack of research done on non-charismatic species and their marginalization. In this study, however, a slightly closer look will be taken on the squirrels that are inhabiting island environments, more specifically, that are endemic to islands. In order to do this, we zoomed into some of the GCMs, and figures created and presented in Chap. 12, and present them here in a comparative manner. Below, Figs. 7.7 illustrate the current and future modeled distribution of some main island groups. These islands are inhabited by many endemic squirrel species. These maps can easily be compared with each other and thereby the forecasted changes can be better observed.

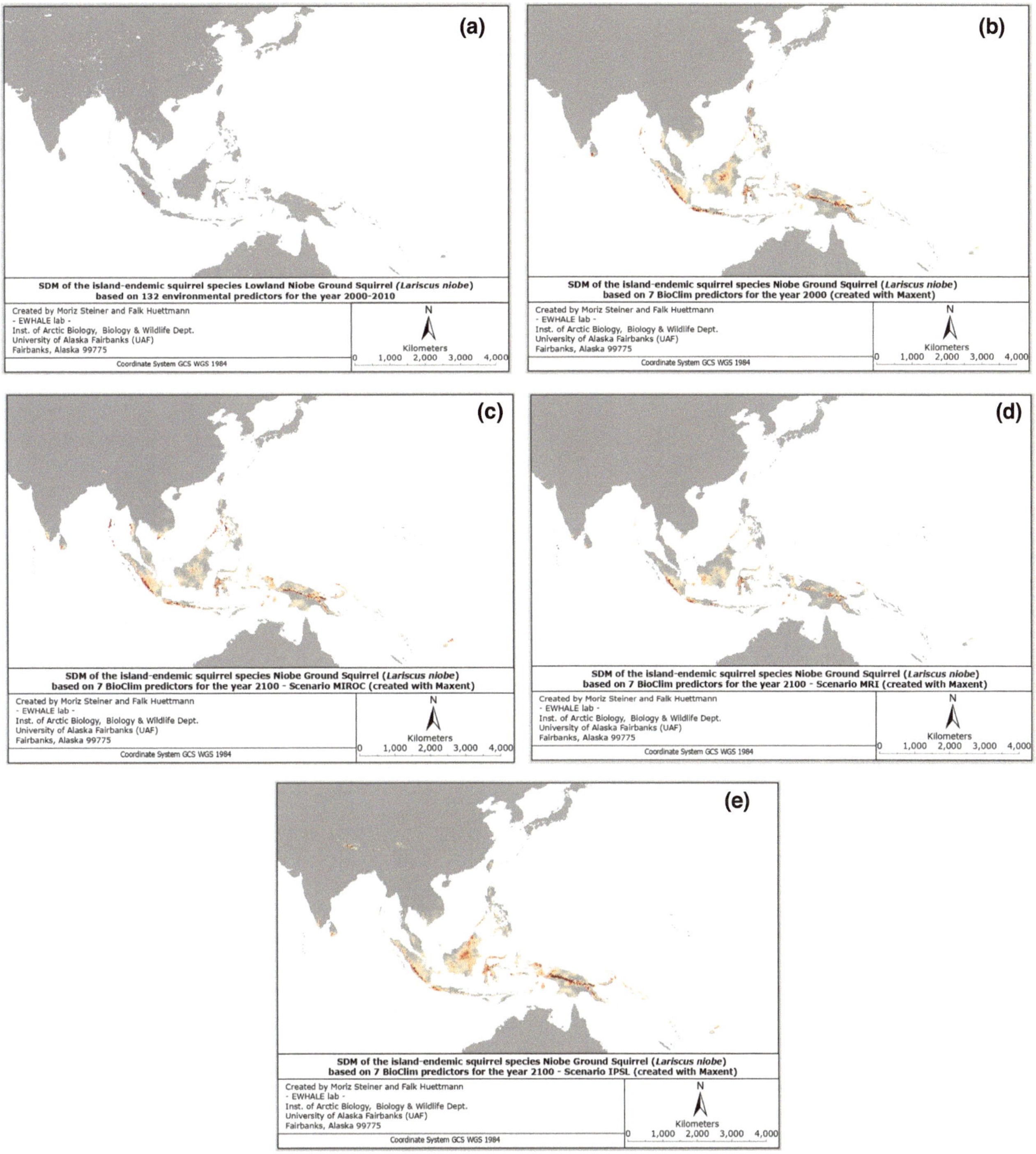

Fig. 7.4 SDM of the island-endemic squirrel species Niobe Ground Squirrel (*Lariscus niobe*) based on (**a**) 132 environmental predictors for the year 2000–2010, (**b**) 7 BioClim predictors for the year 2000, (**c**) 7 BioClim predictors for the year 2100 – Scenario MIROC –, (**d**) 7 BioClim predictors for the year 2100 – Scenario MRI, (**e**) 7 BioClim predictors for the year 2100 – Scenario IPSL

These figures have been included here, as they depict clearly how the global squirrel population will develop in the future. In Chap. 12 (Steiner and Huettmann 2023), the three global climate scenarios (MIROC6, MRI-ESM 2.0, and IPSL-CM6A-LR) have been considered as low/cooling scenario, medium/business-as-usual scenario, and high/hot scenario, respectively. It can clearly be identified that in the two more probable future scenarios (MRI, and IPSL) the squirrel distribution range in the identified region declines dramatically. Only in the scenario of global cooling by 2100, the squirrel population seems to increase in range, however, this is fairly unlikely (Førland et al. 2011; Hall and Fagre 2003; Raftery et al. 2017; Rogelj et al. 2018).

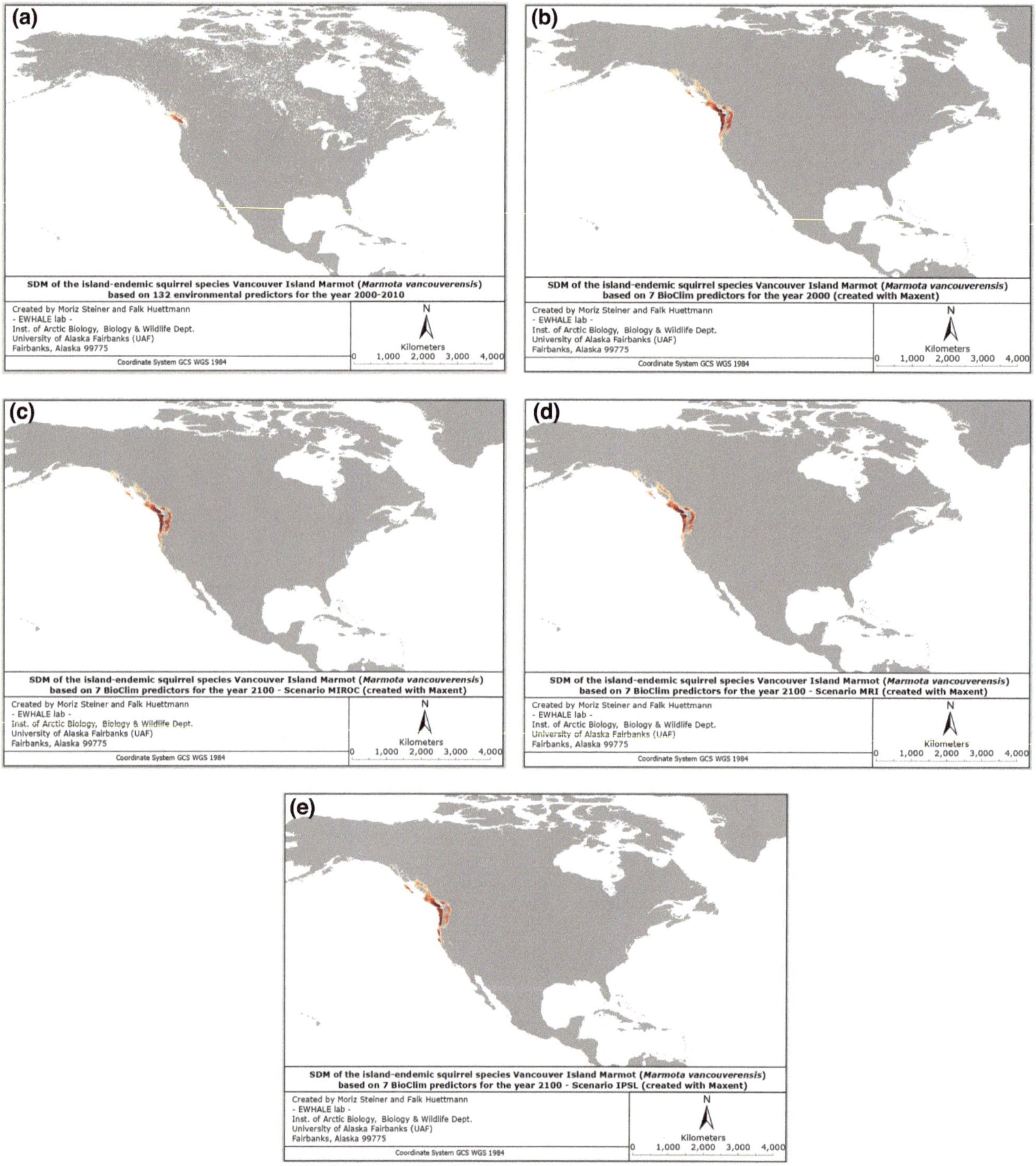

Fig. 7.5 SDM of the island-endemic squirrel species Vancouver Island Marmot (*Marmota vancouverensis*) based on (**a**) 132 environmental predictors for the year 2000–2010, (**b**) 7 BioClim predictors for the year 2000, (**c**) 7 BioClim predictors for the year 2100 – Scenario MIROC, (**d**) 7 BioClim predictors for the year 2100 – Scenario MRI, (**e**) 7 BioClim predictors for the year 2100 – Scenario IPSL

In more detail, for the MIROC scenario, the abundance of the squirrels within the hotspot region increases. However, also a clear shift in their distribution is noticeable. Current core distributions in Japan and South Korea drastically decrease and move southwards. However, for species endemic to Japan (e.g. Japanese Squirrel (*Sciurus lis*), and Japanese Giant Flying Squirrel (*Petaurista leucogenys*)), such a southwards shift is impossible and will thus lead to a significantly increased level

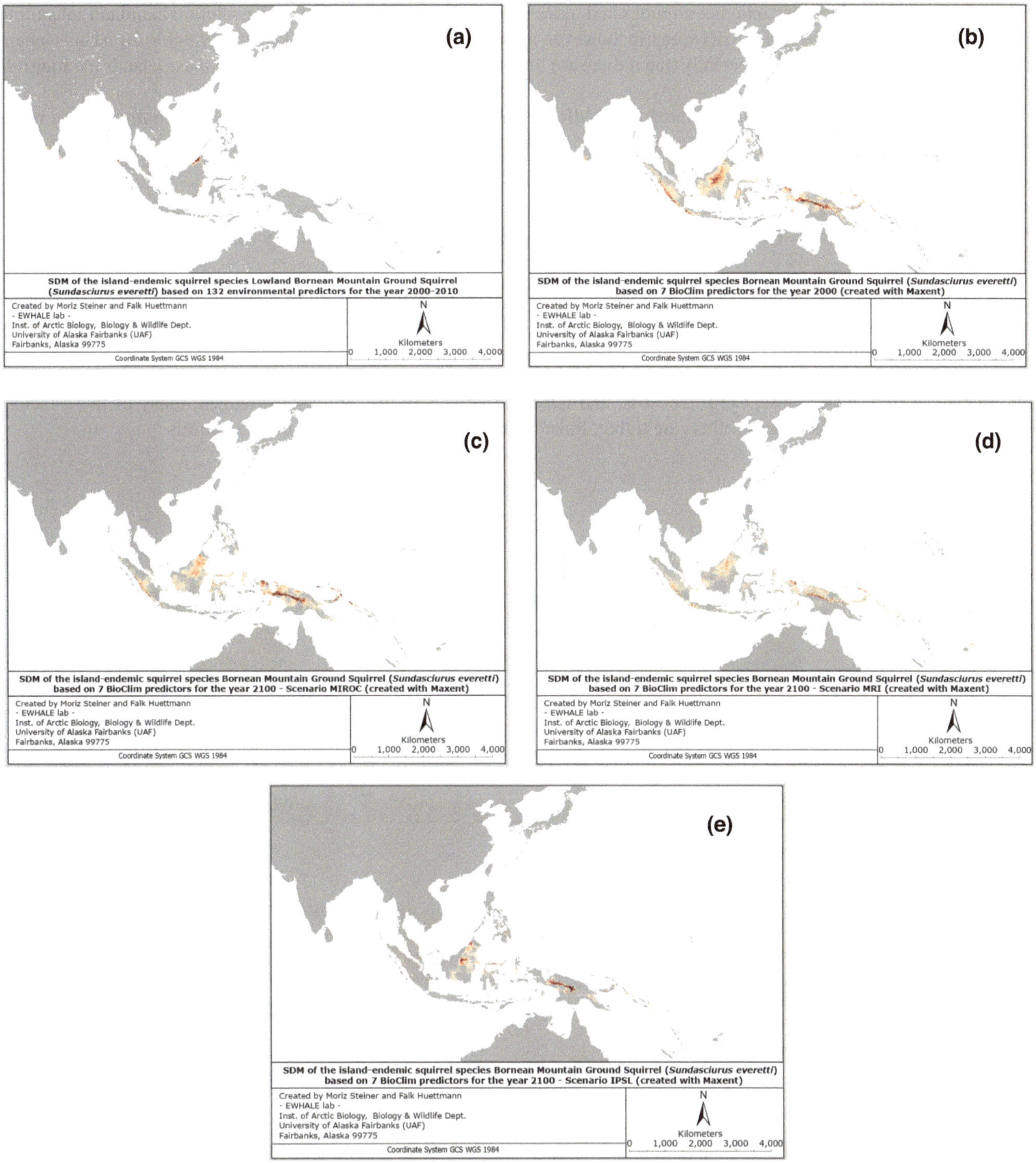

Fig. 7.6 SDM of the island-endemic squirrel species Bornean Mountain Ground Squirrel (*Sundasciurus everetti*) based on (**a**) 132 environmental predictors for the year 2000–2010, (**b**) 7 BioClim predictors for the year 2000, (**c**) 7 BioClim predictors for the year 2100 – Scenario MIROC –, (**d**) 7 BioClim predictors for the year 2100 – Scenario MRI –, (**e**) 7 BioClim predictors for the year 2100 – Scenario IPSL

of stress for these squirrel species, perhaps even extinction. In turn, however, an increase in the distribution ranges, likely population numbers, and abundance of other squirrels in more southern regions is expected.

For the MRI scenario, a drastic decrease in the species range and likely abundance and richness can be observed. From this coarse model, it is difficult to identify which species face the greatest threats and stresses, however, a general decline is clearly noticeable. A core shift to higher altitudes and more northern latitudes seems to be the solution mainland-inhabiting squirrels will choose in case of this MRI scenario, however, as discussed before, this is not always possible for islands inhabiting squirrels. This latter case is especially true if there are little or no hills or mountains present on these islands the squirrels inhabit.

For the IPSL scenario, the same trend as for the MRI scenario can be observed. Here, however, the distribution range declines and overall shift towards more northern latitudes seems to be even more extreme.

7.3.4 Further Human-Caused Climate Aspects Affecting Squirrels Today and In 2100

There are many, even still unknown effects that the climate will have on nature, the Anthropocene, and generally, the planet by 2100. Some major known and researched aspects that will affect the Anthropocene, nature, and thereby also squirrels by 2100 have been summarized in Table 7.1.

In Table 7.1 topics such as Sea level rise, Ocean acidification, pH value drop, Flooding, Extreme hurricanes, Human density increase, Virgin habitat loss, Coral reef loss, and Global Biodiversity loss are acknowledged; many of those factors impact squirrels on islands directly, others are tightly linked with island processes relevant for squirrels.

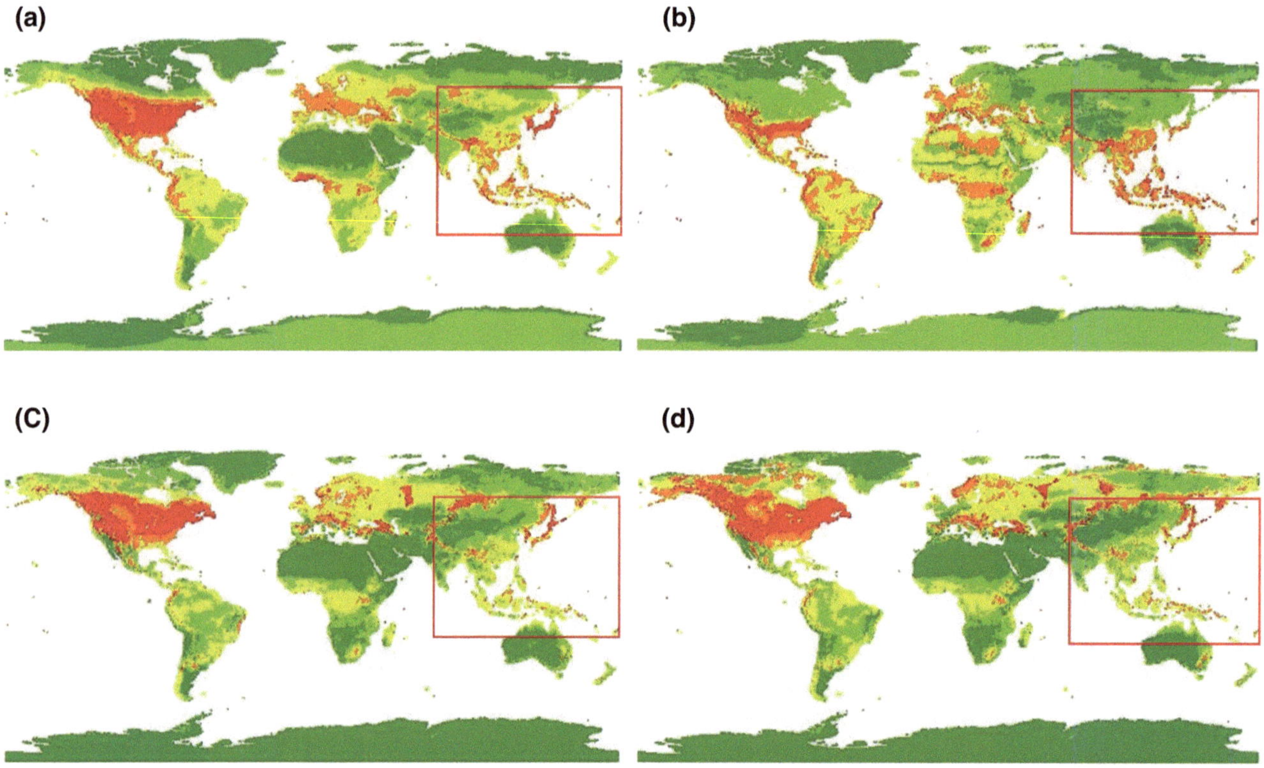

Fig. 7.7 All global squirrel species (**a**) in their current distribution from TreeNet, (**b**) GMC MIROC 2100 from TreeNet, (**c**) GCM MRI 2100 from TreeNet, (**d**) GMC IPSL 2100 from TreeNet

Table 7.1 Further man-caused climate aspects affecting squirrels today and in 2100

Threat	Severity	References
Sea level rise	Increase by at least 1m, likely 3–5 m	Antonioli et al. (2017), Horton et al. (2014), Jevrejeva et al. (2014), Nicholls and Leatherman (1996) and Taherkhani et al. (2020)
Ocean acidification and PH value drop	>0.2 pH decrease	Brander et al. (2012) and Cerrano et al. (2013)
Flooding	5500 km2 in Coastal Italy, up to 20-fold increase in economic risk in damage (England), floodings double approximately every 5 years. Some major coastal metropolis cities are in great danger, all over the world.	Antonioli et al. (2017), Hall et al. (2003), Knowlton et al. (2011) and Taherkhani et al. (2020)
Extreme hurricanes	36% stronger (Florida, USA)	Denamiel et al. (2020), Elsner et al. (2011), Knowlton et al. (2011) and Mudd et al. 2014
Human density increase	Multiplication factor 2–4	Chen and Sun (2019) and Roser 2013
Virgin habitat loss	Severity is globally not assessed and literature lacking	Virkkala (2016) and Vondrák et al. (2019)
Coral reef loss	Reduction by 50%	Brander et al. (2012) and Cerrano et al. (2013)
Global Biodiversity loss	Total bird density declined by >18%, Globally between 40% and 50%	Virkkala (2016) and Wilting et al. (2017)
Heat waves	>0.5 deg. C. temperature/10 years in Egypt, 0.66 °C. increase/10 years in Spain with 1400 deaths/year. Almost 10 °C. increase by 2100 in Alaska (USA), in reality likely more.	Díaz et al. (2019), Kenney et al. (2014), Knowlton et al. (2011), Melvin et al. (2017) and Mostafa et al. 2019
Droughts	>1–1.5 mm decrease of precipitation/year in Egypt, but more severe in other regions of the world	Chen and Sun (2019), Díaz et al. (2019), Mohsenipour et al. (2018), and Siegert et al. (2001)
Forest fires	increase by 10–50% over most of North America	Barbero et al. (2015), Flannigan et al. (2000) and Siegert et al. (2001)
Wars	Approx. 54% increase in armed conflicts with increased warmth	Burke et al. (2009) and Zhang and Lee (2010)

7.4 Discussion & Conclusion

As shown elsewhere (e.g. Connell 2013) and identified in Chap. 3 (Steiner and Huettmann 2023), islands are part of the so-called "regions of/ under high risk". These regions are either at particular risk to squirrels or are at high risk themselves while hosting squirrels, e.g. due to human pressure on a finite isolated landmass. Here, we analyzed this "region of/under high risk", Islands. We used literature and found a high similarity between the documented distribution of the island-endemic squirrel species and the modeled one that has newly been presented here. We also presented additional SDMs which have not been presented in Chap. 3 (Steiner and Huettmann 2023) previously, and an in-depth future forecast analysis of the major island regions of the world by zooming in on the distribution range change of all global squirrel species, caused by climate change. Next, some globally relevant climate change effects will be discussed, how these influence squirrels, and which conservation actions are most likely to lead to success. Lastly, future discussion points are presented which have been included to start a discussion on the there-mentioned topics and the inclusion of squirrels in it. This is accompanied by the identification of possible future research focus points.

7.4.1 Literature Review Analysis of Island-Endemic Squirrel Species

By comparing the upper-presented Literature review (Sect. 7.3.1) and SDMs that have been created in Chap. 3 (Steiner and Huettmann 2023), and above (Sect. 7.3.2), it can be observed that their distribution, both in literature and the models, is highly similar. We found that their distribution is almost identical. Unfortunately, for these less well-known squirrel species only a few literature references are existing which are mostly just short, generalized notations (Wilson and Reeder 2005), however, useful for species where otherwise the information would be absent.

7.4.2 Analysis of the Species Distribution Models (SDMs)

In the SDMs presented above (Figs. 7.1, 7.2, 7.3, 7.4, 7.5), it can be clearly observed that some squirrels are endemic to islands. Island endemism is strongly connected with isolation. The fate of species isolation is not always bright. Scheffer et al. (2006), and Fahrig et al. (2019) suggest that species richness in isolation can often even be significantly increased compared to a vast homogenous landscape. This is because some islands (land-locked, or water-locked) might indeed be considered biodiversity-rich habitats since their niche is heterogeneous compared to a vast homogenous environment that tendentially supports more life forms and species. However, isolation is commonly considered as unbeneficial for species and their integrity (Bailey et al. 2010; Ryser et al. 2019). This is because genetic diversity on a species level is very limited (Furlan et al. 2012; Juan et al. 2004), genetic migration/gene flow is impossible (Furlan et al. 2012; Juan et al. 2004), and the populations are literally "stuck" (Koprowski et al. 2005).

In Sect. 7.3.2, five SDMs have been presented in the addition of the two presented ones in Chap. 3 (Steiner and Huettmann 2023). In Chap. 3, SDMs have been presented for the species Palawan flying squirrel (*Hylopetes nigripes*) which is endemic to the single island Palawan, Philippines, and Black-eared squirrel (*Nannosciurus melanotis*) a species that is under a big climate change-derived threat (sea-level rise) as it only inhabits marginal coastal areas (South-East Asia). These two species are already great examples of squirrels endemically inhabiting this region of/under risk.

Above, additional five species distribution models for the species Kinabalu Squirrel (*Callosciurus baluensis*), Lowland Long-nosed Squirrel (*Hyosciurus ileile*), Niobe Ground Squirrel (*Lariscus niobe*), Vancouver Island Marmot (*Marmota vancouverensis*), and Bornean Mountain Ground Squirrel (*Sundasciurus everetti*) have been presented (Figs. 7.2a, 7.3a, 7.4a, 7.5a, 7.6a). For all of these species, it can be observed that their predicted occurrence is restricted to islands. Four out of the five species occur in the South-East Asian hotspot marked by a red rectangle. Only the species Vancouver Island Marmot (*Marmota vancouverensis*) inhabits an island (Vancouver Island) outside of this identified hotspot. The reason for including so many species from that South-East Asian hotspot is because many squirrel species, in fact, (>100 out of 233) inhabit this hotspot region, see Chap. 3 (Steiner and Huettmann 2023) and its appendix for details). This high occurrence of squirrel species there can possibly be explained by the highly favorable environment, tropical habitats, climate, and feeding sources (Das and van Dijk 2013). Additionally, it is assumed that squirrels have diverged to these islands during times with low tights and decreased sea levels (Mercer and Roth 2003), and since then sea levels have risen and the species have evolved into new species due to fragmentation and evolution (Hawkins et al. 2016b). This is mainly due to the increased threats to the squirrels, which are mainly caused by humans and their consumption, manufacturing pressure, poor governance, and conservation marginalization. In order to predict a low-end estimate of how these squirrel populations in the South-East Asian hotspot will develop over time, global climate models have been performed. These climate models aim to forecast the distribution of all the squirrels occurring in these regions, based on widely acknowledged climate predictors. The results of that can be found in the results section above (7.3.1).

These SDMs fulfill the aim to present the current situation and to raise awareness of these squirrel presences and their need for better governance, to fight conservational marginalization.

Additionally, for each of these SDMs that have been created with 132 predictors, the SDM creation has been repeated with seven BioClim predictors and one altitude predictor, instead of 132 environmental predictors. These have been presented for each species as the second figure (7.2b, 7.3b, 7.4b, 7.5b, 7.6b). There, it can be observed that the distribution remains generally similar, however, the distribution range is much larger for all species. This indicates that the niche of each species cannot be properly described with just 7 (+1) predictors and that the outcome of the models does not represent the actual distribution as accurately as possible. Unfortunately, a full species distribution modeling (using +130 predictors) for the year 2100 cannot be performed yet with such a high number of predictors because they are not available for all the descriptive metrics (e.g. global road and river map – for the proximity to roads and rivers, or the proximity to protected areas, or the future threatened mammal density, etc.). This study therefore also underlines the importance of a high number of environmental predictors to be able to gain better knowledge on the original distribution (based on the occurrence points) and create a more detailed future species distribution niche model.

In order to provide a future outlook for these species, three future scenarios have been presented for each of these species. These 3 global climate scenarios have been broadly discussed in Chap. 12 (Steiner and Huettmann 2023), as well as the BioClim predictors that are composing these scenarios.

For the species *C. baluensis*, the overall distribution for all climate scenarios seems to have decreased compared to the modeled distribution of the current climate. For *H. ileile* a strong variation of its distribution extend can be observed between the prediction of 132 and 7 (+1) predictors. Among the three GCMs, there can no major differences be observed. For this species (*H. ileile*) it seems necessary to include more environmental predictors to properly forecast its modeled future distribution. For the species *L. niobe*, it can be observed that the SDMs, also the future ones (SDFs), capture this squirrel's niche very well. It does not seem to change majorly based on these 7 climate predictors, and it aligns very well with the range and habitat described in literature. For the species *M. vancouverensis*, the future models do not present much of a change of the area of this species (also because it is island-locked), only a small increase in its range size with a potential habitat/ suitable niche on the mainland can be observed. This sounds like great news, however, the SDM for the current climate (with 7 predictors) also shows this trend which indicates that this observation can rather be explained by a decreased SDM quality compared to the one created with 132 predictors. Lastly, for the species *S. everetti,* it can be observed that on all SDMs the core range shifts from the original one presented in Fig. 7.6a. This distribution aligns with the one found in literature, however, the models also present a high occurrence likelihood in New Guinea, which is absent in literature. This might be due to modeling inaccuracies of using only 7 (+1) predictors, or because New Guinea has received too little attention in squirrel research and potential presence occurrence data is absent.

Overall, it can be stated that the SDMs created in Chap. 3 (Steiner and Huettmann 2023), and presented here, align very well with the literature, especially the one for the years 2000–2010 using 132 environmental predictors. The other SDMs deliver similar results, however with poorer accuracy, and thus, it is suggested to compile and use as many descriptive predictors as possible and necessary.

7.4.3 Further Focus Points for The SDMs and its 2100 Forecast with the Maxent Algorithm

In Sect. 7.3.2 several SDMs and SDFs have been presented. There, for each presented squirrel species five maps have been presented, two of the current years (2000–2010) and three as forecast to 2100. As it might be observable, there are not often major, continental changes visible, but rather small changes. These small changes are difficult to grasp from the rather coarse maps presented above. In Chap. 12 (Steiner and Huettmann 2023), major range shifts can be observed, but this is not for all species and genera the case, this must be mentioned. For some species, as observable above, the changes are rather regional, which however can still be major difficulty for a single species to tackle. In order to visualize such changes, Figs. 7.8a till 7.8c are presented below. These figures represent a regional view on the distribution change of the squirrel species Kinabalu Squirrel (*Callosciurus baluensis*) in Northern Borneo.

In Figs. 7.8 it can be observed that when one looks closely at the species' distribution change, a shift can be noticed. In the case of the species *C. baluensis*, a clear shift into higher altitudes can be observed by comparing the current situation (Fig. 7.8a) and the relatively hot GCM scenario IPSL for 2100 (Fig. 7.8c). The opposite can be observed when one compares the current situation (Fig. 7.8a) with the future cooling GCM scenario MIROC (Fig. 7.8b).

By demonstrating these local/regional scale changes it is sought to illustrate that small changes can also be significant, even though they seem minor on global/coarse maps. In this study, it must be said that the approach is pretty coarse and can be improved by using more fine-scale modeling methods and approaches. Such approaches can be chosen for future studies, to study fine-scale, local changes of the species. An example for this can be seen by Imbach et al. (2018) for South America. Also, for future studies, more environmental predictors are suggested to be used if they can be compiled for 2100. By combining a more fine-scale approach with more environmental predictors, it is expected to improve the quality of the SDMs and SDFs.

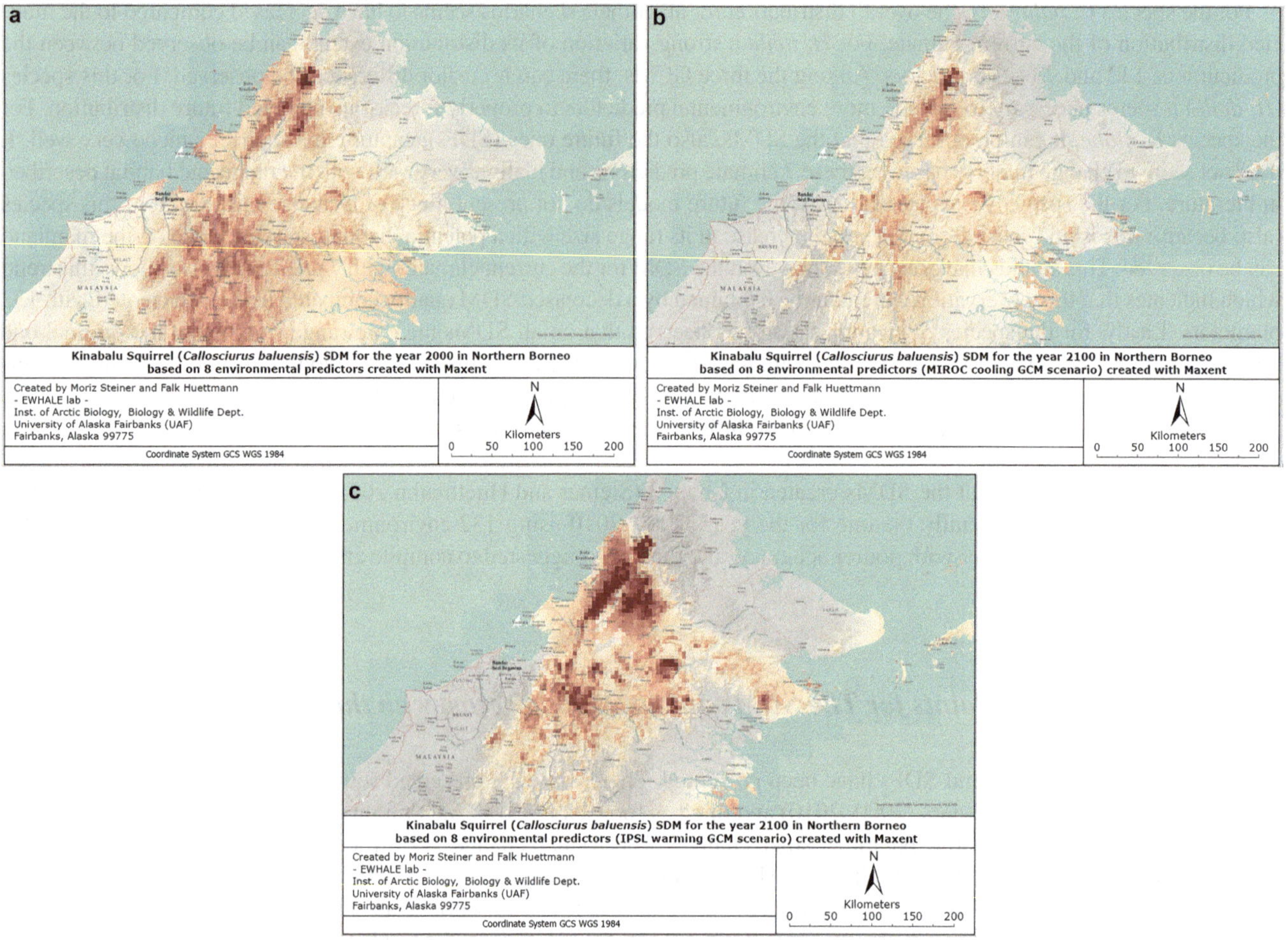

Fig. 7.8 Distribution change of Kinabalu Squirrel (*Callosciurus baluensis*) in Northern Borneo for the year (**a**) 2000 based on 7 BioClim layers (**b**) 2100 based on 7 BioClim layers – MIROC cold GCM scenario (**c**) 2100 based on 7 BioClim layers – IPSL hot GCM scenario

7.4.4 Squirrel Distribution Forecast (SDF) Zoom-In to Squirrel Species-Rich Islands

Figure 7.6a illustrates the current distribution of all global squirrel species using the www.GBIF.org database and 7 BioClim climate layers and one altitude one as the data source. Figure 7.6a represents the current distribution, and thus the baseline of our predictions. From this baseline the Figs. 7.6b-7.6d have been created. How these figures have been created and their methods have been described in detail in Chap. 12 (Steiner and Huettmann 2023). These three latter-mentioned figures illustrate the future prediction of the squirrel distribution in the year 2100, based on three different climate scenarios (MIROC, MRI, and IPSL). It can easily be observed that based on those climate scenarios, the distribution of the squirrels differs. Especially for the identified South-East Asian hotspot, and its islands, a significant range shift can be observed. In the case of the MIROC climate scenario, which has been identified as representing a cooldown of the future climate, the squirrel population within the identified hotspot region significantly increases. Unfortunately, though, this scenario is highly unlikely (Collins et al. 2012; Cubasch et al. 2001; Vecchi and Wittenberg 2010), because a cooldown of the global climate by several degrees is contradicting most climate forecasts (Collins et al. 2012; Cubasch et al. 2001; Vecchi and Wittenberg 2010). On the other hand, the two climate scenarios MRI and IPSL represent the scenarios of business-as-usual, and high but realistic warming forecast, respectively. For the latter two scenarios, the fate of the squirrels does not look bright since their distribution ranges and likely also abundances decline dramatically. These two latter scenarios unfortunately represent with a high probability the future of what squirrels, the Anthropocene, and nature will have to face (Collins et al. 2012; Cubasch et al. 2001; Vecchi and Wittenberg 2010). The reason why the increase of several degrees Celsius of the global climate is such a big threat for squirrels is that it heavily affects their island environment. The main problem for the squirrels there is that they cannot move towards their new best habitable regions because they are literally "stuck" on islands and cannot disperse due to the surrounding waters to move towards regions that support their lifestyles and niches. They are either required to adapt to the new environment or will become extinct. On top of this, the additional threats that climate change brings will heavily impact the squirrels' habitat. Their major threats have been identified and presented above in Table 7.1.

Simply by observing the illustrations in Sect. 7.3.2, the outlook for our planet's squirrels seems grim and depressing. An overall decline can most likely not be escaped. However, the declines can be reduced. Lifestyle approaches and management suggestions to achieve this are being presented below in Sect. 7.4.5.

From these GCMs that have been presented above, a clear shift of the core distribution mainly towards the more northern latitudes and higher altitudes can be observed, as it has been described in Chap. 12 (Steiner and Huettmann 2023).

7.4.5 Further Human-Caused Climate Aspects Affecting Squirrels, the Anthropocene, and Nature Today and in 2100

Table 7.1 includes some major threats caused by climate change that must be faced by the global squirrels, the Anthropocene, and nature by 2100. Here, these major threats will be shortly individually discussed to raise awareness about them, their magnitude, and their influences. In addition, suggested improvements, lifestyle conservation management, and politics are being included in order to decrease their magnitude and severity of negative impacts in the future.

Sea level rise.

This threat is of particular concern for the Anthropocene as many big cities e.g. New York City, Kolkata (Calcutta), Mumbai (Bombay), Dhaka, Guangzhou, Shanghai, Bangkok, even whole states/ countries like Florida and the Netherlands are immediately threatened by a climate-induced rise of the sea level (Nicholls et al. 2007). However, this is also critically important for some squirrel species e.g. Black-eared squirrel (*Nannosciurus melanotis*) which only inhabits marginal coastal areas on some Southeast Asian islands (See Fig. 3.12 in Chap. 3 – Steiner and Huettmann 2023).

Ocean acidification and PH value drop.

This threat is affecting squirrels less directly as squirrels are not sea species. However, some indirect effects might still negatively affect squirrels. This threat is dramatically important for sea species, and especially for the ones that are highly sensitive to the water's pH (e.g. corals, reef-building corals, cold-water corals, crustose coralline algae, *Halimede*, benthic mollusks, echinoderms, coccolithophores, foraminifera, pteropods, seagrasses, jellyfishes, and fishes) (Anthony et al. 2008; Gattuso and Hansson 2011; Guinotte and Fabry 2008; Kleypas and Yates 2009). This even affects humans, especially for the ones that are depending on it e.g. fishers and its consumers (Mathis et al. 2015).

Floods.

This threat goes hand in hand with the upper-mentioned threat of Sea level rise. The higher the sea level rises, the likelier and more frequent floods can be expected (Antonioli et al. 2017; Hall et al. 2003; Taherkhani et al. 2020).

Extreme hurricanes.

Extreme hurricanes are known to increase in frequency and strength, with a predicted increase in strength of up to 36% in regions like Florida (USA) (Denamiel et al. 2020; Elsner et al. 2011; Mudd et al. 2014). Such an increase in the strength and frequency of extreme hurricanes can have severe influences, e.g. through habitat destruction for flora and fauna (including squirrels) (Cely 1991; Gunter and Eleuterius 1971; Pries et al. 2009), economic resources on the local economy (for humans) (Hallegatte 2012; Knowlton et al. 2011; Malmstadt et al. 2009; Padgett et al. 2008), and even human lives (Knowlton et al. 2011).

Human density increase.

Even though here it is referred to humans, it does not solely affect humans, but also other species, under which one can also find squirrels. With an increase in the human population, and thus, also its density, dangers such as an increased frequency of pandemics, housing shortages (Ramankutty et al. 2018; Stycos and Arias 1972; Song et al. 2018; Tian et al. 2020), health care shortages (Marć et al. 2019; Naicker et al. 2010; Wu et al. 2016), drink water shortages (Cassardo and Jones 2011; Hadadin et al. 2010; Pimentel et al. 1994), food shortages (Döös 2002; Ramankutty et al. 2018), outsourcing the planets carrying capacity (Brown and Kane 1995; Meadows and Randers 2012), etc. cannot be avoided.

Virgin/Wilderness habitat loss.

Old-growth/virgin habitats have become rarer and rarer over the last centuries, mainly due to human destruction, and expansion (Virkkala 2016; Vondrák et al. 2019). However, climate change also threatens these valuable habitats worldwide. This is because of increased forest fires (see the section below), biodiversity loss (see the section below), but also a general change of the temperature which will have an impact on the compositing and trophic levels of species inhabiting such systems (e.g. see Aanes et al. 2002; Lauria et al. 2012).

Coral reef loss.

Coral reef loss is affected by climate change by more than just its direct influence of increased temperature. Because it is expected that not only the air's temperature will rise, but also the water's temperature (Ateweberhan and McClanahan 2010; Azra et al. 2020). Additionally, as discussed above, Ocean acidification and PH value drop are also expected to negatively influence coral reefs (Brander et al. 2012; Cerrano et al. 2013). Additionally, human-induced climate change is known to facilitate coral bleaching (Donner et al. 2007).

Global Biodiversity loss.

Biodiversity loss is one of the major threats to nature that is already occurring and predicted to worsen continuously, it is even under discussion that we are currently experiencing the 6th major species extinction event (6th mass extinction) (Agostinho and Gomes 2018; Nazarevich 2015; Stilwell 2018). This event is not only caused by climate change, but also by conservation management failure (Failing and Gregory 2003), unsustainable consumerism (Agostinho and Gomes 2018; Bull and Milner-Gulland 2020), policy failure (Bull and Milner-Gulland 2020; Slingenberg et al. 2009), increased human pressure (Song et al. 2018; Vačkář et al. 2012), habitats loss due to human urban expansion (Song et al. 2018; Vačkář et al. 2012), etc.

Heatwaves.

Due to an increased future temperature and increased events of temperature extremes, heat waves are expected to occur more frequently and thereby also have more severe effects (Ganguly et al., 2009; Knowlton et al. 2011; Kyselý 2010; Schär et al. 2004). Such heatwaves can have significant impacts on the squirrel's lives, their life cycles, and their survival.

Droughts.

There are several cascading events originating from droughts. There is a shortage of available water for humans, animals, and agriculture, leading to arable land declines (Fischer et al. 2005; Larson 2013; Liu et al. 2019; Wang et al. 2019), which in worst-case scenarios can even turn into deaths and wars (Burke et al. 2009; Gill 2000; Rowland et al. 2015; Zhang and Lee 2010).

Forest Fires.

Forest fires are known to be naturally occurring with higher frequency in high latitudes (Chen et al. 2021a, Chen et al. 2021b). Additionally, an increased global temperature, especially a high increase, as it is commonly known to be predicted for the poles, is predicted to significantly increases the frequency and magnitude of forest fires (Chen et al. 2021b). And to worsen the situation even further, due to climate change permafrost melts in the high latitudes (Mack et al. 2012; Schuur et al. 2008), this permafrost liberates methane and carbon, which again, facilitates the creation of new fires and nourishes ongoing ones (Chen et al. 2021a; Mack et al. 2012).

Wars.

Climate change has apart from direct effects also indirect ones. Increased temperatures often lead to droughts (Chen and Sun 2019; Díaz et al. 2019; Mohsenipour et al. 2018; Siegert et al. 2001), this in turn, leads to decreased agricultural yields and therefore starving periods. In order to survive, armed conflicts can arise from such situations (Burke et al. 2009; Zhang and Lee 2010). This topic, in connection with climate change, can become a very "hot" topic in the future. Because with a warming climate, ice will melt, and therefore, new land will become arable (Ramankutty et al. 2018), seaways will become accessible (Borgerson 2008; Smith and Stephenson 2013), and oil fields yieldable (Harsem et al. 2011; Kuzin 1982; Petrick et al. 2017). Here, strong stakeholders and the world's most powerful people and politicians are highly involved and could in the worst-case scenario even lead to future large-scale wars.

All these presented current and future threats to squirrels, humanity, and nature are mainly caused by human-accelerated climate change and its consequences (Hulme et al. 1999; Mitchell et al. 2006). Therefore, the most straight-forward suggestion for futuristic implementation to decrease the rate of the human impact includes the immediate reduction of using fossil fuels and energy sources, decreasing the human footprint, stopping wasting food, moving towards a certain steady-state-economy (www.steadystate.org) – away from a neoliberal capitalistic approach to economy and life. Additionally, the human population increase should be slowed down, consumerism reduced, invested in environmental protection instead of wars, etc. This list almost seems to be never-ending. Many current risks and practices, that have been developed in the past couple of centuries are expected to have fatal influences on the human species and many other species of the world, especially in the near future. Therefore, to live close to and with nature, respecting its natural limits and carrying capacity, and living, therefore, more sustainably, are tasks that are to be followed to live in greater harmony with nature and to allow future generations to enjoy living on this planet (see literature on Buddhism, and Indigenous peoples' approaches e.g. Gethin, 1998; Harvey, 2012).

7.4.6 Further Discussion Points on the Squirrels' Fate on Islands with Suggested Research Focus Points

In this study, the approach to the created/analyzed models (in Sects. 7.3.1, and 7.3.2) is pretty coarse, similar to Ch7 (Steiner and Huettmann 2023). This has the same reasons as in Chap. 12, the authors aim to present the overall critical situation and to initiate discussions to fight the squirrels' marginalization and bring them on scientific agendas, worldwide. However, the issue of such coarse models is that the climate models are very poor in coastal regions and with small islands. Because species might occur on small islands, but these are too small to have a pixel on the map and can thereby not be included in SDMs and SDFs. An attempt to solve this issue is to use vector layers (e.g. shapefiles). Such a vectorized global overview of all islands and the global shoreline has previously been published (e.g. see Sayre et al. (2019)) and should be in the future used as a raster for SDMs. Other islands may also not be included in the data collection projects and provision and thus, there is no data for such islands available, and can thereby not be included in models and predictions. Models still quite fail with that, and a very fine scale approach is needed to be accurate and produce high-quality science. Such fine-scale studies would be a great follow-up to this study presented here, to identify local threats and forecasts, small-scale, and detailed species distribution models. Additional future research following this study is suggested to focus on the process of species divergence on islands, not always at the level of speciation but often at the subspecies level. Because such divergence events can lead to island populations being both phenotypically and genotypically distinct from their corresponding mainland populations (Whittaker and Fernández-Palacios 2007). Generally, approaching the global debate and almost tabu of mammalian subspecies is a topic that requires future attention. However, specifically for squirrels, it can hypothetically be stated that every squirrel population on an island is pretty much an isolated population and could be thereby considered as subspecies. This is because the population is genetically isolated with no genetic flow and follows isolated development and evolution. This in turn bears difficulties for conservation as such sub-populations/subspecies might require different conservation management actions/focus points, and might even be contradicting (Rayner et al. 2021; Suwal et al. 2018; Steiner and Huettmann 2021). It must be mentioned though that research on remote and wilderness islands is difficult to perform, however, it is of crucial conservation importance to move away from charismatic species and places, but rather investing into species where efforts, funding, and policy are highly required and should focus on (all squirrel species except Eurasian red Squirrel (*Sciurus vulgaris*), and Eastern Grey Squirrel (*Sciurus carolinensis*)).

This study and chapter, like several other chapters in this book, mainly aims to include squirrels in more and broader discussions and to present current situations and the affecting threats to/coming from them. The management methods of squirrels on islands regarding their predator-prey relationship and position in trophic levels is still a pretty unknown and unresearched topic which is expected to yield new stunning and conversationally necessary insights. Including squirrels in the discussion on mammal/predator eradication projects on islands (e.g. Rodents in New Zealand etc.) and questioning "What role do squirrels play there?" would be a step onto virgin soil; a research frontier.

Concludingly, in Sect. 7.4.4 several threats have been presented. Approaching these and actively pursuing sustainable management solutions for squirrels, humans, and generally, nature is a major task, but of enormous importance and contemporaneity. Focusing on these, and putting it on the agenda, to seek a life closer to nature, one that is more sustainable is the task that is to be achieved in the upcoming years and decades.

References

Aanes R, Sæther BE, Smith FM, Cooper EJ, Wookey PA, Øritsland NA (2002) The Arctic Oscillation predicts effects of climate change in two trophic levels in a high-arctic ecosystem. Ecol Lett 5(3):445–453

Agostinho AA, Gomes LC (2018) Biodiversity and fisheries management in the Paraná river basin: successes and failures. Repositório Institucional Da Universidade Estadual De Maringá (RI-UEM). http://repositorio.uem.br:8080/jspui/handle/1/5331

Anthony KR, Kline DI, Diaz-Pulido G, Dove S, Hoegh-Guldberg O (2008) Ocean acidification causes bleaching and productivity loss in coral reef builders. Proc Natl Acad Sci 105(45):17442–17446

Antonioli F, Anzidei M, Amorosi A, Presti VL, Mastronuzzi G, Deiana G, Vecchio A et al (2017) Sea-level rise and potential drowning of the Italian coastal plains: flooding risk scenarios for 2100. Quat Sci Rev 158:29–43

Ateweberhan M, McClanahan TR (2010) Relationship between historical sea-surface temperature variability and climate change-induced coral mortality in the western Indian Ocean. Mar Pollut Bull 60(7):964–970

Azra MN, Aaqillah-Amr MA, Ikhwanuddin M, Ma H, Waiho K, Ostrensky A, Abol-Munafi AB et al (2020) Effects of climate-induced water temperature changes on the life history of brachyuran crabs. Rev Aquac 12(2):1211–1216

Bailey D, Schmidt-Entling MH, Eberhart P, Herrmann JD, Hofer G, Kormann U, Herzog F (2010) Effects of habitat amount and isolation on biodiversity in fragmented traditional orchards. J Appl Ecol 47(5):1003–1013

Barbero R, Abatzoglou JT, Larkin NK, Kolden CA, Stocks B (2015) Climate change presents increased potential for very large fires in the contiguous United States. Int J Wildland Fire 24(7):892–899

Borgerson SG (2008) Arctic meltdown: the economic and security implications of global warming. Foreign Aff 87:63

Brander LM, Rehdanz K, Tol RS, Van Beukering PJ (2012) The economic impact of ocean acidification on coral reefs. Clim Chang Econ 3(01):1250002

Brearley G, McAlpine C, Bell S, Bradley A (2012) Influence of urban edges on stress in an arboreal mammal: a case study of squirrel gliders in southeast Queensland. Australia Landsc ecol 27(10):1407–1419

Brown JH (1978) The theory of insular biogeography and the distribution of boreal birds and mammals. In: Great basin naturalist memoirs. pp 209–227

Brown LR, Kane H (1995) Full house: reassessing the earth's population carrying capacity. Earthscan, London

Bull JW, Milner-Gulland EJ (2020) Choosing prevention or cure when mitigating biodiversity loss: trade-offs under 'no net loss' policies. J Appl Ecol 57(2):354–366

Burke MB, Miguel E, Satyanath S, Dykema JA, Lobell DB (2009) Warming increases the risk of civil war in Africa. Proc Natl Acad Sci 106(49):20670–20674

Cassardo C, Jones JAA (2011) Managing water in a changing world. Water 3(2):618–628

Cely JE (1991) Wildlife effects of hurricane Hugo. J Coast Res 8:319–326

Cerrano C, Cardini U, Bianchelli S, Corinaldesi C, Pusceddu A, Danovaro R (2013) Red coral extinction risk enhanced by ocean acidification. Sci Rep 3(1):1–7

Chadwick OA, Derry LA, Vitousek PM, Huebert BJ, Hedin LO (1999) Changing sources of nutrients during four million years of ecosystem development. Nature 397(6719):491–497

Chen H, Sun J (2019) Increased population exposure to extreme droughts in China due to 0.5 C of additional warming. Environ Res Lett 14(6):064011

Chen Y, Romps DM, Seeley JT, Veraverbeke S, Riley WJ, Mekonnen ZA, Randerson JT (2021a) Future increases in Arctic lightning and fire risk for permafrost carbon. Nat Clim Chang 11(5):404–410

Chen Y, Hu FS, Lara MJ (2021b) Divergent shrub-cover responses driven by climate, wildfire, and permafrost interactions in Arctic tundra ecosystems. Glob Chang Biol 27(3):652–663

Chiozza F (2016) Hyosciurus ileile (errata version published in 2017). The IUCN Red List of threatened species 2016: e.T10636A115098495. https://doi.org/10.2305/IUCN.UK.2016-3.RLTS.T10636A22262225.en. Accessed on 23 July 2021

Collins M, Chandler RE, Cox PM, Huthnance JM, Rougier J, Stephenson DB (2012) Quantifying future climate change. Nat Clim Chang 2(6):403–409

Connell J (2013) Islands at risk?: environments, economies and contemporary change. Edward Elgar Publishing, Cheltenham

Cubasch U, Meehl, GA, Boer GJ, Stouffer RJ, Dix M, Noda A, Yap KS, et al. (2001) Projections of future climate change. In: Climate change 2001: The scientific basis. Contribution of WG1 to the Third Assessment Report of the IPCC (TAR). Cambridge University Press, pp 525–582

Das I, van Dijk PP (2013) Species richness and endemicity of the herpetofauna of South and Southeast Asia. Raffles Bulletin of Zoology

De Tores PJ, Hayward MW, Dillon MJ, Brazell RI (2007) Review of the distribution, causes for the decline and recommendations for management of the quokka, Setonix brachyurus (Macropodidae: Marsupialia), an endemic macropodid marsupial from south-west Western Australia. Conserv Sci West Aust 6(1):13

Denamiel C, Tojčić I, Vilibić I (2020) Far future climate (2060–2100) of the northern Adriatic air–sea heat transfers associated with extreme bora events. Clim Dyn 55(11):3043–3066

Department of Conservation (2022) Predator Free 2050. https://www.doc.govt.nz/nature/pests-and-threats/predator-free-2050/. Accessed on 23 July 2021

Di Febbraro M, Martinoli A, Russo D, Preatoni D, Bertolino S (2016) Modelling the effects of climate change on the risk of invasion by alien squirrels. Associazione Teriologica Italiana

Díaz J, Sáez M, Carmona R, Mirón IJ, Barceló MA, Luna MY, Linares C (2019) Mortality attributable to high temperatures over the 2021–2050 and 2051–2100 time horizons in Spain: adaptation and economic estimate. Environ Res 172:475–485

Donner SD, Knutson TR, Oppenheimer M (2007) Model-based assessment of the role of human-induced climate change in the 2005 Caribbean coral bleaching event. Proc Natl Acad Sci 104(13):5483–5488

Döös BR (2002) Population growth and loss of arable land. Glob Environ Chang 12(4):303–311

Elsner JB, Lewers SW, Malmstadt JC, Jagger TH (2011) Estimating contemporary and future wind-damage losses from hurricanes affecting Eglin air force base, Florida. J Appl Meteorol Climatol 50(7):1514–1526

Fahrig L, Arroyo-Rodríguez V, Bennett JR, Boucher-Lalonde V, Cazetta E, Currie DJ, Watling JI et al (2019) Is habitat fragmentation bad for biodiversity? Biol Conserv 230:179–186

Failing L, Gregory R (2003) Ten common mistakes in designing biodiversity indicators for forest policy. J Environ Manag 68(2):121–132

Fischer G, Shah MN, Tubiello F, Van Velhuizen H (2005) Socio-economic and climate change impacts on agriculture: an integrated assessment, 1990–2080. Philos Trans R Soc B: Biol Sci 360(1463):2067–2083

Flannigan MD, Stocks BJ, Wotton BM (2000) Climate change and forest fires. Sci Total Environ 262(3):221–229

Førland EJ, Benestad R, Hanssen-Bauer I, Haugen JE, Skaugen TE (2011) Temperature and precipitation development at Svalbard 1900–2100. Adv Meteorol:1–14

Francis C, Gumal M, Han KH (2016) Nannosciurus melanotis (errata version published in 2017). The IUCN Red List of threatened species 2016: e.T14325A115121817. https://doi.org/10.2305/IUCN.UK.2016-3.RLTS.T14325A22245208.en. Accessed on 23 July 2021

Furlan E, Stoklosa J, Griffiths J, Gust N, Ellis R, Huggins RM, Weeks AR (2012) Small population size and extremely low levels of genetic diversity in island populations of the platypus, Ornithorhynchus anatinus. Ecol Evol 2(4):844–857

Ganguly AR, Steinhaeuser K, Erickson DJ, Branstetter M, Parish ES, Singh N, Buja L et al (2009) Higher trends but larger uncertainty and geographic variability in 21st century temperature and heat waves. Proc Natl Acad Sci 106(37):15555–15559

Gattuso JP, Hansson L (eds) (2011) Ocean acidification. Oxford University Press

Gerrie R, Kennerley R, Koprowski J (2016) Lariscus niobe (errata version published in 2017). The IUCN Red List of threatened species 2016: e.T11306A115101313. https://doi.org/10.2305/IUCN.UK.2016-3.RLTS.T11306A22242433.en. Accessed on 23 July 2021

Gethin R (1998) The foundations of Buddhism. Oxford University Press, Oxford

Gill RB (2000) The great Maya droughts: water, life, and death. UNM Press, Albuquerque

Guinotte JM, Fabry VJ (2008) Ocean acidification and its potential effects on marine ecosystems. Ann N Y Acad Sci 1134(1):320–342

Gunter G, Eleuterius LN (1971) Some effects of hurricanes on the terrestrial biota, with special reference to Camille. Gulf Caribb Res 3(2):283–289

Hadadin N, Qaqish M, Akawwi E, Bdour A (2010) Water shortage in Jordan—sustainable solutions. Desalination 250(1):197–202

Hall MH, Fagre DB (2003) Modeled climate-induced glacier change in Glacier National Park, 1850–2100. Bioscience 53(2):131–140

Hall JW, Evans EP, Penning-Rowsell EC, Sayers PB, Thorne CR, Saul AJ (2003) Quantified scenarios analysis of drivers and impacts of changing flood risk in England and Wales: 2030–2100. Glob Environ Change B 5(2):51–65

Hallegatte S (2012) The rising costs of hurricanes. Nat Clim Chang 2(3):148–149

Harsem Ø, Eide A, Heen K (2011) Factors influencing future oil and gas prospects in the Arctic. Energy Policy 39(12):8037–8045

Harvey P (2012) An introduction to Buddhism: teachings, history and practices. Cambridge University Press, Cambridge

Hawkins MT, Helgen KM, Maldonado JE, Rockwood LL, Tsuchiya MT, Leonard JA (2016a) Phylogeny, biogeography and systematic revision of plain long-nosed squirrels (genus Dremomys, Nannosciurinae). Mol Phylogenet Evol 94:752–764

Hawkins MT, Leonard JA, Helgen KM, McDonough MM, Rockwood LL, Maldonado JE (2016b) Evolutionary history of endemic Sulawesi squirrels constructed from UCEs and mitogenomes sequenced from museum specimens. BMC Evol Biol 16(1):1–16

Horton BP, Rahmstorf S, Engelhart SE, Kemp AC (2014) Expert assessment of sea-level rise by AD 2100 and AD 2300. Quat Sci Rev 84:1–6

Hulme M, Barrow EM, Arnell NW, Harrison PA, Johns TC, Downing TE (1999) Relative impacts of human-induced climate change and natural climate variability. Nature 397(6721):688–691

Imbach P, Chou SC, Lyra A, Rodrigues D, Rodriguez D, Latinovic D, Georgiou S et al (2018) Future climate change scenarios in Central America at high spatial resolution. PLoS One 13(4):e0193570

Jevrejeva S, Grinsted A, Moore JC (2014) Upper limit for sea level projections by 2100. Environ Res Lett 9(10):104008

Juan A, Crespo MB, Cowan RS, Lexer C, Fay MF (2004) Patterns of variability and gene flow in Medicago citrina, an endangered endemic of islands in the western Mediterranean, as revealed by amplified fragment length polymorphism (AFLP). Mol Ecol 13(9):2679–2690

Kennerley R, Ong P (2019) Hylopetes nigripes. The IUCN Red List of threatened species 2019: e.T10604A22243756. https://doi.org/10.2305/IUCN.UK.2019-2.RLTS.T10604A22243756.en. Accessed on 23 July 2021

Kenney WL, Craighead DH, Alexander LM (2014, 1891) Heat waves, aging, and human cardiovascular health. Med Sci Sports Exerc 46(10)

Kleypas JA, Yates KK (2009) Coral reefs and ocean acidification. Oceanography 22(4):108–117

Knowlton K, Rotkin-Ellman M, Geballe L, Max W, Solomon GM (2011) Six climate change–related events in the United States accounted for about $14 billion in lost lives and health costs. Health Aff 30(11):2167–2176

Koprowski JL, Alanen MI, Lynch AM (2005) Nowhere to run and nowhere to hide: response of endemic Mt. Graham red squirrels to catastrophic forest damage. Biol Conserv 126(4):491–498

Kuzin IL (1982) Effect of recent tectonic movements, oscillations of sea level, and climate changes on the formation of oil and gas fields in West Siberia

Kyselý J (2010) Recent severe heat waves in central Europe: how to view them in a long-term prospect? Int J Climatol 30(1):89–109

Larson C (2013) Losing arable land, China faces stark choice: adapt or go hungry. Science 339:644–645

Lauria V, Attrill MJ, Pinnegar JK, Brown A, Edwards M, Votier SC (2012) Influence of climate change and trophic coupling across four trophic levels in the Celtic Sea. PLoS One 7(10):e47408

Liu Y, Duarte GS, Klaus V, Gilgen AK, Hagen EO, Wittwer R, Buchmann N (2019) Ecosystem services of arable land in response to cropping systems and drought. In: Geophysical Research Abstracts. vol 21

Mack MC, Alexander HD, DeMarco J, Melvin A (2012) Wildfire, thermokarst and vegetation change: integrating diverse controls over carbon cycling in arctic and boreal ecosystems. In: AGU Fall Meeting Abstracts. vol 2012, pp B11F–01).

Malmstadt J, Scheitlin K, Elsner J (2009) Florida hurricanes and damage costs. Southeast Geogr 49(2):108–131

Marć M, Bartosiewicz A, Burzyńska J, Chmiel Z, Januszewicz P (2019) A nursing shortage–a prospect of global and local policies. Int Nurs Rev 66(1):9–16

Martin R (ed) (2005) Tree-kangaroos of Australia and New Guinea. CSIRO Publishing, Collingwood

Mathis JT, Cooley SR, Lucey N, Colt S, Ekstrom J, Hurst T, Feely RA et al (2015) Ocean acidification risk assessment for Alaska's fishery sector. Prog Oceanogr 136:71–91

Meadows D, Randers J (2012) The limits to growth: the 30-year update. Routledge

Meijaard E (2016) Callosciurus baluensis. The IUCN Red List of threatened species 2016: e.T3593A22254589. https://doi.org/10.2305/IUCN.UK.2016-2.RLTS.T3593A22254589.en. Accessed on 23 July 2021

Melvin AM, Murray J, Boehlert B, Martinich JA, Rennels L, Rupp TS (2017) Estimating wildfire response costs in Alaska's changing climate. Clim Chang 141(4):783–795

Mercer JM, Roth VL (2003) The effects of Cenozoic global change on squirrel phylogeny. Science 299(5612):1568–1572

Mitchell JF, Lowe J, Wood RA, Vellinga M (2006) Extreme events due to human-induced climate change. Philos Trans R Soc A Math Phys Eng Sci 364(1845):2117–2133

Mohsenipour M, Shahid S, Chung ES, Wang XJ (2018) Changing pattern of droughts during cropping seasons of Bangladesh. Water Resour Manag 32(5):1555–1568

Mostafa AN, Wheida A, El Nazer M, Adel M, El Leithy L, Siour G, Alfaro SC et al (2019) Past (1950–2017) and future (– 2100) temperature and precipitation trends in Egypt. Weather Clim Extremes 26:100225

Mudd L, Wang Y, Letchford C, Rosowsky D (2014) Hurricane wind hazard assessment for a rapidly warming climate scenario. J Wind Eng Ind Aerodyn 133:242–249

Naicker S, Eastwood JB, Plange-Rhule J, Tutt RC (2010) Shortage of healthcare workers in sub-Saharan Africa: a nephrological perspective. Clin Nephrol 74:S129–S133

Nazarevich VJ (2015) The sixth species extinction event by humans. Earth Common J 5(1):61–72

Nicholls RJ, Leatherman SP (1996) Adapting to sea-level rise: relative sea-level trends to 2100 for the United States. Coast Manag 24(4):301–324

Nicholls RJ, Hanson S, Herweijer C, Patmore N, Hallegatte S, Corfee-Morlot J, Muir-Wood R, et al (2007) Ranking of the world's cities most exposed to coastal flooding today and in the future. Organisation for Economic Co-operation and Development. OECD, Paris

Padgett J, DesRoches R, Nielson B, Yashinsky M, Kwon OS, Burdette N, Tavera E (2008) Bridge damage and repair costs from Hurricane Katrina. J Bridg Eng 13(1):6–14

Peacock DE (2009) The grey squirrel Sciurus carolinensis in Adelaide, South Australia: its introduction and eradication. Vic Nat 126(4):150–156

Petrick S, Riemann-Campe K, Hoog S, Growitsch C, Schwind H, Gerdes R, Rehdanz K (2017) Climate change, future Arctic Sea ice, and the competitiveness of European Arctic offshore oil and gas production on world markets. Ambio 46(3):410–422

Pimentel D, Harman R, Pacenza M, Pecarsky J, Pimentel M (1994) Natural resources and an optimum human population. Population and environment, 15, 347–369

Pries AJ, Branch LC, Miller DL (2009) Impact of hurricanes on habitat occupancy and spatial distribution of beach mice. J Mammal 90(4):841–850

Raftery AE, Zimmer A, Frierson DM, Startz R, Liu P (2017) Less than 2 C warming by 2100 unlikely. Nat Clim Chang 7(9):637–641

Ramankutty N, Mehrabi Z, Waha K, Jarvis L, Kremen C, Herrero M, Rieseberg LH (2018) Trends in global agricultural land use: implications for environmental health and food security. Annu Rev Plant Biol 69:789–815

Rayner MJ, Van Loenen AL, Shepherd LD, Cubrinovska I, Scofield RP, Tennyson AJ, Steeves TE et al (2021) Comprehensive evidence for subspecies designations in Cook's Petrel Pterodroma cookii with implications for conservation management. Bird Conserv Int 31(1):1–13

Roach N (2017) Marmota vancouverensis. The IUCN Red List of threatened species 2017: e.T12828A22259184. https://doi.org/10.2305/IUCN.UK.2017-2.RLTS.T12828A22259184.en. Accessed on 23 July 2021

Rogelj J, Popp A, Calvin KV, Luderer G, Emmerling J, Gernaat D, Tavoni M et al (2018) Scenarios towards limiting global mean temperature increase below 1.5 C. Nature. Climate Change 8(4):325–332

Roser M (2013) Future population growth. Our world in data

Rowland L, da Costa ACL, Galbraith DR, Oliveira RS, Binks OJ, Oliveira AAR, Meir P et al (2015) Death from drought in tropical forests is triggered by hydraulics not carbon starvation. Nature 528(7580):119–122

Ryser R, Häussler J, Stark M, Brose U, Rall BC, Guill C (2019) The biggest losers: habitat isolation deconstructs complex food webs from top to bottom. Proc R Soc B 286(1908):20191177

Sayre R, Noble S, Hamann S, Smith R, Wright D, Breyer S, Reed A et al (2019) A new 30 meter resolution global shoreline vector and associated global islands database for the development of standardized ecological coastal units. J Oper Oceanogr 12(sup2):S47–S56

Schär C, Vidale PL, Lüthi D, Frei C, Häberli C, Liniger MA, Appenzeller C (2004) The role of increasing temperature variability in European summer heatwaves. Nature, 427(6972), 332–336

Scheffer M, Van Geest GJ, Zimmer K, Jeppesen E, Søndergaard M, Butler MG, De Meester L et al (2006) Small habitat size and isolation can promote species richness: second-order effects on biodiversity in shallow lakes and ponds. Oikos 112(1):227–231

Schuur EA, Bockheim J, Canadell JG, Euskirchen E, Field CB, Goryachkin SV, Zimov SA et al (2008) Vulnerability of permafrost carbon to climate change: implications for the global carbon cycle. Bioscience 58(8):701–714

Siegert F, Ruecker G, Hinrichs A, Hoffmann AA (2001) Increased damage from fires in logged forests during droughts caused by El Nino. Nature 414(6862):437–440

Slingenberg A, Braat L, Van Der Windt H, Rademaekers K, Eichler L, Turner K (2009) Study on understanding the causes of biodiversity loss and the policy assessment framework. European Commission Directorate-General for Environment, available at http://ec.europa.eu/environment/enveco/biodiversity/pdf/causes_biodiv_loss.pdf. Accessed 23 July 2021

Smith LC, Stephenson SR (2013) New Trans-Arctic shipping routes navigable by midcentury. Proc Natl Acad Sci 110(13):E1191–E1195

Song XP, Hansen MC, Stehman SV, Potapov PV, Tyukavina A, Vermote EF, Townshend JR (2018) Global land change from 1982 to 2016. Nature 560(7720):639–643

Steiner M, Huettmann F (2021) Justification for a taxonomic conservation update of the rodent genus Tamiasciurus: addressing marginalization and mis-prioritization of research efforts and conservation laissez-faire for a sustainability outlook. Eur Zool J 88(1):86–116

Stilwell JD (2018) The next great extinction event. Monash Univerity

Stycos JM, Arias J (1972) Speaking of problems, Cabello shows that housing shortage is critical and increasing, and its solution in a short time is beyond the economic capabilities of the countries in the region. On the other hand, according to Arevalo's projections, future demands in. Milbank Mem Fund Q 45:111–114

Suwal MK, Huettmann F, Regmi GR, Vetaas OR (2018) Parapatric subspecies of Macaca assamensis show a marginal overlap in their predicted potential distribution: some elaborations for modern conservation management. Ecology and evolution 8(19):9712–9727

Taherkhani M, Vitousek S, Barnard PL, Frazer N, Anderson TR, Fletcher CH (2020) Sea-level rise exponentially increases coastal flood frequency. Sci Rep 10(1):1–17

Thorington RW Jr, Koprowski JL, Steele MA, Whatton JF (2012) Squirrels of the world. JHU Press, Baltimore

Tian L, Yan Y, Wu Y, Shao L (2020) Breaking the land monopoly: can collective land reform alleviate the housing shortage in China's mega-cities? Cities 106:102878

Tizard RJ (2016) Dremomys everetti. The IUCN Red List of threatened species 2016: e.T6820A22255505. https://doi.org/10.2305/IUCN.UK.2016-2.RLTS.T6820A22255505.en. Accessed on 23 July 2021

Vačkář D, ten Brink B, Loh J, Baillie JE, Reyers B (2012) Review of multispecies indices for monitoring human impacts on biodiversity. Ecol Indic 17:58–67

Vecchi GA, Wittenberg AT (2010) El Niño and our future climate: where do we stand? Wiley Interdiscip Rev Clim Chang 1(2):260–270

Virkkala R (2016) Long-term decline of southern boreal forest birds: consequence of habitat alteration or climate change? Biodivers Conserv 25(1):151–167

Vondrák J, Urbanavichus G, Palice Z, Malíček J, Urbanavichene I, Kubásek J, Ellis C (2019) The epiphytic lichen biota of Caucasian virgin forests: a comparator for European conservation. Biodivers Conserv 28(12):3257–3276

Wang L, Anna H, Zhang L, Xiao Y, Wang Y, Xiao Y, Ouyang Z et al (2019) Spatial and temporal changes of arable land driven by urbanization and ecological restoration in China. Chin Geogr Sci 29(5):809–819

Whittaker RJ, Fernández-Palacios JM (2007) Island biogeography: ecology, evolution, and conservation. Oxford University Press, Oxford

Wilson DE, Reeder DM (2005) Mammal species of the world: a taxonomic and geographic reference, vol 1. JHU Press, pp 754–818

Wilting HC, Schipper AM, Bakkenes M, Meijer JR, Huijbregts MA (2017) Quantifying biodiversity losses due to human consumption: a global-scale footprint analysis. Environ Sci Technol 51(6):3298–3306

Wu Q, Zhao L, Ye XC (2016) Shortage of healthcare professionals in China. BMJ 354:i4860

Zhang D, Lee H (2010) Climate change, food shortage and war: a quantitative case study in China during 1500–1800. Catrina 5(1):63–71

Chapter 8
BIG DATA for Small Tree Squirrels in Old-Growth Forests? Landscape Metrics, Open Access Field Data, Machine Learning, and GIS Models from Remotely-Sensed Imagery in the Tanana Valley State Forest Wilderness of Alaska

Abstract This study aims to provide a specific predictive example for tree squirrel middens and the corresponding squirrel presence distribution of the North American red squirrel (*Tamiasciurus hudsonicus*) in wilderness areas. Here we use the old-growth forest of interior Alaska as the study area, a vast area (approx. similar to the area of Puerto Rico or Kosovo for comparison) of national relevance with an essential part of indigenous land management regimes. The squirrel distribution has been firstly assessed in the Tanana Valley State Forest (Alaska U.S.) around 60 years ago. Since then, a few distribution data sets have been cumulatively collected, also by the authors in recent years, all in a situation where ongoing habitat fragmentation is altering the environment ever since. However, those data are mostly *ad hoc*/opportunistic and lack a wider research design and modern approaches. With these first indicators, a Species Distribution Model (SDM) is provided in this study in order to depict how such squirrel ranges and populations develop in old-growth forest landscapes, in times of ongoing fragmentation and continuously accelerating climate change (discussed in detail in Steiner and Huettmann 2023 – Chap. 12). It allows for a first strategic and sustainable conservation planning. This study speaks on cutting patterns and old-growth management questions, using squirrels; a conservation scheme that is often and widely ignored and lacks a policy link. The importance of this study and the inclusion of old-growth forest tracts and landscapes in this book has been shown by its classification as a region of/under high risk. This habitat type has been included because of its uniqueness and because it is continuously decreasing in the modern world, simultaneously with the inhabiting species, including squirrels. These latter-mentioned statements are supported by a literature review and newly created GIS maps, depicting the retreat of old-growth forest landscapes and their animal populations using an example from the large interior Alaskan wilderness essential for squirrels.

Keywords North American Red Squirrel (*Tamiasciurus hudsonicus*) · Tanana Valley · Alaska · Boreal forest wilderness · GIS · Open access · BIG DATA · Landscape metrics · Machine learning

8.1 Introduction

Tree squirrels are (mainly) tree-inhabiting small mammals that on occasion show social behavior (Smith 1978). They naturally occur virtually worldwide except in Antarctica and Australia & Oceania (see Black 1972; Koprowski and Nandini 2008; Palmer et al. 2007; Rajaratnam and Redman 2001; Chaps. 3 and 10 in Steiner and Huettmann 2023) and are associated with forest-type landscapes they co-evolved with (Smith 1970).

While squirrels are abundant and hunted, most of them (304 out of 307 → 99%) are relatively poorly studied for valid science-based management, and thus, they are widely underrepresented in the public science agenda (see for instance Chap. 3 – Steiner and Huettmann 2023; Koprowski and Nandini 2008). Worldwide there are virtually no squirrel reserves or protected zones (Koprowski 2005), even though many of them (74 out of 292 see Fig. 1.1 in Chap. 1 – Steiner and Huettmann 2023) are of greater conservation concern or endangered. Despite all this, they are often even seen as "pests" (Dunn et al. 2018; Gilson and Salmon 1990; Holmes 2015; Sheail 1999). All of this also applies to a certain extent to the North American Red Squirrel (*Tamiasciurus hudsonicus*) in Alaska, and North America overall. This lack of knowledge and performed research is not only valid for the general distribution of the species, and their population trends, but also their life-essential behaviors (e.g. feed storage and feeding behavior) (Steiner and Huettmann 2021). In order to provide functional tools, here we attempt to fill such knowledge gaps and focus on such a feeding behavior and its expression by the squirrel species *T. hudsonicus*. A sketch of such a North American Red squirrel can be found illustrated in Fig. 8.1. As broadly discussed previously

Supplementary Information The online version contains supplementary material available at https://doi.org/10.1007/978-3-031-23547-4_8.

M. Steiner, F. Huettmann, *Sustainable Squirrel Conservation*, https://doi.org/10.1007/978-3-031-23547-4_8

Fig. 8.1 North American red squirrel drawing. (by Falk Huettmann)

(Steiner and Huettmann 2021), this squirrel species (*T. hudsonicus*) and its congener Douglas Squirrel (*T. douglasii*) are known to create so-called middens (Elkins 2017; Larsen and Boutin 1994; Ramos-Lara 2012). In the social and life history of *T. hudsonicus* middens play a central role and often can exist for decades, with a regeneration time of >250 years for midden sites (Smith and Mannan 1994). Such middens consist of mostly coniferous cone residuals, small branch ends, and even entire small spruce (*Picea* sp.) and pine (*Pinus sp.*) cones (Kranowski 1969; Wolff and Zasada 1975; Robolt and Huettmann 2021; Steiner and Huettmann 2021). Therefore, in this study, we will focus on the modeling of the distribution of these so-called 'middens' as a proxy of squirrel distribution overall, and then assess how these are spatially influenced by factors such as the surrounding environment, surrounding predominant tree species, etc.

8.1.1 Study Area

In this study, the methods and analyses are focused on the study area of the Tanana Valley State Forest of Alaska (USA) which is generally considered an old-growth forest wilderness landscape with various impacts. It has been listed as one of the top 10 forests in the U.S. to visit (ReserveAmerica 2020). The study area has been mapped and outlined in Fig. 8.2. The Tanana Valley State Forest is a unique and ecologically representative forest tract for interior Alaska, and the North American Boreal forest. It is one of the least affected sections in the largest consistent forests of the world: the boreal forest.

For this large wilderness area (1.81 million acres or 7325 km^2, approx. the area of Puerto Rico, Corsica, or Kosovo for comparison) in interior Alaska centered around Fairbanks with c. 90,000+ inhabitants, not much information exists for North American Red Squirrels or their middens on the larger landscape scale (see Fig. 8.3 for habitat features). The same is true for forest-related habitat information needed to study their life parameters, or for sustainable forest and landscape management, large-scale information is widely absent. Only a few punctual studies were done (Robolt and Huettmann 2021), in addition to some over 60 years ago serving as a proxy baseline (Kranowski 1969; Wolff and Zasada 1975). In that setting, we study in more depth where these squirrel middens occur in the old-growth boreal forest landscape using the latest methods of Geographic Information System (GIS) and Machine Learning (ML) as a role model to progress the understanding of this species and its habitat, *sensu* Ohse et al. (2009).

Figure 8.3 illustrates such a discussed squirrel midden with the corresponding canopy cover for the identification of the surrounding tree species.

Fig. 8.2 Study area of the Tanana Valley State Forest

Fig. 8.3 Photos of (**a**) midden site, and (**b**) associated canopy cover. (Photos by MS)

8.2 Methods

8.2.1 Fieldwork Data for Squirrels

In the summer of 2019, we carried out field surveys for squirrels and their middens in parts of the Tanana Valley State Forest (TVSF) using opportunistic geo-referenced cruising. We focused on two regions in the TVSF, namely due to access: the Murphy Dome area and then the wider Fairbanks municipality region. We were able to find 832 squirrel presences and absences where we characterized midden details (number, diameter), occupancy of squirrels, habitat types, geographical location, and photos. The field protocol used is shown in the appendix. This data set was primarily collected by MS (Fig. 8.4).

A second squirrel field data set covering more of the extent within the TVSF focused on independent assessment locations at accessible sites. Those alternative data were collected, uninformed, and served the purpose to confront and assess a model of squirrel occurrences in the study area. While this is a small data set consisting of 'only' 25 locations it has the strength of wider coverage and a more diverse cross profile in the TSVF. This data set was primarily collected by FH in 2020 and 2021.

A third data set was obtained digitally by MS from the public record for 'squirrels' from the family "Sciuridae". This dataset includes records from www.gbif.org and its contributors within. This dataset was also collected independently from the model, and includes citizen-science information, mostly squirrel presence (not middens), and helps with the ecological niche assessments and prediction assessments.

8.2.2 Other Data for Squirrels and Their Habitat

For the GIS layers, we obtained a copy of the Tanana Valley State Forest inventory data from 1983–2010 (http://forestry.alaska.gov/stateforests). This dataset consists of typical landcover inventory data compiled using remote sensing. It has attributes for forest patches, forest type, canopy density, stage, patch area, etc. This dataset is 'seamless' for the study area. It is in a geodatabase format and was converted into a shapefile (.SHP) for easier handling and operations.

Using the vegetation GIS layer, in fragstats (McGarigal 1995), we then computed the landscape metrics of Edge Density for Spruce Dominated polygons, and the same for coniferous dominated ones, and all patches. This allows to address edginess and fragmentation of the landscape of the TVSF and for a wider landscape assessment for squirrels. The datasets that have been used for the distribution and habitat modeling for this study have been summarized in Table 8.1.

8.2.3 Modeling Steps

We then created an overlay in Open Source QGIS (obtainable via https://www.qgis.org/en/site/forusers/download.html), as well as ArcGIS (obtainable via https://pro.arcgis.com/en/pro-app/2.8/get-started/download-arcgis-pro.htm), of all those squirrel data for a data cube that was used for the subsequent model.

We also created in QGIS a 0.5-degree lattice for the study area which was overlaid for the set of predictors with QGIS and ArcGIS.

We then used TreeNet (Stochastic Boosting) in Salford Predictive Modeler (SPM) (https://www.minitab.com/en-us/products/spm) to explain the response of Midden Yes/No in relation to the seven predictors in the following formula:

$$\begin{aligned}\text{Midden Yes / No} \sim \text{Vegetation Class} + \text{EDallPatches} + \text{Polygon size in hectares} \\ + \text{Size Class of Forest} + \text{Edge Density of Spruce} + \text{Density Class of Forest} \\ + \text{Edge Density of Conifer Dominated Forests}\end{aligned} \tag{8.1}$$

In TreeNet, we used equal balanced weights and 2000 trees to find the optimal solution, while otherwise using default settings and 5 samples for a terminal node with a node depth of 20. This tree was captured and saved as a grove file.

The grove file was then applied in SPM to a lattice for predictions and consecutively saved as a CSV file. Those predictions of the relative index of occurrence (RIO; Humphries et al. 2018) were visualized in GIS using the quantile classification with 5 classes.

We then assessed the model prediction with three pieces of evidence: internally via ROC, externally via alternative field data, and with best-available public open access digital data for interior Alaska.

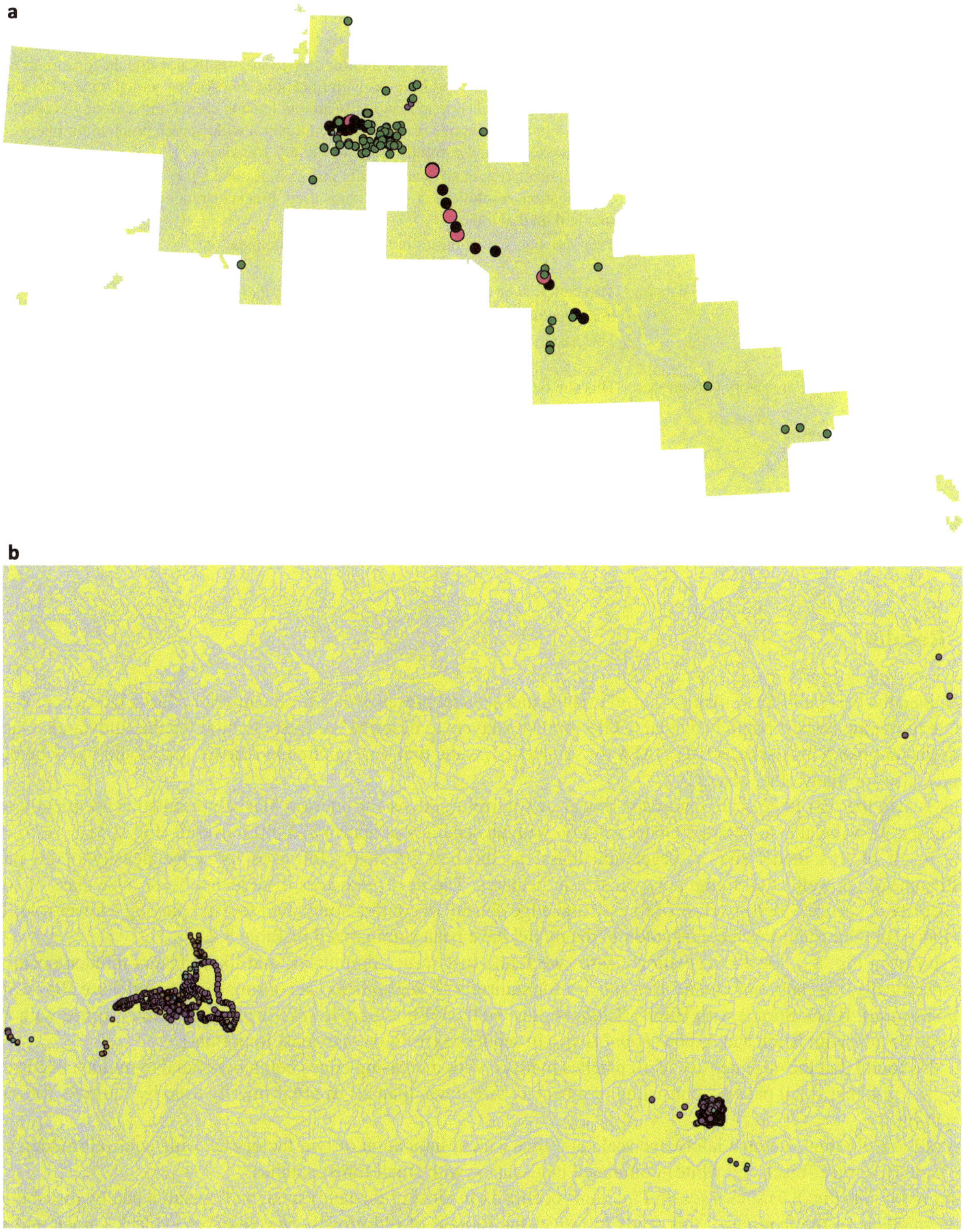

Fig. 8.4 Best publically available compiled field survey locations for squirrel and midden data for the study area (generic vegetation polygon data for the study area overlaid in yellow with a grey outline): (a) This maps shows the three survey data sets 1) Squirrel and Midden presence/absence field survey data (n=824) by MS and FH (upper left cluster in black, for more details see zoom-in in b). 2) (Large-scale field cross profile data points (n=25) for the study area by FH for squirrel midden presence (pink) and absence (black) , and 3) in green, public ocurrence records for 'squirrels' from GBIF, ARCTUS and other sources (n=328). (b) Zoom-in of data set 1 with presence (pink) and other absence categories (black); data details in the data submission package and metadata provided. Other details and applications shown and discussed in the subsequent chapter text

Table 8.1 List of seven data sets used for habitat descriptions of squirrels in interior Alaska, Tanana Valley State Forest

Predictor	Source	Comment
Vegetation class	Tanana Valley State Forest (TVSF) inventory GIS dataset	While vegetation classes appear intuitive, those are usually not straight-forward as a classification process and have many implications. For Alaska, a modern and statistically sound vegetation class map is virtually absent, leaving earlier and currently available schemes in the subjective realm. However, this is a wide-spread problem for many wilderness areas and currently 'state-of-the-art' for Alaska.
EDallPatches	Fragstats	The fragmentation metrics are usually just an index and a function of the map classification, resolution, and algorithms used. Fragstats retains a certain science standard on that topic
Polygon size in hectares	TVSF inventory GIS dataset	There are conceptual biases in forest polygon delineations
Size class of forest	TVSF inventory GIS dataset	This can be seen as forest quality, growth intensity index
		It is often used for identifying valuable timber
Edge density of spruce	Fragstats	This assumes spruce is a distinct and clear unit, which is not always the case
Density class of forest	TVSF inventory GIS dataset	This can be seen as an index and likely is related to soil, nutrients, and microclimate
Edge density of conifer dominated forests	Fragstats	This assumes all forest patches are clearly mapped, which is not always the case (gradients can apply)

8.3 Results

We obtained the best publicly available squirrel midden and squirrel presence data for the interior Alaska landscape. We were also able to obtain the best publicly available GIS predictor layers to describe the forest habitat for the study area. One may refer to that approach in Alaska as BIG DATA for squirrels. It is the first time applied in Alaska, and would likely be of high relevance also for many other places.

This is also the first large-scale landscape model of squirrel middens for interior Alaska presented academically to this day. Further, we were able to obtain a model solution with an accuracy of approx. 75–80% overall (Fig. 8.5).

The result of this model shows 'vegetation class' as the best predictor, followed by' edge density of all patches (EDallPatches)', as well as GIS map polygon size in hectares. The next predictors in sequence were 'size class of forest', 'edge density of spruce', 'density class of forest', and edge density of conifer-dominated forests' playing a lower role. These predictors with their importance/contribution to the model have been summarized in Table 8.2.

As shown in Fig. 8.6, for the vegetation classes we found that spruce dominates, and likely favors midden occurrence. More specifically, the most 'important' tree species for squirrels (*T. hudsonicus*) according to this model are Black Spruce (*Picea mariana*), Black Spruce with Birch (*Betula* sp.), as well as White Spruce (*Picea glauca*), White Spruce with Birch, and White Spruce with Birch and Aspen (*Populus* sp.) in sequence (from most to least important).

We also found that the edge density of all patches matters, with increasing edges being beneficial to midden occurrences. Further, we find that small patches, consisting of a hectare size lower than 20 ha are primarily used by squirrels for midden construction.

Overall, our findings are on a landscape scale and thus a clear indication of two factors favoring squirrel midden occurrence, namely, spruce (black and white)-dominated landscapes, with small forest patches.

Our prediction (Fig. 8.7) shows squirrel middens occurring in most of the study area, and overall a patchy distribution of squirrel middens in the landscape; this tends to reflect the mosaic of old-growth-type reserves – black spruce and white spruce dominated - in the boreal forest landscape of Alaska.

Our model assessment shows an internal performance with a ROC of circa 75%. While this is good but not significantly strong, the alternative shows us a similar pattern widely in favor of the midden predictions. The mean relative index of occurrence (RIO) for presences is 55% compared to a lower (44%) of predicted absence RIOs. While we only have 25 sample plots, not allowing for statistical differences, the trend remains clear. The third assessment using digital squirrel presences,

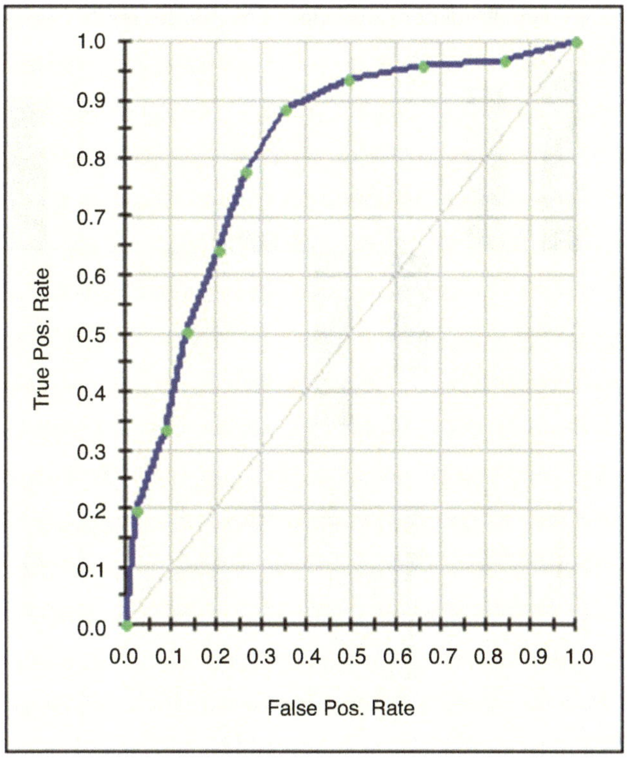

Fig. 8.5 Internal ROC curve of model accuracy based on Treenet

Table 8.2 Ranking of seven predictors in TreeNet showing relative importance values

Predictor	Importance rank in percent	Comment
Vegetation class	100	A categorical predictor dominated in the result by black spruce, black spruce with birch, white spruce, white spruce with birch
EDallPatches	63	Created by fragstats on all polygons
Polygon size in hectares	59	A predictor from GIS in the polygonised landscape of the TSF
Size class of forest	52	A predictor from the TSF classification
Edge density of spruce	41	Created by fragstats on spruce dominated polygons of the TSF
Density class of forest	27	A predictor from the TSF classification
Edge density of conifer dominated forests	18	Created by fragstats on spruce dominated polygons of the TSF

shows a close proximity to spruce stands, favored by middens. In general, this is in support of midden predictions, matching squirrel occurrences. Overall, there is a clear trend in the data, albeit not overwhelmingly strong, indicating some left-over variance of app. 35% that is not well accounted for yet and can deserve further investigation. Likely, and as found with many social mammals, squirrels are complex animals with social and individual behaviors that are not fully described and captured on our scale, resolution, and research design yet. This aspect deserves more attention for the conservation management of this species.

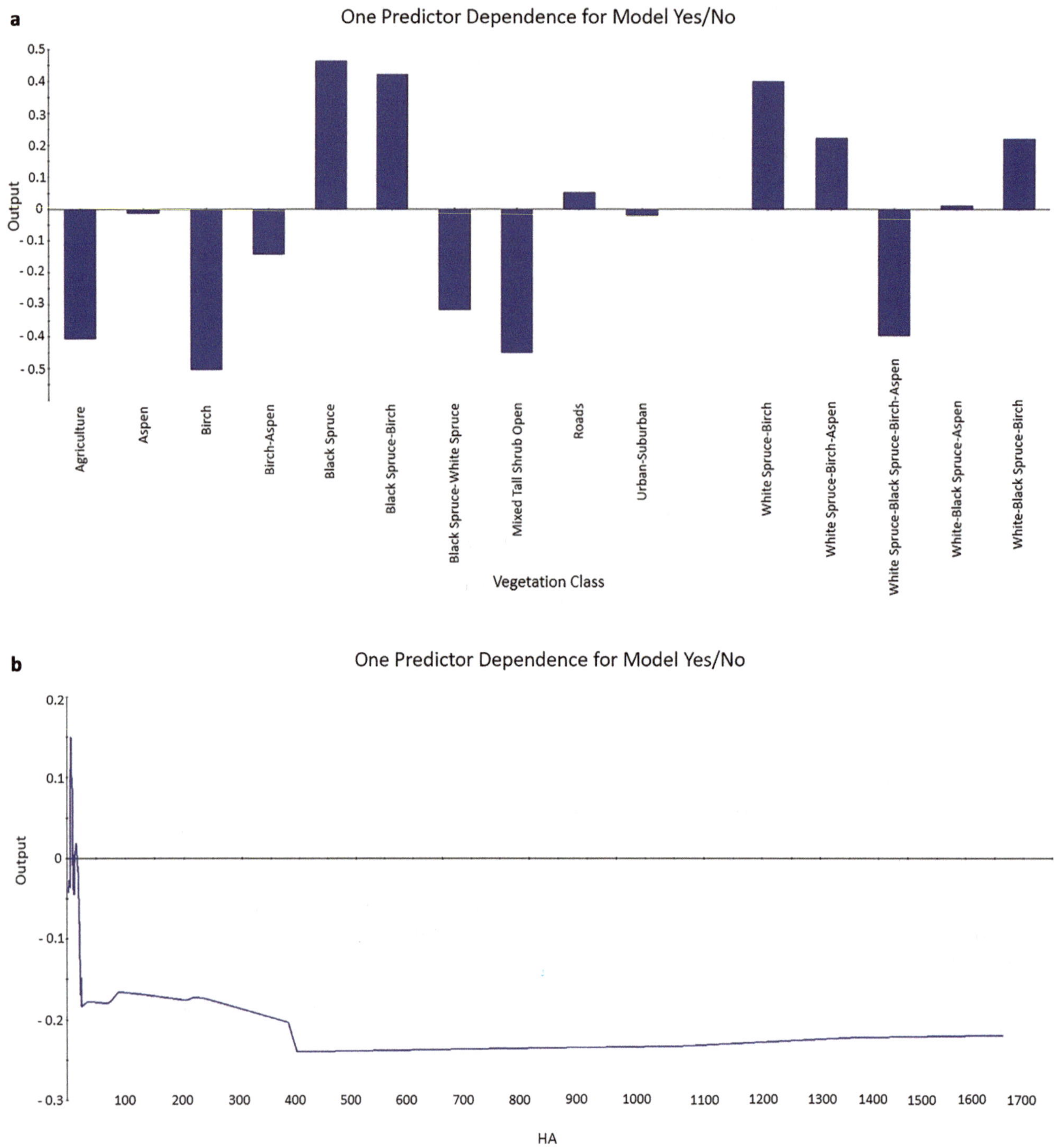

Fig. 8.6 Top 3 predictors using TreeNet algorithm: (**a**) Vegetation Class, (**b**) Edge Density of all patches, (**c**) hectar size class

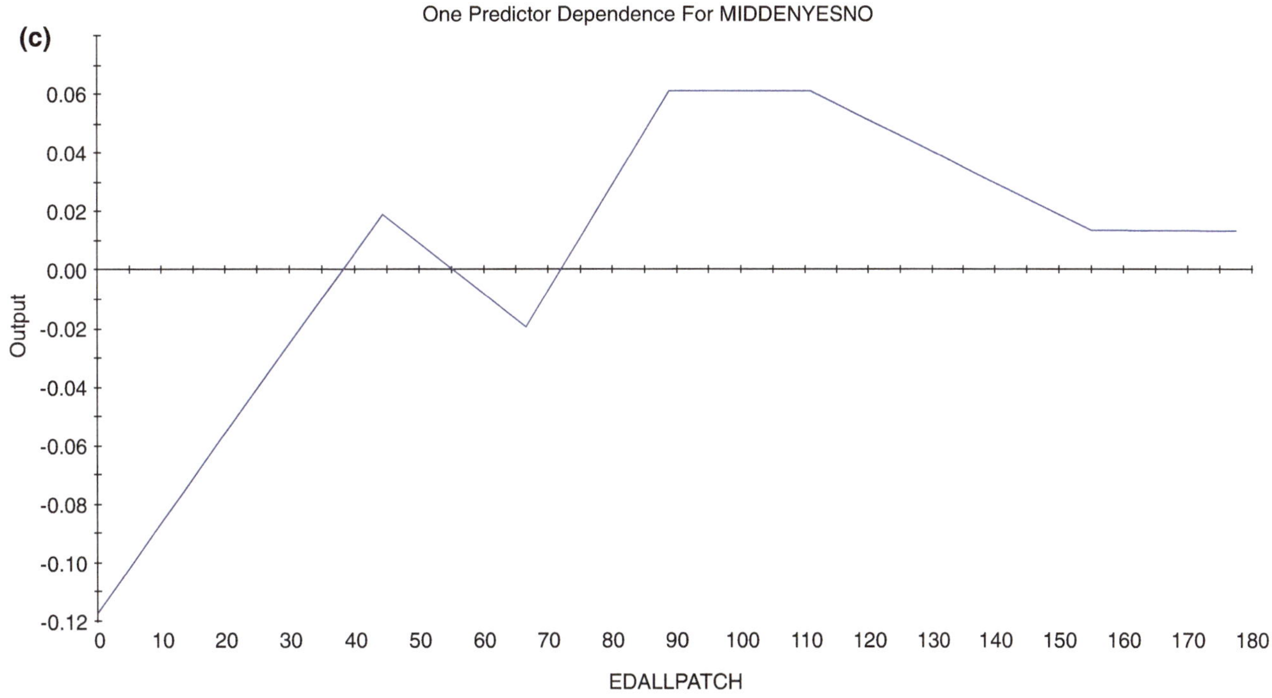

Fig. 8.6 (continued)

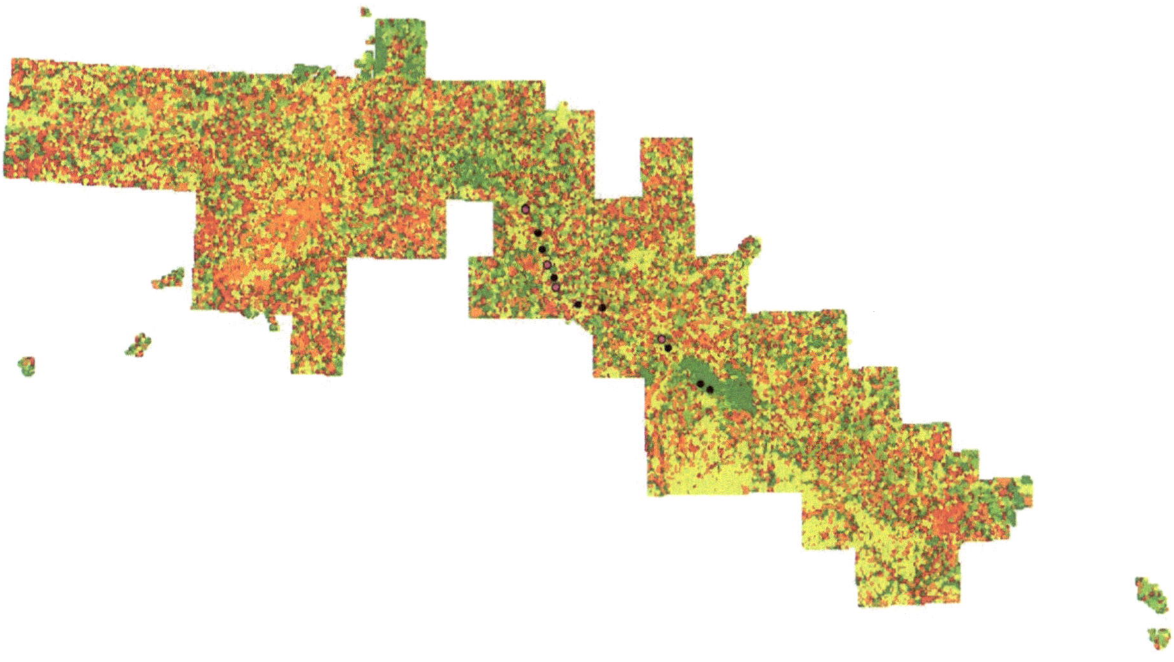

Fig. 8.7 Prediction heatmap map of red squirrel middens in the TVSF using TreeNet. Red indicates a high likelihood of occurrence, green predicts absence

8.4 Discussion

The boreal forest is the largest forest ecosystem in the world (Gauthier et al. 2015; Veraverbeke et al. 2017). While most of its precious wilderness sections occur in Russia (e.g. Bocharnikov and Huettmann 2019), in North America it is heavily cut and used (Hins et al. 2009; Hebblewhite 2017), but still includes sections of old-growth stands, with few substantial areas in Canada and Alaska (Kayes and Mallik 2020). The interior of Alaska, as exemplified by the Tanana Valley State Forest, is part of that old-growth forest landscape (Chapin et al. 2006; Ohse et al. 2009; Young et al. 2017, 2018).

However, this area is widely under-published, and almost unknown, e.g. when compared to the Tongass Forest in southern Alaska or many other areas, e.g. boreal forest in Yukon, northern Alberta, or Labrador, and for instance Russia (Bocharnikov and Huettmann 2019).

The same can be said for the North American Red Squirrel and its midden distribution, specifically in Alaska (the largest state in the U.S. and with the largest protected area holdings). Large-scale distribution studies on this subject are practically absent, especially for Alaska and the TVSF; the latter are areas of wider interest and national concern with squirrels as keystone members.

In this ecosystem, spruce is found at the end of the tree succession stage and it presents a certain old-growth stage in the boreal forest landscape (= habitat climax). Black spruce occurs in wetter areas, and often in higher elevations too, and it is the most widespread habitat type in the study area. It is not much cut for timber and currently carries a low economic value. However, it is of high natural value e.g. global Carbon cycle (Miquelajauregui et al. 2019). White spruce is less abundant but highly in demand, due to its high economic value for timber and rarity e.g. used by high-grading. On appropriate soils, it replaces the mature but younger birch stands, and its patchy distribution can often be explained by continuous, and annual fires, but also due to terrain and human forest cutting/clearing. None of the latter is being practiced/prevented with/by sustainability concepts in Alaska (the Forest Practices Code in Alaska is less than 100 pages strong; growth & yield tables are widely missing, and so does a forest inventory (Division of Forestry Department Of Natural Resources 2018; Ohse et al. 2009; Young et al. 2017, 2018).

North American Red Squirrels are known to occur in Interior Alaska for a long time, with many trapping records, and the first academic records about 60 years ago (Kranowski 1969; Wolff and Zasada 1975) allowing to address some relevant questions for today (Robolt and Huettmann 2021 for suburbanization). Here we can provide an update and use more fine-tuned methods on a wider landscape scale than an ancient landscape. It is not so clear currently what the indigenous land use was before 'western contact' but likely fire management existed and was actively carried out by a relatively small number of nomadic-living human populations in the study are for certain seasons.

In such a context, here we were able to present the first large-scale landscape data and quantitative assessment for North American Red Squirrels in Interior Alaska. Further, we were able to present the forest inventory data as GIS layers, and then run fragstats on it for landscape metrics (as shown in Appendix 8.1).

We also presented the first machine-learning-derived landscape prediction map of squirrel middens with several good evidence for model accuracy (compare with Robolt and Huettmann 2021). This prediction is robust and shows the relevance for squirrel distribution (presence/absence and its predictors) as well as the association of squirrel middens with the surrounding habitat. Arguably, middens play the central role in the squirrel's life history, distribution, and supposedly even survival.

Likely, the true squirrel ranges are larger than the midden ones *per se* and this species is found virtually throughout the entire TVSF and interior Alaska overall. This is reflected in our model assessment where we assess predicted midden occurrences with squirrel presences (as taken from digital data). We find most of the squirrel records around 500 m close to spruce patches in the study area (Figs. 8.8 and 8.9). This distance is easily within the daily squirrel movement range around middens (see Robolt and Huettmann 2021).

Our results showed that "spruce" (black and white) is the dominant/determining factor of middens, and thus for squirrel occurrence (see Wauters et al. 2005), usually occurring in small patches with a subsequent high edge density. It has been shown in other forest landscape studies that the inclusion of patch variables helps to explain predictive power, e.g. Ploton et al. (2020) (Fig. 8.10).

While we found that forest GIS map patches with middens have a size smaller than 20 ha, the size of those patches remains a crucial discussion point. That is because:

(a) The patch size must be seen as a certain artifact of the forest classification itself done by remote sensing and GIS classification methods.
(b) Different forest categories and growth patterns will also determine patch sizes.
(c) Patch size is driven – naturally – by fire and topography, and

Fig. 8.8 Polygons in the Tanana Valley State Forest with dominating black spruce, black spruce with birch, white spruce, and white spruce with birch (shown in solid polygons)

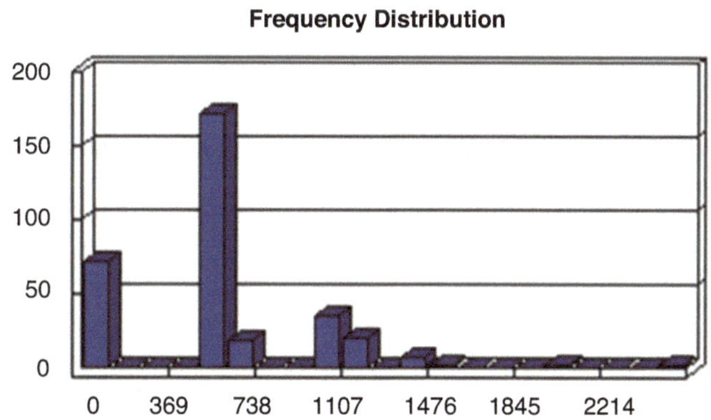

Fig. 8.9 Frequency distribution of 328 red squirrel sightings for the study area by distance (in meters) to spruce patches; the mean distance to spruce patch is c. 534 m

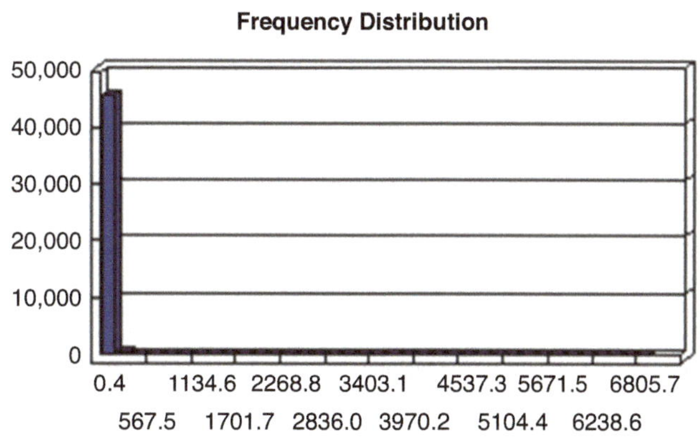

Fig. 8.10 Frequency distribution of spruce patches in hectares (ha); the mean size of a spruce patch is c. 16 ha

(d) Humans lived in the Tanana Valley State Forest for over 30,000 years (Wooller et al. 2012), and cut parts of the forest and its trees, much relevant white spruce and birch stands during the last 100 years, e.g. for mining, timber, and firewood.

The changes during the last 70 years are essential, and a smaller baseline comparison exists with Kranowski (1969), and Wolff and Zasada (1975). For Alaska specifically, parts of the forest have been affected by initial gold mining timber extraction, a railroad, and in part cleared for oil transport (see Trans-Alaska Pipeline System) running 800 miles (1290 km) through Alaska. Another wave of forestry occurred in the study area in the 1990s for Japanese export, as well as for local firewood use, etc. Nowadays, sustainable wood chip production, biofuel, and carbon sequestration play also a role.

Overall, squirrels seem to occur in most parts of that study area. Thus far, we cannot speak to the issue of whether and how squirrels are affected by humans or natural factors. Our study has not yet addressed for instance the relevance of fire or bark beetle outbreaks (but this is now relatively easy to achieve with new GIS model layers available, e.g. Zabihi et al. 2021).

However, squirrel middens occur in small patches and high-edge density areas. This also matches smaller-scale findings by Robolt and Huettmann (2021), and they seem to adjust and co-evolve with human-dominated landscapes during the last 10,000 years, as well as with the patchy landscapes that they currently seem to occur in (see Fig. 8.8 for polygons on black spruce, black spruce with birch, white spruce, and white spruce with birch and aspen).

It should be stated that much of the boreal forest, and thus, the habitat for North American Red Squirrels, are located on indigenous land (see Robolt and Huettmann 2021 for an example of University campus land and www.native-land.ca for a global detailed overview). This invokes many wider questions.

Arguably, the last 50 years have been devastating for the boreal forest as well as for Alaska's landscape wilderness, and some areas have already lost their forest cover in its entirety or become patchworks (see Bocharnikov and Huettmann 2019 for boreal forest fragmentation and patchiness; Hins et al. 2009; Hebblewhite 2017). With climate change and more human intervention, old-growth stands will certainly get reduced and the urban footprint will be on the rise, with many squirrels trying/ having to live in that Anthropocene.

In order to improve the future outlook or decrease the negative impact in the future, it is crucial to make the first step and acknowledge the current situation and threats to the habitat and its species. From there, one can create a better future outlook based on this current situation. Next, it is important to spread knowledge about the current situation and its issues, making them a discussion among scientists and policymakers and bring these topics onto public science agendas for improvement. Only in that way, the future outlook can be improved.

Data Availability Statement: All required data is available on request from the authors.

References

Black CC (1972) Holarctic evolution and dispersal of squirrels (Rodentia: Sciuridae). In: Evolutionary biology. Springer, New York, pp 305–322

Bocharnikov V, Huettmann F (2019) Wilderness condition as a status indicator of Russian Flora and Fauna: implications for future protection initiatives. Int J Wilderness 25:26–39

Chapin FS, Hollingsworth T, Murray DF, Viereck LA, Walker MD (2006) Floristic diversity and vegetation distribution in the Alaskan boreal forest. Oxford University Press, New York, pp 81–99

Division of Forestry Department of Natural Resources (2018) Alaska Forest Resources & Practices Act. http://forestry.alaska.gov/forestpractices. Accessed 29 Aug 2022

Dunn M, Marzano M, Forster J, Gill RM (2018) Public attitudes towards "pest" management: perceptions on squirrel management strategies in the UK. Biol Conserv 222:52–63

Elkins EK (2017) Red squirrel (Tamiasciurus hudsonicus) midden site selection and the influence of conifer species compositions on midden occurrence in the Cooke City Basin of Montana. Doctoral dissertation, Montana State University-Bozeman, College of Agriculture, p 7

Gauthier S, Bernier P, Kuuluvainen T, Shvidenko AZ, Schepaschenko DG (2015) Boreal forest health and global change. Science 349(6250):819–822

Gilson A, Salmon TP (1990) Ground squirrel burrow destruction: control implications. In: Proceedings of the fourteenth vertebrate pest conference 1990, p 33

Hebblewhite M (2017) Billion dollar boreal woodland caribou and the biodiversity impacts of the global oil and gas industry. Biol Conserv 206:102–111

Hins C, Ouellet JP, Dussault C, St-Laurent MH (2009) Habitat selection by forest-dwelling caribou in managed boreal forest of eastern Canada: evidence of a landscape configuration effect. For Ecol Manag 257(2):636–643

Holmes M (2015) The perfect pest: natural history and the red squirrel in nineteenth-century Scotland (William T. Stearn Prize 2014). Arch Nat Hist 42(1):113–125

Humphries GR, Magness DR, Huettmann F (eds) (2018) Machine learning for ecology and sustainable natural resource management. Springer, 3, Cham

Kayes I, Mallik A (2020) Boreal forests: distributions, biodiversity, and management. Springer. https://doi.org/10.1007/978-3-319-71065-5_17-1

Koprowski JL (2005) Management and conservation of tree squirrels: the importance of endemism, species richness, and forest condition. In: Gerald GJ, Brooke GS, Lane EG, Carleton EB (eds) Connecting mountain islands and desert seas: biodiversity and management of the Madrean Archipelago II. Proceedings of RMRS-P-36, vol 36. US Department of Agriculture, Forest Service, Rocky Mountain Research Station, Fort Collins, pp 245–250

Koprowski JL, Nandini R (2008) Global hotspots and knowledge gaps for tree and flying squirrels. Curr Sci, pp 851–856

Kranowski PV (1969) Aspects of red squirrel (*Tamiasciurus Hudsonicus*) population ecology in interior Alaska. University of Alaska Fairbanks (UAF)

Larsen KW, Boutin S (1994) Movements, survival, and settlement of red squirrel (Tamiasciurus hudsonicus) offspring. Ecology 75(1):214–223

McGarigal K (1995) FRAGSTATS: spatial pattern analysis program for quantifying landscape structure, vol 351. US Department of Agriculture, Forest Service, Pacific Northwest Research Station

Miquelajauregui Y, Cumming SG, Gauthier S (2019) Short-term responses of boreal carbon stocks to climate change: a simulation study of black spruce forests. Ecol Model 409:108754

Ohse B, Huettmann F, Ickert-Bond S, Juday G (2009) Modeling the distribution of white spruce (Picea glauca) for Alaska with high accuracy: an open access role-model for predicting tree species in last remaining wilderness areas. Polar Biol 32:1717–1724

Palmer GH, Koprowski J, Pernas T (2007) Tree squirrels as invasive species: conservation and management implications

Ploton P, Mortier F, Réjou-Méchain M, Barbier N, Picard N, Rossi V, Pélissier R et al (2020) Spatial validation reveals poor predictive performance of large-scale ecological mapping models. Nat Commun 11(1):1–11

Rajaratnam SM, Redman JR (2001) Circadian locomotor activity rhythms of the diurnal Indian palm squirrel in constant light. Chronobiol Int 18(1):47–60

Ramos-Lara N (2012) Ecology of the endemic Mearns's squirrel (Tamiasciurus mearnsi) in Baja California, Mexico

ReserveAmerica (2020) 9 fantastic forest regions to explore. ReserveAmerica. https://www.reserveamerica.com/articles/camping/9-fantastic-forest-regions-to-explore. Accessed 29 Aug 2022

Robolt R, Huettmann F (2021) High-resolution prediction of American red squirrel in interior Alaska: a role model for conservation using open access data, machine learning, GIS and LIDAR (2020:03:47121:2:0)

Sheail J (1999) The grey squirrel (Sciurus carolinensis) – a UK historical perspective on a vertebrate pest species. J Environ Manag 55(3):145–156

Smith CC (1970) The coevolution of pine squirrels (Tamiasciurus) and conifers. Ecol Monogr 40(3):349–371

Smith CC (1978) Structure and function of the vocalizations of tree squirrels (Tamiasciurus). J Mammal 59(4):793–808

Smith AA, Mannan RW (1994) Distinguishing characteristics of Mount Graham red squirrel midden sites. J Wildl Manag 58:437–445

Steiner M, Huettmann F (2021) Justification for a taxonomic conservation update of the rodent genus Tamiasciurus: addressing marginalization and mis-prioritization of research efforts and conservation laissez-faire for a sustainability outlook. The European Zoological Journal, 88(1):86–116

Veraverbeke S, Rogers BM, Goulden ML, Jandt RR, Miller CE, Wiggins EB, Randerson JT (2017) Lightning as a major driver of recent large fire years in North American boreal forests. Nat Clim Chang 7(7):529–534

Wauters LA, Bertolino S, Adamo M, Van Dongen S, Tosi G (2005) Food shortage disrupts social organization: the case of red squirrels in conifer forests. Evol Ecol 19(4):375–404

Wolff JO, Zasada JC (1975) Red squirrel response to clearcut and shelterwood systems in interior Alaska, vol 255. Pacific Northwest Forest and Range Experiment Station, Forest Service, US Department of Agriculture

Wooller MJ, Kurek J, Gaglioti BV, Cwynar LC, Bigelow N, Reuther JD, Smol JP et al (2012) An~ 11,200 year paleolimnological perspective for emerging archaeological findings at Quartz Lake, Alaska. J Paleolimnol 48(1):83–99

Young B, Yarie J, Verbyla D, Huettmann F, Herrick K, Chapin FS (2017) Modeling and mapping forest diversity in the boreal forest of interior Alaska. Landsc Ecol 32(2):397–413

Young BD, Yarie J, Verbyla D, Huettmann F, Chapin FS (2018) Mapping aboveground biomass of trees using forest inventory data and public environmental variables within the Alaskan boreal forest. In: Machine learning for ecology and sustainable natural resource management. Springer, Cham, pp 141–160

Zabihi K, Huettmann F, Young BD (2021) Predicting multi-species bark beetle (Coleoptera: Curculionidae: Scolytinae) occurrence in Alaska: open-access big GIS-data mining to provide robust inference. Biodivers Inform 16:1–19

Chapter 9
Can Squirrels Be Used as Indicators to Identify and Protect Old-Growth Forest Reserves? A Worked Strategic Conservation Planning Workflow with Several Optimization Methods for Environmental Impacts and Climate Change, Involving Indigenous Land

Abstract Tree squirrels are commonly acknowledged to be associated with big and mature trees (e.g. spruce), mainly due to their cones. This species group of squirrels inhabits preferably old, mature, and intact forests and associated landscapes; mainly because this species group evolved from, and co-evolved with it. Here we investigate whether their distribution and density can be used to identify such forests. The study area consists of the Tanana Valley State Forest (TVSF) – a representative boreal forest wilderness of interior Alaska (USA) - of international relevance with a perpetual human and industrial history. In this study, we seek to identify scale-related valuable old-growth forest reserve protection for the Interior Alaska Boreal Forest with an optimization application. We employed and tested three optimizer platforms: MARXAN (based on simulated annealing), Gurobi, and Symphony solvers in R language (code). This optimization application is based on strategic conservation planning, using 198 hexagons of 25 km diameter planning units (PUs) in the study area. Using open-source geographic information system (QGIS) software, we attempt a step-wise manual approach, to optimize/classify 50% of the study area for species protection. This is done by using the three variables of (1) high resolution (1 km lattice grid point) prediction locations of American red squirrels (*Tamiasciurus hudsonicus*), (2) 500 ha old-growth forest reserves for spruce (black and white spruce *Picea marinana* and *P. glauca*), and (3) saw timberland polygons. Due to the lack of publicly-available land ownership maps with accepted future model scenarios, our model currently does not take the changing landscape for 2100 into account yet. However, we present the best-optimized baseline data and a workflow to set the stage, making the results more relevant, for a longer time. Our results show that the 50% protection scenario cannot really be achieved in the study area with the three constraining variables; that finding is based on selection frequency, as well as on a singular 'best' solution, and it is due to the geographic non-overlapping constraining arrangement of the predictors. We discuss how such biodiversity-valuable forests can be better protected and what the influences of squirrel conservation and management are on such endemic biodiversity-rich forests of global relevance. Because the study area includes c. 20% indigenous land ownership we show that app. 30% of the 'best' solution planning units (PUs) fall on indigenous land, raising cultural and holistic awareness for the required protection efforts of squirrels, old-growth forest reserves, and indigenous land.

Keywords MARXAN · Gurobi solver · Symphony solver · R language (code) · Strategic conservation planning · Tree squirrels · Interior Alaska · Nature reserve modeling · Tanana Valley State Forest · Wilderness area · Indigenous land

9.1 Introduction

Many squirrel species, especially tree squirrels and flying squirrels, are naturally linked to trees, mainly mature trees. They need established trees and their composition (forests) as feed sources (cones), shelter, and lifetime habitat. Similar to the woodpecker group (Baral and Huettmann 2020), (big) trees are a certain anchor for squirrels to live and survive. Squirrels are social animals and trees are an inherent part of the ecological niche for squirrels as a basic life requirement.

Tree quality, species richness, and vegetation age are linked to the well-being of the earth and how it is managed by humans (Doimo et al. 2020; Hedblom et al. 2017; Loureiro and Veloso 2017; Panagopoulos et al. 2016), but certainly, it also matters for the well-being and distribution of squirrels, their ecosystem (Robold and Huettmann 2021), and wider ecological services. Mature old trees present the natural link for squirrels throughout earth's recent history, and they can usually provide such services best to squirrels. These services include feed provision (cones and similar) (Brink and Dean 1966; Gurnell 1983; Patton and Vahle 1986; Rusch and Reeder 1978; Wauters et al. 2002), shelter (from predators) & nesting (Fitzwater

---

Supplementary Information The online version contains supplementary material available at https://doi.org/10.1007/978-3-031-23547-4_9.

M. Steiner, F. Huettmann, *Sustainable Squirrel Conservation*, https://doi.org/10.1007/978-3-031-23547-4_9

Fig. 9.1 Study area of the interior Alaskan wilderness, Tanana Valley State Forest (USA)

and Frank 1944; Hackett and Pagels 2003; Koprowski 2005; Perkins et al. 2008), lifetime habitat (Patton and Vahle 1986), and present centers of their territory (Dantzer et al. 2012).

In interior Alaska, the Tanana Valley State Forest (TVSF) is a representative wilderness area near Fairbanks that features old-growth forest reserves (see details in Steiner and Huettmann in 2023 – Chap. 8). Those are typically characterized by white spruce (*Picea glauca*) or black spruce (*Picea mariana*) accumulations (polygons), with some birch (*Betula sp.*) or aspen/poplar (*Populus sp.*). Black spruce is relatively widespread, occurs in wet areas, and is mostly of lower commercial value, and thus, often untouched by humans with very old ages and of high interest for the squirrels and other wildlife. Whereas white spruce is described as 'timber', so are birch and poplar. This area has been heavily harvested in the last 100 years with early gold mining, railroads, World War 2 (WW2), and statehood, including during the 1990s for Japan (Knapp 1992; Wurtz et al. 2006). However, some old-growth (white spruce timber) reserves are left in the TVSF e.g. on steep slopes or other inaccessible areas (see Ohse et al. 2009; Young et al. 2017, 2018 for White Spruce, biomass, biodiversity, and forest inventory). In interior Alaska, those areas are few. Based on our own fieldwork in the study area, we found North American Red squirrels (*Tamiasciurus hudsonicus*) within all of those areas in the wider landscape of the TVSF but with a preference for those old-growth coniferous patches (Robold and Huettmann 2021), which are rather small in size and carry a high edge ratio but are extremely valuable for biodiversity. Some of those patches are in urbanized areas even. An overview of the study area can be seen in Fig. 9.1.

We are all now living in the Anthropocene; landscapes are fully used and anthropologically exploited (see the concept of the full world in Daly (2005)), relevant ecological cycles are man-made or severely affected by industrial activities, and the wilderness shrinks. That also applies to interior Alaska, one of the last wilderness areas left in North America (Alaska carries the bulk of the remaining U.S. protected and wild areas). Using the North American Red squirrel in interior Alaska as an example, here we look into the prioritization of relevant habitats for this species. The study area is one of the few remaining wilderness areas in interior Alaska, as part of the boreal forest – the largest intact forest block left in the world (Gauthier et al. 2015; Veraverbeke et al. 2017). This study area is also covered by indigenous land, the Dene culture, and a complex set of landownerships (see https://native-land.ca/; Garnett et al. 2018).

Another squirrel species that could have been used for this study is the Northern Flying Squirrel (*Glaucomys sabrinus*), which also occurs in Alaska, but is widespread to a lesser extent and has little data available (see details in Textbox 9.1). Arguably, this is a squirrel community but needs more research. A sketch of such a flying squirrel can be seen in Fig. 9.2.

Textbox 9.1: Flying Squirrels in Alaska? What Flying Squirrels?

Falk Huettmann, Moriz Steiner

Northern Flying squirrels (*Glaucomys sabrinus*) occur in the Boreal and Temperate Mixed Forest Zones of Canada, Alaska, and the lower 48 (Smith 2007; Steiner and Huettmann 2023 – Chap. 3). In Alaska, they are managed by the Alaska Department of Fish and Game (ADFG) (ADFG 2022). According to them, Northern Flying Squirrels are covered in the game management laws, and poaching would essentially not occur, or matter. The actual harvest of Northern Flying Squirrels is to be minuscule, but they are shot at to a significant extent. That is because people do not like squirrels around bird feeders and their homes/farms/properties raiding bird food etc. Northern Flying Squirrels are also attractive as they are 'cute' and 'fly' (glide) and thus are a hard/interesting/ "playful" target to shoot at. The hunting laws are all-inclusive ('liberal'), making poaching virtually impossible to occur as the hunting laws are difficult to violate. And thus, when it comes to pursuing Northern Flying Squirrels, everything seems to be ok: trapping, shooting, no bag limits, or a relevant 'take' by urbanization and forestry in the wide absence of population monitoring. Now, that is a '*laissez-faire*' approach to conservation management in the year 2023 (Steiner and Huettmann 2021). Consider squirrels then as game but being treated like non-game. And that is unlikely 'modern', scientific or proactive, or even meaningful or effective for a public trust resource governance. It has been widely criticized for other wildlife in Alaska, e.g. bears or wolves (Ripple et al. 2019). And it certainly is initially pretty cheap and time-saving for efforts, and so the governance and the public appear to be well served with it, whereas Flying Squirrels are not really included in that equation.

While Climate Change unfolds, the genetics of those species are already changing (Garroway et al. 2010 for hybridization). And many other issues in Alaska are unfolding too (Ripple et al. 2019).

Instead, game species are to be managed, e.g. with science-based quotas, underlying information, and harvest rates; they need 'a plan' based on life history and modern principles (e.g. Mahoney and Geist 2019). But even the most basic life history metrics for these sophisticated social wildlife species in the wide public eye are absent. Beyond how Flying Squirrels can detect truffles (Pyare and Longland 2001), the management science on Flying Squirrels in Alaska draws few citations whatsoever, and not much detail on them is mutually agreed upon, including population trends and estimates with confidence intervals forward looking (see Bakker and Hastings 2002 for Alaskan den trees). There is no Northern Flying Squirrel manager *per se*, nor really such a job description focusing on Northern Flying Squirrels in Alaska (other than furbearers), while Alaska holds likely the largest wilderness area in the U.S., much of it is public land or a part of the conservation areas.

Arguably, Flying Squirrels are affected by forestry, but this topic is also widely left unresolved for any relevant forestry practices, e.g. Alaska has the largest temperate rainforest in the world with a massive harvest within (e.g. Schoen 2021). And the Forest Practice Code in Alaska has very few pages (considering this is one of the largest forest resources on the continent (Huettmann and Young 2022) (Textbox Fig. 9.1). As bird feeders are still a major habitat feature for flying squirrels, is that a management by bird food and peanut butter?

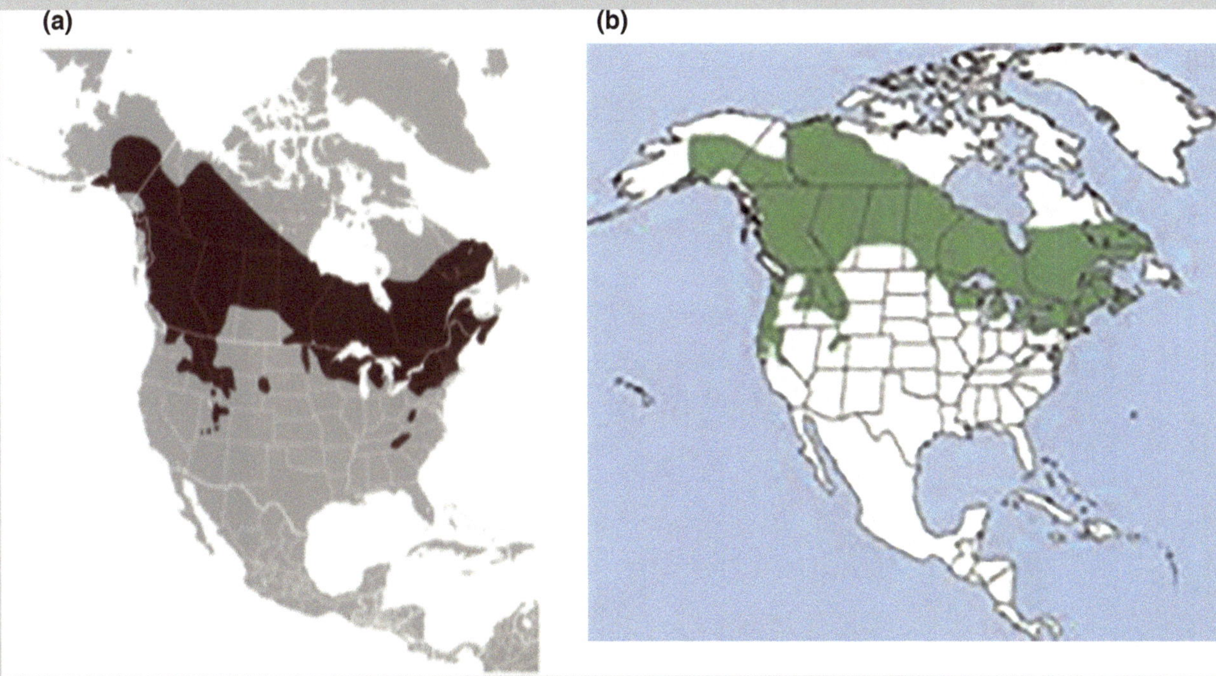

Textbox Fig. 9.1 Range maps of Flying Squirrels (**a**) source: Wikipedia, (**b**) source: NatureWorks

A typical example is already found in the range maps of *G. sabrinus*. Knowing the distribution of a species is an essential item for managing any species. But looking up the literature records, several of those range maps can be found. Some are hand-drawn, some are conflicting, others are blurry, and none show the underlying data or truly agree with each other, including for the Alaskan part and assumed range limits, e.g. in the north and across borders (Canada). Discrepancies are thus widely found, e.g. see the figures above and the displayed distribution on the west coast of the US. So here we show a model based on the latest Alaska public occurrence data for this species (Textbox Figs. 9.2), it indicates that this species really likes the urban and suburban regions, including road corridors (where people live), distant from the wider wilderness.

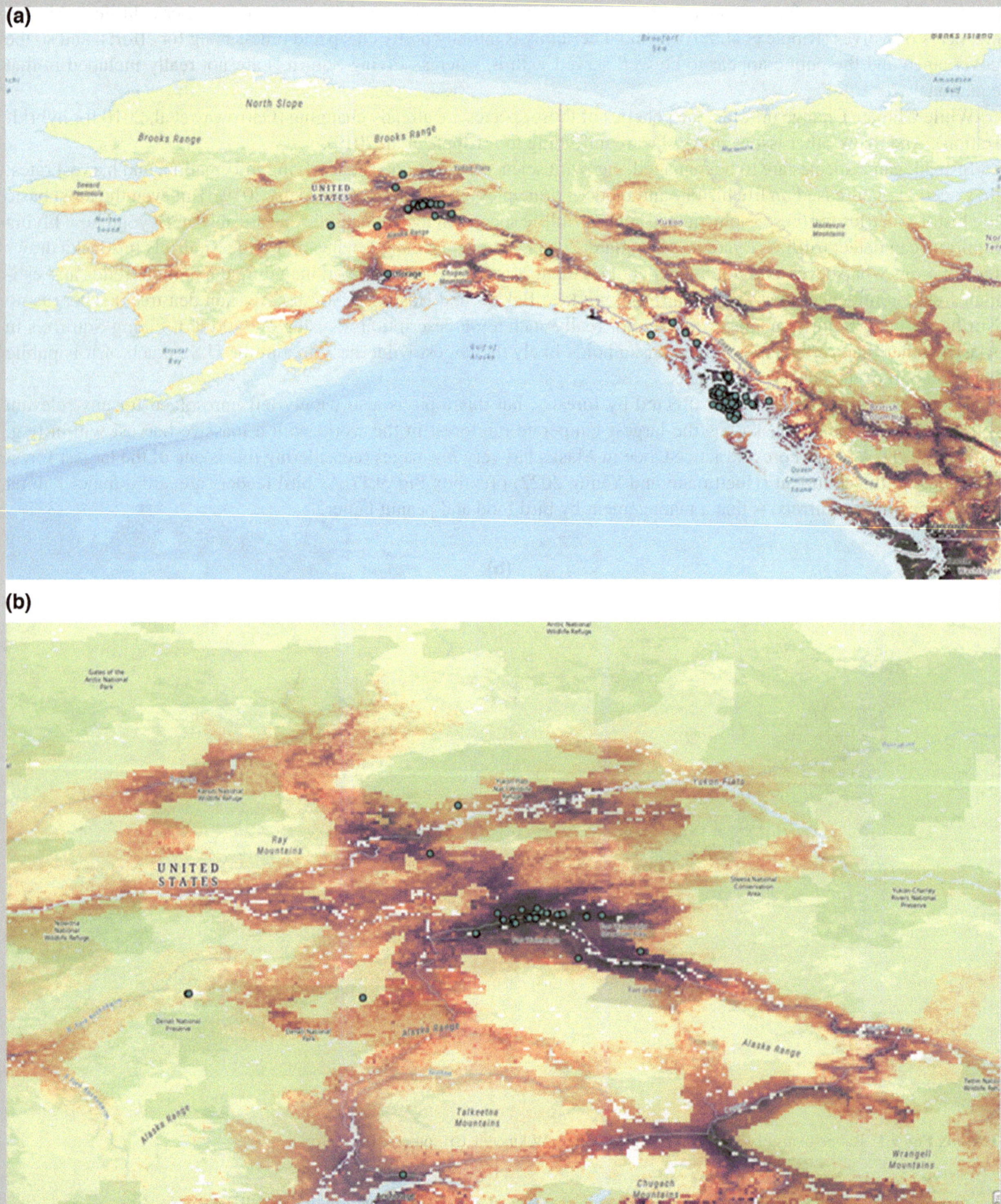

Textbox Figs. 9.2 Predicted map for Northern Flying Squirrels in Alaska and adjacent Canada using Maxent and 132 open-Access predictors (**a**) Alaska – Western Canada overview, (**b**) Fairbanks overview. (Source Steiner and Huettmann 2023)

Why is that?

Likely because in the absence of a proper management - any management and coordination - bird feeders, suet, peanut butter, and other human processed food provided, the amount of those 'subsidies' that are put in the yards – for the birds – remains mind-boggling and is on the rise even (www.feederwatch.org). In Alaska, one may happily add dog food to that equation, easily over 100 large sled dog kennels exist in Alaska, and many have dog food accessible. Flying Squirrels take good advantage of those subsidies, and the distribution reflects that well (Textbox Fig. 9.3).

Textbox Fig. 9.3 Photo of a Flying Squirrel foraging at a typical dog food box during winter in interior Alaska's harvested Old-Growth Forest region of the Tanana Valley State Forest. (Source: FH)

Textbox Fig. 9.4 Flying squirrel sketch. (Drawn by Enrika Gataveckaite)

Northern Flying Squirrels are found in many parts of Northern North America still. They are classified by the IUCN Red List as Least Concerned. However, their habitats are heavily fragmented, large trees are on the decline and urbanization is widely on the rise. In Alaska, they are living on the climate range edge, and much habitat loss and change have already occurred, e.g. in Wasilla regions and Tongass (Kessler 1979; Mills and Andres 2004; Sitka 2022). The Boreal Forest zone remains vast and wild, but the management is not proactive or has any relevant data and science to convince. Compared with other species, one can easily do better, and likely should! (Textbox Fig. 9.4)

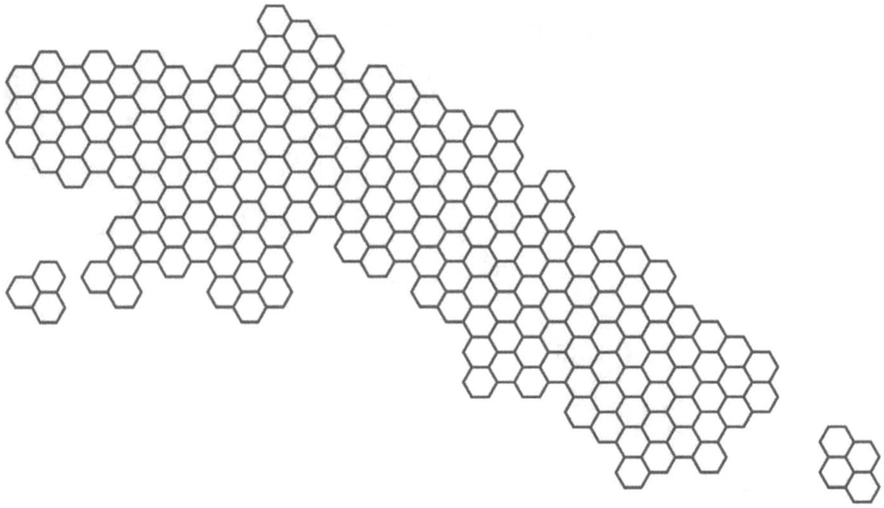

Fig. 9.2 198 Planning Units (PUs) in the study area

Anticipating more changes in the landscape to come, sacrificing sustainability and ecological service decay, and upcoming tensions about land and forest resources in the future, here we try to elaborate on it and present methods that can be used to better address those conflicts among all stakeholders.

9.2 Methods

9.2.1 North American Red Squirrel (Tamiasciurus hudsonicus) Data

We used the best-publicly available *T. hudsonicus* prediction for the study area, the Tanana Valley State Forest (TVSF). This prediction layer is found online, and comes from Steiner and Huettmann (2023 - Chap. 8) and has been created with the best publicly available data in a GIS format, presenting a relative index of occurrence (RIO) for a pixel of a 1 km-spaced lattice (45,419 lattice points). Details and raw data are described in the references provided above. As high-quality squirrel locations in the study area, we used the 0.9 RIO as a cut-off resulting in 98 top squirrel habitat locations from the prediction lattice entering the analysis (see Appendix).

9.2.2 Habitat Data and GIS Preparation

While the study area is of wider, national, and international relevance, there are just a few GIS habitat layers available on that scale (see Ohse et al. 2009 for Alaska; Sriram and Huettmann (unpublished), Steiner and Huettmann 2023 – Chap. 3 for over 132 GIS layers; Young et al. 2017, 2018). Here we use spruce (White and Black Spruce) polygons (n = 46,080) and timber saw wood (n = 11,373) polygons from the TVFS vegetation GIS database layers. Because we are interested in old-growth spruce reserves, we selected polygons larger than 500 ha (n = 73). To address land ownership issues more realistically and beyond 'just' timber sales, we also infer and refer to indigenous land layers (created by MS from the API) with the information from www.native-land.ca.

We then used a hexagon GIS layer to group and stratify hexagon areas. This was done with the MMGIS plugin in QGIS for ~25 km diameter hexagons in the study area resulting in 198 Planning Units (PUs; Fig. 9.2). We extracted the underlying GIS layers for each hexagon polygon in QGIS based on the UTM North Zone 6 projection in meters. This resulted in 135 PUs for saw timber, 60 PUs that had spruce old-growth forests larger than 500 ha, and 71 PUs with squirrel predictions higher than an ROI of 0.9 (the best prediction lattice points).

9.2.3 *Optimization*

We generally followed Marxan as the tool of choice for strategic conservation planning. We then added the Marxan approach in R with optimizers of GUROBI and Symphony (Ardron et al. 2008; Schuster et al. 2020). We tried to optimize for 'the singular best' solution as well as selection frequencies. This allowed us to get at the question of strategic conservation planning for North American Red squirrels in the study area based on the constraint questions set by the user. As we are not aware of any strategic conservation planning exercises for North American Red squirrels, or virtually any squirrel species in the world (Koprowski and Nandini 2008), with exception of a hand full of (unpublished) conservation attempts and indirect by-products, this is among the first applications for squirrels of this concept we are aware of. Thus, this study can present a role model and template for best-possible conservation decisions for squirrels, and other small mammals and their ecosystems, anywhere (for Alaska see Murphy et al. 2010, 2012).

There are no good and high-resolution 'future' landscape and climate models available for the study area (compare with Huettmann et al. 2005 and Nielsen et al. 2008). Thus, we apply our strategic conservation planning for the current time period, where we have data.

We tried to find 'the best' – singular and selection frequency 0–100 – solution for 50% area coverage of squirrel predictions in the PUs with the following maximization Formula 9.1:

$$\begin{aligned}&50\% \text{ of the Planning Units in the study area protected} \sim \text{high Red Squirrel predictions}\\&+\text{Old Growth reserve } (500\text{ha}) \text{ presence} + \text{Timber Saw land presence}\end{aligned} \tag{9.1}$$

For a more holistic land concept involving more stakeholders, the resulting optimization map will then be overlaid with and assessed by, indigenous land holdings in the study area. Ideally, and in future attempts, stakeholders should be 'on board' and participate for re-iterative updates.

Figure 9.3 illustrates a workflow overview of strategic conservation planning, which has also been utilized in this study.

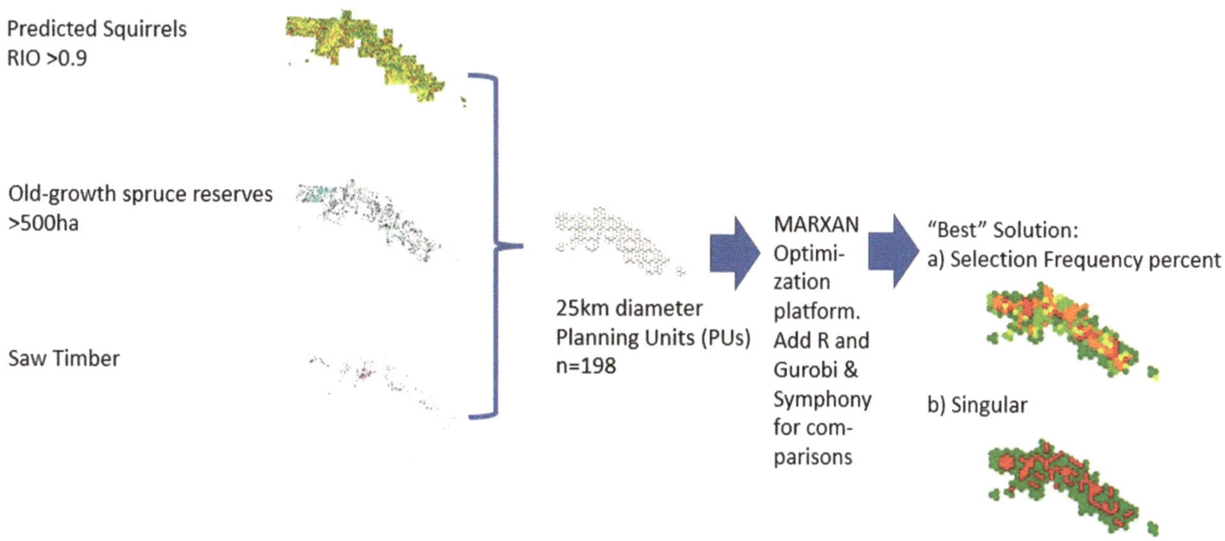

Fig. 9.3 Workflow of strategic conservation planning

9.3 Results

We were able to compile the best-publicly available GIS data for strategic conservation planning using North American Red squirrels as indicators. We did that for the study area of the Tanana Valley State Forest (TVSF), a vast tract of the wilderness of international importance in interior Alaska. It is an area around the city of Fairbanks where urbanization, fragmentation, and climate change is known to occur in high magnitude and where indigenous land ownerships can be found.

Our results show that outcomes of optimization techniques somewhat differ across platforms; that is true for singular solutions as well as selection frequency (Figs. 9.4, 9.5, 9.6 and 9.7). The selection frequency percentage for MARXAN shows only 22 PUs as the top solution (app. 11% of the study area). The best singular solution selected 67 PUs (app 30%) of the study area also indicating that a 50% protection goal within the 3 variable constraints is hardly feasible. Results for rprioritizer, gurobi, and Rsymphony confirm that trend. It shows that the study area is already relatively compromised and that a decent 50% protection is not possible within the framework of squirrels, old-growth, and saw timber; indicating that these three components can hardly co-exist sustainably in modern days.

Simply judged by the constraint of saw timber, the rsymphony solution – selection frequency – seems to perform best (see Fig. 9.7b for a visual assessment).

Our findings also show that most old-growth forests in the study area consist of smaller patches but that human impact can already be found in most parts of the study area, specifically for the 500 ha patches (73 out of 40,000 GIS polygons in the study area overall). Virtually all of the accessible and commercially viable stands in the study area are already accessed or cut down for white spruce; whereas black spruce is mostly left untouched but has natural fragmentation, e.g. due to rivers

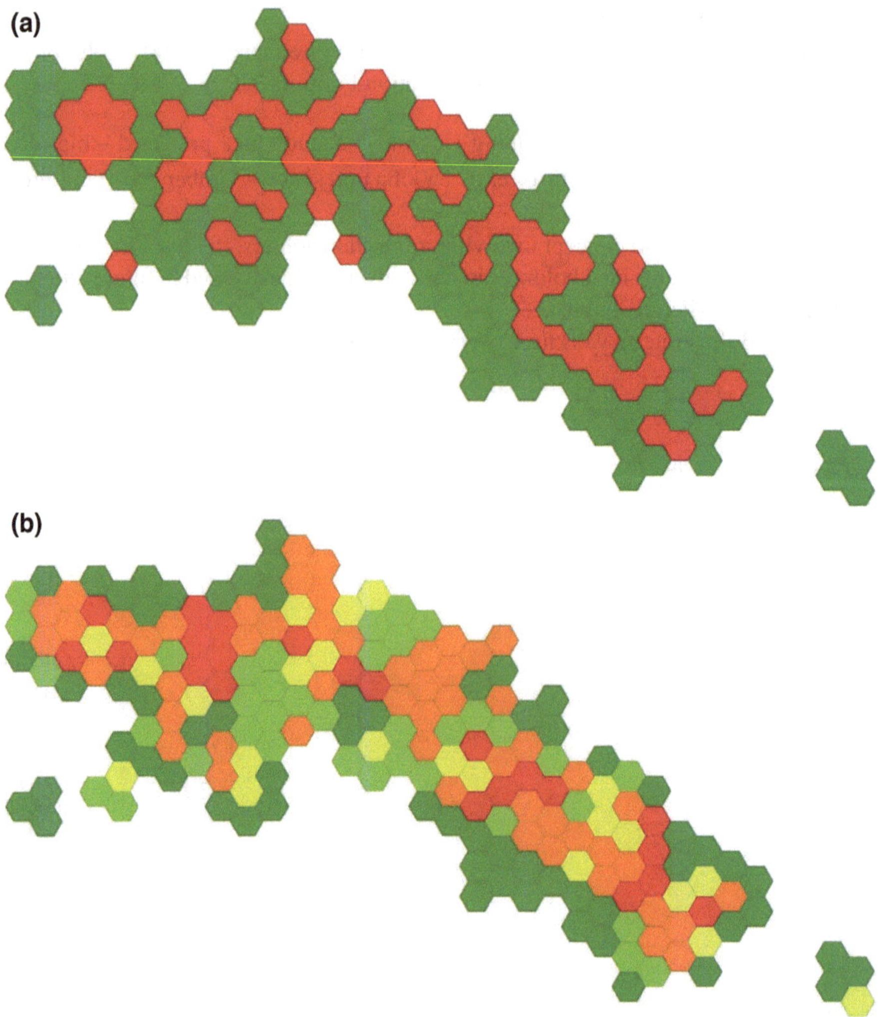

Fig. 9.4 Result of 'the best' solution (**a**) singular solution, and (**b**) selection frequency percent for Planning Units (PUs) using MARXAN

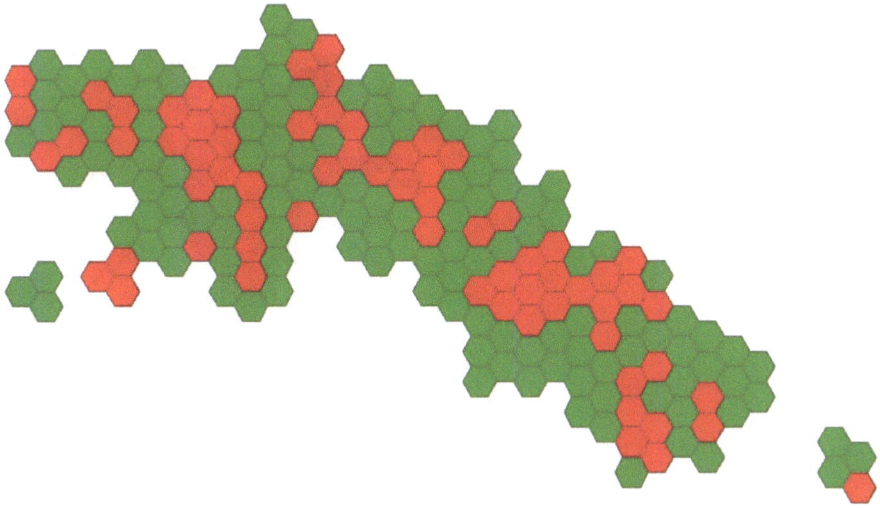

Fig. 9.5 Result of 'the best' solution (singular solution) for Planning Units (PUs) using rprioritizer

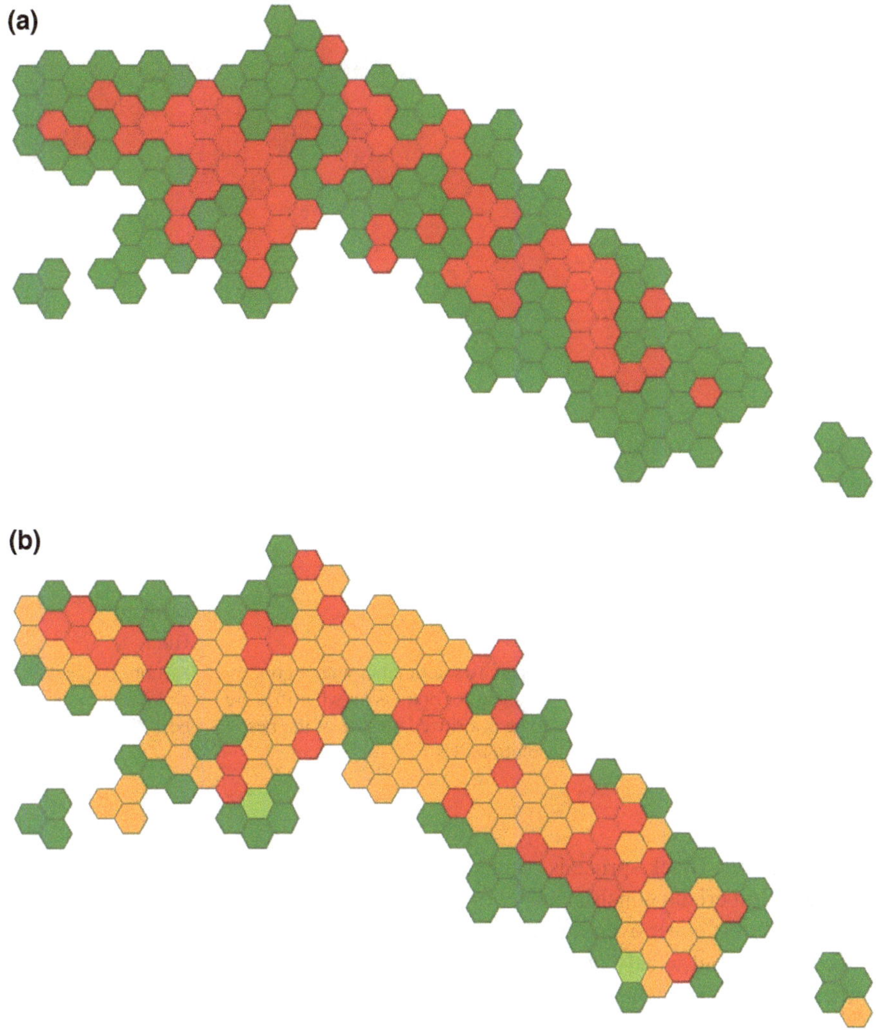

Fig. 9.6 Result of 'the best' solution (**a**) singular solution, and (**b**) selection frequency percent for Planning Units (PUs) using gurobi package in R

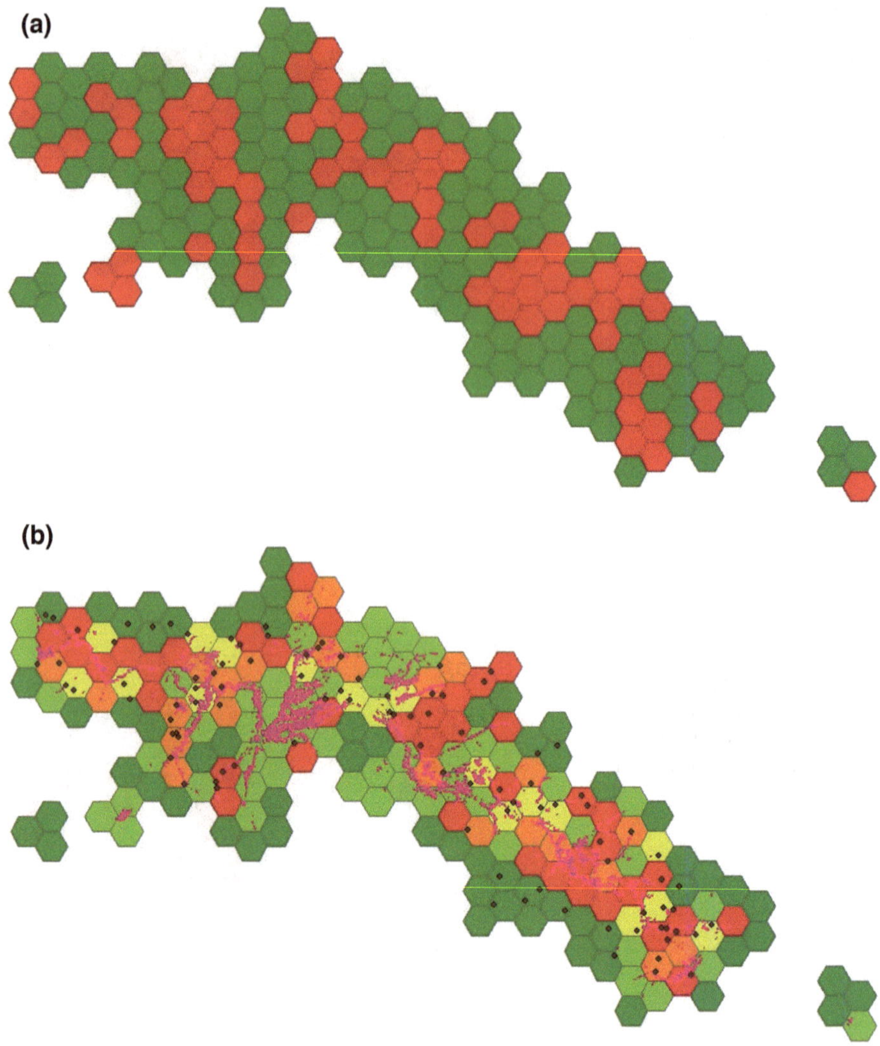

Fig. 9.7 Result of 'the best' solution (**a**) singular solution, and (**b**) selection frequency percent for Planning Units (PUs) using Rsymphony package in R

Table 9.1 Characteristics of the two 'best' solutions using MARXAN for selection frequency percent and singular best

Criteria	Selection frequency	Singular best
PUs selected as 'best'	22%	67%
Cover percent of the area achieved	App. 11%	App. 30%
Connectivity	Little	Medium
Clumsiness	Low	High
General characteristics of the solution	A patchy solution but with a good gradient for options on what PUs are to be prioritized outside of the best solution PUs	A more connected, spread-out solution but with little choices and freedom on what PUs are to be prioritized other than the best PU solution itself

and topography. Black spruce is commercially less viable and occurs in alpine and/or wet areas; often with standing water. Secondly, timberland consists mostly of poplar trees and thus does not impact cone production needed for squirrels but contributes to forest cover overall (Table 9.1).

Grasslands and widely open forests seem not to be really relevant for North American Red squirrels (Boon et al. 2008; Ferner 1974; Ransome and Sullivan 1997; Robold and Huettmann 2021), and thus, are mostly excluded from the resulting good priority planning units. The TVSF has few open grass areas and open forest land, e.g. Fairbanks municipality, railways, and airports. Those are recent landscape artifacts created by mankind with no big value for squirrels. It represents a typical example of increased habitat transition and loss of habitat for squirrels, mostly occurring in just the last 60 years, with mining-related fuel wood overuse during the last 100 years (usually near rivers and roads).

We also found for 'the best' solutions that some urban areas are included. That is primarily because Fairbanks' sprawl area covers over 120 km in diameter (aided by steam boats and planes to over 500km) and is covered by forest that has many isolated, old-growth patches and areas within (Robold and Huettmann 2021). Squirrels seem to strive there, e.g. on edges and frequent trash bins, etc., but exact life history details such as disease transmissions (to other animals including dogs, and humans - zoonosis), stress, behavior patterns, and life expectancy are mostly unknown. Urban areas could be sinks sustained by surrounding wilderness sources (this is well known for other species such as coyotes, bears, ravens, etc). Other urban squirrel populations exist, and those warrant more research in the future (see Chap. 5 in Steiner and Huettmann 2023).

9.4 Discussion

This is the first strategic conservation planning work for interior Alaska's wildlife and for squirrels using GIS data. While this work should be improved, in the Anthropocene and its pressures (see Murphy et al. 2010, 2012 for Alaska), we find that a 50% protection goal is easily warranted but hardly possible while the study area is of major conservation concern and under threat of development, e.g. mining and housing.

Following the general MARXAN platform, we find that the optimizers used in this study show diversity in the outcome, but can improve and provide more options for planning purposes, e.g. the best commercially friendly solution is found with Rsymphony (selection frequency). However, it would still miss some predicted squirrel hotspots (Fig. 9.7b) and thus compromise habitat integrity. The use of more ecologically meaningful GIS layers would also be useful.

Thus far, our optimization models also fall short on the climate change application to optimize a future state of the study area (compare with Murphy et al. 2010, 2012), and it needs to be improved with a good future outlook set of GIS layers for interior Alaska (compare with model methods in Huettmann et al. 2005; Nielsen et al. 2008).

We use selection frequency for our 'best' model approach. It allows for a gradient and perhaps some planning options. However, using the singular 'best' model shows a similar but already somewhat different optimization for PUs (see Figs. 9.4, 9.5, 9.6 and 9.7), and they cannot reach 50% of the study area as protected with an overall Planning Units (PUs)scale of 25 km. This means the study area has no good protection options for the specified aims (e.g. wildlife, forestry, and old-growth forests). Scale matters.

Unfortunately, one must not use the best-available indigenous land ownership map for academic research and GIS. Using the source of Garnett et al. (2018), on a continental scale when eyeballing the study area it might show approx. 35% percent of our MARXAN solution is indigenous land. In our 'best' squirrel solution, we find approx. 55% of the PUs are covered by indigenous land regardless (as per recent legal definition, because before contact all land was probably used by indigenous groups). This indicates that the role and relevance of indigenous land ownership and management require more attention (Schuster et al. 2018). This indigenous land data cannot be shown and shared due to ethical and copyrighted reasons, where the data providers aim to respect the indigenous communities from misusing the detailed data. This puts a massive constraint on the outcome of this study and should be addressed better by all parties involved. The original data can however be requested and obtained from the original authors of this report (Garnett et al. 2018, and UNEP-WCMC and ICCA Consortium 2021). Another indigenous land data set was intended to be used, provided by Native Land Digital (www.native-land.ca), however, this is difficult to use, technically (the data were not shared as GIS files and could only be observed on their interactive map (see link above)). Upon inquiring, GIS file data sharing has been denied and the team has guided the inquiring author (MS) to their API section, where the corner points of the polygons constructing their map are being displayed. Without any other option, MS created manually a hand-traced GIS file using these geographical corner points provided. This has only been done for this wider study area (Alaska, USA) because this manual technique requires extensive time efforts precluding its detailed use. This newly but coarsely created data file is now freely available, and accessible and can be found as Shapefile (.shp), and CSV file (.csv) in the appendix of this study. This latter data set could also not be used for the fine-scale model of this study since details such as land ownership, etc. are lacking in this dataset. For indigenous research, science, and the general well-being of humankind, we encourage the agreed and continuous open-access data sharing approach for especially such data are to be followed and pursued in a constructive science-based fashion. This aims to move away from the business-as-usual conservative approach, in the direction of open data sharing science and research, as it is best-professional practice and common/mandated policy by public funders and in the sciences (Zuckerberg et al. 2011; Huettmann 2015). It provides progress, transparency, reproducibility, and sustainability.

Within the constraints of wilderness and Alaska, here we were able to set up the first workflow for strategic conservation planning using North American red squirrels. We think this concept can be improved from the results at hand and then applied wider to many other applications of interior Alaska, its squirrels, and related conservation management problems. However, it should also apply to other species and countries/states. Noteworthy remains the already impossible aim of trying to combine economic and environmental values (aka 'win-win') in this landscape affected by industrialization.

Acknowledgments We thank H. Berrios and E. Huettmann for their input and advice. The incredible Chrome team deserves our deep gratitude also. We acknowledge the data received from (Native Land Digital – https://native-land.ca/, and UNEP-WCMC and ICCA Consortium (2021) for indigenous land data, as well as all other open access data for this research. This is EWHALE publication # 271.

References

ADFG (2022) Northern flying squirrel (Glaucomys sabrinus yukonensis). Alaska Department of Fish and Game. https://www.adfg.alaska.gov/index.cfm?adfg=northernflyingsquirrel.main. Accessed 31 Aug 2022

Ardron JA, Possingham HP, Klein CJ (2008) Marxan good practices handbook. Pacific Marine Analysis and Research Association, Vancouver, p 149

Bakker VJ, Hastings K (2002) Den trees used by northern flying squirrels (Glaucomys sabrinus) in southeastern Alaska. Can J Zool 80:1623–1633

Baral HS, Huettmann F (2020) Chapter 35. The fate of the great woodpeckers and hornbills in Nepal: no big trees, no life. In: Regmi GR, Huettmann F (eds) Hindu Kush-Himalaya watersheds downhill: landscape ecology and conservation perspectives. Springer Gland, Switzerland, pp 685–694

Boon AK, Réale D, Boutin S (2008) Personality, habitat use, and their consequences for survival in North American red squirrels Tamiasciurus hudsonicus. Oikos 117(9):1321–1328

Brink CH, Dean FC (1966) Spruce seed as a food of red squirrels and flying squirrels in interior Alaska. J Wildl Manag 30:503–512

Daly HE (2005) Economics in a full world. Sci Am 293(3):100–107

Dantzer B, Boutin S, Humphries MM, McAdam AG (2012) Behavioral responses of territorial red squirrels to natural and experimental variation in population density. Behav Ecol Sociobiol 66(6):865–878

Doimo I, Masiero M, Gatto P (2020) Forest and wellbeing: bridging medical and forest research for effective forest-based initiatives. Forests 11(8):791

Ferner JW (1974) Habitat relationships of Tamiasciurus hudsonicus and Sciurus aberti in the Rocky Mountains. Southwest Nat 18(4):470–473

Fitzwater WD, Frank WJ (1944) Leaf nests of gray squirrel in Connecticut. J Mammal 25(2):160–170

Garnett ST, Burgess ND, Fa JE, Fernández-Llamazares Á, Molnár Z, Robinson CJ, Leiper I et al (2018) A spatial overview of the global importance of indigenous lands for conservation. Nat Sustain 1(7):369–374

Garroway CJ, Bowman J, Cascaden TJ, Holloway Gi L, Mahan CG, Malcolm JR, Steele MA, Turner G, Wilson PJ (2010) Climate change induced hybridization in flying squirrels. Glob Chang Biol 16(1):113–121

Gauthier S, Bernier P, Kuuluvainen T, Shvidenko AZ, Schepaschenko DG (2015) Boreal forest health and global change. Science 349(6250):819–822

Gurnell J (1983) Squirrel numbers and the abundance of tree seeds. Mammal Rev 13(2–4):133–148

Hackett HM, Pagels JF (2003) Nest site characteristics of the endangered northern flying squirrel (Glaucomys sabrinus coloratus) in Southwest Virginia. Am Midl Nat 150(2):321–331

Hedblom M, Knez I, Gunnarsson B (2017) Bird diversity improves the well-being of city residents. In: Ecology and conservation of birds in urban environments. Springer, Cham, pp 287–306

Huettmann F (2015) On the relevance and moral impediment of digital data management, data sharing, and public open access and open source code in (tropical) research: the Rio convention revisited towards mega science and best professional research practices. In: Huettmann F (ed) Central American biodiversity: conservation, ecology, and a sustainable future. Springer, New York, pp 391–418

Huettmann F, Young BD (2022) The so-called modern 'sustainable forestry' destroys wilderness, old-growth forest landscapes and ecological services worldwide: a short first-hand review and global narrative on the use of 'growth-and-yield'as a destructive and even impossible goal. In: Forest dynamics and conservation. Springer, Singapore, pp 53–82

Huettmann F, Franklin SE, Stenhouse GB (2005) Predictive spatial modelling of landscape change in the foothills model Forest. For Chron 81(4):525–537

Kessler WB (1979) Bird population responses to clearcutting in the Tongass National Forest of southeast Alaska (No. 71). US Department of Agriculture, Forest Service, Alaska Region, Tongass National Forest, Ketchikan Area

Knapp G (1992) Native timber harvests in southeast Alaska, vol 284. US Department of Agriculture, Forest Service, Pacific Northwest Research Station

Koprowski JL (2005) The response of tree squirrels to fragmentation: a review and synthesis. In Animal conservation forum (8, 4, pp. 369-376). Cambridge University Press

Koprowski JL, Nandini R (2008) Global hotspots and knowledge gaps for tree and flying squirrels. Curr Sci:851–856

Loureiro A, Veloso S (2017) Green exercise, health and well-being. In: Handbook of environmental psychology and quality of life research. Springer, Cham, pp 149–169

Mahoney SP, Geist V (eds) (2019) The North American model of wildlife conservation. Johns Hopkins University Press

Mills TK, Andres BA (2004) Changes in loon (Gavia spp.) and red-necked grebe (Podiceps grisegena) populations in the lower Matanuska-Susitna Valley, Alaska. Can Field-Nat 118(2):210–214

Murphy K, Huettmann F, Fresco N, Morton J (2010) Connecting Alaska landscapes into the future. U.S. Fish and Wildlife Service, and the University of Alaska. http://www.snap.uaf.edu/downloads/connecting-alaska-landscapes-future

Murphy K, Reynolds J, Jenkins J, Whitten E, Fresco N, Lindgren M, Huettmann F (2012) Predicting Future Potential Climate-Biomes for the Yukon, Northwest Territories, and Alaska: A climate-linked cluster analysis approach to analyzing possible ecological refugia and areas of greatest change. Prepared by the Scenarios Network for Arctic Planning (SNAP) and the EWHALE lab, University of Alaska-Fairbanks on

behalf of The Nature Conservancy Canada., Government Northwest Territories. http://snap.uaf.edu/webshared/Nancy%20Fresco/Cliomes/Cliomes%20FINAL%20March%2031%202012.pdf
Nielsen SE, Stenhouse GB, Beyer HL, Huettmann F, Boyce MS (2008) Can natural disturbance-based forestry rescue a declining population of grizzly bears? Biol Conserv 141(9):2193–2207
Ohse B, Huettmann F, Ickert-Bond SM, Juday GP (2009) Modeling the distribution of white spruce (Picea glauca) for Alaska with high accuracy: an open access role-model for predicting tree species in last remaining wilderness areas. Polar Biol 32(12):1717–1729
Panagopoulos T, Duque JAG, Dan MB (2016) Urban planning with respect to environmental quality and human well-being. Environ Pollut 208:137–144
Patton DR, Vahle JR (1986) Cache and nest characteristics of the red squirrel in an Arizona mixed-conifer forest. West J Appl For 1(2):48–51
Perkins MW, Conner LM, Howze MB (2008) The importance of hardwood trees in the longleaf pine forest ecosystem for Sherman's fox squirrels. For Ecol Manag 255(5–6):1618–1625
Pyare S, Longland WS (2001) Mechanisms of truffle detection by northern flying squirrels. Can J Zool 79:1007–1015
Ransome DB, Sullivan TP (1997) Food limitation and habitat preference of Glaucomys sabrinus and Tamiasciurus hudsonicus. J Mammal 78(2):538–549
Ripple WJ, Miller SD, Schoen JW, Rabinowitch SP (2019) Large carnivores under assault in Alaska. PLoS Biol 17(1):e3000090. https://doi.org/10.1371/journal.pbio.3000090
Robold RB, Huettmann F (2021) High-resolution prediction of American red squirrel in interior Alaska: a role model for conservation using open access data, machine learning, GIS and LIDAR. Peer J 9:e11830
Rusch DA, Reeder WG (1978) Population ecology of Alberta red squirrels. Ecology 59:400–420
Schoen J (2021) Tongass odyssey: seeing the Forest ecosystem through the politics of trees. University of Alaska Press, Fairbanks
Schuster R, Germain RR, Bennett JR, Reo NJ, Secord DL, Arcese P (2018) Biodiversity on Indigenous lands equals that in protected areas. BioRxiv:321935. https://doi.org/10.1101/321935
Schuster R, Hanson JO, Strimas-Mackey M, Bennett JR (2020) Exact integer linear programming solvers outperform simulated annealing for solving conservation planning problems. PeerJ 8:e9258. https://doi.org/10.7717/peerj.9258
Sitka (2022) Threats to the Tongass. Sitka Conservation Society. https://www.sitkawild.org/threats_to_the_tongass. Accessed 31 Aug 2022
Smith WP (2007) Ecology of Glaucomys sabrinus: habitat, demography, and community relations. J Mammal 88(4):862–881
Steiner M, Huettmann F (2021) Justification for a taxonomic conservation update of the rodent genus Tamiasciurus: addressing marginalization and mis-prioritization of research efforts and conservation laissez-faire for a sustainability outlook. Eu Zoologic J 88(1):86–116
UNEP-WCMC and ICCA Consortium (2021) A global spatial analysis of the estimated extent of territories and areas conserved by Indigenous peoples and local communities, Territories of Life: 2021 Report. UNEP-WCMC (Cambridge, UK) and ICCA Consortium (worldwide). Accessible via https://report.territoriesoflife.org/wp-content/uploads/2021/05/ICCA-Territories-of-Life-2021-Report-GLOBAL-ENG.pdf
Veraverbeke S, Rogers BM, Goulden ML, Jandt RR, Miller CE, Wiggins EB, Randerson JT (2017) Lightning as a major driver of recent large fire years in North American boreal forests. Nat Clim Chang 7(7):529–534
Wauters LA, Tosi G, Gurnell J (2002) Interspecific competition in tree squirrels: do introduced grey squirrels (Sciurus carolinensis) deplete tree seeds hoarded by red squirrels (S. vulgaris)? Behav Ecol Sociobiol 51(4):360–367
Wurtz TL, Ott RA, Maisch JC (2006) In: Chapin FS III, Oswood MW, Van Cleve K, Viereck LA, Verbyla DL (eds) Timber harvest in interior Alaska. Alaska's Changing Boreal Forest, pp 302–308
Young B, Yarie J, Verbyla D, Huettmann F, Herrick K, Chapin FS (2017) Modeling and mapping forest diversity in the boreal forest of interior Alaska. Landsc Ecol 32(2):397–413
Young BD, Yarie J, Verbyla D, Huettmann F, Chapin FS (2018) Mapping aboveground biomass of trees using forest inventory data and public environmental variables within the Alaskan boreal forest. In: Machine learning for ecology and sustainable natural resource management. Springer, Cham, pp 141–160
Zuckerberg B, Huettmann F, Frair J (2011) Proper data management as a scientific foundation for reliable species distribution modeling. In: Predictive species and habitat modeling in landscape ecology. Springer, New York, pp 45–70

Chapter 10
Squirrel Economics: A Global and National Cross-Scale Assessment of GDP vs Conservation Status in Regards to What Type of Human Economy the Squirrels Would Choose

Abstract We present a holistic perspective discussing how the squirrel conservation status is related to the world's economy. Using the Gross Domestic Product (GDP) on two scales, global and national (for the United States), we analyzed the socio-economic and environmental correlation with the IUCN conservation status and population trends of squirrels. This allows us to express the interactions between humans and the global environment with the aim to conserve ecosystems around the globe for sustainable life on earth overall. We base this assessment on presenting state-of-the-art GIS maps and squirrel species centers linked with the GDP distribution. We find that while GDP-rich countries often label endangered species as Endangered, GDP-poor countries mostly list them as Data Deficient, likely due to a lack of research resources. We present that the economy is globally connected – telecoupled – and we additionally provide an outlook on how the future will look like if the economy is continuing in this unsustainable booming and busting way and how that affects wildlife conservation embedded in the wider universal processes. Additional suggestions are included on how a more sustainable approach to the economy can change the future of the world's wildlife conservation. We also show in which projects the global efforts and financial resources should be invested, and demonstrate how landscape ecology principles support our suggestions. Conclusively, we propose other concepts, including a widely improved human-wildlife-habitat relationship allowing mankind and wildlife to flourish well side-by-side and together.

Keywords Economy · Society · GDP · IUCN · Economic growth · Wildlife conservation · Telecoupling

10.1 Introduction

Squirrels of the world and their wilderness are among the most marginalized species and habitat groups (Steiner and Huettmann 2021 and references within). The research effort dedicated to them is very limited, with few publications in high-impact journals, comparatively small research and management budgets for all the 300+ species, and a poor conservation status (both in terms of the overall percentage of threatened species, and the degree of actually properly assessed species), and widely ignored (Koprowski and Nandini 2008; Steiner and Huettmann 2021). Even most basic life history data are widely missing for the vast majority of squirrel species (approx. 302/307) (see Chaps. 1, 3, and Table 12.1 in Chap. 12 in Steiner and Huettmann 2023; see also Koprowski and Nandini 2008).

In a finite space and with limited resources on earth, arguably, squirrels rank rather low 'on the totem pole'. Squirrels are (wrongly) perceived as simply ubiquitous and abundant, or as irrelevant to mankind and get widely ignored in policy and global agendas (Marshall et al. 2020; Roth and Thorington 1982). This is also shown by the fact that they are used as target practice by teenagers and adults in the hunting sector (see Steiner and Huettmann 2023 – Chaps. 5 and 11 for details).

At best, squirrels rank as part of biodiversity, or mammals. But the research policies that mention squirrels are few (Koprowski and Nandini 2008). We only know of five squirrel species that really receive any relevant attention (Eastern Gray Squirrel (*Sciurus carolinensis*), Eurasian Red Squirrel (*S. vulgaris*), North American Red Squirrel (*Tamiasciurus hudsonicus*), Fox squirrel (*S. niger*), and Eastern Chipmunk (*Tamias striatus*). Instead, and like tried for amphibians, one may think of a global conservation listing for the entire species group (Family). This effort is however lacking for squirrels, they remain widely ignored and a *laissez-faire* approach remains the *status quo* (Steiner and Huettmann 2021).

While squirrels are not given the attention they deserve, other aspects apparently more dear to the heart of modern humans – money and its economy – are promoted beyond reason (=greed) and that proportion receives consistently a boost, inflation,

Supplementary Information The online version contains supplementary material available at https://doi.org/10.1007/978-3-031-23547-4_10.

M. Steiner, F. Huettmann, *Sustainable Squirrel Conservation*, https://doi.org/10.1007/978-3-031-23547-4_10

and subsidy (see Gaffney 2007). The recent budget and its 'innovation' spent by billionaires flying to Mars and into Space is a typical example of many (e.g. budgeted and spent on the military, oil business, etc.). Another example includes politics and voting campaigns, how often is it that politicians attempt to gain votes by promoting their GDP boosting plans? Your answer is likely, "Pretty much all the time". On the contrary, how often are nature or squirrel conservation plans included in such campaigns? Here your answer is probably "Never". This is a culture to be changed, because who can eat, breath, or drink money? We rather have to take care of nature and its resources, including squirrels, so we can continue breathing fresh air, drinking fresh spring water, and admire squirrels jumping around in the trees. In this study, we perform a first assessment on how economy is overall linked with squirrels, and investigate how different nations approach this topic.

Here we assess whether and how this one-sided approach to the economy has not done much good and how it affected squirrels. There are only a few such studies ever undertaken (see for instance Resendiz-Infante and Huettmann 2015; Huettmann 2015a, b for conservation), and to our knowledge none on squirrels. And so here we attempt such a first assessment and we try to provide baseline data and approaches across scales.

10.2 Methods

Here we use GDP as a predictor for the global squirrel conservation statuses. As a justification for our approach, in the Anthropocene, the GDP and its behavior, as fueled by economic growth, presents a clearly accepted driver for wildlife conservation, human well-being, and Mother Earth (Daly and Farley 2011; Rich 2013). It is often forgotten and left out of the discussion, that without nature, a basic agricultural production, the economy cannot remain stable nor increase (see the trophic theory of money (Czech 2019)). In order to globally sustain the economy of a growing human population, agriculture and the close interaction with nature are of core importance (see ecological economics Common and Stagl 2005; Daly and Farley 2011). That also affects the squirrels of the world. It is even more unfortunate to see that not even the basic and "origin" of the economy appreciates squirrels, where squirrels and generally rodents are eradicated on big scales as they are considered "*pests*", "*harmful for the business*", or "*unprofitable*" (see mass eradication of squirrels and other rodents in agriculture Palmer et al. 2007; Singleton and Petch 1994). Here we assess this relationship at two wider policy scales: Global and National (U.S.).

10.2.1 Global Analysis

We obtained GDP data of the year 2020 for all nations of the world from the United Nations (UN) using public information (https://data.un.org/Data; https://www.imf.org/en/Data; see data presentations in Wikipedia (2022a)). We then used the IUCN Red List species database and plotted it in OpenSource (QGIS) and ESRI GIS (ArcGIS). This is the best-available data and allowed us to select the nations that were in the range of squirrels, as well as squirrel species that are listed as Endangered and Critically Endangered (see Appendix 10.3 for the complete list; reprint from Steiner and Huettmann 2023 – Chap. 12 – Table 12.1).

We then summarized the GDP for those nations with IUCN Red-Listed Squirrels (classified as "Endangered" and "Critically Endangered") and compared them with other nations of the world within the squirrel range (Fig. 10.1). Here the green-red coloration gradient indicates the number of endangered species within that nation (green = low, red = high). Every black dot indicates the centered occurrence of a squirrel species (see details in Chapter 1 - Steiner and Huettmann 2023).

10.2.2 National Analysis (U.S.)

We obtained GDP data for the US and its States from the United Nations UN and online using public information (see Bea (2022), for a presentation see also Wikipedia (2022b)). We then used again the IUCN Red list species database and plotted it in OpenSource QGIS and ESRI GIS (ArcGIS). This allowed us to select the states that were in the range of squirrels, as well as squirrel species that are listed as Endangered and Critically Endangered (see Appendix 10.3 for list) (Fig. 10.2). Here the green-red coloration gradient indicates the number of endangered species within that state (green = low, red = high). Every black dot indicates the centered occurrence of a squirrel species (see details in Chapter 1 - Steiner and Huettmann 2023).

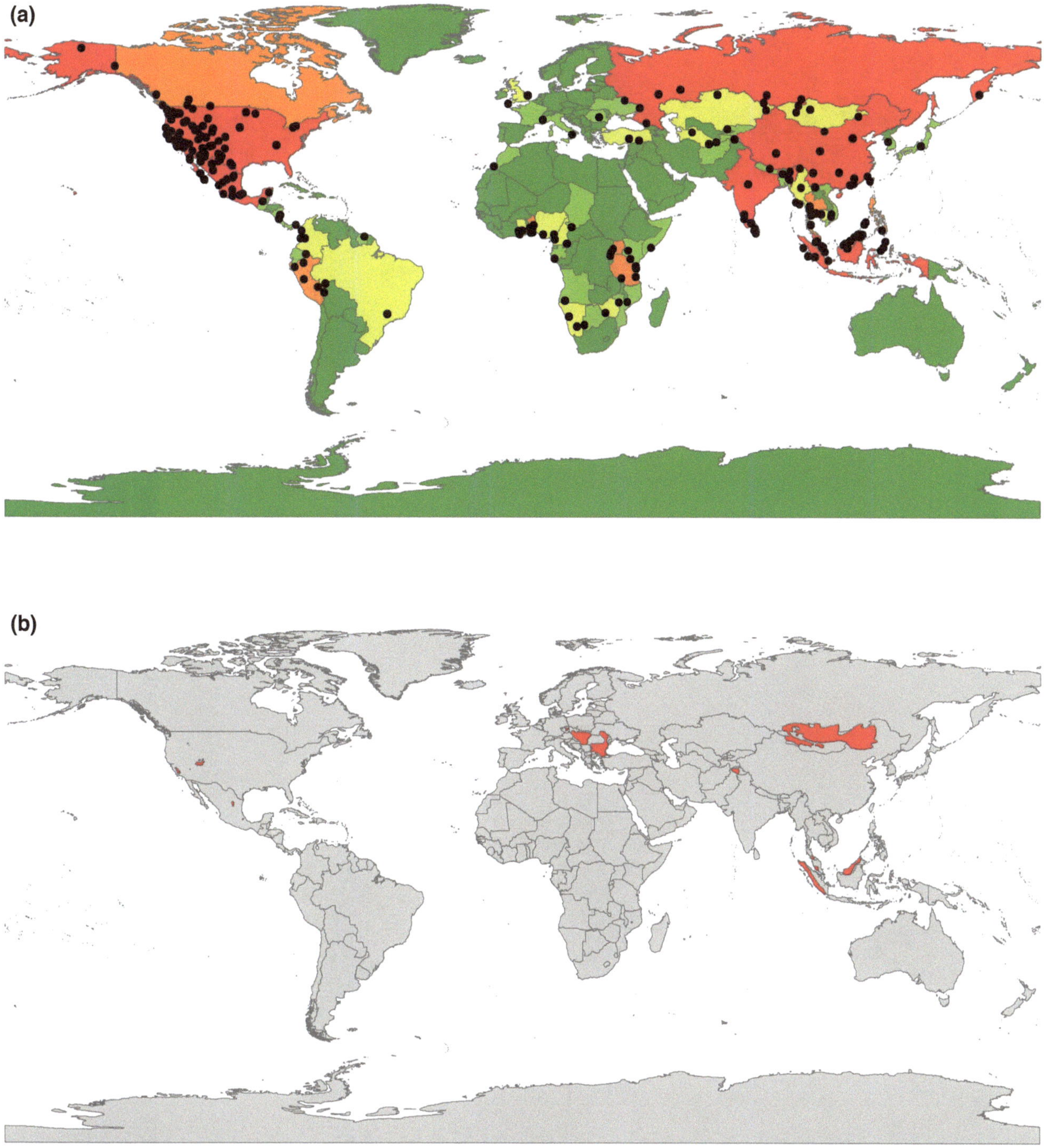

Fig. 10.1 Global map of (**a**) approximate centered squirrel occurrences with (green = low number of endangered species, red = high number of endangered species), and (**b**) known IUCN Red-listed squirrel species ranges (in red)

10.3 Results

10.3.1 Global Analysis

The global GDP consists currently of approx. $ 100+ trillion (https://steadystate.org/), growing by more than 1,25% annually. From that, the nations that have squirrels in their range have a combined GDP of approx. $ 95 trillion. The GDP is not equally distributed throughout the world, and the majority of nations and their citizens show a relatively low and poor GPD (Fig. 10.3).

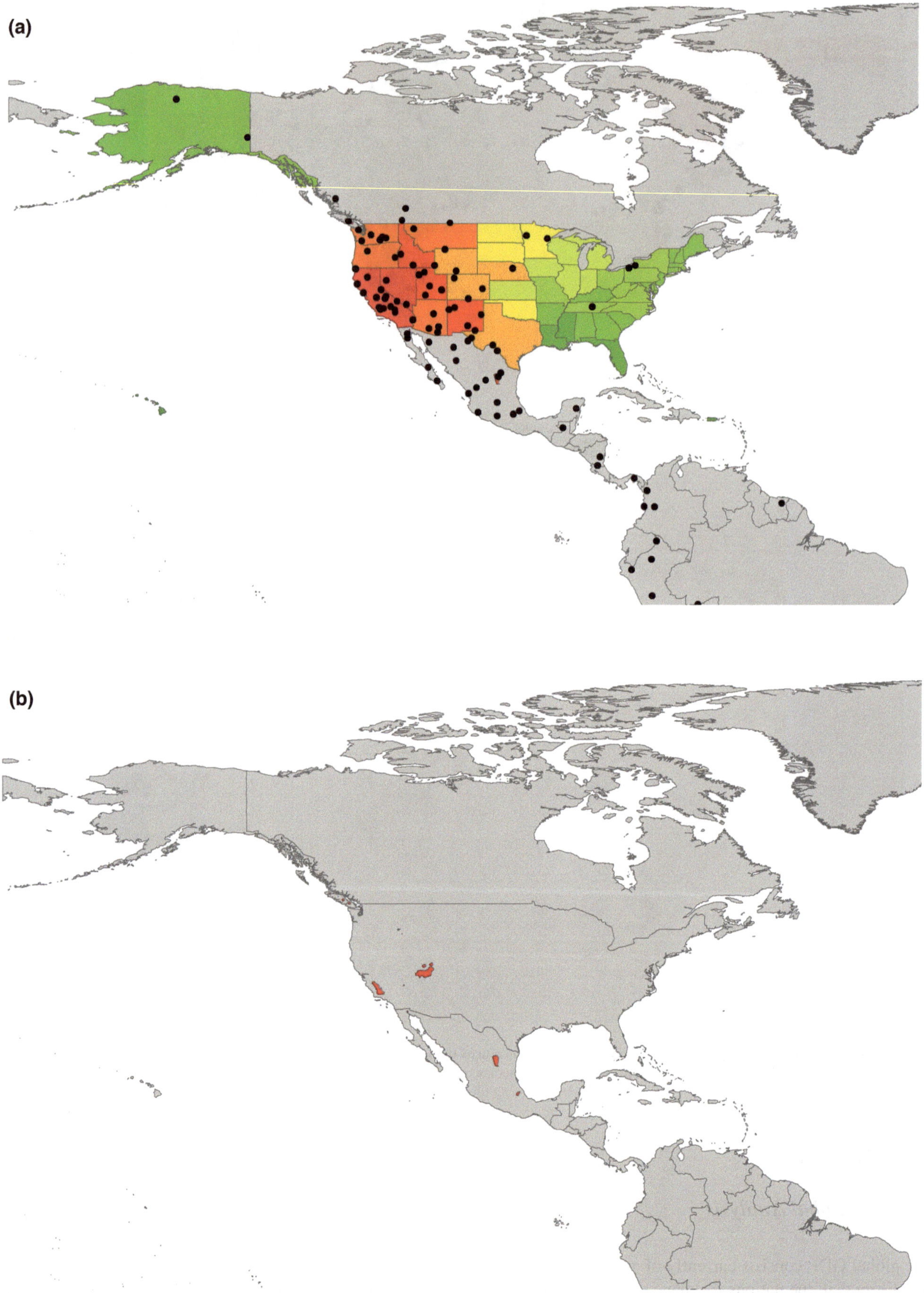

Fig. 10.2 National map for the U.S. with (**a**) approximate centered squirrel occurrences with (green = low number of endangered species, red = high number of endangered species), and (**b**) IUCN Red-listed squirrel ranges (in red)

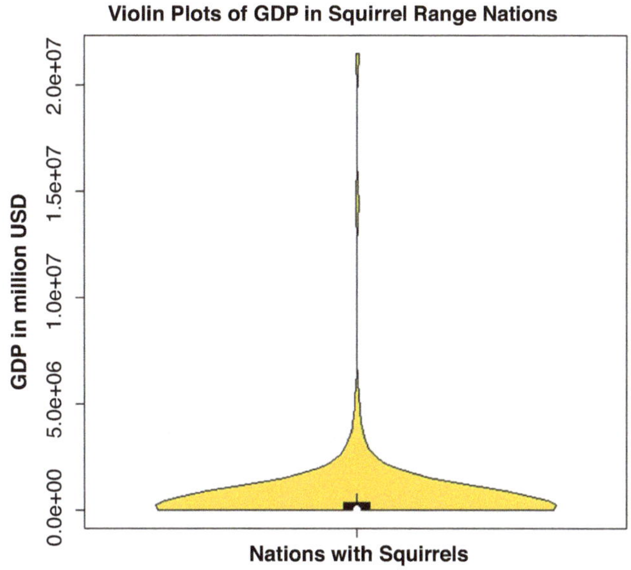

Fig. 10.3 Violin plot of the World's National GDPs where squirrels occur

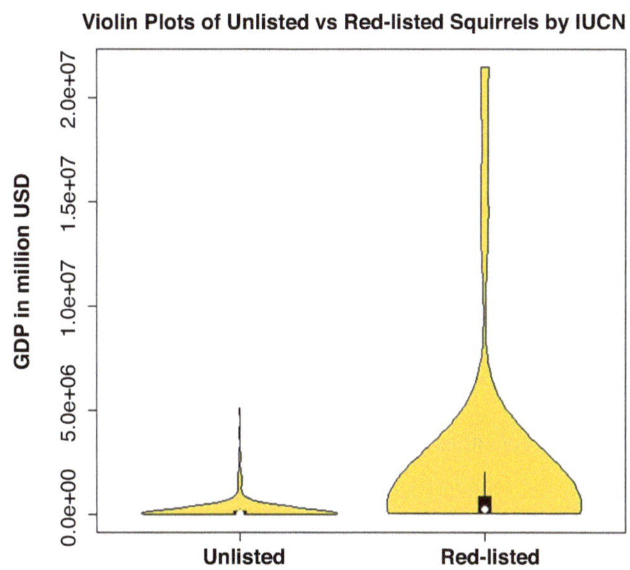

Fig. 10.4 Violin plot of the World's GDP for nations with and without IUCN red-listed squirrels

As it can be seen in Fig. 10.4, our data show that endangered red-listed squirrel species link with the GDP. Only a few nations have a relatively high GDP, and those are nations where squirrels are red-listed. The GDP is based on the consumption of goods and services, driven by industrial activities with oil, gas, and mining exploration at its core. Our data from the global analysis show that on a finite planet with finite resources a big GDP is in direct contradiction to a rather concerning squirrel conservation status.

By observing these figures, and generally comparing GDP and the endemic squirrel species inhabiting the corresponding countries, one can observe, that GDP-rich countries often have a high number of species classified as Endangered and Critically Endangered, whereas GDP-poor countries have a much higher percentage of Data Deficient rather than Endangered species, likely due to the lack of avaialable research resources (see Chap. 1 Steiner and Huettmann (2023) for details on Data Deficiency and its corresponding conservation status class).

10.3.2 National Analysis (U.S.)

The U.S. has one of the globally highest GDPs, approx. $ 25 trillion; and virtually all 50 states of the U.S. are inhabited by squirrels, except Hawaii. The GDP is not equally distributed within the nation (Fig. 10.5), and many U.S. states actually have a larger GDP than other individual nations (e.g. California has a larger GDP than France, the United Kingdom, and even India (as of 2021; that is even more so at the per-capita basis)). The U.S. GDP is growing by c. 2.3% annually. The states that have IUCN red-listed squirrels in the U.S. are California (GDP of $3,290,170 in Mio$), Idaho (GDP of $92,300 in Mio$), Nevada (GDP of $187,394 in Mio$), and Utah (GDP of $212,855 in Mio$) as of 2020. All of those have a larger GDP than many other U.S. States and nations even. Already the State of California itself ranks among the top 15 business entities in the world carrying approx. 1/4th of the overall US GDP. In Figure 10.6 we can observe that the GDP of states with red-listed species is significantly higher than the GDP of states with no red-listed soecies.

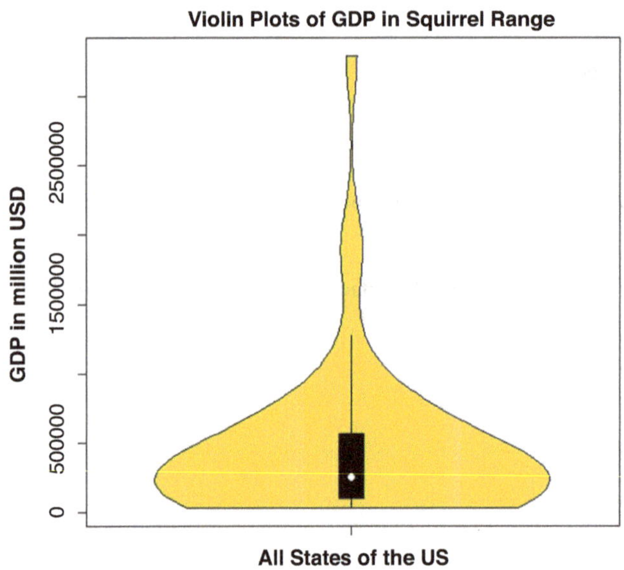

Fig. 10.5 Violin plot of the U.S. States and their GDP distribution

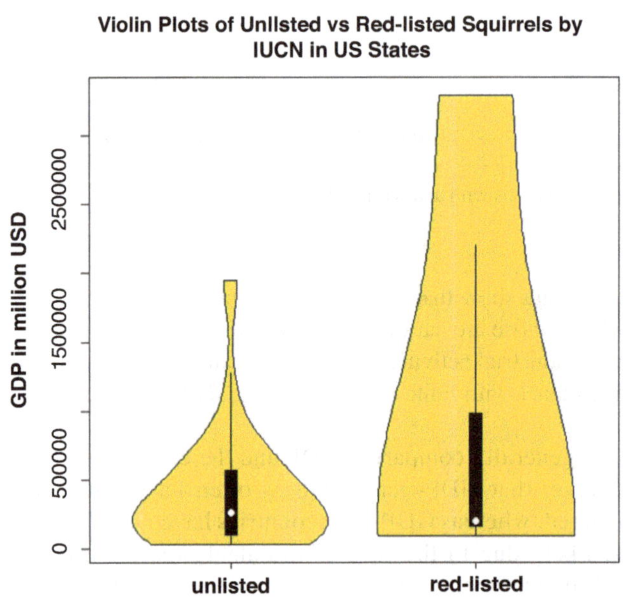

Fig. 10.6 Violin plot of the U.S. States GDP for States with and without IUCN red-listed squirrels

10.4 Discussion

There are over 300 species of squirrels (Burgin et al. 2020), and they are widely distributed in the world (see Steiner and Huettmann 2023 – Fig. 3.3, Chap. 3). Like any other living species now in the Anthropocene, squirrels are affected by humans and their activities, including the economy ("it's the economy, stupid"). It is widely recognized that we all live in the Anthropocene by now, where virtually all relevant processes in the world are directly ruled and driven by human society. While squirrels are no exception, virtually no research publication focuses on it, or any of the mandated and trusted conservation agencies, NGOs included.

The GDP is an international and cultural metric that is widely boosted, worldwide. It makes for an unprecedented approach in the global history with agencies in charge! It can be correlated with the status of wildlife, (e.g. Czech 2000, 2002, 2014), and it shows that much of the wilderness and wildlife species are in direct conflict with a growing GDP (Czech and Krausman 2001; see also Huettmann (2015a) for the ocean and islands, and Resendiz-Infante and Huettmann (2015) for parrots).

Here we showed on a global level, as well as on a national state level, using GIS and open-access data, how GDP correlates with squirrels and their conservation status, as judged by the IUCN Red List conservation status classes.

We find that squirrels are usually not red-listed in most of the low GDP nations and states. These are areas where 'modernity' is less intense. Whereas the red-listed species are found primarily in nations and states with a high or the highest GDPs and/or growth rates in the world; California and Mongolia stand out as good examples; both of them are fueled by industry, mining, and growth sectors. The status of squirrels somewhat reflects that with the highest numbers of endangered species there. Nations, governance, and agencies in charge of GDP and boosting it must recognize that!

However, it is noteworthy that the GDP-rich countries approach and express conservation assessments differently and find that their local species are Endangered. In contrast to low-GDP countries, which have their local Endangered species predominately labeled as Data Deficient (see details in Steiner and Huettmann 2023 – Chap. 1). To explain this conservation schism in greater detail, according to the IUCN Red List (2022), 1 out of the 34 Data deficient species (= 3%), and 5 out of the 17 Endangered and Critically Endangered species occur in Western countries (= almost 30%) (information retrieved: 22 Feb 2022). By looking at these numbers, there is a clear pattern visible, western countries seem to have a ten-fold higher likelihood to be inhabited by endangered species than data deficient species. Arguably, this situation is a man-made cultural feature, a labeling trick that comes with a profile favoring the ignorance of serious industrial impacts and 'greed'. From a conservation point of view, neither of those statuses is 'good', and all require increased efforts and resources to secure the species from becoming extinct.

It should be mentioned here that industrialization cannot be disconnected – divorced or uncoupled – from economic growth and that it tends to fuel man-made climate change, e.g. via CO_2 release and consumption of products and goods, across borders and globally (Brenner 2019; Cherniwchan 2012; Giddens 1994; Gregory 1977). With that, the Kuznets curve is not a realistic function, the opposite patterns tend to apply (see Huettmann 2015a; Resendiz-Infante and Huettmann 2015; Stern 2017).

The world and its economy are widely and deeply connected across scales. The world is telecoupled, and spill-over effects can be seen in the GDP and beyond (e.g. Liu et al. 2018). For a more in-depth study and thorough conservation in the public eye, this should be taken better and more directly into account.

Our study is relatively coarse but shows already clear patterns and trends across scales and locations, many of them equal to what has been shown in the literature earlier and for other species (Huettmann 2015a, b; Resendiz-Infante and Huettmann 2015), following basic ecological principles and the underlying laws of thermodynamics (Jorgensen 2018; Ruth 2013).

When compared to a few very rich nations, in our assessment we find a globally dominating low GDP in nations, indicating little true progress during the last 400 years of western industrialization (Rich 2013); a new form of poverty still dominates and the gap between rich and poor is actually known to be still rising (Flores 2017; Tritch 2006). Relevant progress brought by the modern approach is widely missing there, while essentially making things worse.

The notion of climate change is not directly addressed yet in this study. But arguably, a large GDP reflects human-driven climate change, and its change will put major pressure on the GDP with even the rich nations carrying a large debt. The subsequent outlook for national governance – business as usual – remains gloomy at best (Kompas et al. 2018).

'The economy' has many factors and remains very complex; many schools of thought contribute to and shape it, but not all are good and sustainable or provide for a better society or environment (Bennis and O'Toole 2005; Cheit 1985; Gaffney 2007; Hirsch Jr 2010; Lavy et al. 2012; Rich 2013). Clearly, a one-sided promotion of economic growth will not be beneficial to humans and living beings alike, as shown here with IUCN red-listed squirrels. Many more species and their ecosystems will likely show the same pattern, and face the same issues (now and even more so in the future).

10.5 Conclusion

The impacted conservation status of squirrels and their degree of being Endangered is positively connected with the size of the GDP, globally and for the U.S. states (as an example of a high GDP governance entity). The GDP is a constructed but rather powerful human-made metric to express the state of conservation, certainly for squirrels. We suggest that the GDP as a national goal and metric needs to be changed to better serve squirrels, ecological services, conservation, humans, and the global biotic world alike. Namely, in the current global crisis a different live, culture, and subsequent distribution of wealth and consumption pattern is needed on a finite planet.

Acknowledgments We acknowledge the data delivery for carrying out this study. FH appreciates the help of H. Berrios and E. Huettmann as well as the grand Chrome team. This is EWHALE lab publication # 260.

References

Bea (2022) GDP by State. https://www.bea.gov/data/gdp/gdp-state. Accessed 19 July 2022
Bennis WG, O'Toole J (2005) How business schools have lost their way. Harv Bus Rev 83(5):96–104
Brenner MH (2019) Industrialization and economic growth: estimates of their effects on the health of populations. In: Assessing the contributions of the social sciences to health. Routledge, pp 65–115
Burgin CJ, Wilson DE, Mittermeier RA, Rylands AB, Lacher TE, Sechrest W (eds) (2020) Illustrated checklist of the mammals of the world. Lynx
Cheit EF (1985) Business schools and their critics. Calif Manag Rev 27(3):43–62
Cherniwchan J (2012) Economic growth, industrialization, and the environment. Resour Energy Econ 34(4):442–467
Common M, Stagl S (2005) Ecological economics: an introduction. Cambridge University Press
Czech B (2000) Economic growth as the limiting factor for wildlife conservation. Wildl Soc Bull 28:4–14
Czech B (2002) Shoveling fuel for a runaway train: errant economists, shameful spenders, and a plan to stop them all. Island Press
Czech B (2014) The conflict between economic growth and wildlife conservation. In: Peak oil, economic growth, and wildlife conservation. Springer, New York, pp 99–117
Czech B (2019) The trophic theory of money: principles, corollaries, and policy implications. J Proc R Soc NSW 152(471/472):66–81
Czech B, Krausman PR (2001) The endangered species act: history, conservation biology, and public policy. JHU Press
Daly HE, Farley J (2011) Ecological economics: principles and applications. Island Press
Flores RL (2017) The rising gap between rich and poor: a look at the persistence of educational disparities in the United States and why we should worry. Cogent Soc Sci 3(1):1323698
Gaffney M (2007) The corruption of economics, Georgist Paradigm series. Shepheard-Walwyn Publishers, Chicago
Giddens A (1994) Industrialization, ecology, and the development of life politics. In: Ecology, society and the quality of social life. Transactions, New Brunswick, pp 1–10
Gregory PR (1977) Russian industrialization and economic growth: results and perspectives of western research. Jahrb Gesch Osteur H. 2:200–218
Hirsch ED Jr (2010) The schools we need: and why we don't have them. Anchor
Huettmann F (2015a) Reversed Kuznets curves in the Caribbean: linking marine biodiversity endangerment with the National Gross Domestic Product (GDP) and assessed with human metrics (life expectancy, infant mortality, population growth rate). In: Huettmann F (ed) Central American biodiversity conservation, ecology, and a sustainable future. Springer, New York, pp 105–134
Huettmann F (2015b) An oceanography conservation view of Central America: Caribbean wealth, Pacific wilderness, plunder and mismanagement meets seafloor mining, Deep-Sea drilling, climate change, and human population explosion in the EEZs and beyond. In: Huettmann F (ed) Central American biodiversity conservation, ecology, and a sustainable future. Springer, New York, pp 135–158
IUCN Red List (2022) Family sciuridae. https://www.iucnredlist.org/search?taxonomies=101297&searchType=species. Accessed 19 July 2022
Jorgensen SE (2018) Thermodynamics and ecological modelling. CRC Press
Kompas T, Pham VH, Che TN (2018) The effects of climate change on GDP by country and the global economic gains from complying with the Paris climate accord. Earth's Future 6(8):1153–1173
Koprowski JL, Nandini R (2008) Global hotspots and knowledge gaps for tree and flying squirrels, vol 95. Curr Sci, pp 851–856
Lavy V, Silva O, Weinhardt F (2012) The good, the bad, and the average: evidence on ability peer effects in schools. J Labor Econ 30(2):367–414
Liu J, Dou Y, Batistella M, Challies E, Connor T, Friis C, Sun J et al (2018) Spillover systems in a telecoupled Anthropocene: typology, methods, and governance for global sustainability. Curr Opin Environ Sustain 33:58–69
Marshall AJ, Meijaard E, Leighton M (2020) Extreme ecological specialization in a rainforest mammal, the Bornean tufted ground squirrel, Rheithrosciurus macrotis. bioRxiv. https://doi.org/10.1101/2020.08.03.233999
Palmer GH, Koprowski J, Pernas T (2007) Tree squirrels as invasive species: conservation and management implications. In: Witmer GW, Pitt WC, Fagerstone KA (eds) Managing vertebrate invasive species: proceeding of an international symposium, Fagerstone, USDA/APHIS/WS. National Wildlife Research Center, Fort Collins
Resendiz-Infante C, Huettmann F (2015) Bird conservation status and meaningful socioeconomic correlates in Central America: results from an open access data-mining approach for parrots using machine learning indicate serious economic problems. In: Central American biodiversity. Springer, New York, pp 77–104
Rich B (2013) Mortaging the earth: the World Bank, environmental impoverishment, and the crisis of development, 2nd edn. Island Press
Roth VL, Thorington RW (1982) Relative brain size among African squirrels. J Mammal 63(1):168–173

Ruth M (2013) Integrating economics, ecology and thermodynamics, vol 3. Springer Science & Business Media

Singleton GR, Petch DA (1994) A review of the biology and management of rodent pests in Southeast Asia. Australian Centre for International Agricultural Research

Steiner M, Huettmann F (2021) Justification for a taxonomic conservation update of the rodent genus Tamiasciurus: addressing marginalization and mis-prioritization of research efforts and conservation laissez-faire for a sustainability outlook. Eur Zool J 88(1):86–116

Stern DI (2017) The environmental Kuznets curve. In: Oxford research encyclopedia of environmental science. Oxford University Press, Oxford

Tritch T (2006) The rise of the super-rich. The New York Times, 19

Wikipedia (2022a) List of countries by GDP (nominal). https://en.wikipedia.org/wiki/List_of_countries_by_GDP_(nominal). Accessed 19 July 2022

Wikipedia (2022b) List of U.S. states and territories by GDP. https://en.wikipedia.org/wiki/List_of_U.S._states_and_territories_by_GDP. Accessed 19 July 2022

Part III
Problems and Governance in the Squirrel World

Chapter 11
The Global Squirrel Hunting Status and Its Marginalized Governance and Law Enforcement

I have already started the Squirrel Wars for the season. I'll consider it even when the pile of squirrels equals the amount of damage they have been doing. Haven't had a squirrel war for a few years, time to thin the herd!

Source: Anonymous Online

Abstract For millennia squirrels are hunted, eaten, poached, traded for fur, killed as shooting targets, and trapped all around the world; they are an inherent part of bushmeat (Akinnifesi and Veloso 2014; Djagoun et al. 2018; Ripple et al. 2016; Wolfe et al. 2005). And unfortunately, almost everywhere it is unmanaged and far away from being harvested sustainably (Steiner and Huettmann 2021; see multiple chapters in Steiner and Huettmann 2023). Even the most basic demographic metrics for squirrels – as needed for conservation management in the public eye – are not well researched, e.g. annual harvest rates, reproductive rates, life expectancy, fecundity, etc. While they might be generically assumed, local estimates across meaningful scales are virtually left unresearched, carry no relevant confidence intervals, or remain totally unknown. Often even the taxonomic species status is also widely unclear and in great flux (Steiner and Huettmann 2021, 2023 - Chapter 1), leaving the essential question unresolved of what actually is managed? In this study, we are taking a closer look at the global squirrel hunting, poaching, fur-bearing, and trapping status. This is done by starting a global interview (inquiry done through written mailing letters) and subsequent email follow-ups to ask global governmental authorities on issues of (1) Assigned budget for wildlife conservation actions in each mandated area for all squirrel species present; (2) Laws and regulations regarding squirrel hunting, fur-bearing, and trapping; (3) Reports of known and documented poaching activities; and (4) Whether the current squirrel management is aligned with the 17 UN Sustainable Development Goals (SDGs)? The aim of this study consists of compiling for the first time, for squirrels, a consistent management status database, gathering the above-mentioned information that is usually not known or provided for the public nor well analyzed for most species. This study will help to fill that embarassing gap in a constructive fashion: it allows for the first time to start a synthesis on squirrel hunting, fur-bearing, and trapping regulations on a global scale. The responses to these mailings are analyzed with state-of-the-art data mining techniques. They are to determine to which extent the mailed information corresponds to the one published online, and even if responses are given (response rates were rather small, app. 44%). This is then discussed here to illustrate the importance of data transparency and repeatability for wildlife conservation, survival, the interactions between humans and animals, and the general well-being of nature (including resource extraction and associated climate change politics). It all seems to be interlinked on a global scale (*sensu* telecoupling). This study aims to set an example of the importance of data transparency with which we would like to invite the mandated governments and agencies to follow this approach online to enlighten the broad population with the necessary information to guarantee future sustainable conservation management for squirrels and other species.

Keywords Squirrels · Hunting · Governance · Intransparency · Wildlife · Data mining · Telecoupling · Letter survey

11.1 Introduction

In most parts of the world where squirrels occur, they are traditionally either hunted, poached, traded for fur, or trapped for various reasons. Such reasons include gaining a financial benefit, provision of food, pest control, or acting as a leisure time activity for teenagers and young adults (mostly males; Alaska Department of Fish and Game (2021) – see Alaskan Trappers report for the largest wilderness in the US as the nation with the largest budget potentially available to achieve a squirrel

Supplementary Information The online version contains supplementary material available at https://doi.org/10.1007/978-3-031-23547-4_11.

M. Steiner, F. Huettmann, *Sustainable Squirrel Conservation*, https://doi.org/10.1007/978-3-031-23547-4_11

wildlife governance; Herring et al. 2021). Animals have been killed by humans in order to survive and be used as food, and clothing sources for easily more than 60,000 years (Cordain et al. 1998; Frassetto et al. 2009). And this remains the case until this very day for many remote countries and communities, despite national welfare systems in place turning some subsistence lives into perverted subsidies.

Traditional and initial subsistence has widely been abandoned in the modern world with legal control (management), where the hunter-gatherer lifestyle has little space in an urbanized world, and is even rejected and widely perceived as primitive, cruel, or barbaric. In Appendix 11.3, a short but representative story can be found which describes such a rural/tribal lifestyle in modern times narrated by a friend of the authors which grew up in a small town in Sierra Leone (West Africa) where this kind of hunter-gatherer lifestyle is still practiced with a focus on tropical bushmeat hunting (Larson et al. 2016). This story – as one example of many – also describes the importance of these hunter-gatherer individuals in the communities as a crucial part of the communal food provision. This is assumed to be part of mankind and its development for centuries, millennia, or even millions of years (Cordain et al. 1998; Frassetto et al. 2009). Assumingly, since it has always been practiced with steady pressure on the local squirrel (and other bushmeat) populations, the local human community must have had some kind of unwritten management for those species. Because if no sustainable management would have happened, those communities would not have been able to pursue this lifestyle for thousands/millions of years. Taboos in indigenous communities for instance help to maintain populations and not to overharvest them. However, suddenly, this mindset in human beings must have changed or been lost throughout the last few decades, greatly facilitated by industrialized tools, push markets, a one-sided globalization and the internet.

The onset of industrialization brought a massive and fundamental change. That is because this original mindset of survival, pursuing food, ideally in a re-occurring – sustainable – fashion, maintaining a balance with nature, and having taboos and respect, has essentially been completely lost in the modern Anthropocene, in a neoliberal world. It is now where most people are living and feel detached from nature (Kesebir and Kesebir 2017). Without this inherent respect, and regulation from people, the balanced relationship and awareness between humans and other species – wilderness – is widely gone (see the concept of "nature-deficit disorder"; Fletcher 2017; Sandry 2013).

This can also be observed in the co-evolved global squirrel species population and community setup. Without this natural sense of being somewhat sustainable – in balance with nature – and acting respectfully, the humans' actions have shown in the recent past a major failure in maintaining the natural balance between taking from nature and providing space and time to regenerate the taken/used natural resources (e.g. see Huettmann 2020; Taber and Payne 2003). In modern days, the rate at which we humans are taking from nature is highly unsustainable and by far out of any kind of sustainable balance (e.g. see the Club of Rome; Botkin et al. 2014). This can be observed in many examples worldwide and for projects like the Lake Owens (Wikipedia (2021), as part of many others, e.g. for forest habitat destructions like the ones that have been described by Eshetu (2014), Rudel and Roper (1997), Stearns (1997), and Williams (1992).

In addition to this relatively new unbalanced acting of an increasing number of humans and the dominant consuming industrial human societies and cultures in the earth's history, now virtually fully detached from nature, there are also no, or very few conservation measures in place by governments for small mammals such as squirrels. In order to better assess and quantify this global lack of conservation measures and its policy reinforcement, in this study, we are investigating the global squirrel hunting governance and its influence on squirrels.

Squirrels are one of the most hunted species worldwide and they are predominantly taken as bushmeat (Akinnifesi and Veloso 2014; Djagoun et al. 2018; Ripple et al. 2016; Wolfe et al. 2005). One assumes that the squirrel population is still conservationally regulated with vast insensitive underestimates and the number of animals that can be hunted. Whereas regulations on a species level are widely absent, as well as harvest numbers and estimates. This is especially assumed after understanding how many of the squirrel species are hunted (e.g. see harvesting reports from the U.S. state of Georgia, where over 800,000 squirrels have reportedly been harvested within just 2 years (The Georgia Department of Natural Resources Wildlife Resources Division (2021))) – likely a minimum estimate. However, this assumed conservation science-driven decision and policy making is unfortunately not in place, nowhere really (compare for instance with the elaborate quota setting procedure and annual public debate for moose, caribou, or deer). Squirrel governance is widely absent and ignored.

To further outline this discrepancy, for many squirrel species it is not known whether their population trend is increasing, stable, or decreasing even though the species are actively hunted and thus managed with a tax-paid public funding scheme underneath, e.g. often no assigned or transparent budget for squirrels is known (see survey responses for question 1 discussed in Chap. 13 – Steiner and Huettmann 2023; and regarding unknown population, trends see Chap. 1 – Steiner and Huettmann 2023). The issue with this is that, naturally, whenever a population is overhunted and diminished to a grade where it can no longer sustain itself, this species will likely decline and then head to extinction sooner or later. It is not a recommended management framework, nor best practices, and rather a negative spiral that has no check and benchmark, or any limit (except the point where the number of extant individuals has dropped too low, to 0). The fact that many species are actively hunted even though their conservation status, number of extant mature individuals, life history metrics, and population trends are

Fig. 11.1 Squirrel hunting practiced in the USA; how about bag limits? (Source: McBroom 2021)

Fig. 11.2 Squirrel/bushmeat hunting practiced in Central/Western Africa (Source: InfoCongo 2016)

completely unknown, raises the question of whether hunting is actually sustainable and respecting the natural boundaries of the global squirrel population and what they can sustain. We see no answer in the textbooks on those topics (e.g. see Silvy 2020).

In this study, we are focusing on sampling the current hunting situation of the global squirrel species, where we want to elaborate on the current law enforcement for squirrel hunting, poaching activities, bushmeat hunting, trapping, and how conservation success can still be achieved even though hunting is practiced. Additionally, we want to discuss the current issues with hunting squirrels, and how these issues influence the global squirrel population and their long-term conservation success.

In order to visualize hunting practices enforced on squirrels, Figs. 11.1, 11.2, 11.3 and 11.4 illustrate squirrel hunting/trapping practices in different countries and situations around the globe.

Fig. 11.3 Squirrel/bushmeat hunting practiced in Southwestern Asia (Source: Kimbrough 2013)

Fig. 11.4 An example of inhumane squirrel trapping and furbearing practices (the bloody tracks show an intense struggle and no immediate, but cruel death, e.g. a meaningful trap check is misssing) (Source: The Wildlife Whisperer 2021)

This study addresses several arguments originating from well-known but malfunctioning sustainable wildlife governance, in specific here, squirrel-centric governance in the Anthropocene.

Squirrels are known to be mesopredators, contributing significantly to the ecosystem, ecosystem services, fungi dispersal, soil properties, forest, landscape set up through seed dispersal, etc. (Callahan 1993; Steiner and Huettmann 2021 and references within; Swaisgood et al. 1999; Willson et al. 2003). They are ecosystem engineers. Thereupon, this study tries to assess how conservation, specifically squirrel conservation, is affected and limited by current governance and policies. Further, it inquires whether this governance and policy marginalize small mammals wholesale and how this affects environmental and hunting regulations, thus, population regulators, landscape setup, and ecological services. For the hunting situation assessment, it is determined in which nations squirrels are hunted and poached, to which extent, and to which extent data is available, as presented by mandated governments in charge of this public trust resource. Here we inquire about this through a mailing interview query to all state governmental entities in North America -U.S., Canada, and Mexico -, as well as for the governmental entities of many other countries that are inhabited by squirrels, e.g. Russia, Germany, Italy, Netherlands, UK, Malaysia, etc.).

11.2 Methods

We employed mail-in letter inquiries on a list of management details for squirrels in North America and around the world. This was done in the absence of easily accessible management information for squirrel species.

Like virtually all wildlife, squirrels are a public trust resource, in most jurisdictions, the public pays tax and in exchange has the right and need to have insight into the lawful relations of squirrel hunting, trapping, fur-bearing, etc. By publicly announcing the hunting regulations for squirrels and other small mammals, the citizens know which guidelines are to be followed. However, often such regulations are not so meaningful and are nowhere to be found online, most of the time they are non-existing. This is especially true for not-so-famous and charismatic species. On the contrary, Eastern grey squirrel (*Callosciurus carolinensis*), and North American red squirrel (*Tamiasciurus hudsonicus*) are mentioned in many hunting regulations forms and statements (see Table 11.4 below, and Steiner and Huettmann 2021), however, species with smaller population sizes and population ranges – the vast majority – are widely missing in the hunting/trapping report statements and conservation management. This is fatal for these less-prominent species since their populations are already smaller by nature, with no guidelines and laws to follow, those populations are unmanaged, ignored, and their future seems to be doomed. If any species needs such regulations, its rather those understudied ones with small populations rather than the over-studied omnipresent species. This is the reason for sending out the letter surveys to all countries/states/provinces that are inhabited by squirrels, to receive first-hand information and a baseline.

In order to assess first-hand legally valid information for this study, we sent out a total of 107 letters and 96 follow-up emails to the global governmental hunting authorities. Using a personalized physical and electronic letter survey where all USA states, Canadian provinces, and many other countries such as Italy, France, Nepal, Indonesia, Colombia, etc. have been individually contacted to utilize this first-hand state-of-the-art information. The physical letter inquiries have been sent out during the last week of 2021. The follow-up emails have been sent out to all the countries/states/provinces on January 25th that have not responded to the letter survey until that date, by sending an electronic response to MS. The physical letters for North America have been sent out by FH (based in Alaska, USA) and all the remaining letters have been sent out by MS (temporarily based in Italy). Also, all email follow-ups have been sent out by MS.

The mail-in letter contained the following four basic inquiry topics.

1. Assigned budget for wildlife conservation actions in your mandated area for all squirrel species present.
2. Lawful regulations in your mandate area regarding squirrel hunting, fur-bearing, and trapping.
3. Reports of known and documented poaching activities.
4. Is the current squirrel management aligned with the 17 UN Sustainable Development Goals (SDGs)? If yes, with which ones specifically, and how does the current management fulfill SDG 11 (Sustainable cities and communities), 12 (Responsible consumption and production), 13 (Climate action), and 15 (Life on land)?

In Appendix 11.2 the full letter of the survey can be found.

11.2.1 Literature Review on the Current Squirrel Hunting Regulations and Practices

In order to start with the first baseline of species conservation and management (also for hunting), here we conduct a literature review on the basic regulations in place for terrestrial small mammals. Such basic regulations include the review of who is responsible for the species, if they are endangered or not (= whether the state/nation, the country, or a certain agency is responsible for them). This jurisdictional review focused on identifying the legal correspondents and responsibility on different levels of the country for the inhabiting species and their assigned conservation classes.

Our review aims to identify the general laws that apply to squirrel management, hunting, and conservation. This means that we are summarizing who is legally responsible for squirrels, and subsequently, we compare it with what we actually observe in practice, and how squirrels are effectively treated. This literature review performed on the hunting regulations for all countries, states, and provinces is an update and extension of the previously performed work in Steiner and Huettmann (2021), where a similar work approach has been applied for the USA only and all its states for one genus (*Tamiasciurus*).

This literature review also aids in understanding and observing possible mismatches with the existing data provided online, and the data inquired first-hand from each country's hunting authority. Therefore, we aim to compare the results of this review, and the results from the letter survey, and analyze them for their match. Ideally, we assume the information provided online to the public and received by email/letter as a response to the letter survey should be approximately identical.

11.2.2 Study Design of the Mail Survey

In order to avoid all predictable biases, we have designed the letter survey in a way that all entities that are supposedly responsible for the squirrels' management and conservation are included in the survey. Also, we sent out both a physical and electronic letter (email) to all the included entities. By doing so, we aimed to avoid cases where the physical letters did not arrive in time, or at all. Also, it provides a backup in case the email did not arrive at the desired destination.

The questions in the survey have been reduced to the four questions mentioned in Sect. 11.2 as these have been identified by us to be the most important topics to ask the responsible authorities in order to successfully report on the squirrels' hunting regulations, their conservation, poaching, their assigned conservation budget, and their correlation with the UN-initiated 17 SDG's.

An overview of all the states, provinces, and countries to which the mailing interview has been sent can be found in Table 11.1. Table 11.2, later on, shows an overview of all squirrel species included here.

11.2.3 Analyzing the Survey Responses

Lastly, here we also analyze the responses to the surveys. This is done by compiling information on whether an entity replied and to which of the four questions from the survey answers have been provided (Table 11.3). Additionally, to gain a deeper understanding of the actual hunting regulations, Table 11.4 has been created.

The addresses for the best-available letter survey and the email addresses for the email survey as a backup can be found in Appendix 11.1.

11.3 Results

11.3.1 Literature Review on the General Laws in Place for Wildlife and Squirrels in the U.S. and Other Places around the Globe, Reviewing Legislative Levels and Law Enforcement on the Ground

On a global level, it can be observed that most wildlife laws, including the ones for squirrels, are country-specific. This actually means they are not coordinated and have no wider plan or vision. It is highly fragmented and means that each country has its own wildlife regulations in place for (ideally) all species inhabiting the corresponding country (for Europe see Schmeller et al. 2008 and references within, for the U.S. and the rest of the world see Croteau and Mott 2013). Neighboring

Table 11.1 Overview of all countries/states/provinces included in the letter survey (list was compiled based on nations/ provinces that are knowingly and heavily populated by squirrel species)

European countries	Non-European countries			United States				Canadian provinces
Austria	Benin	Indonesia	Rwanda	Alabama	Indiana	Nebraska	South Carolina	Alberta
Finland	Bhutan	Ivory Coast	South Korea	Alaska	Iowa	Nevada	South Dakota	British Columbia
France	Brazil	Japan	Sri Lanka	Arizona	Kansas	New Hampshire	Tennessee	Manitoba
Germany	Burundi	Kenya	Togo	Arkansas	Kentucky	New Jersey	Texas	New Brunswick
Italy	Cameroon	Malaysia	Uganda	California	Louisiana	New Mexico	Utah	Newfoundland and Labrador
Netherlands	China	Mexico		Colorado	Maine	New York	Vermont	Northern Territory
Norway	Columbia	Nepal		Connecticut	Maryland	North Carolina	Virginia	Nova Scotia
Poland	Costa Rica	Nicaragua		Delaware	Massachusetts	North Dakota	Washington	Nunavut
Spain	Ecuador	Nigeria		Florida	Michigan	Ohio	West Virginia	Ontario
Sweden	Ghana	Panama		Georgia	Minnesota	Oklahoma	Wisconsin	Prince Edward Island
Switzerland	Guam	Peru		Hawaii	Mississippi	Oregon	Wyoming	Quebec
Turkey	Guatemala	Puerto Rico		Idaho	Missouri	Pennsylvania		Saskatchewan
United Kingdom	India	Russia		Illinois	Montana	Rhode Island		Yukon

states and countries hardly ever communicate with each other and mutually agree on wildlife laws, regulations, and especially law enforcement. The Federal Model dominates, where federal agencies and laws interact with local state laws and entities. While this Federal Model is seen as a leading and very dominant democratic framework, arguably it is currently not well implemented for squirrels and does not really serve them well, or with a meaningful sustainable business model. Unfortunately though, rarely all species inhabiting a single country or state are also lawfully regulated and organized (Qian et al. 2018). The reason for this is that some species for which no wider public interest exists are mostly marginalized and left out of wildlife legislations and thus, law enforcement (Koprowski and Nandini 2008; Ramos-Lara and Koprowski 2014). Squirrels have no lobby to put them onto the agenda, fairly. For squirrels, this general trend can also be widely observed (Steiner and Huettmann 2021 and references within). This is especially true for countries with very high species richness (e.g. see Malaysia, Philippines, Indonesia, United States, India, and Mexico). To provide evidence and back this up, please see Steiner and Huettmann (2021), and specifically Table 11.2 below where these countries and their reported managed squirrel species are presented.

By looking at the U.S. only, – a major landholder in North America and holding most squirrel species and management power – one can observe a three-level law enforcement system. These laws can be observed on a federal, state, and local level (Mahoney and Geist 2019; Press et al. 1996). Many hunting regulations are developed and enforced on a state level, which is occasionally subdivided into local areas and also interfering with Federal laws, such as the Endangered Species Act, and other laws and policies, e.g. National Park policies, air quality laws, and infrastructure bills (see Code of Federal Regulations 2021; Croteau and Mott 2013; National Park Service 2021; Press et al. 1996). For species that are considered endangered on a federal level, conservation actions are also supposed to be enforced federally (see Croteau and Mott 2013; and references above). Unfortunately, law enforcement is not linked and reported to a complete extent to the local entities and published online (Schmeller et al. 2008 and references within). Already getting a complete picture of those details remains very difficult to compile, virtually impossible for all squirrel species in the US.

11.3.2 Analysis of the Received Letter Responses with a Focus on Hunting & Trapping/ Fur-Bearing Regulations

A total of 107 letter surveys have been sent out by the authors, of which 11 entities replied via email within the first 30 days. 30 days after sending the letters, 96 follow-up emails have been sent by MS to all entities that have not replied yet. From these follow-up emails, 23 replied within another 30 days. Ten replied after the cut-off date which has been indicated by using red font color in Tables 11.3 and 11.4.

Overall, and if taken all together, the majority of our questions are left unanswered. It can be noted that approximately only half of the entities in the U.S. have replied (32/50–64%), and slightly more European entities have replied (8/13–67%), leaving a vast amount without a reply. Canadian states/ provinces have just replied partially (5/13–42%), and countries outside of Europe and North America have barely replied at all (2/31–6%). Again, this is essentially a public request from citizens with a research interest.

Looking at the answers to the individual questions, assigned budgets are present to the largest extent in Europe with 2/7 (29%) out of the ones that responded. Hunting/trapping/fur-bearing regulations have been provided to the largest extent by the U.S. with 28/32 (87.5%), and to the lowest extent by the countries outside of Europe and North America with 0/31 (0%).

Table 11.2 Summary of squirrel species included in management plans in some countries with the highest squirrel species richness

Country	Number of species mentioned in legislations	Number of species occurring	Overall percentage of inclusion	Source
Malaysia	4	43	9.3%	CITES (2014)
Phillipines	1–2	13	7–15	Legacy.senate.gov (2010) and US AID (2021)
Indonesia	6 species and 3 genera	54	12- approx. 30%	Maryanto et al. (2019) and US AID Indonesia (2004)
United States	No numbers on a federal level	69	Lost likely < 50% see results of letter survey	
India	Approx. 1-2	31	3–6%	No public documents available for the whole country – at least in English language
Mexico	Approx. 3	33	9%	IUCN Small Mammal Specialist Group (SMSG) (2021)

Table 11.3 Response table squirrel survey

Countries	Responded	Assigned budget provided	Lawful Regulations	Poaching known	SDG based manageme nt	Comments	Email follow-up sent
Austria	Yes	Yes	Yes	No known poaching activities.	Yes	Survey squirrels with drones	Yes
Finland	Yes	No	Yes	No reports available.	Not directly		Yes
France	No						Yes
Germany	Yes	No	Partly	No details provided	No	Very vague response, no specific responses to the questions asked	Yes
Italy	No						Yes
Netherlands	Yes	No	Yes	No indications of significant poaching levels,	No		Yes
Norway	Yes	No	Yes	No poaching activities known.	Yes		Yes
Poland	No						Yes
Spain	Yes	No Budget available	Yes	Poaching is Not a conservatio n issue for red squirrel populations in Spain.	Yes	Very professional response	No
Sweden	Yes					Forwarded to https://www.naturvardsverket.se/en/	Yes
Switzerland	Yes	Yes	Yes	Very rarely reported.	Not directly	Phone call recorded with MS	Yes
Turkey	No						Yes
United Kingdom	No						Yes
Yes count ratio	**(8/12) 0.67**	**0.13**	**0.47**	**0**	**0.2**		**0.93**

(continued)

Table 11.3 (continued)

Countries	Responded	Assigned budget provided	Lawful Regulations	Poaching known	SDG based management	Comments	Email follow-up sent
Benin	No						Yes
Bhutan	No						Yes
Brazil	No						Yes
Burundi	No						Yes
Cameroon	No						Yes
China	No						Yes
Columbia	Pending						Yes
Costa Rica	No						Yes
Ecuador	No						Yes
Ghana	No						Yes
Guam	No						Yes
Guatemala	Yes	No	No	No	No	No information available	Yes
India	No						Yes
Indonesia	No						Yes
Ivory Coast	No						Yes
Japan	No						Yes
Kenya	No						Yes
Malaysia	No						Yes
Mexico	No						Yes
Nepal	No						Yes
Nicaragua	No						Yes
Nigeria	No						Yes
Panama	No						Yes
Peru	No						Yes
Puerto Rico	Yes	No	No	No	No	No squirrels present	Yes
Russia	No						Yes
Rwanda	Pending						Yes
South Korea	No						Yes
Sri Lanka	No						Yes
Togo	No						Yes
Uganda	No						Yes
Yes count ratio	**(2/31) 0.06**	**0**	**0**	**0**	**0**	**0**	**1**

(continued)

Table 11.3 (continued)

States	Responded	Assigned budget provided	Lawful Regulations	Poaching known	SDG based manage ment	Comments	Email follow-up sent
Alabama	No						Yes
Alaska	Yes	No	Yes	No bag limit for either hunting or trapping, so poaching is Not an issue.	No	No bag limit. Year-round season	No
Arizona	Yes	Yes	Yes	Yes	Aligned but not reported		Yes
Arkansas	Pending						Yes
California	No						Yes
Colorado	Yes	No	Yes	Information request requires paying an invoice	No	Ask to pay for poaching reports	No
Connecticut	Yes	No	Yes	Not aware of any documented poaching activities in this state.	No	Do Not know what the SDG's are	No
Delaware	No						Yes
Florida	Yes	Not available	Yes	1 squirrel hunt in 3 years. No other incidents reported	No		Yes
Georgia	Yes	Partly	Yes	Not provided. Forwarded to their website. Nothing mentioned there.	No		Yes
Hawaii	Yes	No	No	No	No	No squirrels present	No
Idaho	No						Yes
IlliNois	No						Yes
Indiana	Yes	Yes	Yes	Pending	Yes		Yes
Iowa	No						Yes
Kansas	Yes	No budget assigned to squirrels	Yes	No	No		Yes
Kentucky	No						Yes
Louisiana	Yes	No	Yes	Yes, these include hunting out of season and most common is bag limit violations	No		Yes

(continued)

Table 11.3 (continued)

Maine	Yes	Na for squirrels	Yes	12 poaching records since 2017.	No but similar	Received a second reply. Very odd (Becky Orff)	Yes
Maryland	Yes	No	Yes	Poaching is Not known to be an issue of concern in regard to small-game management within Maryland. Poaching incidents involving squirrels are relatively uncommon, and likely unrelated events that do Not have population-level impacts.	Yes	Additionally added squirrel observation data.	Yes
Massachusetts	Yes	No	No	No	No	The contacted authority had No information to answer any of our questions	Yes
Michigan	Yes	No	Partly	Forwarded to someone else	No	"Unfamiliar with the UN SDG's"	Yes
Minnesota	No						Yes
Mississippi	No						Yes
Missouri	No						Yes
Montana	Yes	No	Yes but no limits	No	No		Yes
Nebraska	Yes	No	Yes	Seems to be occurring from time to time.	No	Unfamiliar with SDG's	No
Nevada	No						Yes
New Hampshire	Yes	No	Vaguely	No known reports	No		Yes
New Jersey	No						Yes
New Mexico	Yes	Yes	Yes	There have been No squirrel poaching reports in the state, to the best of their knowledge.	Yes		Yes
New York	Yes	No	Yes	Forwarded to submitting a request to the	No, but their managem ent is		Yes

(continued)

Table 11.3 (continued)

				law enforcement division.	relatively sustainabl e.		
North Carolina	No						Yes
North Dakota	Yes	Yes	Yes	No reports of known poaching activities in ND.	No		Yes
Ohio	Yes	No	Yes	14 recorded poaching violations since 2017,	No		No
Oklahoma	Yes	No	Yes	No tracking metric in place to quantify the day-to-day poaching enforcement effort or time spent.	No		Yes
Oregon	Pending		Yes				Yes
Pennsylvania	No						Yes
Rhode Island	Yes	No	Yes	No poaching activities reported or documented by the Division of Fish and Wildlife.	No	Very extensive response	No
South Carolina	No						Yes
South Dakota	Yes	No	Partly	No known poaching cases or unknown.	No	Short and unprofessio nal response	No
Tennessee	Yes	No	Yes	TWRA wildlife officers have issued 45 citations and 20 warnings for violations related to squirrels since 2018. These include hunting violations, live wildlife possession, and taxidermy.	No		No
Texas	Yes	No	Yes	No details provided	"Unfamili ar with the SDGs".		Yes
Utah	No						Yes
Vermont	No						Yes
Virginia	No						Yes
Washington	No						Yes
West Virginia	Yes	No	Yes	WV DNR has no record of investigation of	Tracked but Not reported		Yes

(continued)

Table 11.3 (continued)

				any squirrel poaching cases.			
Wisconsin	Yes	No	Yes	Annually there are relatively few incidents of unlawful take or poaching activities of squirrels. However, those that occur are a result of harvest outside of the legal season and bagging limits (daily and possession)	No	No guidelines for some species	Yes
Wyoming	Yes	No	Yes	No	No		Yes
Yes count ratio	**(32/50) 0.64**	**0.1**	**0.54**	**0.12**	**0.06**		**0.82**

Territories	Responded	Assigned budget provided	Lawful Regulations	Poaching known	SDG based management	Comments	Email follow-up sent
Alberta	No						Yes
British Columbia	Yes	No	Vaguely	No information provided in the survey response.	No		No
Manitoba	No						Yes
New Brunswick	Yes	No	Yes	There are very few if any reports of known and documented poaching activities regarding squirrels.	No		Yes
Newfoundland and Labrador	No						Yes
Northern Territory	No						Yes
Nova Scotia	No						Yes
Nunavut	No						Yes
Ontario	Yes	No	Yes	No, not a significant	Management based on sust.		Yes

(continued)

Table 11.3 (continued)

				source of illegal activities	Goals but Not reported to the UN.		
Prince Edward Island	No						Yes
Quebec	Yes	No	Yes	"No information"	Management based on sust. Goals but Not reported to the UN.	"Will respond in a few weeks"	Yes
Saskatchewan	Yes	No	Yes	No	Not directly		Yes
Yukon	No						Yes
Yes count ratio	**(5/12) 0.42**	**0**	**0.33**	**0**	**0**		**0.92**
Total Yes count ratio	**(47/107) 0.41**	**0.065**	**0.355**	**0.056**	**0.056**		**0.9**

Table 11.4 Hunting Regulations table from squirrel survey, based on best-available data received by authors

Countries	Lawful Regulations in place	Huntable	Huntable species	Daily Bag limit	Possession limit	Hunting season	Furbearing and trapping laws?	Hunting Law details	Comments
Austria	Yes	Partly	Eurasian Red Squirrel (*Sciurus vulgaris*)	Information lacking on the national level	Information lacking on the national level	Information lacking on the national level	Not available	Viennese Hunting law § 3	European ground squirrel (*Spermophilus citellus*) fully protected.
Finland	Yes	Yes. Allowed to hunt with shotguns and bows.	Eurasian Red squirrel (*Sciurus vulgaris*)	Information lacking on the national level	Information lacking on the national level	1st November to 28th February (response letter survey). 1st December to 31st January (Hunting decree).	Iron traps may be used up to a diameter of 20 cm. Use of restraining traps and foot snares is allowed. A leaf blower for the capturing or chasing of squirrels is allowed.	- Finnish Hunting Law (615/1993) - Finnish Hunting Decree (666/1993).	The hunting season of survey response and online Hunting decree (666/1993) do not overlap.
Germany	No	Not complete, but 'no' for *S. vulgaris*	None indicated	No information provided	No information provided	No information provided	No information provided	No references provided	Very incomplete response
Netherlands	Yes	Native species not, invasive species yes.	Pallas's Squirrel (*Callosciurus erythraeus*), Eastern Gray Squirrel (*Sciurus carolinensis*)	Information lacking on the national level	Information lacking on the national level	No closed season for invasive species.	Only used for invasive species.	Article 3.10 of the Nature conservation act (Wet Natuurbescherming - BWBR0037552). Nature conservation act. Article 3.19. EU list of Invasive Alien Species of Union concern (EU Regulation 2019/1262).	It is forbidden to trade in squirrels or parts of squirrels.
Norway	Yes	Yes	Eurasian Red squirrel (*Sciurus vulgaris*)	Information lacking on the national level	Information lacking on the national level	1st November to 15th March in all of the country (4,5 months in the grey fur winter season). Hunting and trapping are not permitted in the period from and including 24 December to and including 31 December and not on Good Friday, Easter Eve, and the first day of Easter.	It is illegal to use any traps.	- Act on hunting and trapping of game (Wildlife Act). - Regulations on the practice of hunting, felling, and trapping. - Regulations on hunting and hunting times as well as collecting eggs and down for the hunting seasons from and including 1 April 2022 to and including 31 March 2028	The use of chemicals or poisons to kill game other than small rodents and reptiles is prohibited.
Spain	Yes	No	None	None	None	None	Not practiced nor authorized	Law 42/2007, of December 13, on natural heritage and biodiversity (article 54).	The red squirrel is protected in Spain and it is not possible to kill, hunt, disturb, or destroy its habitat
Sweden	NA	NA	NA	NA	NA	NA	NA	NA	NA
Switzerland	Yes	Partly	Alpine Marmot (*Marmota marmota*)	Defined by the regions	Defined by the regions	Defined by the regions	Practiced by farmers to protect their fields, sanctioned in the most extreme cases	Bundesgesetz über die Jagd und den Schutz wildlebender Säugetiere und Vögel (https://www.fedlex.admin.ch/eli/cc/1988/506_506_506/de) Bundesgesetz über den Natur- und Heimatschutz	Most of the laws are only available in German, Italian, and French (no English version available).

(continued)

Table 11.4 (continued)

								(https://www.fedlex.admin.ch/eli/cc/1966/1637_1694_1679/de)	
Guatemala	No	No	No	No	No	No	No	No	No squirrels present
Puerto Rico	No	No	No	No	No	No	No	No	No squirrels present

Countries	Lawful Regulations in place	Huntable	Huntable species	Daily Bag limit	Possession limit	Hunting season	Furbearing and trapping laws?	Hunting Law details	Comments
Alaska	Yes	Yes	North American Red Squirrel (*Tamiasciurus hudsonicus*)	No limit	No limit	Huntable year-round, with no closed season.	Same as hunting, with no closed season with no limits.	Link 1, Link 2	"There is no biological reason to limit harvest." It is not required to report shot or trapped squirrels to the Wildlife Division. Northern Flying Squirrel (*Glaucomys sabrinus*) is not hunted because it is nocturnal.
Arizona	Yes	Partly	Any tree squirrel except the Mount Graham red Squirrel (*Tamiasciurus fremonti grahamensis*), Tassel-eared tree squirrel (*Sciurus aberti*).	Five (5) tree squirrels per day.	Fifteen (15) tree squirrels of which no more than five (5) may be taken in any one day.	Oct 1 - Dec 31, 2021 (*T. fremonti grahamensis*). Jul 1, 2021 - Jun 30, 2022 & Sep 1, 2021- May 31, 2022 (*S. aberti*). See more details under Link 3 for other weapons.	See more details under Link 3 for all kinds of weapons/ traps.	Link 3	To hunt tree squirrels in Arizona, you need a valid hunting or combination license.
Colorado	Yes	Yes	Fox Squirrel (*Sciurus niger*), Pine Squirrel (*Tamiasciurus spp.*), Marmot (*Marmota spp.*), Abert's Squirrel (*Sciurus aberti*), Black-tailed (*Cynomys ludovicianus*), white-tailed (*Cynomys leucurus*) and Gunnison prairie dogs (*Cynomys gunnisoni*), Wyoming (Richardson's) ground squirrel (*Urocitellus richardsonii*).	- *S. aberti* (Two (2) squirrels) - *Tamiasciurus spp. & S. niger* (Five (5) squirrels) - *Cynomys ssp. & Urocitellus r.* (no bag limits) - *Marmota spp.* (Two (2) marmots)	- *S. aberti* (Four (4) squirrels) - *Tamiasciurus spp. & S. niger* (Ten (10) squirrels) - *Cynomys ssp. & Urocitellus r.* (no possession limits) - *Marmota spp.* (Four (4) marmots)	- *S. aberti* (Statewide: November 15 - January 15 annually) - *Tamiasciurus spp. & S. niger* (Statewide: October 1 - end of February annually) - *Cynomys ssp.* (Public Land: June 15 - end of February annually & Private Land: January 1 - December 31 annually) - *U. richardsonii* (Statewide: January 1 - December 31 annually) - *Marmota spp.* (Statewide: August 10 - October 15 annually)	No significant limitations, the following arms, are allowed: (1. Any rifle or handgun. 2. Any shotgun. 3. Handheld bows and crossbows.)	Link 4	It is unknown which species are exactly meant by "Marmots" since more species inhabit Colorado. Some species have special falconry seasons.
Connecticut	Yes	Yes	Eastern Gray Squirrel (*Sciurus carolinensis*), Woodchuck (*Marmota monax*)	- *S. carolinensis* (8 squirrels) - *M. monax* (no limits)	- *S. carolinensis* (season limit 40 squirrels) - *M. monax* (no limits)	- S. *carolinensis* (Jan. 1 – Feb. 28 & Sep. 1 – Dec. 31) - *M. monax* (March 15 – Nov. 15).	CT does not have a trapping season for squirrels	Link 5, Link 6, Link 7	
Florida	Yes	Yes	Eastern Gray Squirrel (*Sciurus carolinensis*)	Daily bag limit: 12	Possession limit: 24	Statewide Oct. 9 – March 6	*Sciurus c.* Falconry (Oct. 1 – March 31) Bag limit (12), Possession limit (24). No trapping/ fur-bearing.	Link 8	*Sciurus c.* may be taken during archery, crossbow, and muzzleloading gun seasons. Fox Squirrel (*Sciurus niger*) may not be taken
Georgia	Yes	Yes	NA (Fox Squirrel (*Sciurus niger*), others not specified)	In general (12); *S. niger* 1/hunter	NA	Statewide Aug. 15-last day in Feb.	NA	Link 9	S. niger hunting in some areas is more restricted/ forbidden.
Hawaii	No	No	No	No	No	No	No	No	No squirrels present

(continued)

Table 11.4 (continued)

Indiana	Yes	Yes	Eastern Gray Squirrel (*Sciurus carolinensis*), Fox Squirrel (*Sciurus niger*)	5 Squirrels	NA	August 15 through January 31 of the following year.	Seems to be allowed. Not strictly mentioned.	State law in IC 14-22 (Link 10); 312 IAC 9-3-17 (Link 11)	See more details under Link 11. Franklin's ground squirrel (*Spermophilus franklinii*) is protected and not huntable. Southern Flying Squirrel (*Glaucomys volans*) is only briefly mentioned.
Kansas	Yes	Yes	No species specified	5 squirrels	20 squirrels	June 1 through the last day of February	No details provided	No links provided. Article paragraphs attached to email.	
Louisiana	Yes	Yes	No species specified	8 squirrels per person per day during the Fall/Winter season and 3 squirrels per person per day during the Spring season.	No information provided	1st Saturday in October through the last day of February (Fall/Winter Season) and opens again the 1st Saturday in May for 23 days (Spring Season)	No information provided	No references provided	
Maine	Yes	Partly	Eastern Gray Squirrel (*Sciurus carolinensis*), Woodchuck (*Marmota monax*), North American Red Squirrel (*Tamiasciurus hudsonicus*)	*S. carolinensis* (4), *M. monax* & *T. hudsonicus* (no limits)	*S. carolinensis* (8), *M. monax* & *T. hudsonicus* (no limits)	*S. carolinensis* (September 25 - December 31), *M. monax* & *T. hudsonicus* (No closed season)	Statewide trapping season *T. hudsonicus* (October 31 - December 31).	Summary of Maine Hunting Laws (Link 12), Summary of Maine Trapping Laws (Link 13)	Chipmunks and flying squirrels are non-game species.
Maryland	Yes	Partly	Eastern Gray Squirrel (*Sciurus carolinensis*), Woodchuck (*Marmota monax*), North American Red Squirrel (*Tamiasciurus hudsonicus*), Southern Flying Squirrel (*Glaucomys volans*), Chipmunks (*Tamias spp.*)	NA	NA	NA	NA	Link 14	Delmarva fox squirrel (*Sciurus niger cinereus*) may not be hunted. Barely any details on hunting limitations are available.
Massachusetts	NA	NA	NA	NA	NA	NA	NA	NA	The corresponding division has "no information or involvement in the topics"
Michigan	Yes	Partly	Eastern Gray Squirrel (*Sciurus carolinensis*), Woodchuck (*Marmota monax*), North, Fox Squirrel (*Sciurus niger*)	*M. monax* (No limits) *S. carolinensis* & *S. niger* (5)	*M. monax* (No limits) *S. carolinensis* & *S. niger* (10)	- *M. monax* (no closed season except for state park and recreation areas which shall be closed April 1 to Sep. 14) - *S. carolinensis* & *S. niger* (Sep. 15 to March 31)	Trapping is statewide year-around allowed, except within state park and recreation areas from April 1 to September 14.	Link 15	
Montana	Yes	Yes	All squirrels can be hunted. For residents no license is needed for non-residents a hunting/ trapping license is needed. Limits are equal for residents and non-residents.	No limits for squirrels hunted	No limits for squirrels hunted	No closed season for squirrels hunted	No limits for squirrels hunted	Link 32	Uinta ground squirrel (*Urocitellus armatus*), and Wyoming ground squirrel (*Urocitellus elegans*) are considered potential species of concern.

(continued)

Table 11.4 (continued)

Nebraska	Yes	Yes	Eastern Gray Squirrel (*Sciurus carolinensis*), North, Fox Squirrel (*Sciurus niger*)	7 squirrels	28 squirrels	August 1 - January 31	Trapping follows the same season and limits as hunting.	Link 16	Southern Flying Squirrel (*Glaucomys volans*) may not be taken.
New Hampshire	Yes	Yes	Eastern Gray Squirrel (*Sciurus carolinensis*),	5 squirrels	NA	Hunting season for gray squirrels shall open on September 1 and close on January 31	NA	Link 17	
New Mexico	Yes	Yes	Abert's Squirrel (*Sciurus aberti*), North American Red Squirrel (*Tamiasciurus hudsonicus*), Arizona Gray Squirrel (*Sciurus arizonensis*), Eastern Gray Squirrel (*Sciurus carolinensis*), Fox Squirrel (*Sciurus niger*)	8 (singly or in aggregate)	16 (singly or in aggregate)	Sep. 1 – Nov 30.	NA	Link 18	
New York	Yes	Yes	Eastern Gray Squirrel (*Sciurus carolinensis*), Fox Squirrel (*Sciurus niger*), North American Red Squirrel (*Tamiasciurus hudsonicus*)	6 squirrels, regardless of species, *T. hudsonicus* has no limits	No information was provided. *T. hudsonicus* has no limits	Sept. 1 – Feb. 28 (almost the entire state), Nov. 1 – Feb. 28 (New York City and Long Island), *T. hudsonicus* has no closed season	No trapping season for squirrels	Link 25, Link 26	
North Dakota	Yes	Yes	Tree squirrels	4 squirrels	12 squirrels	September 11 through February 28	Seems to be allowed, but no details were provided	No reference provided	No details were provided regarding which squirrels are meant by "Tree squirrels".
Ohio	Yes	Yes	Eastern Gray Squirrel (*Sciurus carolinensis*), Fox Squirrel (*Sciurus niger*), North American Red Squirrel (*Tamiasciurus hudsonicus*)	6 squirrels	NA	September 1 to January 31	Commonly not considered trapping/ fur-bearing animals	Link 19	Groundhog hunting has no closed season.
Oklahoma	Yes	Yes	Prairie Dog (*Cynomys spp.*), Eastern Gray Squirrel (*Sciurus carolinensis*), Fox Squirrel (*Sciurus niger*)	*Cynomys spp.* (No limits), *S. carolinensis & S. niger* (25)	*Cynomys spp.* (No limits), *S. carolinensis & S. niger* (50)	*Cynomys spp.* (No closed season), *S. carolinensis & S. niger* (May 15 - Jan. 31; statewide)	Furbearing and trapping seem not to be applied to squirrels.	Link 20	
Rhode Island	Yes	Yes	Eastern Gray Squirrel (*Sciurus carolinensis*), North American Red Squirrel (*Tamiasciurus hudsonicus*)	5, singly or in aggregate	NA	October 16 - February 28	Not performed on squirrels	Link 21	
South Dakota	Yes	Yes	Eastern Gray Squirrel (*Sciurus carolinensis*), North American Red Squirrel (*Tamiasciurus hudsonicus*), Fox Squirrel (*Sciurus niger*)	5 squirrels	15 squirrels	September 1 through the last day of February	No information available.	Link 22	
Tennessee	Yes	Yes	"Squirrels"	10 squirrels	NA	Aug. 28 – Feb. 28 & May 14. – June 12	NA	Link 23	No specification of what is meant by "squirrels".

(continued)

Table 11.4 (continued)

Texas	Yes	Yes	Presumably Eastern Gray Squirrel (*Sciurus carolinensis*)	East Texas (10), All Other Counties (no limits)	No details provided.	East Texas (Oct. 1 - Feb. 27 & May 1 – 31), All Other Counties (Sept. 1 - Aug. 31)	"Only homeowners are allowed to trap squirrels that are causing damage to private properties and these animals must be released alive"	Link 27, Link 28	
West Virginia	Yes	Yes	Eastern Gray Squirrel (*Sciurus carolinensis*), Black, Albino, Fox Squirrel (*Sciurus niger*)	6 squirrels	24 squirrels are the possession limit, No season limit	September 11 through February 28	No information available	No reference provided	
Wisconsin	Yes	Yes	Eastern Gray Squirrel (*Sciurus carolinensis*), North American Red Squirrel (*Tamiasciurus hudsonicus*), Fox Squirrel (*Sciurus niger*)	*T. hudsonicus* (no limits), *S. carolinensis & S. niger* (5)	*T. hudsonicus* (no limits), *S. carolinensis & S. niger* (15)	T. hudsonicus (no closed season), S. carolinensis & S. niger (Sept. 18th – Jan. 31st)	"Tails, skins, and skulls may be bought and sold if severed from the rest of the carcass."	No reference provided	"Red Squirrels have unprotected status in Wisconsin so they can be harvested year-round by any means."
Wyoming	Yes	Yes	Eastern Gray Squirrel (*Sciurus carolinensis*), North American Red Squirrel (*Tamiasciurus hudsonicus*), Fox Squirrel (*Sciurus niger*)	10 squirrels	20 squirrels	Sep. 1 – Mar. 31	Squirrels are commonly not trapped.	Link 29, Link 30, Link 31	

Countries	Lawful Regulations in place	Huntable	Huntable species	Daily Bag limit	Possession limit	Hunting season	Furbearing and trapping laws?	Hunting Law details	Comments
British Columbia	Yes, but vague details provided	Partly	Eastern Gray Squirrel (*Sciurus carolinensis*), North American Red Squirrel (*Tamiasciurus hudsonicus*)	Barely hunted, no details provided.	Barely hunted, no details provided.	No details were provided.	Barely	No references provided	Vancouver Island Marmot (*Marmota vancouverensis*) is a fully protected species
New Brunswick	Yes	Yes	North American Red Squirrel (*Tamiasciurus hudsonicus*), the Eastern Grey Squirrel (*Sciurus carolinensis*), and the Northern Flying Squirrel (*Glaucomys sabrinus*)	No details were provided.	No details provided.	October 1st to the last day of February	October 30th to the last day of February	No references provided	
Ontario	Yes	Partly	North American Red Squirrel (*Tamiasciurus hudsonicus*), Fox Squirrel (*Sciurus niger*), Eastern Gray (Black) Squirrel (*Sciurus carolinensis*)	*S. carolinensis & S. niger* (Combined 5 squirrels)	*S. carolinensis & S. niger* (Combined 15 squirrels)	*S. carolinensis & S. niger* (Sep. 15 to Dec. 31 in some regions, Sep. 25 to Dec. 31 in other regions)	*Tamiasciurus hudsonicus* only trapped, not hunted. Commonly with a leg-hold trap (Oct. 25 to end of Feb.), no other squirrels	Link 33, Link 34, Link 35	Eastern Chipmunk (*Tamias striatus*), Least Chipmunk (*Tamias minimus*), Northern Flying Squirrel (*Glaucomys sabrinus*), and Southern Flying Squirrel (*Glaucomys volans*) are specially protected species.
Quebec	Yes	Partly	Woodchuck (*Marmota monax*) for hunting, Eastern gray squirrel (*Sciurus carolinensis*) & North American Red Squirrel (*Tamiasciurus hudsonicus*) for trapping & furbearing	No details provided	No details provided	*M. monax* (1 April to 31 March)	No details provided	Link 36 Link 37 Link 38 Link 39	Least chipmunk (*Tamias minimus*), Eastern chipmunk (*Tamias striatus*), Southern flying squirrel (*Glaucomys volans*), Northern flying squirrel (*Glaucomys sabrinus*) Are not exploited

(continued)

Table 11.4 (continued)

Saskatchewan	Yes	Yes	North American Red Squirrel (*Tamiasciurus hudsonicus*), Black-tailed Prairie dog (*Cynomys ludovicianus*)	No details were provided.	No details were provided.	November 1 to March 15	November 1 to March 15	The Wildlife Regulations, 1981, The Wildlife Act, 1998 and the Fur Animals Open Seasons Regulations; Link 24	

Links Bibliography

Link 1

https://www.adfg.alaska.gov/static/hunting/trapping/pdfs/trap2020.pdf

Link 2

https://www.adfg.alaska.gov/static/applications/web/nocache/regulations/wildliferegulations/pdfs/smgame.pdfBCF74A546CDD1EE2DDA101083D667E41/smgame.pdf

Link 3

https://azgfd-portal-wordpress-pantheon.s3.us-west-2.amazonaws.com/wp-content/uploads/archive/2021-22-Arizona-Hunting-Regulations_211012.pdf

Link 4

https://cpw.state.co.us/Documents/RulesRegs/Regulations/Ch03.pdf

Link 5

https://portal.ct.gov/DEEP/Hunting/CT-Hunting-and-Trapping

Link 6

https://portal.ct.gov/-/media/DEEP/hunting_trapping/pdf_files/SeasonSummary2022.pdf

Link 7

https://portal.ct.gov/-/media/DEEP/hunting_trapping/pdf_files/2022-CT-Hunting-Guide.pdf

Link 8

https://www.eregulations.com/assets/docs/resources/FL/21FLHD-LR.pdf

Link 9

http://rules.sos.ga.gov/gac/391-4-2

Link 10

http://iga.in.gov/legislative/laws/2021/ic/titles/014/#14-22

Link 11

http://iac.iga.in.gov/iac/iac_title?iact=312

Link 12

https://www.maine.gov/ifw/hunting-trapping/hunting-laws/index.html

Link 13

https://www.maine.gov/ifw/hunting-trapping/trapping-laws/index.html

Link 14

https://www.eregulations.com/maryland/hunting/small-game-hunting

Link 15

https://www.michigan.gov/documents/dnr/WCO_458867_7.pdf

(continued)

Table 11.4 (continued)

Link 16

http://digital.outdoornebraska.gov/i/1397794-2021-small-game-and-waterfowl-guide-web/15?

Link 17

https://casetext.com/regulation/new-hampshire-administrative-code/title-fis-executive-director-fish-and-game-department/chapter-fis-300-wildlife-seasons-and-rules/part-fis-301-game-animals/section-fis-30111-gray-squirrel

Link 18

https://www.srca.nm.gov/wp-content/uploads/attachments/19.031.0005.pdf

Link 19

https://codes.ohio.gov/ohio-administrative-code/1501:31

Link 20

https://www.wildlifedepartment.com/hunting/hunting-resources/small-game/squirrel-resources

Link 21

http://www.dem.ri.gov/programs/bnatres/fishwild/pdf/huntabs.pdf

Link 22

https://sdlegislature.gov/Rules/Administrative/16213

Link 23

https://www.eregulations.com/tennessee/hunting/small-game-seasons-bag-limits

Link 24

https://publications.saskatchewan.ca/

Link 25

https://www.dec.ny.gov/outdoor/29460.html

Link 26

https://govt.westlaw.com/nycrr/Document/I21bf137dc22211ddb7c8fb397c5bd26b?viewType=FullText&originationContext=documenttoc&transitionType=CategoryPageItem&contextData=%28sc.Default%29

Link 27

https://tpwd.texas.gov/regulations/outdoor-annual/regs/animals/squirrel/#autotoc-item-autotoc-0

Link 28

https://texreg.sos.state.tx.us/public/readtac$ext.TacPage?sl=R&app=9&p_dir=&p_rloc=&p_tloc=&p_ploc=&pg=1&p_tac=&ti=31&pt=2&ch=65&rl=116

Link 29

https://wyoleg.gov/statutes/compress/title23.pdf

Link 30

https://wgfd.wyo.gov/regulations#Trapping

Link 31

https://wgfd.wyo.gov/Regulations/Regulation-PDFs/REGULATIONS_CH11

Link 32

https://fwp.mt.gov/binaries/content/assets/fwp/hunt/regulations/2021/2021-furbearer-final-for-web.pdf

Link 33

https://www.ontario.ca/laws/regulation/980669

Link 34

https://files.ontario.ca/books/mnrf-2021-hunting-regulations-summary-en-2021-04-01-v2.pdf

Link 35

https://furmanagers.com/wp-content/uploads/2021/09/Regulations-English-2021.pdf

Link 36

https://www.legisquebec.gouv.qc.ca/en/document/cr/C-61.1,%20r.%2012

Link 37

https://www.legisquebec.gouv.qc.ca/en/document/cr/C-61.1,%20r.%201%20/

Link 38

https://www.legisquebec.gouv.qc.ca/en/document/cr/C-61.1,%20r.%2021

Link 39

https://www.legisquebec.gouv.qc.ca/en/document/cr/C-61.1,%20r.%203%20/

Poaching has been reported to the largest extent by the U.S. with 6/50 (12%), whereas no entities outside of the U.S. have reported any poaching activity on squirrels (see contrasting facts posted with photos on many public media sites like Facebook/ Meta).

Finally, for the last question which concerns the 17 SDGs, European countries include them to the largest extent in their squirrel management with 3/13 (23%). The individual responses for all countries/states can be seen in Table 11.3. Arguably SDGs are not linked to squirrels from the U.N nor properly implemented locally, not even reported by the responsible entities.

In order to have a greater understanding of the hunting regulations for the global squirrels, we have taken a closer look at the responses concerning the hunting regulations and the management regulations for trapping and fur-bearing. In Table 11.4, MS created an overview of all the responses with details on huntable species, bag limits, season restrictions, etc. for all countries/states that have replied to the survey.

Overall it can be observed that in European countries there are rarely hunting regulations in place for the entire nation, besides regulations that fully protect a species. Also, it is surprising that in some countries/states a species can be fully protected, whereas in other neighboring countries/states – on the other side of the border – it can be hunted without any limits. This can be seen for example with the species The North American Red Squirrel (*Tamiasciurus hudsonicus*). It gets even more convoluted when subspecies of such 'no limits' species are endangered but their taxonomy is chaotic and understudied. With no proper hunting laws in place and no law enforcement on the ground, such endangered subspecies are likely to be wiped out by unaware citizens, a clear sign of '*laissez-faire*', and ignorance. Additionally, it is noteworthy to have a closer look at the huntable species and the total of species mentioned in the regulations. This number is surprisingly small. Less than 20 species have in total been mentioned in the hunting regulations which includes species protected from hunting. Considering that the global squirrel population consists of approx. 300 species, which means that less than 10% are included in these regulations. This appears far from achieving convincing squirrel management, assuming citizens pay tax for such services.

This overview of the squirrel hunting regulations can be seen as a first-hand update and expansion of the work from the authors of a couple of years ago (Steiner and Huettmann 2021), where the squirrel hunting regulations from all U.S. states have been summarized. A detailed overveiw of squirrel hunting in Alaska has been presented in Textbox 11.1 below.

Textbox 11.1: Hunting and Trapping of Squirrels Are Not Really Part of the Commercial Alaskan Subsistence Lifestyle

Falk Huettmann, Moriz Steiner

Like many wilderness nations, there is a romantic view imposed about subsistence lifestyles and the wilderness of Alaska. It often involves urban perspectives of 'getting away from it all' and being oneself in 'the wild'. It is close to the ideal of the noble savage (Fairchild 1928; Ellingson 2001), living in great nature, being in harmony with the world and Mother Earth, and using natural resources in a good, sustainable way. It is a great concept if it is true, applicable, and lived well. But in the Anthropocene, it is difficult to scale this up to a global level and provide such opportunities for every human being in all social classes with good governance. And so, the life of hunting and trapping remains a traditional and very powerful ideal though, rarely achieved and less lived by many people. Similar applies to many indigenous concepts. It sits strong in an urbanized world, and in political circles where it gets promoted by the media and vested stakeholders. It is a widely politicized and created view. In the modern world, one might start thinking "what does all this modernity bring us?", is it increased happiness? "Were ancient humans less happy because they did not have the comforts of modern life, despite being synergized with nature?" "Is modernity any good?". The use of wildlife for own gain and income, and sustainable subsistence, is part of this concept. It applies to trapping and hunting and 'clean' nature, unspoiled by 'man' and free of contaminants and impacts. But there are a few problems with this narrative: (a) The subsistence lifestyle is pretty rough for most modern urbanized humans if one has to rely on it for the most part of the year. (b) There is also hardly enough wildlife left and available to live year-round for modern people , e.g. most indigenous people were nomadic and had to frequently adjust to migratory and available wildlife resources on a large landscape. (c) Squirrels either played little part in this lifestyle or were often seen as spiritual animals, hardly eaten. At best, they were meager emergency food items used on occasion, during winter. To our knowledge, there is no indigenous society, trapper, or fur reseller, where squirrels played a relevant part in the diet, the food culture, or the income. Squirrels take much effort to pursue, they were token animals, tryouts for kids, hardly more. With few exceptions, this is mostly true for the northern temperate and subarctic zone. If one has to rely on squirrels, one reached a low-end point. In more southern countries and regions, squirrels are often trapped as bushmeat until this day. However, that is widely done without any kind of holistic management plan. Examples for this have been seen on a global scale e.g. the detection of a new squirrel species (Laotian giant flying squirrel (*Biswamoyopterus laoensis*)) on the food/bushmeat market in Laos (Jenkins et al. 2005; Sanamxay et al. 2013). In the case of North America, in the northern regions, squirrels played no relevant role in indigenous people's income or most trappers. They were a marginal and rel. small additional resource and could not be relied on, at best. Compared to other resources, the skin of squirrels is too small, sells not well (who has or wants a squirrel pelts ?) and the meat proportion is too little nor good for eating. Diseases might occur too. The only source for which squirrels have been used and are still used today is for some specific and recent western fur trades (Borlik 2020; Howard-Johnston 2020; Martin 2004; see examples from the Hutson Bay Company HBC Obbard et al. 1987; Ray 1998 – noteworthy is that the HBC went bankrupt early on). But this fur trade was widely on the East coast, little in Alaska, or Russian territories. Trappers just did not make a living off squirrels, nor did indigenous societies; the squirrel resource did not provide…much. Now referring to squirrels as part of the traditional and subsistence ancient lifestyle, using them accordingly, and promoting them (e.g. Taras 2017 not just for kids') and in high abundance and for commercial reasons to a widely uninformed public, is a modern artifact and not so factual. It is widely misleading, tries to celebrate a false narrative and image, and just shows how eager the modern world and governance became for resources and money. Squirre;s became a 'scape goat'. However, the squirrel subsistence scheme is mentioned in several book stories and myths, as such a cute animal is close to people's hearts with high importance of preserving them in the portfolio (Dugger 2021; Smelcer et al. 1992). This subsistence artifact and narrative for Alaska sits well though with some people and ideologies in the absence of fully pristine nature and space left where one could live a subsistence lifestyle. And thus, squirrels now get used as a replacement item for big game and such an ideal. While climate change, the internet, and the human population unfold to record heights, squirrels should not be promoted on the wrong side of the equation. No Alaska, Yukon, or Canadian indigenous person, hunter, or trapper and families can really live from the squirrel pursuit; nor should a tourist company promote and rely on them with hunting guides and advertise them to a global audience for a shooting event. It's just not Alaskan. Instead, if we have witnessed one thing during the last decades then that squirrels are pressured and stressed more and with many populations locally extinct or on the wider decline. It is not sustainable or subsistence. Just another untouched icon of the wild here is getting pushed out, urbanized, politicized, and changing its entire ecology and DNA; its very identity

11.3.3 Analysis of the Coherence Between Literature Records and Regulations with the Survey Responses

We observed that the survey responses were overall relatively coherent with the literature records and the online government, however, some of the online references provided by the respondents were not coherent with other websites of the same mandate area found on the world wide web. A few examples have been pointed out in Table 11.4.

11.4 Discussion

Squirrels have been hunted for millennia and have been in the public interest ever since. Presumably, squirrels have already been hunted in hunter-gatherer times, by indigenous groups, in tribal settings, even until this day (Lew-Levy et al. 2017 and see Appendix 11.3 where such a modern hunter-gatherer lifestyle is still practiced and described first-hand in Sierra Leone – West Africa). However, in the modern days, where humans are fairly detached from nature and wildlife management, with large consumption and a very high human population, wildlife management is no longer based on communal decisions by simultaneously respecting natural limits (Keane et al. 2008; Martin et al. 2016). Nowadays, it is largely based on the assigned budget and financial interests of the stakeholders involved, often with a lack of communication and meaningful coordination among neighboring countries/states/legal entities (Keppel et al. 2012; Miles et al. 2020). This can be directly observed by the extreme differences in the regulations of neighboring states/countries (e.g. see also the response of Switzerland vs Germany, France, or Italy to the letter survey).

Considering the response rate, it was pretty decent for European countries and the U.S. (with approx. 65%), ok for Canadian provinces (with 42%), and very low for other countries outside of Europe and North America (circa 6%). Ideally, a response rate of over 90% we considered 'good', however, overall that is far away from the response rate of our survey.

To move forward from here, the authors acknowledge the complexities of wildlife management in the Anthropocene but also suggest considering finite resources and good management during crisis times and climate change. Best management practices can be followed in the public eye. Arguably, management entities should ideally communicate better with the public and with each other across jurisdictions to align their policies and regulations for all species in a sustainable fashion.

Also, it appears from our work that the vast majority of squirrel species (>90%) are currently left out completely from hunting/trapping/fur-bearing regulations (see this topic addressed also in Aycrigg et al. 2015 – with suggested research needs). For a sustainable, holistic future conservation approach, all squirrel species should be included in the hunting regulations, especially the ones that are prohibited to be hunted.

Tackling wider and more inclusive, holistic approaches to natural resources – Mother Earth – is unlikely harmful (Chapin et al. 2011; *sensu* Huettmann 2020; Suzuki 1993, and https://www.wisdomweavers.world/) but instead brings good progress and public buy-in into policy. So why not pursuing it more?

Acknowledgements We acknowledge the great help from Jeneba Cise with the hunting story from Sierra Leone. FH acknowledges discussions with,and insights from, governance entities contacted, as well as E. Huettmann, S. Linke, H. Berrios and the incredible Chrome team. This is EWHALE lab# publication 299.

References

Akinnifesi FK, Veloso N (2014) Bringing Brazil's school feeding best practices to scale in Africa and Latin America. Nat Faune 28(2):75–77

Alaska Department of Fish and Game (2021) 2020 Alaska trapper report: 1 July 2020–30 June 2021. https://www.adfg.alaska.gov/static/hunting/trapping/pdfs/trap2020.pdf. Accessed 31 Dec 2021

Aycrigg J, Beauvais G, Gotthardt T, Huettmann F, Pyare S, Andersen M, Walton K et al (2015) Novel approaches to modeling and mapping terrestrial vertebrate occurrence in the northwest and Alaska: an evaluation. Northwest Sci 89(4):355–381

Borlik T. A (2020) Performing The Winter's Tale in the "Open": Bear Plays, Skinners' Pageants, and the Early Modern Fur Trade. In The Routledge Handbook of Shakespeare and Animals (pp. 190-203). Routledge

Botkin JW, Elmandjra M, Malitza M (2014) No limits to learning: bridging the human gap: the report to the club of Rome. Elsevier

Callahan JR (1993) Squirrels as predators. The Great Basin Naturalist, pp 137–144

Chapin FS, Power ME, Pickett ST, Freitag A, Reynolds JA, Jackson RB, et al., Bartuska A (2011) Earth Stewardship: science for action to sustain the human-earth system. Ecosphere, 2(8), 1–20

CITES (2014) International trade in endangered species act 2008. https://cites.org/sites/default/files/projects/NLP/Malaysia_wildllife_Act686-5_8_2014.pdf. Accessed 31 Dec 2021

Code of Federal Regulations (2021) Parks, forests, and public property. www.ecfr.gov/current/title-36. Accessed 31 Dec 2021

Cordain L, Gotshall RW, Eaton SB (1998) Physical activity, energy expenditure and fitness: an evolutionary perspective. Int J Sports Med 19(05):328–335

Croteau E, Mott CL (2013) Saving endangered species: a case study using global amphibian declines. Nat Educ Knowl 4(4):9

Djagoun CA, Sogbohossou EA, Kassa B, Ahouandjinou CB, Akpona HA, Sinsin B (2018) Effectiveness of protected areas in conserving the highly hunted mammal species as bushmeat in Southern Benin. The Open Ecology Journal, 11(1)

Dugger JM (2021) Lewis and Clarke in the Caves. Mythlore, 40(1), 63-83. Mythlore, Vol. 40, No. 1 (Fall/Winter 2021), pp. 63–83

Ellingson T (2001) The myth of the noble savage. Univ of California Press

Eshetu AA (2014) Forest resource management systems in Ethiopia: historical perspective. Int J Biodivers Conserv 6(2):121–131

Fairchild HN (1928) The noble savage. In The Noble Savage. Columbia University Press

Fletcher R (2017) Connection with nature is an oxymoron: a political ecology of "nature-deficit disorder". J Environ Educ 48(4):226–233

Frassetto LA, Schloetter M, Mietus-Synder M, Morris RC, Sebastian A (2009) Metabolic and physiologic improvements from consuming a paleolithic, hunter-gatherer type diet. Eur J Clin Nutr 63(8):947–955

Herring G, Eagles-Smith CA, Goodell J, Buck JA, Willacker JJ (2021) Small-mammal shooting as a conduit for Lead exposure in avian scavengers. Environ Sci Technol 55(18):12272–12280

Howard-Johnston J (2020) The fur trade in the early middle ages. In Viking-Age Trade (pp. 57–74). Routledge

Huettmann F (2020) "The gods are angry": a first-hand environmental account and an experience of the 2015 Nepal earthquake in hindsight towards a new culture on how to approach and live well with mother earth. In: Hindu Kush-Himalaya watersheds downhill: landscape ecology and conservation perspectives. Springer, Cham, pp 611–632

InfoCongo (2016) Consumption Patterns Decimating Africa's wildlife. Info Congo. https://infocongo.org/en/consumption-patterns-decimating-africas-wildlife-3/. Accessed Dec 2021

IUCN Small Mammal Specialist Group (SMSG) (2021) Mexico. https://small-mammals.org/portfolio/mexico/. Accessed 31 Dec 2021

Jenkins PD, Kilpatrick CW, Robinson MF, Timmins RJ (2005) Morphological and molecular investigations of a new family, genus and species of rodent (Mammalia: Rodentia: Hystricognatha) from Lao PDR. Systematics and Biodiversity, 2(4), 419–454

Keane A, Jones JP, Edwards-Jones G, Milner-Gulland EJ (2008) The sleeping policeman: understanding issues of enforcement and compliance in conservation. Anim Conserv 11(2):75–82

Keppel G, Morrison C, Watling D, Tuiwawa MV, Rounds IA (2012) Conservation in tropical Pacific Island countries: why most current approaches are failing. Conserv Lett 5(4):256–265

Kesebir S, Kesebir P (2017) How modern life became disconnected from nature. Greater Good

Kimbrough L (2013) Scientists discover new flying mammal in bushmeat market. Mongabay. https://news.mongabay.com/2013/08/scientists-discover-new-flying-mammal-in-bushmeat-market/. Accessed Dec 2021

Koprowski JL, Nandini R (2008) Global hotspots and knowledge gaps for tree and flying squirrels. Curr Sci 95:851–856

Larson LR, Conway AL, Hernandez SM, Carroll JP (2016) Human-wildlife conflict, conservation attitudes, and a potential role for citizen science in Sierra Leone, Africa. Conserv Soc 14(3):205–217

Legacy.senate.gov (2010) Fifteenth Congress of The Republic of the Philippines. http://legacy.senate.gov.ph/lisdata/99798560!.pdf. Accessed 31 Dec 2021

Lew-Levy S, Reckin R, Lavi N, Cristóbal-Azkarate J, Ellis-Davies K. (2017) How do hunter-gatherer children learn subsistence skills? A meta-ethnographic review. Human Nature, 28, 367–394

Mahoney SP, Geist V (eds) (2019) The North American model of wildlife conservation. Johns Hopkins University Press

Martin J (2004) Treasure of the land of darkness: The fur trade and its significance for medieval Russia. Cambridge University Press

Martin JL, Maris V, Simberloff DS (2016) The need to respect nature and its limits challenges society and conservation science. Proc Natl Acad Sci 113(22):6105–6112

Maryanto I, Maharadatunkamsi AA, Achmadi SW, Sulistyadi E, Yoneda M, Suyanto A, Sugardjito J (2019) Checklist of the mammals of Indonesia: scientific, English, Indonesia name and distribution area table in Indonesia including CITES, IUCN and Indonesian category for conservation, vol 3. Lembaga Ilmu Pengetahuan Indonesia, Bogor

McBroom K (2021) Wintertime Squirrel Hunting. Henry USA. https://www.henryusa.com/the-henry-repeater/wintertime-squirrel-hunting/. Accessed Dec 2021

Miles A, Muñoz JMP, Bayle-Sempere JT (2020) Low satisfaction and failed relational coordination among relevant stakeholders in Spanish Mediterranean marine protected areas. J Environ Manag 272:111003

National Park Service (2021) Laws, policies & regulations. www.nps.gov/aboutus/lawsandpolicies.htm. Accessed Dec 2021

Obbard ME, Jones JG, Newman R, Booth A, Satterthwaite AJ, Linscombe G (1987) Furbearer harvests in North America. Wild Furbearer Management and Conservation in North America. Ontario Ministry of Natural Resources and the Ontario Trappers Association, 1007–1034

Press D, Doak DF, Steinberg P (1996) The role of local government in the conservation of rare species. Conserv Biol 10(6):1538–1548

Qian H, Deng T, Beck J, Sun H, Xiao C, Jin Y, Ma K (2018) Incomplete species lists derived from global and regional specimen-record databases affect macroecological analyses: a case study on the vascular plants of China. J Biogeogr 45(12):2718–2729

Ramos-Lara N, Koprowski JL (2014) Deforestation and knowledge gaps threaten conservation of less charismatic species: status of the arboreal squirrels of Mexico. Mammalia 78(4):417–427

Ray AJ (1998) Indians in the fur trade: their role as trappers, hunters, and middlemen in the lands southwest of Hudson Bay, 1660–1870: with a new introduction. University of Toronto Press

Ripple WJ, Abernethy K, Betts MG, Chapron G, Dirzo R, Galetti M, Young H et al (2016) Bushmeat hunting and extinction risk to the world's mammals. R Soc Open Sci 3(10):160498

Rudel T, Roper J (1997) The paths to rain forest destruction: crossnational patterns of tropical deforestation, 1975–1990. World Dev 25(1):53–65

Sanamxay D, Douangboubpha B, Bumrungsri S, Xayavong S, Xayaphet V, Satasook C, Bates PJ (2013) Rediscovery of Biswamoyopterus (Mammalia: Rodentia: Sciuridae: Pteromyini) in Asia, with the description of a new species from Lao PDR. Zootaxa, 3686(4), 471–481

Sandry N (2013) Nature deficit disorder. Educ Young Child Learn Teach Early Child Years 19(2):32–34

Schmeller DS, Gruber B, Bauch B, Lanno K, Budrys E, Babij V, Henle K et al (2008) Determination of national conservation responsibilities for species conservation in regions with multiple political jurisdictions. Biodivers Conserv 17(14):3607–3622

Silvy NJ (ed) (2020) The wildlife techniques manual: volume 1: research. Volume 2: management (Vol. 1). JHU Press

Smelcer JE (2011) The complete Ahtna poems. State University of New York at Binghamton

Stearns FW (1997) History of the Lake states forests: natural and human impacts, General technical report NC-189. U.S. Department of Agriculture, Forest Service, North Central Forest Experiment Station, St. Paul

Steiner M, Huettmann F (2021) Justification for a taxonomic conservation update of the rodent genus Tamiasciurus: addressing marginalization and mis-prioritization of research efforts and conservation laissez-faire for a sustainability outlook. Eur Zool J 88(1):86–116

Suzuki D (1993) Wisdom of the elders: sacred native stories of nature. BoD–Books on Demand

Swaisgood RR, Owings DH, Rowe MP (1999) Conflict and assessment in a predator–prey system: ground squirrels versus rattlesnakes. Anim Behav 57(5):1033–1044

Taber RD, Payne NF (2003) Wildlife, conservation, and human welfare: a United States and Canadian perspective. Krieger Publishing Company, Malabar

The Georgia Department of Natural Resources Wildlife Resources Division (2021) Squirrel Fact Sheet. https://georgiawildlife.com/sites/default/files/wrd/pdf/fact-sheets/2005_squirrel.pdf. Accessed 31 Dec 2021

The Wildlife Whisperer (2021) Squirrel trap disadvantages. The Wildlife Whisperer. https://totalwildlifecontrol.com/critter-facts-control/squirrel-control-attic-removal/squirrel-trap-disadvantages/. Accessed Dec 2021

US AID (2021) Protect wildlife final report. https://www.dai.com/uploads/Protect%20Wildlife%20Final%20Report.2021-5-14.pdf. Accessed 31 Dec 2021

US AID Indonesia (2004) Report on biodiversity and tropical forests in Indonesia. https://pdf.usaid.gov/pdf_docs/pnada949.pdf. Accessed 31 Dec 2021

Wikipedia (2021) Owens Lake. https://en.wikipedia.org/wiki/Owens_Lake. Accessed 31 Dec 2021

Williams M (1992) Americans and their forests: a historical geography. Cambridge University Press

Willson MF, Santo TLD, Sieving KE (2003) Red squirrels and predation risk to bird nests in northern forests. Can J Zool 81(7):1202–1208

Wolfe ND, Daszak P, Kilpatrick AM, Burke DS (2005) Bushmeat hunting, deforestation, and prediction of zoonotic disease. Emerg Infect Dis 11(12):1822

Chapter 12
Where Do the World's Squirrel Hotspots and Coldspots of 230+ Species Go with Climate Change in 2100? A First BIG DATA Minimum Estimate from an Open Access Climate Niche Rapid Model Assessment

Abstract Man-made climate change and its impact on the living world remain the problem of our time waiting for a good science-based resolution. Here, we focus on forecasting scenarios for the global squirrel population as a representative and powerful but overlooked species group for the year 2100. This approach was possible by using app. 230 publicly available Species Distribution Model (SDM) prediction maps for the world's squirrels (233 out of 307; 75% – with the remaining 25% virtually absent and thus seemingly impossible to properly manage). These distribution forecasts are originating from 132 GIS predictors, implemented with an ensemble of three machine learning algorithms (TreeNet, RandomForest, and Maxent). We found that most of the world's squirrel ranges will be shifting (usually towards higher altitudes and latitudes) and remain/ become more fragmented; some species extend their range, and many can 'spill' into new landscapes. Considering that here we just ran a first Rapid Assessment of Big Data, dealing with a climate niche envelope scenario of the future but not the entire more holistic perspective of climate change and 2100, we assume wider serious changes will occur for squirrels, their habitats, and the world in the future Anthropocene of 2100. These changes can lead to more stress, genetic mixing and loss, extinction, and increased zoonotic disease transmissions, and this process will occur with an increased gradient over time, while man-made climate change remains widely unabated.

Keywords Human-driven climate change · Man-made CO_2 · Squirrels (Sciuridae) · Ecological Niche · Geographic Information Systems (GIS) · Climate models · Zoonotic diseases · Rapid Model and Risk assessments

12.1 Introduction

Man-made climate change caused by the unsustainable consumption of materials and goods, and subsequent CO_2 release remains the problem of our time, for the living world and the universe (e.g. IPPC.org). It is widely unresolved in a scientific matter, with useful data for conservation widely missing. While temperatures rise and impacts increase on our planet Earth almost beyond human comprehension, e.g. zoonotic diseases, many species and habitats are still not even well assessed or get marginalized for risks, trends, and the effort of science-based management. The squirrel family (Sciuridae) consists according to Thorington et al. (2012) of 285 species, and according to the tables presented previously Steiner and Huettmann (2023 – Chap. 1) of 307 species (see Steiner and Huettmann (2021) for a generic taxonomic review, validity, and lack of mutual agreement). This study attempts to include all globally known squirrel species (approx. 307), however, due to a lack of open-access data only 233 (~75%) can be utilized thus far (Steiner and Huettmann 2023 - Chapters 1 and 3). The remaining 25% -1/4th- are virtually in a chaos state and can therefore pretty much not be used, nor managed.

Of the over 300 known squirrel species, not all are taxonomically agreed upon or carry meaningful data (see Steiner and Huettmann 2021; Steiner and Huettmann in press – Chap. 1). Thus far, only very few squirrel species are well-studied and present in the research literature, e.g. Eurasian Red Squirrel (*Sciurus vulgaris*), Eastern Grey Squirrel (*Sciurus carolinensis*), Eastern Fox Squirrel (*Sciurus niger*), and North American Red Squirrel (*Tamiasciurus hudsonicus*) (Lombardia et al. 2019). The rest of the squirrel species (approx. 99% -303/307) are widely understudied, and literature is often absent on many squirrel-related questions; hardly exist in English , to be accessible to a global audience. Science-based conservation management is thus hardly possible.

In previous work Steiner and Huettmann (2023 – Chap. 3), best-available distribution data for over 230 squirrels has been obtained from the public record via GBIF.org and made available as model-predicted distribution maps. Here, we make use

Supplementary Information The online version contains supplementary material available at https://doi.org/10.1007/978-3-031-23547-4_12.

M. Steiner, F. Huettmann, *Sustainable Squirrel Conservation*, https://doi.org/10.1007/978-3-031-23547-4_12

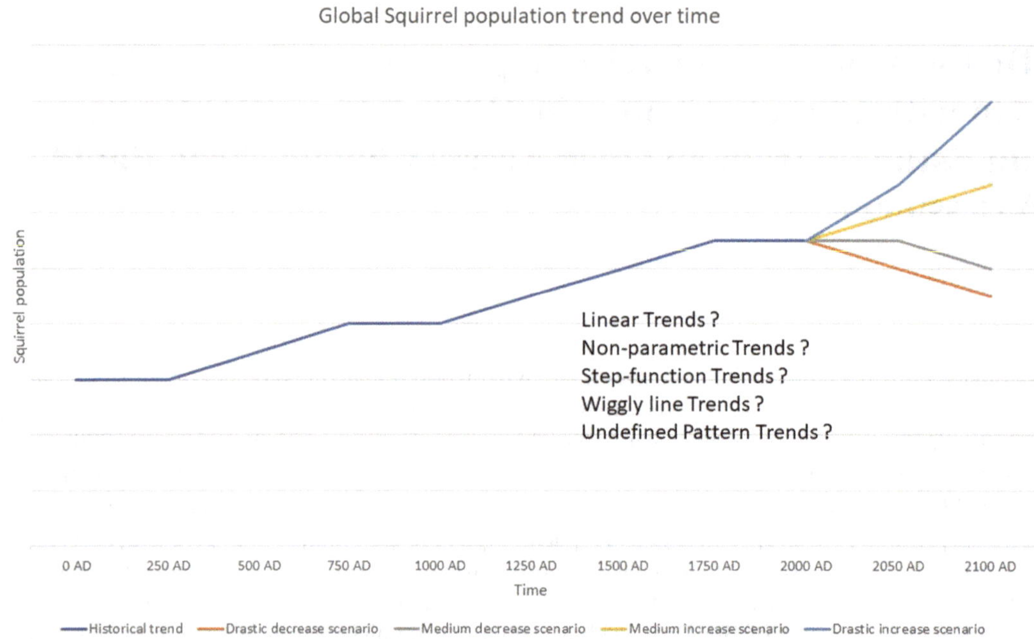

Fig. 12.1 Conceptual squirrel population scenarios over time with forecast to 2100

of that public data set and its summary of hot-spots and cold-spots (biological prioritization scenarios, *sensu* Lawler et al. 2011).

Moving beyond coarse global model predictions one can look at IUCN's Top 10 most endangered squirrel species. So we identified in previous work that the genera *Geosciurus*, *Heliosciurus* & *Paraxerus* were responding strongly to the climate in their distribution modeling (Steiner and Huettmann 2023 – Chap. 4). We therefore used those four cohorts (1. all global squirrels, 2. IUCN's Top 10 most endangered squirrel species, 3. genus *Geosciurus*, 4. genera *Heliosciurus* & *Paraxerus* merged) for a Meta-Analysis summary, testing how these squirrel groups respond in the absence of detailed studies to future climate scenarios (1. cooling, 2. business as usual, and 3. high warming) for a first and generic trend. To illustrate such scenarios, a hypothetical trend model can be seen in Fig. 12.1. This model presents the estimated evolution of the squirrel population during the last approx. 2000 years, with slowly rising population metrics. From the current day until 2100 and further, the future is unknown. Therefore, four possible population trend scenarios have been presented in Fig. 12.1 which are considered 'likely' for the future. Those models can be done with different algorithms – aspatially and spatially (e.g. for methods and approach see Huettmann et al. 2005; Nielsen et al. 2008; compare also with Peters et al. 2020).

Throughout this study, such scenarios of climate and future squirrel population changes (as presented in Fig. 12.1) will be discussed.

Here we aim to present the first distribution forecast scenarios for the global squirrel population and four cohorts – usually an overlooked group in such assessments and legal policies (Steiner and Huettmann 2021). This aims to set a baseline, start a discussion, and outline the assessed severity of climate change on squirrels, as one aspect of the living world. Also, we include a short assessment of the future trends towards a higher predicted risk of zoonotic disease transmission frequency.

12.2 Methods

12.2.1 Species Model and Cohort Prediction Layers

We were able to obtain over 230 SDM layers which are based on 132 environmental predictors (Steiner and Huettmann (2023 – Chap. 3); *sensu* Sriram and Huettmann unpublished), as ASC (.asc ASCII) files which were created with Maxent (Elith et al. 2006; Elith and Leathwick 2009; Milanovich et al. 2012). These 233 files contain for each squirrel species in the world the individual species distribution model (SDM) (see details on modern climate-predictor-based SDMs by Taheri et al.

Table 12.1 List of the 10 (17) most endangered squirrel species

Species	Taxonomic species numbers (TSN)	Conservation status (IUCN)	General occurrence	Occurrence reference	Included in the analysis
Ammospermophilus nelson	180182	Endangered	California, USA	SDM – Appendix Chap. 3	Yes
Biswamoyopterus biswasi	632491	Critically endangered	Namdapha National Park, Arunachal Pradesh, India	IUCN Red List (Molur 2016)	No
Cynomys mexicanus	632340	Endangered	Central Mexico	SDM – Appendix Chap. 3	Yes
Cynomys parvidens	180187	Endangered	Southern Utah, Norther Arizona, USA	SDM – Appendix Chap. 3	Yes
Eupetaurus cinereus	632492	Endangered	Karakoram Range	IUCN Red List (Zahler 2010)	No
Hylopetes sipora	632500	Endangered	Palau Sipora, Sumatra, Indonesia	IUCN Red List (Lee 2016a)	No
Iomys sipora	632504	Endangered	Mentawai, Sumatra, Indonesia	IUCN Red List (Lee 2016b)	No
Marmota sibirica	632385	Endangered	Mongolia	SDM – Appendix Chap. 3	Yes
Marmota vancouverensis	180142	Critically endangered	Vancouver Island, Canada	SDM – Appendix Chap. 3	Yes
Neotamias palmeri	180198	Endangered	Las Vegas, Nevada, USA	SDM – Appendix Chap. 3	Yes
Paraxerus vincenti	632403	Endangered	Northern Zambezia, Mozambique	IUCN Red List (Kennerley and Peterhans 2016)	No
Prosciurillus weberi	632407	Endangered	Malangke, South Sulawesi, Indonesia	IUCN Red List (Musser et al. 2019)	No
Pteromyscus pulverulentus	632526	Endangered	Malaysia, Sumatra, Indonesia	IUCN Red List (Clayton 2016)	No
Spermophilus citellus	632446	Endangered	Eastern Europe	SDM – Appendix Chap. 3	Yes
Tamiasciurus mearnsi	632478	Endangered	Baja California, Mexico	SDM – Appendix Chap. 3	Yes
Urocitellus brunneus	930316	Endangered	Idaho, Oregon, Washington State, USA	SDM –Appendix Chap. 3	Yes
Xerospermophilus perotensis	930311	Endangered	Mexico City, Mexico	SDM – Appendix Chap. 3	Yes

(2021)). Those were then converted into geo-TIFF (.tif) files, and summarized (merged) in open-source QGIS and ArcGIS using the Raster Calculator analysis tool to create Global distribution hotspots and coldspots by MS. In principle, this tool merges all SDMs into one file which then allows summarizing the distribution of all its componential species at once (as attempted previously by Koprowski and Nandini (2008) with other methods and a smaller sample size).

We then selected the 10 most endangered squirrel species by using the IUCN Red List as a reference (www.iucnredlist.org). In principle, we selected all squirrel species that have been classified as Critically Endangered and Endangered (See Table 1.4 in Steiner and Huettmann (2023 – Chap. 1), and IUCN Red List. These two conservation classes contain in a total of 17 squirrel species, however, since for 7 of them (41%) no distribution data is available from GBIF.org (download DOI: https://doi.org/10.15468/dl.665b59), only 10 squirrels were able to be included. Table 12.1 presents these 17 squirrel species and indicates which ones have been used for this and further analyses. Those 10 were extracted from the SDM set (from Steiner and Huettmann 2023 – Chap. 3) and summarized as hotspot and coldspot maps by using the raster calculator tool in ArcGIS.

Further, we also used the genera *Geosciurus* as one group, as well as *Heliosciurus & Paraxerus* combined as another species group, from the global set of 233 squirrel species. These three species groups have been selected for a specific reason: namely, in our assessment (see Steiner and Huettmann 2023 – Chap. 3) they responded most significantly to climate predictors produced previously (compared to all remaining environmental predictors). The species from the genus *Geosciurus* responded most significantly to the IUCN conservation status classes analysis, and the genera *Heliosciurus* and *Paraxerus* responded most significantly to the IUCN population trend metric of the analysis. The species from the latter two genera have been merged since they occur in the same regions and very similar environments and have responded highly similar to the climate predictors. Similarly, as above, those were then extracted from the SDM set and their hotspot and coldspot maps have been created using the raster calculator tool in ArcGIS.

12.2.2 Climate Scenario Predictor Data

The state of the climate in the year 2100 is uncertain, and not well-agreed upon for a commonly used approach, namely, what models and future scenarios to employ and how to approximate future conditions such as for 2100 (Baltensperger and Huettmann 2015; Chunrong et al. 2016; Reygondeau and Huettmann 2014). Worldclim.org (BioClim) offers an assumed data platform with options to do so, and here we used seven BioClim predictor layers and one elevation predictor to describe an assumed 2100 (Fick and Hijmans 2017). Noteworthy here is that BioClim follows IPCC.org but is not officially approved or endorsed by IPCC (which has its own datasets across the world and with different scales and aims). The elevation layer is also obtained from the Worldclim.org compilation, however, here it is often left out in the discussions as it is a layer of reference that will likely not noticeably change between 2000 and 2100. We then implemented the three scenarios as described by the following authors (Boucher et al. 2020 for IPSL, Tatebe et al. 2019 for MIROC, Yukimoto et al. 2019 for MRI). These climate scenario predictors for the year 2100 have been downloaded in late 2021 (no version number available).

12.2.3 Climate Modeling with Bioclim Predictors and for 2100

The hotspot and coldspot maps were derived from SDMs based on 132 environmental predictors. However, those predictors do not all exist for 2100. Therefore, we used instead agreed-upon 2100 matching proxy predictors from Worldclim.org (BioClim) to transfer models in the climate space. Namely, we used BIO1, BIO7, BIO 10, BIO11, BIO12, BIO16, and BIO 17 (see predictor overview in the table of Appendix 12.3). We recognize their limits while making models coarser when using 7 predictors instead of 133 but in the absence of better information on a global scale for 2100, that is what has been used, as commonly done elsewhere across locations and disciplines (Baltensperger and Huettmann 2015; Huettmann et al. 2017; Chunrong et al. 2016; Reygondeau and Huettmann 2014). In order to present different climate scenarios, we have used three Global Climate Models (GCMs) that have also been utilized and provided by WorldClim.org (MIROC6, MRI-ESM2-0, and IPSL-CM6A-LR). For this study, the MIROC6 scenario may be considered as the temperature decrease scenario, even as a certain cooling scenario, with a decreased global temperature of approx. 1 °C (Tatebe et al. 2019). The MRI-ESM2-0 scenario is considered as a low-medium temperature increase of an approximate global increase of 2 °C (Yukimoto et al. 2019). Although problematic, one might refer to it as a '*business as usual*' model (we assume that the status quo is more severe). Lastly, the IPSL-CM6A-LR scenario is considered as a medium-high increase of temperature by approx. 3 °C (Boucher et al. 2020). While this is perceived as a high/extreme scenario, it should be stated that climate change – and when left unabated – has no real limits thus far and can probably increase way beyond ten degrees Celsius (see Meltofte et al. 2008 for parts of the Arctic easily reaching 12 degrees Celsius and more), also around the three poles (Huettmann 2012; Infante 2012). As a matter of fact, the poles hosted dinosaurs in earth's history and water levels were much higher. Temperatures so hot that such areas were widely inhospitable for mammals and something relatively common for the earth's history, or the universe's history.

We then modeled the four squirrel cohorts (Global Squirrels, Top10 Endangered Squirrels, *Geosciurus*, as well as *Heliosciurus* with *Paraxerus*), with those seven Bioclim predictor layers (see predictor overview in the table of Appendix 12.3) for the three scenarios of 2100 (IPSL, MRI, and MIROC).

To be more robust and stable, and to increase the models' accuracy/quality, we used three ensemble-based leading machine learning (ML) algorithms to do so: Random Forest & TreeNet (https://www.minitab.com/en-us/products/spm/), and Maxent (https://biodiversityinformatics.amnh.org/open_source/maxent/) (details in Hegel et al. 2010; Humphries et al. 2018).

For the mapping visualization, we used the Jenks (Natural Break) Legend with 5 categories in ArcGIS as this shows sufficient details of the changes, the major breaks, and it suits the step-function nature of the (tree-based) ML algorithms (Breiman et al. 1984).

It is not our intention to focus on individual model details and differences here but instead let the models predict to their abilities and then overall infer in concert (sensu Breiman et al. 1984; Breiman 2001a), using the common trends within the model predictions and scenarios and infer on those for prioritization and progress. As model predictions are made available, this can be fine-tuned as needed.

We then summarized generic evidence trends from those models and present them in a summary table (for an approach see for instance Infante 2012). It must be mentioned that the predictions are based on a future scenario and carry no known quantitative certainty and serve only as an estimate/index based on our models. To our knowledge, it is the best one available though. The better the models, the higher the predictive accuracy, but even with a high number of predictors and robust models, they can hardly predict the real future distribution.

12.3 Results

12.3.1 TreeNet Model 2000 vs 2100

12.3.1.1 Global Squirrels

The TreeNet Model of Global Squirrels reproduces well but somewhat underpredicts (Appendix 12.1). For the medium scenario (MRI), the generic trend shows a more fragmented pattern for Europe and a decline in West Africa. Additionally, it shows a range extension in Central Asia, whereas Southeast Asia declines. For the hot scenario (IPSL), the generic trend from the medium scenario continues. For the cold scenario (MIROC), the generic trend shows a widening of the distribution worldwide, but the cold regions remain free of squirrels (Fig. 12.2).

12.3.1.2 World's Top 10 Endangered Squirrels

Here again, the TreeNet Model of the World's Top 10 Endangered Squirrels reproduces well but underpredicts for Europe and North America (Appendix 12.1). For the medium scenario (MRI), the generic trend shows an increase in Europe and North America but a decline in Central America. For the hot scenario (IPSL), the generic trend shows a wider increase. For the cold scenario (MIROC), the generic trend shows a southern movement and decline for Central Asia, next to a punctual increase in North America (Fig. 12.3).

12.3.1.3 *Geosciurus*

The TreeNet Model of the *Geosciurus* squirrels reproduces well but also generally underpredicts (Appendix 12.1). For the medium scenario (MRI), the generic trend shows a generic decline. For the hot scenario (IPSL), the generic trend of decline prevails. For the cold scenario (MIROC), the generic trend also shows a decline (Fig. 12.4).

12.3.1.4 *Heliosciurus* Merged with *Paraxerus*

The TreeNet Model of the *Heliosciurus* and *Paraxerus* squirrels reproduces generally well (Appendix 12.1). For the medium scenario (MRI) the generic trend shows an increase for Africa and a slight increase in Latin America. For the hot scenario (IPSL) the generic trend remains and also shows an increase for Central America. For the cold scenario (MIROC) the generic trend shows a decline from the status quo (Fig. 12.5).

12.3.2 Random Forest Model Predictions 2000 vs 2100

12.3.2.1 Global Squirrels

The Random Forest model of Global Squirrels reproduces well but shows a general underprediction (Appendix 12.1). For the medium scenario (MRI), the generic trend shows an extension for North America and also into Central and South-East Asia. For the hot scenario (IPSL), the generic trend shows a similar increase with a more pronounced pattern towards higher altitude regions. For the cold scenario (MIROC), the generic trend shows the status quo but shows a high increase in its initial distribution patches.

12.3.2.2 World's Top 10 Endangered Squirrels

The Random Forest model of the World's Top 10 Endangered Squirrels reproduces well overall but with an underprediction (Appendix 12.1). For the medium scenario (MRI), the generic trend shows a retraction for Europe and Central Asia. For the hot scenario (IPSL), the generic trend shows a strong move northward, specifically into high arctic regions. For the cold scenario (MIROC), the generic trend shows a move southward towards the equator regions.

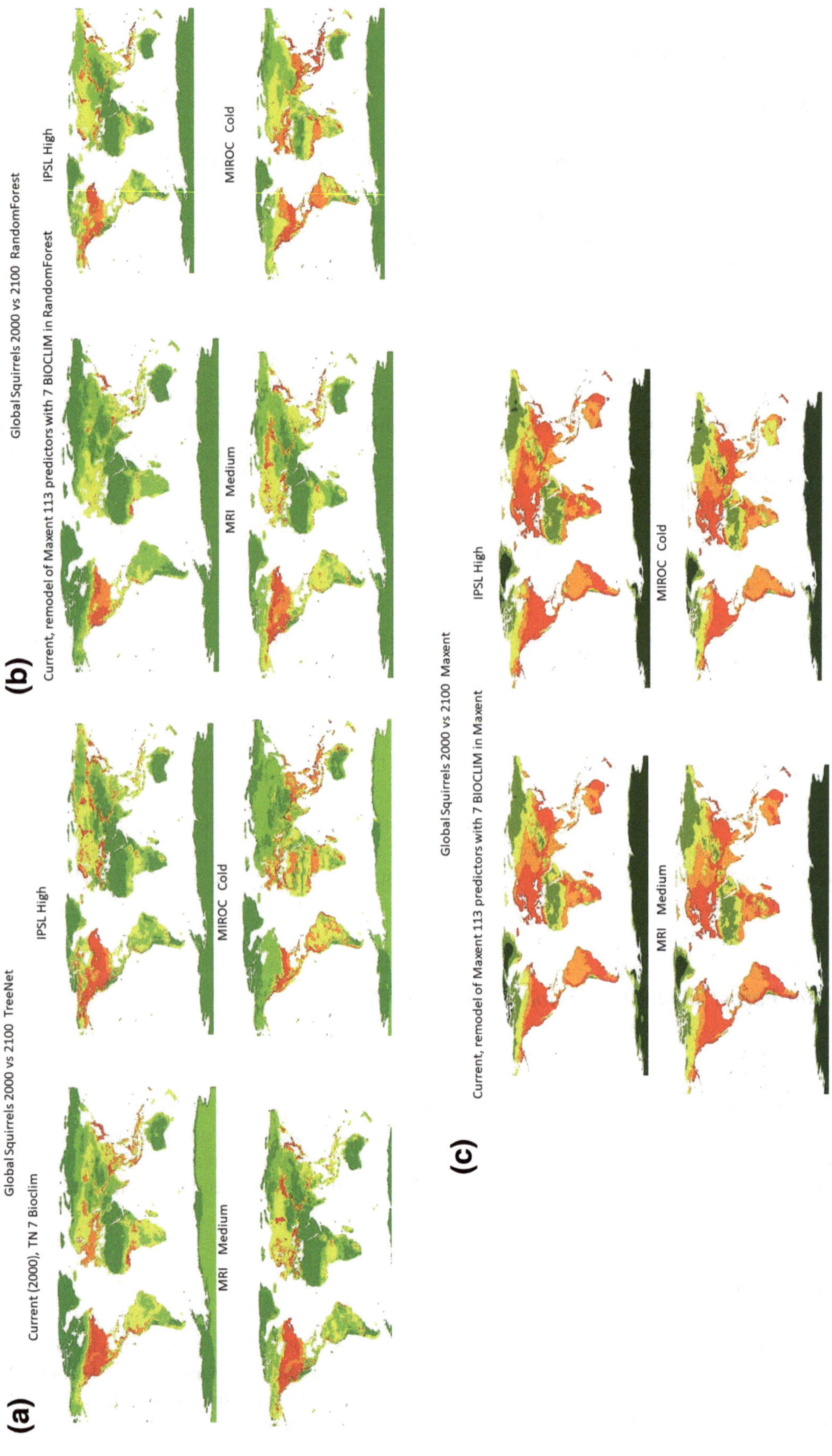

Fig. 12.2 Three climate change scenarios for Global Squirrel hotspots and coldspots for 2100 **(a)** Treenet prediction, **(b)** RandomForest, and **(c)** Maxent

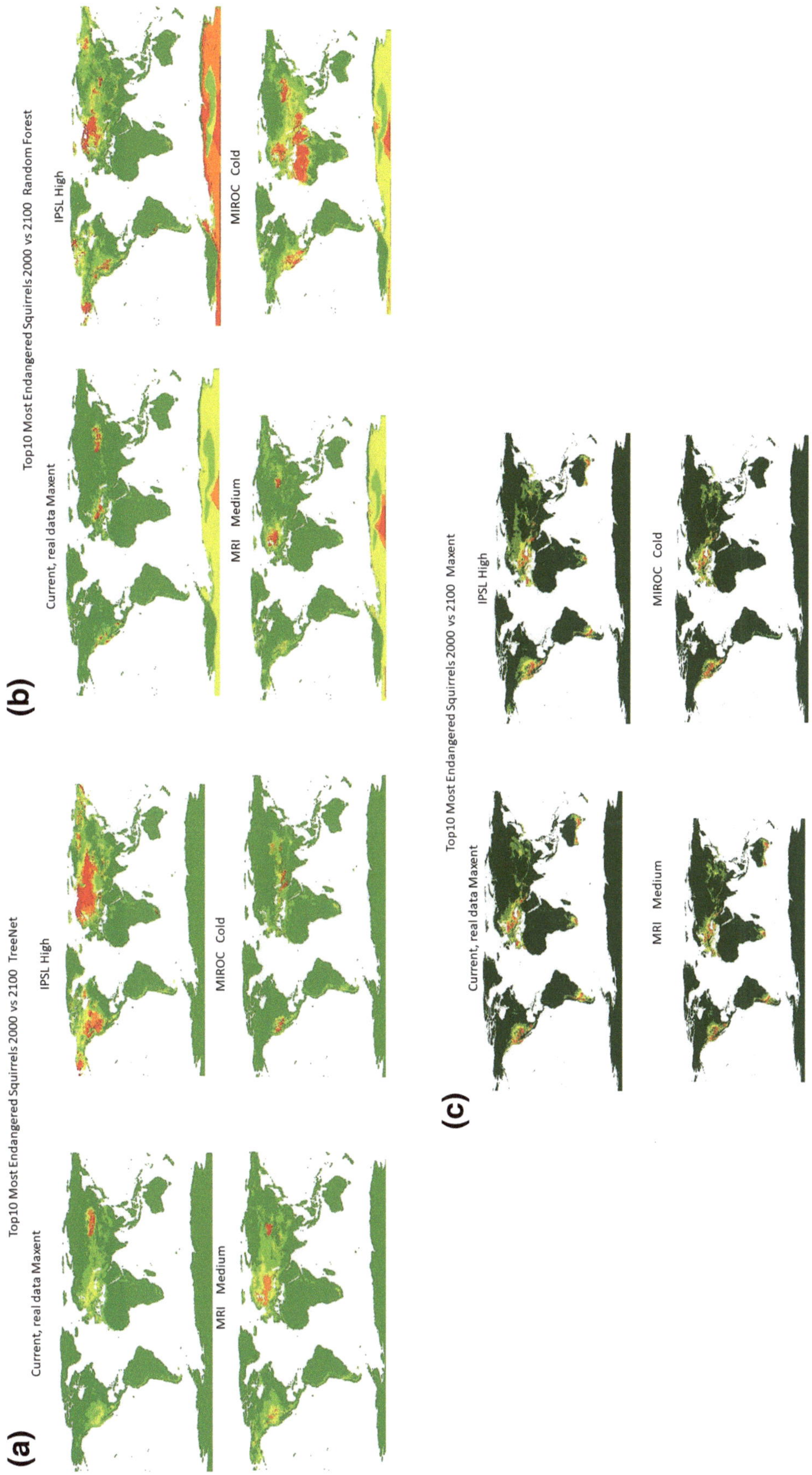

Fig. 12.3 Three climate change scenarios for the World's Top Ten Endangered Squirrel hotspots and coldspots for 2100 **(a)** Treenet prediction, **(b)** RandomForest, and **(c)** Maxent

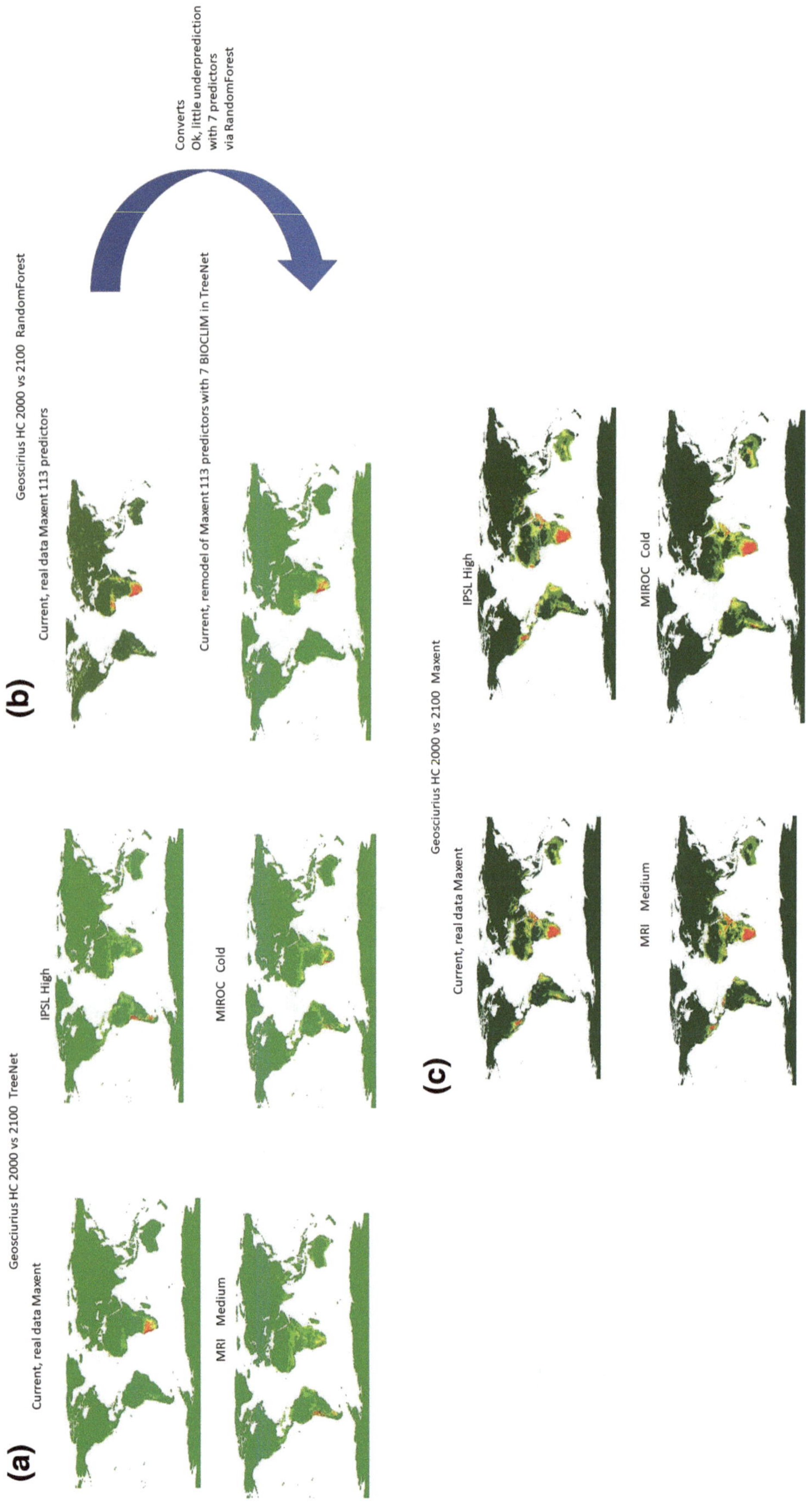

Fig. 12.4 Three climate change scenarios for *Geosciurus* hotspots and coldspots for 2100 (**a**) Treenet prediction, (**b**) RandomForest, and (**c**) Maxent

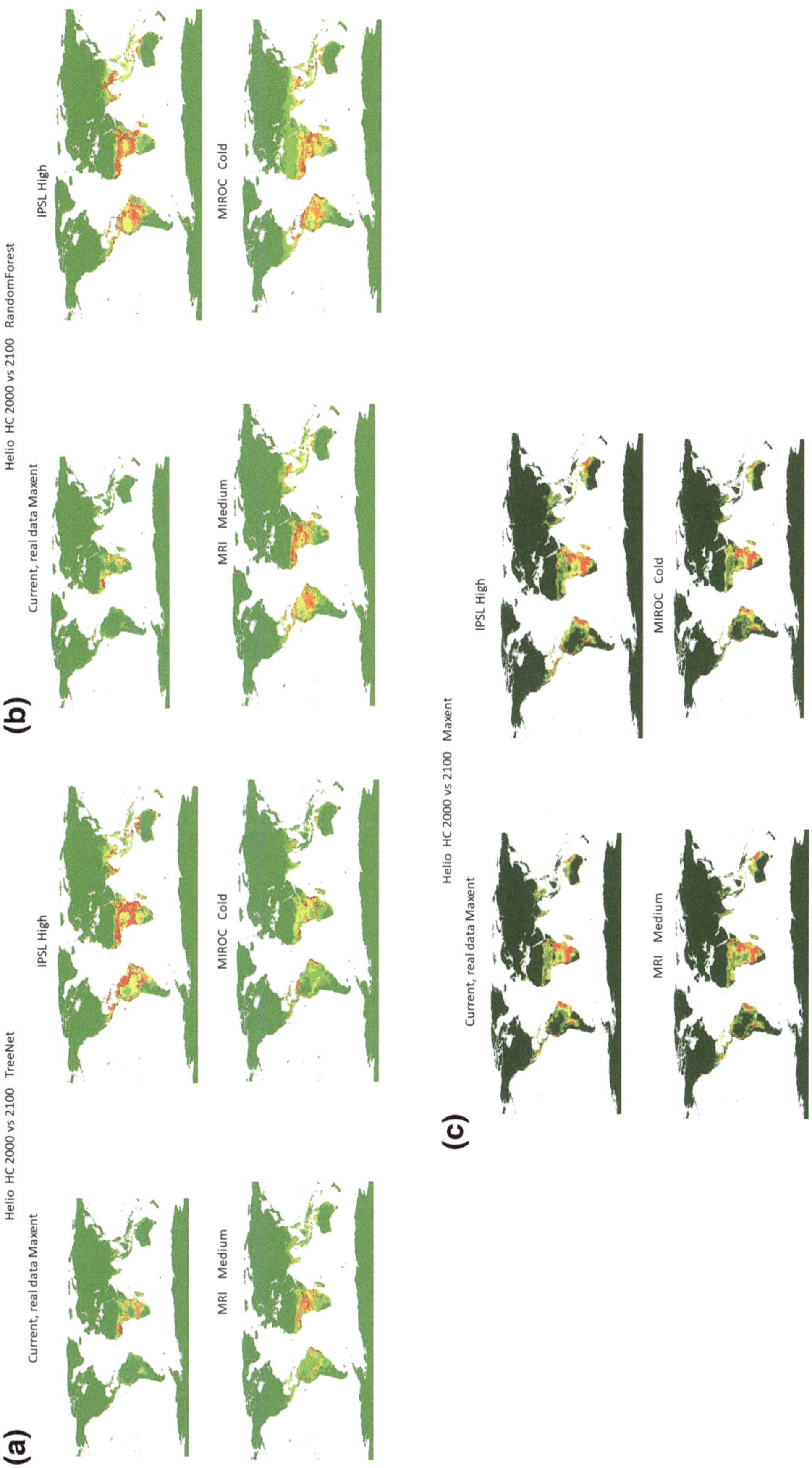

Fig. 12.5 Three climate change scenarios for *Heliosciurus* and *Paraxerus* hotspots and coldspots for 2100 **(a)** Treenet prediction, **(b)** RandomForest, and **(c)** Maxent

12.3.2.3 *Geosciurus*

The Random Forest model of the *Geosciurus* squirrels reproduces well, however, with an underprediction (Appendix 12.1). For the medium scenario (MRI), the generic trend shows an extension into West Africa, but also for parts of Latin America and northern Australia (where they currently do not occur, but such a niche is predicted there, e.g. for sister taxa, vacant, etc). For the hot scenario (IPSL), the generic medium trend is less pronounced. For the cold scenario (MIROC), the generic trend shows a strong increase for North and South Africa.

12.3.2.4 *Heliosciurus* merged with *Paraxerus*

The Random Forest model of the *Heliosciurus* and *Paraxerus* squirrels reproduces well with a small underprediction (Appendix 12.1). For the medium scenario (MRI) the generic trend shows an increase in existing areas and a new distribution patch in Central and Latin America. For the hot scenario (IPSL), the generic trend shows a similar pattern to the medium regime but is slightly increased. For the cold scenario (MIROC), one can observe a generic trend moving southwards and an increase in the distribution range.

12.3.3 Maxent Model Prediction 2000 vs 2100

12.3.3.1 Global Squirrels

The Maxent Model of Global Squirrels reproduces ok but shows some overpredictions (Appendix 12.1). For the medium scenario (MRI), the generic trend shows no relevant changes. For the hot scenario (IPSL), the generic pattern from the 2000s also remains highly similar. For the cold (MIROC) scenario, it shows some contraction.

12.3.3.2 World's Top 10 Endangered Squirrels

The Maxent Model of the World's Top 10 Endangered Squirrels reproduces not so well (Appendix 12.1). For the medium scenario (MRI), there is no big change in the trend. For the hot scenario (IPSL), the main patterns also remain and show no relevant changes. The same applies to the cold scenario (MIROC).

12.3.3.3 *Geosciurus*

The Maxent Model of the *Geosciurus* squirrels reproduces in an acceptable fashion overall but shows some overpredictions and some underpredictions as well (Appendix 12.1). For the medium scenario (MRI), the generic pattern remains. For the hot scenario (IPSL), the generic trend also remains. For the cold scenario (MIROC), the generic trend shows a small increase around existing hotspots.

12.3.3.4 *Heliosciurus* and *Paraxerus*

The Maxent model of the *Heliosciurus* and *Paraxerus* squirrels reproduces ok but underpredicts (Appendix 12.1). For the medium scenario (MRI), the generic trend shows a wider and concentrated range patch. For the hot scenario (IPSL), the generic trend shows an extension. For the cold scenario (MIROC), the generic trend remains the status quo.

12.3.4 Forecast Distribution Meta-analysis Summary

Table 12.2 summarizes our findings presented in Sect. 12.3. This summary indicates that across the three scenarios and the three algorithms used, a drastic (tendentially negative) change in the global distribution range of squirrels if we continue to pursue the 'business-as-usual' approach. One always must keep in mind that species such as small mammals (squirrels) can

Table 12.2 Meta-Analysis summary of the Distribution Metrics for hotpots and coldspots 2100 for Global Squirrels, Top 10 Endangered Squirrels, *Geosciurus* and *Heliosciurus* & *Paraxerus*

	Feature of distribution								
Model type	Core zone enlarge-ment	Core zone Shrinkage	General core shift	Core zone shift North-wards	Core zone shift upwards in elevation	Shift towards flat area	Core zone fragmen-tation	Range de-clines	**Overall average per row**
All squirrels	0.44	0.44	0.67	0.44	0.56	0.00	0.56	0.67	**0.47**
Ten most threatened Squirrels	0.44	0.44	0.78	0.67	0.44	0.11	0.56	0.44	**0.49**
Geosciurus	0.11	0.67	0.67	0.22	0.33	0.11	0.56	0.67	**0.42**
Helioscius & *Paraxerus*	0.67	0.22	0.22	0.00	0.00	0.22	0.00	0.11	**0.18**
Overall average per column	**0.42**	**0.44**	**0.59**	**0.33**	**0.33**	**0.11**	**0.42**	**0.47**	**0.39/0.39**

hardly move (and establish new suitable habitat sites) 100+km within the upcoming 80 years. So even small niche changes on the map can be a big hurdle to overcome and cope with for small animals like squirrels. We use a somewhat parsimonious approach and while most of our models underpredict reality, when compared with the initial hotspot and coldspot maps, many metrics of the distribution will still change dramatically either way. Specifically, the 'ten Most Threatened Squirrels' will be affected strongly. Generally, the most changes that are predicted to happen are observed for the metric "General core shift" which indicates a general shift of the core population/distribution. Followed by the metric "Range decline", which indicates a general decline in the squirrel group's range. After these metrics, the most changes can be observed for the two metrics "Core zone shrinkage", and "Core zone fragmentation". "Core zone shrinkage" indicates in contrast to "Range declines", the decrease in size of the core distribution, compared to the overall distribution. The least amount of changes are observed for the metric "Shift towards flat areas". Thereby, a general shift towards flatter areas is not so likely, but rather towards higher altitudes, following the presented models, and this study's hypothesis.

Table 12.2 represents a meta-analysis summary of the distribution metrics for hotpots and coldspots for the year 2100 for Global Squirrels, Top 10 Endangered Squirrels, *Geosciurus*, *Heliosciurus* & *Paraxerus*. This summary has been created based on the table presented in Appendix 12.2. There, more details can be observed, where for each created model the changes from the current distribution and the modeled 2100 distribution are being described by the same metrics as in Table 12.2. The actual shifts could be smaller than predicted due to lagged responses, which in turn can increase stress factors for the species, leading possibly earlier to extinction than predicted (see theory on niche tracking e.g. Lee 2021).

The calibration results of the TreeNet and RandomForest model runs can be found in Appendix 12.1. These indicate the model results for each of the four squirrel groups and the predicted changes, illustrated as maps for 2100. The accuracy of these models is determined by the quality and the number of the predictors used, as well as the modeling algorithm of each software. In order to avoid any kinds of software biases, these models have been created and presented in a comparative manner, which allows anyone to compare them to each other and form their own conclusion from these models (let the models speak for themselves – *sensu* Breiman 2001b).

12.4 Discussion

Man-made climate change remains widely unabated, and the associated CO_2 release is widely not controlled, with often poor, lacking, or failing future outlooks (Arar and Southgate 2009; Baes et al. 1977; Hale 2010; Jamieson 2014). Using the best available open access BIG DATA, here we were able to look at the best predicted SDM summary for over 230 squirrel species – a group that is somewhat ignored and marginalized with lacking science-based management, funding, and efforts; even the taxonomy is not agreed upon (Steiner and Huettmann 2023 – Chap. 1; Steiner and Huettmann 2021).

We created the first globally important hotspot and coldspot maps and modeled them forward with bioclimatic variables from Worldclim.org (BIOCLIM), using 3 machine learning algorithms (for TreeNet, Random Forest, and Maxent), globally with 0.5-degree pixel accuracy.

Table 12.3 Selection of landscapes and habitats affected for squirrels by our 2100 models

Landscape/habitat	Selection of countries affected	Comment
Europe	Germany, France, Switzerland, UK, Italy, etc.	A heavy fragmented squirrel habitat and clear shift to Scandinavian/North-Eastern countries
Western North America	U.S.	A key region for squirrel biodiversity with high abundances
Latin America	Brazil, Ecuador	A key region for squirrel biodiversity
Central America	Costa Rica, Panama	A key region for squirrel biodiversity
Central Asia	Kazakhstan	A key region for squirrel range expansion with higher squirrel density and diversity
North Africa	Algeria, Morocco, Tunisia	Becomes less interesting for an increased warming scenario (MRI, IPSL), but more interesting for cooling scenarios (MIROC)
South Africa	South Africa	Becomes a habitat less desired and populated by squirrels
S.E. Asia	Indonesia, Malaysia	A very diverse landscape with high squirrel biodiversity, but also extinction
Islands	Globally	A key area affected by climate change shifts
Mid-elevation mountains	Italy, Austria, Switzerland, Alps	A key region for squirrel refugium
Tropics	Brazil, Congo, Indonesia, Malaysia	A key region for squirrel biodiversity
Boreal Forest	Russia, Canada, Alaska	A key region for northern squirrel species

We then used three climate scenarios (Global Climate Models – GCM), namely MRI, IPSL, and MIROC. Those come from a wide variety of possible climate scenario options. To start the rapid assessment here, we then tried to show three scientifically accepted climate scenario models and apply them to the world's squirrels and some genera belonging to them. Our results indicate underpredictions but already show a generic distribution shift for the majority of squirrel species, especially for the World's Top Ten most endangered Squirrel species (carrying the conservation status classes Critically Endangered, and Endangered). Most importantly, we see a shift in the core ranges, as well as a fragmentation of the squirrel distributions. Such patterns are known to result in additional population stress factors, often extinction, especially in island environments (Fitak et al. 2013; Hanski 2015; Koprowski et al. 2005; Mimura 1999; Seltmann et al. 2017). Arguably, the ecosystems change as well, and thus with human-driven CO_2 and Greenhouse Gases (GHGs) one may expect vast shifts overall globally, with squirrels as the collateral. To summarize, an overview of some globally observed trends has been created (Table 12.3).

As seen in the maps and a selection presented in Table 12.3, for landscapes and habitats affected, it is clear that Central America, as well as Latin America, are future conservation hotspots for squirrels, even genera that currently do not occur in this part of the world would flourish there well. The same can be said for Central Europe, parts of western North America, Central Asia, parts of North and South Africa, and the entirety of South-East Asia. Islands should receive the most attention, as well as some mid-elevation mountain areas, the tropics overall, and also the boreal forest and parts of Patagonia.

These indicated squirrel hotspot regions correlate not surprisingly with the global hotspots of zoonotic disease transmission recently published by (Han et al. 2016). Especially the disease transmission for rodents correlates with the squirrel hotspots. Within these squirrel hotspots, one can find rural, and suburban areas, but also urban areas with a high human density. All this together indicates that the frequency of zoonotic disease transmission between rodents (squirrels) and humans is on the rise, and likely negatively influencing both parties (Gibb et al. 2020; Estrada-Peña et al. 2014; Wolfe et al. 2005).

The approach presented by us here aims to utilize holistic assessment methods and to initiate/present a workflow with data (Jenkins et al. 2013). We here tried to present the global species trend and some rough subdivisions in order to publish a global big-picture of the situation for rapid assessment actions. In-depth analyses are always a follow-up option that can be achieved starting with the data and methods used and presented here, e.g. by using a regional or species-specific approach (see the Tropics in Steiner and Huettmann 2023 – Chap. 6 and Islands Steiner and Huettmann 2023 – Chap. 7, etc.). The rapid assessment methods used here, primarily aim to present and start such views and initialize debates and discussions on this topic. Without acknowledging a marginalized and undesired scenario/outlook, no betterments can be expected.

We would also like to highlight the fact that these SDFs could only be created for 233 species instead of all global squirrel species. As mentioned above, this is due to the missing occurrence data available in open-source databases. It is a result of our work to highlight those species with and without data for improvements and further research efforts; it sets a new research agenda for the world's squirrels, which did not really have one before. Ideally, research efforts and funding should directly flow into minimizing the number of Data Deficient species and collecting occurrence data of these Data Deficient and endangered species. If that can be achieved in the near future, the future of the global squirrels is a bit less uncertain, and more research can be conducted to finally sustainably conserve the endangered squirrel species. Once there is enough data available for all squirrel species, it can be evaluated which species are most threatened and, thus, have the highest conservation needs. From there onwards, conservation scientists and projects should ideally focus on implementing sustainable conservation measures to guarantee the long-term survival of these Data Deficient and endangered species. With all species well assessed, the conservation priorities amongst the whole family might change drastically. Even one properly assessed species can change local and regional conservation properties. An example in the Western world with one of the richest nations and states in the world is the Arizona Grey Squirrel (*Sciurus arizonensis*), which actually is considered Data Deficient (Linzey et al. 2019). It is one of the only Data Deficient squirrel species in the Southwestern part of the U.S., where other than that species there are a few other endangered species (Palmer's Chipmunk (*Neotamias palmeri*), Nelson's Antelope Squirrel (*Ammospermophilus nelsoni*), Utah Prairie Dog (*Cynomys parvidens*), and Mearns' Squirrel (*Tamiasciurus mearnsi*)). Proper assessment of *S. arizonensis* might indicate that it is heavily endangered, meaning that conservation efforts should be reinforced and likely reprioritized over some of the current projects ongoing for the other endangered species. Therefore, we would urge to focus on global conservation assessment initiatives and basic research on the life history of all the understudied squirrel species (approximately >95% of all squirrel species) in an efficient policy manner focusing on habitat protection and maintenance.

While our models just deal with bioclimatic predictors as proxies for the future, the real-world changes in the next 100 years are likely much bigger, more complex, and more severe. For instance, a human population increase is expected, higher consumption of natural resources (UN 2011), increased contamination (Fuller et al. 2022), more pandemics (Smith 2021), and loss of wilderness (Li et al. 2022). We believe that our models represent a scenario and are a minimum estimate of what is to come and what squirrels are facing, and those findings should present a good foundation for sustainable actions. It should start a discussion and a template from where it can be worked from with all relevant data in the open access realm documented with metadata to do so.

We acknowledge that the true future remains unknown; there is no single-bullet solution to knowing what 2100 will be like. Here we had to use a narrow and parsimonious approach still and start the process for squirrels of the world. But arguably, the patterns and trends we see are robust, and they are already a concern and most of them are not in favor of a good future for these species in the Anthropocene (Baker and Harris 2015; Ellis 2013; Huettmann 2012; Ramankutty et al. 2018; Shorohova et al. 2011; Song et al. 2018; Taheri et al. 2021).

Data Availability: All data generated or analyzed during this study are included in this published study (and its Supplementary Information files).

References

Arar JI, Southgate D (2009) Evaluating CO2 reduction strategies in the US. Ecol Model 220:582–588

Baes CF, Goeller HE, Olson JS, Rotty RM (1977) Carbon Dioxide and Climate: the uncontrolled experiment: possibly severe consequences of growing CO_2 release from fossil fuels require a much better understanding of the carbon cycle, climate change, and the resulting impacts on the atmosphere. Am Sci 65:310–320

Baker PJ, Harris S (2015) Urban mammals: what does the future hold? An analysis of the factors affecting patterns of use of residential gardens in Great Britain. Mamm Rev 37:297–315

Baltensperger AP, Huettmann F (2015) Predictive spatial niche and biodiversity hotspot models for small mammal communities in Alaska: applying machine-learning to conservation planning. Landscape Ecology 30(4):681–697

Boucher O, Servonnat J, Albright AL, Aumont O, Balkanski Y, Bastrikov V, Vuichard N et al (2020) Presentation and evaluation of the IPSL-CM6A-LR climate model. J Adv Model Earth Syst 12(7):e2019MS002010

Breiman L (2001a) Random forests. Mach Learn 45:5–32

Breiman L (2001b) Statistical modeling: the two cultures (with comments and a rejoinder by the author). Stat Sci 16(3):199–231

Breiman L, Friedman J, Olshen R, Stone C (1984) Classification and regression trees. Wadsworth Int. Group. 37:237–251

Chunrong M, Huettmann F, Guo Y (2016) Climate envelope predictions indicate an enlarged suitable wintering distribution for Great Bustards (Otis tarda dybowskii) in China for the 21st century. PeerJ 4:e1630

Clayton E (2016) Pteromyscus pulverulentus. The IUCN Red List of threatened species 2016: e.T18703A22245307. https://doi.org/10.2305/IUCN.UK.2016-2.RLTS.T18703A22245307.en. Accessed 22 July 2021

Elith J, Leathwick JR (2009) Species distribution models: ecological explanation and prediction across space and time. Annu Rev Ecol Evol Syst 40:677–697

Elith J, Graham CH, Anderson RP, Dudík M, Ferrier S, Guisan A, Zimmermann EN et al (2006) Novel methods improve prediction of species' distributions from occurrence data. Ecography 29(2):129–151

Ellis CJ (2013) A risk-based model of climate change threat: hazard, exposure, and vulnerability in the ecology of lichen epiphytes. Botany 91:1–11

Estrada-Peña A, Ostfeld RS, Peterson AT, Poulin R, de la Fuente J (2014) Effects of environmental change on zoonotic disease risk: an ecological primer. Trends Parasitol 30:205–214

Fick SE, Hijmans RJ (2017) WorldClim 2: new 1-km spatial resolution climate surfaces for global land areas. Int J Climatol 37:4302–4315

Fitak RR, Koprowski JL, Culver M (2013) Severe reduction in genetic variation in a montane isolate: the endangered Mount Graham red squirrel (Tamiasciurus hudsonicus grahamensis). Conserv Genet 14:1233–1241

Fuller R, Landrigan PJ, Balakrishnan K, Bathan G, Bose-O'Reilly S, Brauer M, Yan C (2022) Pollution and health: a progress update. Lancet Planet Health

Gibb R, Redding DW, Chin KQ, Donnelly CA, Blackburn TM, Newbold T, Jones KE (2020) Zoonotic host diversity increases in human-dominated ecosystems. Nature 584(7821):398–402

Hale S (2010) The new politics of climate change: why we are failing and how we will succeed. Environ Polit 19:255–275

Han BA, Kramer AM, Drake JM (2016) Global patterns of zoonotic disease in mammals. Trends Parasitol 32:565–577

Hanski I (2015) Habitat fragmentation and species richness. J Biogeogr 42:989–993

Hegel TM, Cushman SA, Evans J, Huettmann F (2010) Current state of the art for statistical modelling of species distributions. In: Spatial complexity, informatics, and wildlife conservation. Springer, Tokyo, pp 273–311

Huettmann F (ed) (2012) Protection of the three poles. Springer

Huettmann F, Franklin SE, Stenhouse GB (2005) Predictive spatial modelling of landscape change in the Foothills Model Forest. For Chron 81:525–537

Huettmann F, Magnuson EE, Hueffer K (2017) Ecological niche modeling of rabies in the changing Arctic of Alaska. Acta Veterinaria Scandinavica 59:1–11

Humphries GR, Magness DW, Huettmann F (2018) Machine learning for ecology and sustainable natural resource management. Springer, Cham

Infante CGR (2012) Meta-analysis on the effects of global economic growth on birds in the nations of the three poles. Doctoral dissertation

Jamieson D (2014) Reason in a dark time: why the struggle against climate change failed – and what it means for our future. Oxford University Press

Jenkins CN, Pimm SL, Joppa LN (2013) Global patterns of terrestrial vertebrate diversity and conservation. Proc Natl Acad Sci 110:E2602–E2610

Kennerley R, Peterhans KJ (2016) Paraxerus vincenti (errata version published in 2017). The IUCN Red List of threatened species 2016: e.T16212A115132722. https://doi.org/10.2305/IUCN.UK.2016-3.RLTS.T16212A22243323.en. Accessed 22 July 2021

Koprowski JL, Nandini R (2008) Global hotspots and knowledge gaps for tree and flying squirrels. Curr Sci, pp 851–856

Koprowski JL, Alanen MI, Lynch AM (2005) Nowhere to run and nowhere to hide: response of endemic Mt. Graham red squirrels to catastrophic forest damage. Biol Conserv 126:491–498

Lawler JJ, Wiersma YF, Huettmann F (2011) Using species distribution models for conservation planning and ecological forecasting. In: Predictive species and habitat modeling in landscape ecology. Springer, New York, pp 271–290

Lee B (2016a) Hylopetes sipora. The IUCN Red List of threatened species 2016: e.T10606A22243951. https://doi.org/10.2305/IUCN.UK.2016-2.RLTS.T10606A22243951.en. Accessed 22 July 2021

Lee B (2016b) Iomys sipora (errata version published in 2017). The IUCN Red List of threatened species 2016: e.T10846A115099877. https://doi.org/10.2305/IUCN.UK.2016-3.RLTS.T10846A22249119.en. Accessed 22 July 2021

Lee H (2021) Climate change biology. Academic

Li F, Li W, Li F, Long Y, Guo S, Li X, Li J et al (2022) Global projections of future wilderness decline under multiple IPCC Special Report on Emissions Scenarios. Resour Conserv Recycl 177:105983

Linzey AV, Timm R, Álvarez-Castañeda ST, Castro-Arellano I, Lacher T (2019) Sciurus arizonensis. The IUCN Red List of threatened species 2019: e.T20005A22247935. https://doi.org/10.2305/IUCN.UK.2019-3.RLTS.T20005A22247935.en. Accessed 29 Aug 2022

Lombardia R, Piemonte R, Liguria R, Bertolino S (2019) Preventing grey squirrel spread in northwest Italy. Natural Resources Wales (NRW)

Meltofte H, Christensen TR, Elberling B, Forchhammer MC, Rasch M (2008) High-arctic ecosystem dynamics in a changing climate. Academic Press Elsevier

Milanovich JR, Peterman WE, Barrett K, Hopton ME (2012) Do species distribution models predict species richness in urban and natural green spaces? A case study using amphibians. Landsc Urban Plan 107:409–418

Mimura N (1999) Vulnerability of island countries in the South Pacific to sea level rise and climate change. Clim Res 12:137–143

Molur S (2016) Biswamoyopterus biswasi (errata version published in 2017). The IUCN Red List of threatened species 2016: e.T2816A115063959. https://doi.org/10.2305/IUCN.UK.2016-3.RLTS.T2816A22271554.en. Accessed 22 July 2021

Musser G, Dando T Kennerley R (2019) Prosciurillus weberi. The IUCN Red List of Threatened Species 2019: e.T18365A22256206. https://doi.org/10.2305/IUCN.UK.2019-1.RLTS.T18365A22256206.en. Accessed on 14 March 2023

Nielsen SE, Stenhouse GB, Beyer HL, Huettmann F, Boyce MS (2008) Can natural disturbance-based forestry rescue a declining population of grizzly bears? Biol Conserv 141:2193–2207

Peters MP, Prasad AM, Matthews SN, Iverson LR (2020) Climate change tree atlas, Version 4. U.S. Forest Service, Northern Research Station and Northern Institute of Applied Climate Science, Delaware. https://www.fs.usda.gov/nrs/atlas

Ramankutty N, Mehrabi Z, Waha K, Kremen C, Herrero M, Rieseberg LH (2018) Trends in global agricultural land use: implications for environmental health and food security

Reygondeau G, Huettmann F (2014) Past, present and future state of pelagic habitats in the Antarctic Ocean. Biogeographic Atlas of the Southern Ocean, pp 397–403

Seltmann A, Czirják GÁ, Courtiol A, Bernard H, Struebig MJ, Voigt CC (2017) Habitat disturbance results in chronic stress and impaired health status in forest-dwelling paleotropical bats. Conserv Physiol 5(1)

Shorohova E, Kneeshaw D, Kuuluvainen T, Gauthier S (2011) Variability and dynamics of old-growth forests in the circumbolear zone: implications for conservation, restoration and management

Smith J (2021) Q&A: future pandemics are inevitable, but we can reduce the risk. EU Commission. https://ec.europa.eu/research-and-innovation/en/horizon-magazine/qa-future-pandemics-are-inevitable-we-can-reduce-risk#:~:text=Coronavirus%20will%20not%20be%20the,higher%20now%20than%20ever%20before. Accessed 29 Aug 2022

Song XP, Hansen MC, Stehman SV, Potapov PV, Tyukavina A, Vermote EF, Townshend JR (2018) Global land change from 1982 to 2016. Nature 560(7720):639–643

Sriram S, Huettmann F. A global model of predicted Peregrine Falcon (Falco peregrinus) distribution with open source GIS code and 104 Open Access Layers for use by the global public. Earth System Science Data Discussions, 1–39 (Unpublished)

Steiner M, Huettmann F (2021) Justification for a taxonomic conservation update of the rodent genus Tamiasciurus: addressing marginalization and mis-prioritization of research efforts and conservation laissez-faire for a sustainability outlook. Eur Zool J 88:86–116

Taheri S, Naimi B, Rahbek C, Araújo MB (2021) Improvements in reports of species redistribution under climate change are required. Sci Adv 7(15):eabe1110

Tatebe H, Ogura T, Nitta T, Komuro Y, Ogochi K, Takemura T, Kimoto M et al (2019) Description and basic evaluation of simulated mean state, internal variability, and climate sensitivity in MIROC6. Geosci Model Dev 12(7):2727–2765

Thorington RW Jr, Koprowski JL, Steele MA, Whatton JF (2012) Squirrels of the world. JHU Press

UN (2011) Humanity's voracious consumption of natural resources unsustainable – UN report. United Nations. https://news.un.org/en/story/2011/05/374942#:~:text=According%20to%20the%20report%20by,decoupled%E2%80%9D%20from%20natural%20resource%20use. Accessed 29 Aug 2022

Wolfe ND, Daszak P, Kilpatrick AM, Burke DS (2005) Bushmeat hunting, deforestation, and prediction of zoonotic disease. Emerg Infect Dis 11:1822

Yukimoto S, Kawai H, Koshiro T, Oshima N, Yoshida K, Urakawa S, Ishii M et al (2019) The Meteorological Research Institute Earth System Model version 2.0, MRI-ESM2. 0: description and basic evaluation of the physical component. J Meteorol Soc Jpn Ser II

Zahler P (2010) Eupetaurus cinereus. The IUCN Red List of threatened species 2010: e.T8269A12904144. https://doi.org/10.2305/IUCN.UK.2010-2.RLTS.T8269A12904144.en. Accessed 22 July 2021

Chapter 13
Squirrel's Marginalization and Modern Lack of Conservation, and a Poor Sustainability Outlook as a Call to Good Action

Abstract Judged by all assessment data and modern governance options at hand, the conservation status of the world's wild squirrels is arguably in an undesirable, marginalized state, with an even worse outlook for modernity and its trends predicted for 2100. Very few nations around the globe include squirrels in their conservation management, and if so, often with minuscule or even no assigned budgets. This study attempts to provide a real leap forward on this topic as it assesses, acknowledges, shows, and identifies this lack of conservation efforts with relevant conclusions. This is done by presenting contemporary GIS maps including the respective conservation status for every squirrel species with direct policy implications. These representations are mapped by nations of the world and include details about the available information regarding conservation budgets and squirrel protection policies. We also assess the sustainability of the global squirrel conservation management strategies by surveying the actual implementations of the 17 United Nations Sustainable Development Goals (SDGs) in squirrel management. Here we aim to identify conservation status trends and data availabilities to present conservation efforts invested by each nation. For these latter investigations/ assessments, the necessary information has been compiled from the governmental official webpages and directly requested by official letters & emails to the corresponding departments to deliver a 'state-of-the-art' overview of the global conservation efforts for squirrels. Such data are otherwise impossible to compile for budget assignments for squirrel conservation and governance actions or even small mammals since data is simply not publicly available. So what is the governance based on? Because these data are widely absent and not transparently shared by the governments, scientists, and institutions with the public, it hinders conservation and sustainable land and resource stewardship nationally and across the borders, internationally. Here we emphasize the concept of transparent wildlife and environmental management and outline the importance of data-sharing for combined conservation success and effectiveness; good governance principles. Additionally, to eradicate this older culture and to set a good example, here we share all the obtained information.

Keywords Squirrels · Conservation budgets · Mis-prioritization · Governmental failures in conservation · Human-nature interaction · U.N. sustainable development goals (SDG) · Data mining · Letter survey

13.1 Introduction

By following the news, especially on social media, it is often mentioned how many species are endangered world-wide, how many species became extinct in recent times, and how we as humans are heavily contributing to this problem. The species Great Auk (*Pinguinus impennis*) (Banks and Hochuli 2017; Thomas 2018), Passenger Pigeon (*Ectopistes migratorius*) (Cuker 2020; Takahashi et al. 2017), Dodo (*Raphus cucullatus*) (Angst et al. 2017; Ceballos et al. 2015; Fulton 2017), Woolly Mammoth (*Mammuthus primigenius*) (Nogues-Bravos et al. 2008; Salsberg 2000), Steller's Sea Cow (*Hydrodamalis gigas*) (Ceballos et al. 2015; Malhi et al. 2016), and many more species are famous to have become extinct due to human presence and unsustainable management/ hunting (Allendorf and Hard 2009; Koch and Barnosky 2006; Wang et al. 2017). Other species e.g. Asian Tiger and several subspecies (*Panthera tigris*) (Kolipakam et al. 2019; Wodak 2020), Asiatic Lion (*Panthera leo persica*), Black Softshell Turtle (*Nilssonia nigricans*) (He et al. 2021; Purkayastha 2018), Vaquita (*Phocoena sinus*) (Flessa et al. 2019; Munguia-Vega et al. 2007), Spix's Macaw (*Cyanopsitta spixii*) (Butchart et al. 2018; Parr and Juniper 2010), Javan Rhino (*Rhinoceros sondaicus*) (Goossens and Ambu 2012; Matschiner 2021), Hawaiian Crow (*Corvus hawaiiensis*) (Fleischer and McIntosh 2001; Milberg and Tyrberg 1993), Socorro Dove (*Zenaida graysoni*) (Ortiz-Alcaraz et al. 2017; Lennard 1997), and many more less-charismatic species are on the brink of extinction in modern days (Ceballos et al. 2015; Kirch 1996; Waller et al. 2017; Wardle and Bardgett 2004). In addition, other mammals were associated to have experienced human-induced drastic declines or extinctions. These species include

Supplementary Information The online version contains supplementary material available at https://doi.org/10.1007/978-3-031-23547-4_13.

M. Steiner, F. Huettmann, *Sustainable Squirrel Conservation*, https://doi.org/10.1007/978-3-031-23547-4_13

Ground sloths & Giant Ground Sloths (*Megatherium americanum*) (Bednekoff 2011; Turvey and Crees 2019), several Elephant species (Elephantidae)(Burney and Flannery 2005; Nogués-Bravo et al. 2008; Sumanarathna et al. 2017), subspecies of Wapiti (Gippoliti et al. 2018; Van Wieren 2012), Red Wolf (*Canis lupus rufus/Canis rufus*) (Agan et al. 2021; Agan et al. 2020), Eastern Cougar (*Puma concolor couguar*) (Cardoza and Langlois 2002; Goodman 2018), etc. With unsustainable and non-science-based management of these species and many more to come (of which many are probably unknown), more will likely become extinct. Therefore, conservation is of massive importance and has become more and more a topic of general awareness, relevance, and well-being (Taliaferro 2019). That is not only for the species *per se* but also for ecological services and mankind.

In this study, the conservation status of the global squirrels is being assessed. Across the world, many conservation organizations and governmental entities are mandated with, and in charge of, species conservation. A vast set of laws, legal concepts and courts are occupied with it. But one might wonder how these governmental entities actually operate and assure holistic science-based sustainable species conservation, homogenized across landscapes and nations? In recent years, the United Nations (UN) has developed 17 development goals (SDGs) that are supposed to help with creating a sustainable future globally. An introduction to the 17 SDGs can be found in Textbox 13.1.

Textbox 13.1: SDGs "for the Birds" and Squirrels?

The sustainable development goals (SDGs) have been introduced by the U.N. and adopted by all United Nations Member States in 2015 (see details www.sdgs.un.org/goals). While this appears to make sense, it actually does not, ecologically, and there are numerous conceptual and reality problems with the SDGs making them on a large scale harmful for the world, certainly for squirrels.

Already the underlying concept of win-win for everybody is problematic, and widely unrealistic when rolled out on a global scale with finite resources.

These SDGs mainly aim to improve sustainability for all major parts of our society and in this world, by applying a so-called form of 'sustainability', imposed globally, and breaking it down into 17 goals. Once all these 17 goals are supposed to be achieved, the planet is then assumed to be sustainable, and democracy is to fare well; squirrels would be safe…

Unfortunately, these SDGs are not really sustainable; they are an arbitrary and parsimonious number, and top-down biased, come with a dubious not-widely-shared world view and approach to life and governance, and are still lacking crucial components in order to actually achieve global and overall sustainability, or democracy and good governance, spirituality and a happy life. They are lacking to be holistic and are far from complete when it comes to dealing with relevant aspects of life, well-being, or the environment. One of these problems is global acceptance and coverage (see Koenig-Archibugi 2011; Marchetti 2008; Rich 2001 for a global democracy approach). SDGs are widely popular in European governance and some other thriving developing countries that feature strong administrations, e.g. top-down. However, they are widely unaccepted, poorly executed and paid for, and even ignored by a few but major nations in real life. Such nations include China, the USA, Russia, Australia, Canada, etc. (Ali et al. 2018; Petersmann 2021). There are also similar concerns for member nations of the U.N. that are in violation of democratic principles and human rights (Bartels 2013; Börzel and Risse 2004) but still striving for SDGs. Global sustainability, and sustainability for squirrels, can hardly be achieved without a common will and synergized efforts under the basic pre-amble of equity, fairness, inclusion, and human rights including all aspects of life, while considering finite resources. These are discussed in our study and have been included in the letter survey to all major nations/ countries/ states that are inhabited by squirrels because they are assumed to be the most developed sustainable approach to this date. However, as it can be seen for the majority of the survey responses we received, the SDGs are widely excluded, or ignored for the management of the local, and generally, global squirrel population. It is essentially a wide disregard for SDGs to start with. SDGs seem to play no role in local governance and carry little buy-in. Therefore, it can be safely concluded that the model of the 17 SDGs developed by the UN is globally still widely failing, certainly in the short run (Coscieme et al. 2020; Kloke-Lesch 2018; Pogge and Sengupta 2015), and for squirrels.

Another component that indicates that these 17 SDGs are not yet fully applicable to the real world is that these "sustainable" development goals are still widely based on the increase of financial wealth and an increase in the local and global economy (GDP as the measure), all to happen on a finite planet and atmosphere. The crucial point – distribution of wealth – is widely ignored, and unlikely to settle even with a good agreement for all citizens of the world – it was never achieved by the western world governance, but instead created warfare and a deep rift among mankind. Nature paid the bill. Without including an approach of economical long-term sustainability – a realistic poverty reduction and a more even distribution of wealth - in combination with the concept of our finite earth, e.g. resources are finite and can run out -, such development goals cannot be "sustainable". SDGs are to acknowledge that, but they do not, by design. By including concepts of a steady-state economy, or similar, e.g. ecological economics, such issues of the mismatch between a finite planet and a continuously growing economy can be addressed and minimized (see details in Coscieme et al. 2020; Czech 2017; Daly 1993; Dietz and O'Neill 2013 and references within). This includes the notion of human population growth, abortion, and equal consumption per capita (all issues that are avoided/ ignored for decades). All of those are issues known for a long time and essentially ignored in the SDGs.

And there is a long list of other inherent problems with the SDGs by the UN:

(a) It clearly sets a stage for a global top-down world government, one that is not even democratically voted and agreed upon, nor funded and with an equal burden for all. Who wants a global government, for instance, the world court or a world currency? All efforts in that direction went widely underwhelming, thus far.
(b) The creation of a world government involves world institutions. This must include a global election and funding system, such as a world tax. The overall environmental, social, and economic realities of a world agency must remain mind-boggling, setting up the stage for giant social issues centered around such giant agencies, e.g. a world pension fund. The spectacular failures of the IUCN (species declines and extinctions unabated), the IPCC (now in its 8th version while CO_2 and climate change still being on the rise, totally unabated while IPCC prepares for yet another version), or the so-called peace and humanitarian missions like in Rwanda (acknowledged genocides) or Syria, Afghanistan, Iraq, Sudan, and Yemen with the 'Blue Helmets' show us no other. Is it a scam (see Krishna et al. 2022 for pandemics and basic oversights on even simple diseases to manage by such global bodies)?

Now, how are squirrels faring in such a world design, and when considering they are already widely ignored and marginalized currently?

Therefore, for a more successful and actually sustainable future, one that is not negatively affecting Mother Earth or/ and squirrels, the authors suggest revising the underlying business model of SDGs, including several additional goals to the existing 17, and modifying many others that address those issues, and therefore, increase the potential global sustainability and meaningful goals that nations should strive for with a buy-in by its citizens. It is a wish list for the birds (and the squirrels).

In order to assure successful species conservation, it must not only be sustainable but also meaningful, lawfully binding, and overarching amongst all areas/ regions in which a certain species occurs. To assess this on a global scale, the authors sent 107 surveys to countries/ states all around the world, and to all major regions where squirrels occur. This has been done to obtain first-hand information on the two upper-mentioned conservation management components (sustainable management & holistic collaboration capacity among entities for successful species management).

13.2 Methods

13.2.1 Species Richness Overview of the World's Squirrels

In order to create the global species richness overview, the occurrence data from GBIF (www.gbif.org) have been used (download DOI: https://doi.org/10.15468/dl.665b59), similarly as in several other chapters of this book (Steiner and Huettmann 2023). All the occurrence data points from each species have been pulled to one average hotspot location using the QGIS function "Mean Coordinates…" for each species individually. This can be found in QGIS under the tab "Vector", and then "Analysis Tools". This returns the mean latitudinal (y) and longitudinal (x) coordinates of all occurrences of a single species as one averaged single x and y coordinate. In the absence of detailed range maps for squirrels available open-access, it allows showing approximate location hotspots for each species. This has been performed for all global squirrel species. This yielded a global overview of all species' mean coordinates and can be found in Appendix 13.3 as an illustration. Consecutively, in order to present the species richness for each country, the symbology has been adjusted in a way that in each country the number of mean coordinates (represented by one dot for each species) is being counted. This can be achieved by using the analysis tool called "Count Points in Polygon…". This analysis tool can be found in QGIS under "Vector" in the toolbar and then under "Analysis Tools". Consecutively, the symbology can be set up in a way to follow a color ramp based on the number of dots (species mean coordinates) occurring within this country (Shapefile – .shp – polygon). As a result of this analysis, Fig. 13.1a has been obtained. How this can be replicated in ArcGIS is explained in detail in Appendix 13.4.

To perform the same analysis for just the United States, all the steps have been repeated, except for presenting the data per country, it has been presented separately for each state. This was possible by using a different base map (shapefile (.shp)) which identified each state as an individual polygon and therefore the dots, and thus, species can be counted per polygon and presented thereafter (raw files can be obtained on request from MS).

13.2.2 Conservation Status Overview of the World's Squirrel Species

In order to create an overview of the conservation status classes of the world's squirrels, data from Steiner and Huettmann (Chap. 1 – 2023) has been used. This data also entails the conservation status of each species. These conservation status classes have been retrieved from the IUCN red list (www.iucnredlist.org). For this study, these conservation status class data have been associated with the centered occurrence of each squirrel species created previously. Using this method, we produced an illustration of all global countries and the centered occurrence for all squirrel species presented by their conservation status. This has again been repeated for the United States only as it represents a global squirrel species richness hotspot region.

13.2.3 Online and Physical Survey of the Squirrels' Assigned Conservation Budget and Management Sustainability of All US States and Squirrel-Inhabited Nations

In order to use first-hand and state-of-the-art information to analyze the budget assigned to squirrel conservation and the management sustainability, 107 physical letters, and 96 emails have been sent to each legislative entity/ agency of each state of the United States, as well as to a number of countries, and all Canadian territories (see Table 13.1 for details). The original survey, including the questions that have been asked to the entities, can be found in Appendix 13.1. The individual addresses, and physical and email addresses, of the entities that have been selected for the survey, can be found and Appendix 13.2.

Consecutively, the answers to these online and physical surveys/ letters have been analyzed and presented in the results in Sects. 13.3.2 and 13.3.3. In this study, complementary to Chap. 11 (Steiner and Huettmann 2023), we focus on the survey responses concerning the remaining two questions about the assigned conservation management budget and the 17 Sustainable development goals (SDGs).

13.2.4 Squirrel Conservation Management (Theory)

Squirrels are small mammals, often considered game species, with an almost global distribution. Approximately 300 species exist, however, very few are properly managed (Koprowski and Nandini 2008). For the very few species for which some data is available, it has been summarized in Table 13.2 below. According to management theory, humans should look after their local species, protect the ones that are endangered and recover them, and diminish the ones that became dominant as a pest, or invasive with a negative effect on the other local species; abundant species can be carefully harvested (Silvy 2020; Thomson 1992). The key to holistic sustainable management practice consists of - but is not limited to - efficiency, monitoring, transparency, communication, and communal agreement among all included parties upon executed practices. This seems to be a simple concept but is widely absent in conservation management. For example, in Switzerland, the Eurasian Red Squirrel (*Sciurus vulgaris*) is fully protected, whereas in many other European neighboring countries the same species just across the border is not protected at all (see Graf and Fischer 2021, and see survey responses). It is noteworthy that the distribution is not assumed to be much different in Switzerland compared to the neighboring countries nor is the life history (Shar 2016). Other countries, such as the USA have other policies in place, e.g. see Organ et al. (2012), but the issues remain, which are poor holistic management of species and a lack of collaboration amongst entities across borders.

13.3 Results

13.3.1 Richness Overview Grouped by Country/Nation

The map illustrating the global species richness grouped by nation/country for the world's squirrels is presented and Fig. 13.1.

In Fig. 13.1a, we can observe the high density of squirrel species in the United States, Mexico, Malaysia, Indonesia, and China. In Fig. 13.1b, the highest squirrel hotspot counts can be observed in California, New Mexico, Arizona, Utah, and Washington State. A clear west tendency can be observed in Fig. 13.1b.

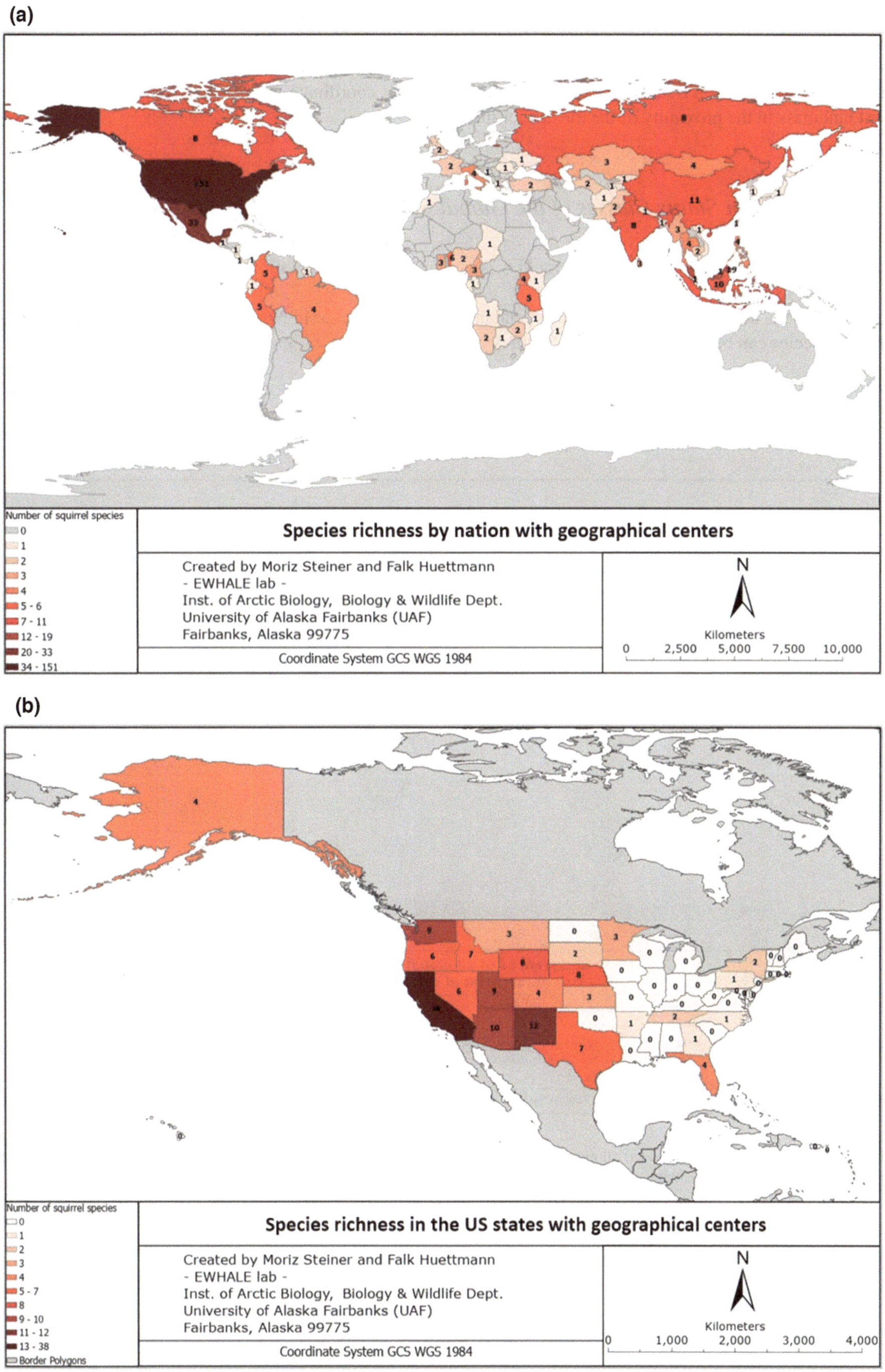

Fig. 13.1 Species richness overview of (**a**) all global squirrel species by nation (**b**) all species within the states of the U.S

Two species occur – according to the GBIF data – on more than one continent (Eastern Gray Squirrel (*Sciurus carolinensis*) – occurring in the eastern USA and has been introduced into some parts of Europe –, and Arctic Ground Squirrel (*Urocitellus parryii*) – occurring in Northwestern North America and Northeastern Russia –. These species and the ones occurring on several islands in Southeast Asia that displayed a mean coordinate in the sea have been manually adjusted to the closest landmass in the proximity of the mean coordinates.

13.3.2 Conservation Status Classes of the Global Squirrels Mapped by Nations

The figures illustrating the global and national (for U.S.) overview of the conservation status classes of the world squirrels is presented in Figs. 13.2.

In Figs. 13.2 one is able to observe that the most endangered species occur in North America. And the highest number of data deficient species can be found in Northern South America, tropical Africa, and Southeast Asia.

(a)

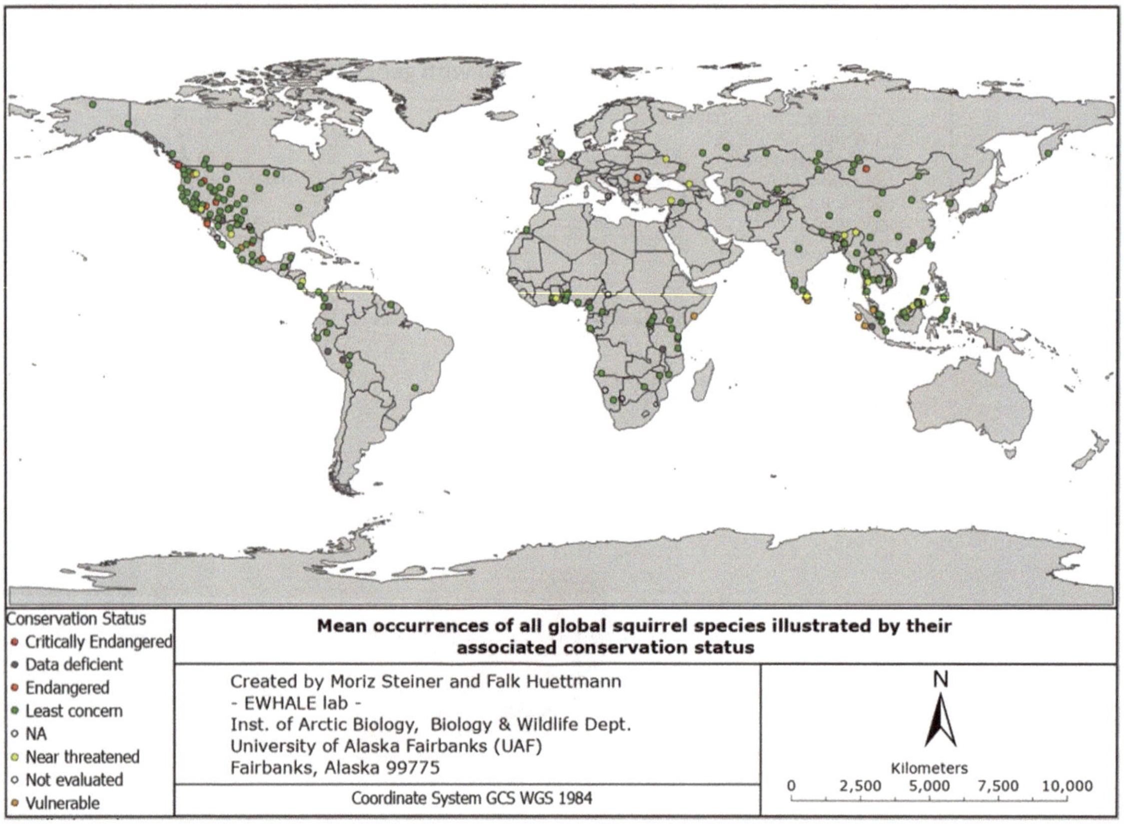

Fig. 13.2 Conservation status overview of (**a**) all global squirrel species (**b**) all species within the U.S. with centered squirrel species occurrence points (one dot represents the approximate center of one squirrel species)

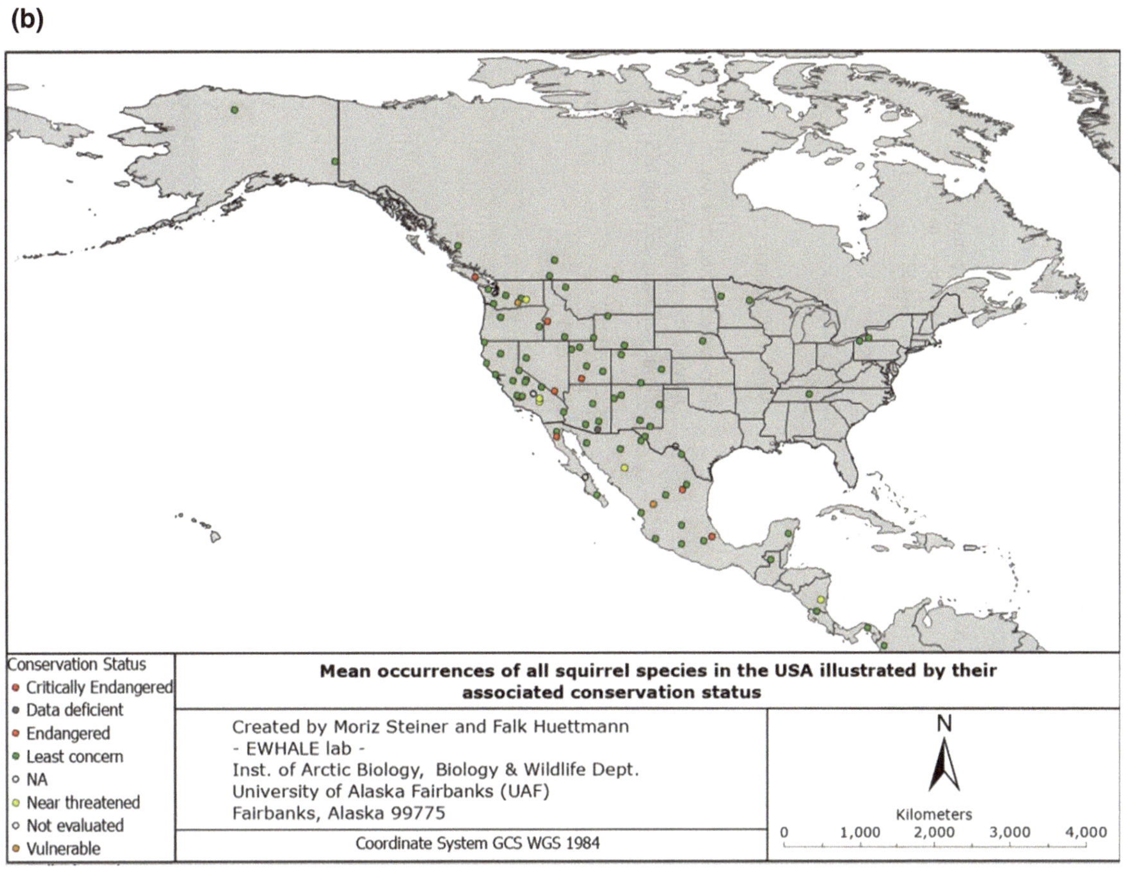

Fig. 13.2 (continued)

13.3.3 Results for the Survey on the Budget Assigned to Squirrel Conservation

The results of the online survey using emails, and the physical survey using letters, are presented in Table 13.1.

From these surveys and the meta-analysis performed on the responses (see meta-analysis theory by Moher and Olkin 1995; Van de Kaa et al. 2007), we were able to observe that surprisingly few entities have management budgets specifically assigned to squirrels/ small mammals. Only 7 out of the 107 entities included in the survey have reported their assigned budget (= 6.5%). In many entities, the conservation budget is not split up and administrated on a species level, but rather "where it's needed", as observed multiple times in the survey responses. The most common total budget mentioned in the survey responses was approximately 25,000–30,000 Euro/ US Dollar. This has been mentioned by Austria (for monitoring with drones), Arizona (for removal of invasive species middens), and North Dakota (for monitoring). The other entities that indicated an assigned budget were Indiana (with 3,300 USD), and New Mexico (with less than 14,000 USD). Switzerland and Georgia (U.S.) also indicated that they have assigned budgets for squirrels, but the numbers were not available to the authors. All other entities, which do not have any budget assigned to squirrel conservation management, or which do not have their budgets administrated on a species level, invest likely significantly less than the upper-mentioned entities that provided numbers.

Table 13.1 Meta-analysis of the physical and online survey on the budgeting of squirrel conservation and the SDG-based conservation management (**a**) for European countries, (**b**) for non-European countries, (**c**) for North American states, (**d**) for Canadian provinces

Countries	Responded	Assigned budget provided	Lawful regulations	Poaching known	SDG based management	Comments	Email follow-up sent
Austria	YES	YES	YES	No known poaching activities	YES	Survey squirrels with drones	YES
Finland	YES	NO	YES	No reports available	Not directly		YES
France	NO						YES
Germany	YES	NO	Partly	No details provided	NO	Very vague response, no specific responses to the questions asked	YES
Italy	NO						YES
Netherlands	YES	NO	YES	No indications of significant poaching levels	NO		YES
Norway	YES	NO	YES	No poaching activities known	YES		YES
Poland	NO						YES
Spain	YES	NO Budget available	YES	Poaching is not a conservation issue for red squirrel populations in Spain	YES	Very professional response	NO
Sweden	YES					Forwarded to https://www.naturvardsverket.se/en/	YES
Switzerland	YES	YES	YES	Very rarely reported.	Not directly	Phone call recorded with MS	YES
Turkey	NO						YES
United Kingdom	NO						YES
Yes count ratio	**(8/12) 0.67**	**0.13**	**0.47**	**0**	**0.2**		**0.93**
(b) Non-European countries							
Benin	NO						YES
Bhutan	NO						YES
Brazil	NO						YES
Burundi	NO						YES
Cameroon	NO						YES
China	NO						YES
Columbia	Pending						YES
Costa Rica	NO						YES
Ecuador	NO						YES
Ghana	NO						YES
Guam	NO						YES
Guatemala	YES	NO	NO	NO	NO	No information available	YES
India	NO						YES
Indonesia	NO						YES
Ivory Coast	NO						YES
Japan	NO						YES
Kenya	NO						YES
Malaysia	NO						YES
Mexico	NO						YES

Nepal	NO						YES
Nicaragua	NO						YES
Nigeria	NO						YES
Panama	NO						YES
Peru	NO						YES
Puerto Rico	YES	NO	NO	NO	NO	No squirrels present	YES
Russia	NO						YES
Rwanda	Pending						YES
South Korea	NO						YES
Sri Lanka	NO						YES
Togo	NO						YES
Uganda	NO						YES
Yes count ratio	**(2/31) 0.06**	**0**	**0**	**0**	**0**	**0**	**1**
(c) U.S. States							
Alabama	NO						YES
Alaska	YES	NO	YES	No bag limit for either hunting or trapping, so poaching is not an issue	NO	No bag limit. Year-round season	NO
Arizona	YES	YES	YES	YES	Aligned but not reported		YES
Arkansas	Pending						YES
California	NO						YES
Colorado	YES	NO	YES	Information request requires to pay an invoice	NO	Ask to pay for poaching reports	NO
Connecticut	YES	NO	YES	Not aware of any documented poaching activities in this state	NO	Do not know what the SDG's are	NO
Delaware	NO						YES
Florida	YES	Not available	YES	1 squirrel hunt in 3 years. No other incidents reported	NO		YES
Georgia	YES	Partly	YES	Not provided. Forwarded to their website. Nothing mentioned there	NO		YES
Hawaii	YES	NO	NO	NO	NO	No squirrels present	NO
Idaho	NO						YES
Illinois	NO						YES
Indiana	YES	YES	YES	Pending	YES		YES
Iowa	NO						YES
Kansas	YES	No budget assigned to squirrels	YES	NO	NO		YES
Kentucky	NO						YES
Louisiana	YES	NO	YES	Yes, these include hunting out of season and most common is bag limit violations	NO		YES

(continued)

Table 13.1 (continued)

Countries	Responded	Assigned budget provided	Lawful regulations	Poaching known	SDG based management	Comments	Email follow-up sent
Maine	YES	NA for squirrels	YES	12 poaching records since 2017	No but similar	Received a second reply. Very odd	YES
Maryland	YES	NO	YES	Poaching is not known to be an issue of concern in regards to small game management within Maryland. Poaching incidents involving squirrels are relatively uncommon, and likely unrelated events that do not have population level impacts	YES	Additionally added squirrel observation data	YES
Massachusetts	YES	NO	NO	NO	NO	The contacted authority had no information to answer any of our questions	YES
Michigan	YES	NO	Partly	Forwarded to someone else	NO	"Unfamiliar with the UN SDG's"	YES
Minnesota	NO						YES
Mississippi	NO						YES
Missouri	NO						YES
Montana	YES	NO	YES but no limits	NO	NO		YES
Nebraska	YES	NO	YES	Seems to be occurring from time to time	NO	Unfamiliar with SDG's	NO
Nevada	NO						YES
New Hampshire	YES	NO	Vaguely	No known reports	NO		YES
New Jersey	NO						YES
New Mexico	YES	YES	YES	There have been no squirrel poaching reports in the state, to the best of their knowledge	YES		YES
New York	YES	NO	YES	Forwarded to submitting a request to the law enforcement division	No, but their management is relatively sustainable		YES
North Carolina	NO						YES
North Dakota	YES	YES	YES	No reports of known poaching activities in ND.	NO		YES
Ohio	YES	NO	YES	14 recorded poaching violations since 2017	NO		NO
Oklahoma	YES	NO	YES	No tracking metric in place to quantify the day to day poaching enforcement effort or time spent	NO		YES
Oregon	Pending		YES				YES
Pennsylvania	NO						YES
Rhode Island	YES	NO	YES	No poaching activities reported or documented by the Division of Fish and Wildlife	NO	Very extensive response	NO
South Carolina	NO						YES
South Dakota	YES	NO	Partly	No known poaching cases or unknown	NO	Short and unprofessional response	NO

Tennessee	YES	NO	YES	TWRA wildlife officers have issued 45 citations and 20 warnings for violations related to squirrels since 2018. These include hunting violations, live wildlife possession, and taxidermy	NO		NO
Texas	YES	NO	YES	No details provided	"Unfamiliar with the SDGs"		YES
Utah	NO						YES
Vermont	NO						YES
Virginia	NO						YES
Washington	NO						YES
West Virginia	YES	NO	YES	WV DNR has no record of investigation of any squirrel poaching cases	Tracked but not reported		YES
Wisconsin	YES	NO	YES	Annually there are relatively few incidents of unlawful take or poaching activities of squirrels. However those that occur are a result of harvest outside of the legal season and over-bagging (daily and possession)	NO	No guidelines for some species	YES
Wyoming	YES	NO	YES	NO	NO		YES
Yes count ratio	**(32/50) 0.64**	**0.1**	**0.54**	**0.12**	**0.06**		**0.82**
(d) Canadian Provinces and Territories							
Alberta	NO						YES
British Columbia	YES	NO	Vaguely	No information provided as survey response	NO		NO
Manitoba	NO						YES
New Brunswick	YES	NO	YES	There are very few if any reports of known and documented poaching activities regarding squirrels	NO		YES
Newfoundland and Labrador	NO						YES
Northern Territory	NO						YES
Nova Scotia	NO						YES
Nunavut	NO						YES
Ontario	YES	NO	YES	No, not a significant source of illegal activities	Management based on sust. Goals but not reported to the UN		YES
Prince Edward Island	NO						YES
Quebec	YES	NO	YES	"No information"	Management based on sust. Goals but not reported to the UN	"Will respond in a few weeks"	YES
Saskatchewan	YES	NO	YES	NO	Not directly		YES
Yukon	NO						YES
Yes count ratio	**(5/12) 0.42**	**0**	**0.33**	**0**	**0**		**0.92**
Total Yes count ratio	**(47/107) 0.41**	**0.065**	**0.355**	**0.056**	**0.056**		**0.9**

13.3.4 Results of the Survey on the SDG-Based Squirrel Conservation Management

The results of the online survey using emails, and the physical survey using letters, is presented in Table 13.1. Overall, 6/ 107 (5.6%) entities have responded that their conservation management is based upon at least one of the 17 SDGs. European countries include the SDGs to the largest extent in their squirrel management with 3/13 (23%).

A trend that was observable amongst the responses from North American entities is that surprisingly many respondents were unaware of the 17 SDGs, which likely indicates that their conservation management is far away from the approached principles of the 17 UN-based SDGs. Another observable trend was that several conservation plans are considered sustainable and thus, indirectly follow the 17 SDGs, but they do not report to the UN. This is likely because the 17 SDGs are not fit to be used on a species level but rather on a much larger scale (as discussed with the Federal Department of the Environment of Switzerland). All these mentioned inconveniences with the 17 UN SDGs highlight the current drawbacks and shortcomings mentioned in Textbox 13.1.

13.4 Discussion

The conservation status of many squirrel species worldwide is classified as 'Endangered' according to the IUCN Red List and Butchart et al. (2010); Clausen and York (2008); Lanz et al. (2018); Malcolm and Markham (2000); Stork (1997). It can often be observed that the closer a species is being assessed, the more insight is gained into how endangered a species is and how small the actual species numbers are (see Anekwe 2019). For the world's squirrels, officially 34 species are listed as endangered (17 Vulnerable, 15 Endangered, and 2 Critically Endangered – IUCN Red List February 2022). However, also 35 species are listed as Dated Deficient (IUCN Red List February 2022). This means that for these 35 species there is not even enough data available to assess the conservation status of the species at stake, suggesting that the numbers and research efforts invested into the species are minimal, almost non-existing. Additionally, it seems appropriate to reference Chap. 1 (Steiner and Huettmann 2023), where the authors mention that the IUCN Red List classes are assigned in some cases overly positive to species listed on the IUCN Red List (see also Nijman 2022).

In this study, a global overview of the squirrel species richness for each country is presented (Fig. 13.1a), in addition to a zoomed-in version for the United States only (Fig. 13.1b). The same is being presented for the conservation status classes of all global squirrel species, globally (Fig. 13.2a), and in the USA only (Fig. 13.2b).

Taking a closer look at Fig. 13.1a, it is possible to observe that the global squirrel species hotspots in terms of species richness, occur approximately in the tropical, subtropical, and temperate zone. This is known to be a very attractive climate for the thriving of many species (Thorington et al. 2012; Steiner and Huettmann 2021; Steiner and Huettmann 2023). Not only plant species but also animal species have been observed to occur in higher species numbers in the tropical region compared to other climate regions (see Chaps. 3, 6, and 7 in Steiner and Huettmann ; Brown 2014; Buzas et al. 2002; Laurance et al. 2012). Therefore, the climate is one of the reasons that might explain many aspects of the high species diversity in the Southeast Asian region. Another possible reason is the high number of islands there with many species that are endemic to only one or a few islands. This is a very typical example of the topics of speciation on islands, discussed in Chap. 2 – Steiner and Huettmann 2023. Other speciation causes involve spatial and temporal differentiation and habitat use (Spatiotemporal Diversification – Abreu-Jr et al. 2020). This can be observed for several tropics-endemic species, where multiple squirrel species inhabit the same special area, but in different canopy heights, allowing them to occupy different specific microhabitat niches (see Wikipedia 2022). Another one is the co-evolution with humans, which is difficult to trace back but has been recorded for similar speciation causes (Rosshart 2019).

Another major hotspot of squirrel species can be observed in parts of North America, namely southwestern U.S. (~Rocky Mountains, California region etc). Here, a possible explanation for the high species richness might be the early genesis of the entire squirrel family (Olson 1976; Safran and Nosil 2012). As discussed in Chap. 2, according to the literature, it is suspected that the squirrel evolution originally started in the modern USA as a cradle (Mercer and Roth 2003). Since the species have had the most time in their originating grounds to evolve, they had the longest time for diversification and specification into certain ecological niches (Mercer and Roth 2003; Olson 1976), this possibly explains the high species richness there.

Figures 13.2 illustrate an overview of the global distribution of endangered squirrel species. By observing these figures, one can easily identify a trend, or better, hotspots of endangered species. It seems like, these hotspots of the endangered species can be observed in Western North America, Mexico, and Southeast Asia. This aligns with the high numbers of species and generally high species richness in these regions. For more insights into this specific topic, please see Textbox 13.2, addressing squirrels in the Global South.

Textbox 13.2: Like with Most Biodiversity, the Majority of (Squirrel) Species Occur in the ("Global) South": Implications for an effective Conservation Management Such as Including Poverty and Distribution of Wealth
Falk Huettmann, Moriz Steiner

Squirrels are small mammals, and most of them are living in, and are directly associated with, trees (Stapanian and Smith 1978; Smith 1970; Steele 2008). Squirrels co-evolved with this habitat and later with humans, making it a case of species radiated. Squirrels are virtually found worldwide, but the highest species diversity is usually found in the equator region, the tropics, and the 'global south' (see Chap. 3 Steiner and Huettmann 2023; Burgin et al. 2020; Thorington et al. 2012). One reason is that much land mass is found in 'the tropics', and another one is that this region is closest to the sun and thus, was never really glaciated. Besides many other features, this region is a very diverse landmass and offers not only ancient tropical and temperate forests but also rugged terrains, deserts and mountains, coastal zones, and a large human population. All of those actors are to fit together on limited pixels, and so something has to give, e.g. becomes extinct. And it is the species such as squirrels that have a difficult time defending themselves in such a competitive world; they get marginalized, lose their co-evolved habitat, and suffer from the decay of their ecological niche (Huettmann 2007; see Taber and Payne 2003 for small mammals). Climate change has made that more apparent than ever (Koskimäki et al. 2014; Van Horne 2008).

But the remote 'Global North' has much to do with the current global problem. Many species became already extinct in the Global North. And while most squirrel species are located in the Global South, the impacts and destructions of this zone are widely set up and are caused by, the Global North; also referred to as the 'Western World' and its governance and culture (see the concept of Telecoupling and Diamond 2011; Lees 2021; Mitlin and Satterthwaite 2012).

This is a process that was started by Christoph Columbus and James Cook for instance ('Enlightenment') (Munck 2000; Outram 2019; Porter 1990), and got further entrenched by Charles Darwin and the Western Science dominance, culminating in colonialism, and then the subsequent World Wars 1 and 2, and now, Globalization (Cabecinhas and Brasil 2019; Connell and Dados 2014; López 2007; Willems 2014). It is an ongoing trend that seems to have no good and sustainable end in sight. For squirrels, those are very problematic times indeed. Until then, squirrels did not suffer much from human pressure or habitat degradation; that has been like that for millennia. Man-made Global Climate change was virtually none-existing for most parts of the squirrel evolution.

However, by now, we live in the Anthropocene. And in this regime, where humans, e.g. the *homo economicus* (Gintis 2000; Levitt and List 2008; Thaler 2000), the *homo consumericus* (Saad 2011, 2014), as well as *homo electricus* (see Michael 2007, 2014; Huettmann et al. 2020; Neutrino Energy Group 2018), drive the wholesale set-up of landscapes and the atmosphere, and the demand of resources worldwide, this is no laughing matter for squirrels, hardly for any other biotic species including mankind.

Squirrels – again with the majority located in the south – get still described with a rather naïve and antiquated taxonomy scheme from the North (C. Linnaeus, Sweden). Arguably that fails (Steiner and Huettmann 2021). Apart from missing life history data and information, squirrels are not even assessed for their behavior traits (as a major scheme for most mammals).

In the absence of a good, agreed and applied wildlife management concept, the actual squirrel management remains minuscule while most species in the south are affected by humans, usually due to the rich north. The poverty distribution shows a clear peak in the 'global south'; most people earn less than $4 per day. While this must be seen as being absurd – many resources are found near the equator but poverty prevails there instead – the economic and administrative realities are flipped, delivering products to the North. It comes from the global governance design. And those resources come from the south, where most squirrel species are located. Examples are oil development in tropical nations, diamond mining by Chinese companies in countries like Sierra Leone (Conteh and Maconachie 2021; Maconachie 2009), and gold and other precious metal mining in tropical South America (Crundwell et al. 2011; Malm 1998; Nriagu 1994), and more (see Kirsch 2014 for mining in Papua New Guinea).

Our current global conservation management does not tackle these issues well and hardly acknowledges them. It still has not even accepted them and/or ignores them, globally.

Here we propose to acknowledge the *status quo*, put more attention to this problem, and better resolve the facts and the conservation management gaps for meaningful global sustainability; squirrels included. Many tools to do so exist, e.g. applying global models, using Open Access data, using holistic approaches, including ecology and telecoupling analysis on a finite planet.

Countries/ states with higher species richness, especially endangered ones, are supposedly expected to have a higher assigned budget for squirrel conservation. Unfortunately, however, that is not necessarily the case as it can be observed in greater detail in Chap. 10 (Steiner and Huettmann 2023). This assumption is also proven wrong by the results of the surveys. The majority of the entities that have replied to our surveys have no assigned budgets for squirrel conservation. There has also been no pattern in the survey responses that suggest an increased assigned budget for wealthier countries/ states (measured by the country-wide/ state-wide GDP).

From this, we can conclude that the overall distribution of wealth and endangered species richness does not imply that assigned conservation budgets are higher compared to countries/ states with lower GDP and endangered species richness.

So if wealth and the richness of threatened species are not determining the assigned budget for species conservation, one might wonder what does affect it. Assumingly, though, it is not a single factor that determines the financial conservation efforts, but rather the combination of several factors.

Throughout the survey responses, many different management methods have been observed. However, unfortunately, many of those management strategies indicate a high degree of conservation marginalization and '*laissez-faire*'. As an example of this, some governmental and law enforcement entities seem to have no incentive to make hunted squirrels mandatory to be reported, let alone to protect species that are threatened in nearby regions (contact MS for details about this).

For the modern world, it might be difficult to imagine how the livelihood between humans and animals (small mammals) should ideally be approached sustainably (Majumder 2021). An example of how some communities in the Kailash Mountain region (Nepal) interact with a local species has been described by Koju et al. (2021), which describes the interaction between the Pika's in the Himalayas and the local people (see also projects like Ukali in Nepal (https://www.ukali.org/)).

There, the wilderness is in relatively good shape. In that region and around the Kailash Mountain (India, Nepal, Tibet/ China), the animals are sacred and highly respected by virtually all its inhabitants and visitors. It is a prime example of how human-nature interaction can be happening, how it has always been there, and how it is even in modern days still viewed, respected, and approached in this way by the local people, in synergy with nature (e.g. see Dalai Lama on nature conservation and human-nature interactions; Buckley and Lama 2021; Johnson 1992). More insights into squirrel management science and its directing system can be found in Textbox 13.3.

Textbox 13.3: Do Squirrels Want Science Denial, Science-Based Conservation Management, Ecosystem Management, Co-management, or Science Dictatorship?

The call for more science is ubiquitous in the modern world. But what science is and how done, and on what platform, remains blurry, at best. Science – as it is usually understood in the public eye – tends to be done at institutions, and by people with a Ph.D., as tenured professors with freedom of speech. It involves a decade of training and then would lead the public.

Well, so goes the theory. Other science perspectives and views are widely dismissed in the western world, e.g. indigenous views and cosmologies, or so-called untrained perspectives by 'amateurs'.

As we see with climate change, or with the conservation of wildlife and habitats, the traditional science scheme has no relevant track record on this and has virtually not achieved anything for the world's squirrel conservation. Of course, one will find many dead squirrels and their skeletons in the world's museums, with even more dead squirrels found as road kills on the world's streets, millions of squirrels poached all over the world, and many other millions shot and trapped for game or fur but often just for fun. Squirrels are moribund topics to publish 'curiosity papers'. Likely billions of squirrels (population numbers are globally unknown, without any scientific estimation or true confidence intervals) get stressed by humans and barely survive on a daily basis and the western 'business-as-usual' in the long run.

But science? How does science contribute to a better world for squirrels, or for mankind and nature even? Thus far, it is fair to say, it failed dramatically, so did most of its institutions and NGOs.

Science remains a vague concept, has a strong cultural and administrative aspect to it, and it is difficult to achieve well and to sustain in the world we live in with funding-driven institutions without a sustainable business model. Most people in the world make less than $4 a day and lack the means for western-style university education. And if they get such an education, it is not squirrel biology they study or want to see money to be spent on.

Squirrel science remains an elusive beast, even in the wildlife biology sciences and its publications, as well as in the wealthy western world, and certainly in Ivy League Schools (who are to lead society).

While most people know easily over 10 car or expensive clothing brands, who really can name a squirrel research publication or a squirrel data set or 3 squirrel species names, let alone their conservation status?

Scientists argue they are to be in the driver's seat, leading society, making the best decisions for squirrels and about squirrels; all with reason. But that is not democratic and not happening either way. In a democracy, everybody should vote and be part of the decision process; but that hardly happens either, certainly not in the tropics (where most of the world's squirrels live). Consider here the structure and funding of 'science' in nations, e.g. Academies of Science, paid by tax payers and/or industries, and how nations and their sciences organize themselves on a global level, e.g. in the United Nations, or with The World Bank funding. Do we have squirrels and their science there anywhere in the driver's seat, or at least sitting at the table and part of the agenda?

Science can be done in many ways and it remains very vague; the outcome can be seen in the status of squirrels, data, conservation status, and future outlook Science can focus on an ecosystem, but the authors are not aware of any policy by the nations of the world that have implemented it for squirrels really in courts and legal texts, nor has it defined well what an ecosystem really is. We are also not aware of how science really contributes to a voting democracy, certainly not for squirrels. Perhaps science could contribute to the old notion of a 'doomed harvest surplus' ('spilled milk') and how big it really is, or of the noble but never achieved concept of maximum sustainable yield (MSY)? But we are not aware of any relevant or modern science being devoted to population estimates of most squirrel species nor of reported annual yields or computations and a formula for what MSY is for squirrels. For virtually none of the squirrels of the world, we have meaningful demographic metrics even, e.g. fecundity, longevity estimates, population numbers, population age structures cohorts or male/female ratios, predation rates, and overall life history details (see details for the few species for which parts of this data is available in Table 13.2 above). Those are needed for basic wildlife managemen thought. What do the Ivy Leagues and their funders - as the promoters and beneficiaries of world's science - argue for squirrels?

Clearly, the science of squirrels remains elusive and hardly relevant for their conservation management. Further, there is no good science model and institute really studying on a relevant scale how squirrel science contributes to their conservation or our global well-being and climate change. Arguably, a squirrel science based on experiments, lab work, control sites, linear regressions and parsimony has not achieved, and likely never will. And if climate change is included, then mostly with non-progressive questions like: "*Is there really an impact?*" (of course there is one, no need to focus much on it as climate change impacts are known to occur to wildlife for easily over 50 years. The bigger and more essential question is instead: what to do about it and how to reduce consumption, man-made CO_2, and Methane emissions)? The exception (like with a few Endangered Species), confirms the rule. If one would ask squirrels, they likely would not be in favor of science, or at least how science is being carried out these days. And why would they?

Table 13.2 Basic life history of the few relatively extensive studied squirrels

Taxonomic group	Life expectancy	Gestation period	Litter size	Natural predation rate	Male–female ratio	Annual reproduction	References
Eastern grey squirrel (*Sciurus carolinensis*)	Average life span adults 6 years; 12–24 years	40–44 days	1–9 (average 2 or 3)	Approx. 15.6%	59.1 one year old males to 40.9 one year old females; 110.3 males per 100 females	1–18 (average 4 to 6)	Kline (1964), Nixon and McClain (1975), Pest Strategies (2022), Saunders (1988), Shorten and Elton (1951) and Twining et al. (2020)
North American red squirrels (*Tamiasciurus hudsonicus*)	Life span adults 2–8 years	31–40 days (shorter in higher latitudes)	3–4 offspring	Approx. 39%	1: 1	Average 3–8 (depending on lat. & long)	Alaska Department of Fish & Game (2008), Goheen and Swihart (2005), Haughland and Larsen (2004), Hurly (1987), Lair (1985), Pearson (2000), Pest Strategies (2022) and Wishart et al. (2018)
Fox squirrels (*Sciurus niger*)	Life span adults 13–18 years	44 days	Average 2–4 (max. 7).	Approx. 35%	119.2 males per 100 females (54% male dominance)	Average 2–8 (max. 14).	Hansen and Nixon (1985), Kline (1964), Moore (1957), Pest Strategies (2022) and University of Wisconsin-Stevens Point NA
Southern flying squirrel (*Glaucomys volans*)	Life span adults 3–5 years	40 days	3 or 4 (extremes are 2–7)	Approx. 39%	Close to 1:1	On average 4–9	Atwood (2016), Jacques et al. (2017), Laves and Loeb (2006), Pest Strategies (2022) and Saunders (1988)

(continued)

Table 13.2 (continued)

Taxonomic group	Life expectancy	Gestation period	Litter size	Natural predation rate	Male–female ratio	Annual reproduction	References
Northern flying squirrel (*Glaucomys sabrinus*)	Potential longevity is not known; some adults survive at least 4 years	40 days	2–4 (extremes 1–6)	1–56%	0.6–1.7 depending on habitat	Average: 2–9	Malamuth and Mulheisen (2011), New Hampshire PBS NA; Patterson and Patterson (2010), Saunders (1988), Smith (2007) and Smith et al. (2011)
Eurasian red squirrel (*Sciurus vulgaris*)	7–12 years	38–39 days	1–7 young	Approx. 7–37%	Slightly male-biased (50.9%)	Average: 2–14	Kenward et al. (1998), Magris and Gurnell (2002), Mari et al. (2008), Seinfeld (1999), Turkia et al. (2016) and World Life Expectancy NA
Overall	6–18 years	31–44 days (average 40 days)	1–7 (average 4)	Approx. 30–35%	Slightly male dominant	2 breeding cycles (spring and autumn) in general 2–14 (average 7–8)	

In addition to the upper-mentioned 'obvious' marginalization of the global squirrels, here we also performed a cloud analysis to identify to which extent squirrels, or small mammals are included in the names of institutions and wildlife departments. More on this topic can be found in Textbox 13.4.

Textbox 13.4: What's in a Name? Squirrels Are Not Fish: An Assessment of Agency Names Governing North American Squirrels in 2022

In North America squirrels are part of 'game'; it is a plentiful terrestrial non-water living natural resource that is covered by the North American Model of Wildlife Management (Mahone and Geist 2019). But considering a '*laissez-faire*' approach to squirrels (Steiner and Huettmann 2021), how well does it really fare?

Here we look at agencies that are to provide the legal and management coverage for those species. As those squirrels are usually managed by the states and provinces in a wider federal framework, there are 50 states in the U.S. and 10 provinces and 3 territories in Canada in charge of those species of squirrels (in North America).

Simply judged from the outside, text mining the agency names, and expressing the governance structure through the agency title, done on the state and provincial levels show us surprisingly diverse and complex governance concepts, all done essentially for the same species set. If words are having a meaning and value, the names of those agencies can be used to give a first indication of how those wildlife resources are governed, and how the agencies perceive themselves, e.g. in the wider administrative governance context of 2023.

On a federal level, there are two government entities, the US and Canada; both are 'stable' and widely unchanged e.g. U.S. Fish & Wildlife Service (USFWS), but they have frequent elections and subsequent changing concepts and trends within. Recent changes in the ESA or SARAH show those changes strongly.

And on the state/provincial level, where the daily management action for squirrels sits, there then is a bewildering array of departments, divisions, commissions, comitees, and combined terms used for the same thing: management of squirrels. Three languages can be found as part of the administrative set: English and some French, little Spanish (while this is the largest rising language group in North America), and virtually no indigenous one whatsoever.

Then, there is an even bigger combination set of terms used in the agency name itself as shown in Textbox Table 13.1. Please note the relevance of 'fish' lumped in concert with game and non-game management such as for squirrels. Often fish even dominates the governance scheme where squirrels are fully embedded!

One may easily agree that squirrels should deserve specific localized attention; but why would they be managed by a commission vs a department vs a division or be widely linked with fish (a species group that lives in the water)? The concept that endemic state game, wildlife, and fish are all treated the same, by similar and same administrative units, in the same national governance apparatus, must appear very odd and lacking sophistication and care for detail. It is not really adequate and lacks a modern structure and awareness. Interestingly, fish are still combined with other wildlife/ game divisions, but one might wonder why birds are completely left out or organized differently then. Or amphibians and reptiles, let's say? Neither mammals nor birds are in the agency title, but fish are, and the latter occurs even as

Textbox Table 13.1 Combinations of terms, titles and names used for agencies in charge of squirrels on a 'state' level in North America

Name of the state agency	How often found in the U.S.
Department of natural resources	10
Division of fish and wildlife	3
Game and fish department	2
Department of fish and wildlife	2
Department of fish and game	2
Department of conservation and natural resources	2
Terms like fish & game, or conservation occur	Just once each

the leading species group. This is especially interesting because some squirrel species are only monitored as a convenient side project of monitoring bird species (see for instance monitoring efforts of the Cornell Lab (www.feederwatch.org), whereas such feeders are a major food source and relevant for the backyard, and the federal monitoring project of avian species in Switzerland – see squirrel survey responses). At a minimum, finite management resources should likely not be divided or shared with water-living creatures which are entirely disconnected from the tree-living squirrel world, but yet all treated the same (Textbox Fig. 13.1).

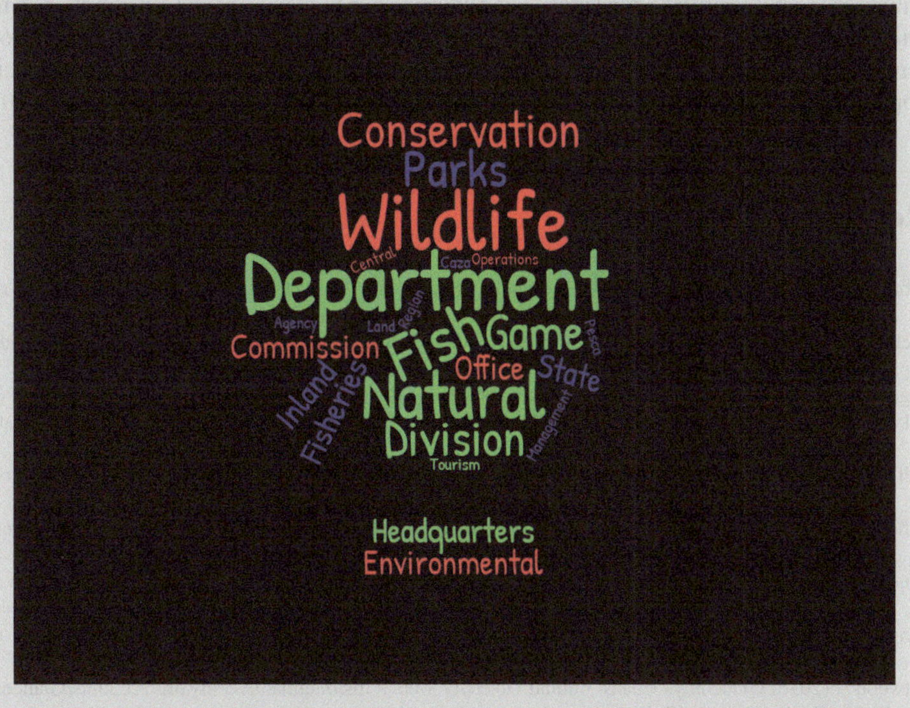

Textbox Fig. 13.1 'Name cloud' obtained as a visualization when using terms, titles, and names used by conservation agencies in charge of squirrels for North America on a state level. This presents one example for the squirrel governance in the world, e.g. North America as one of the richest entities to do so

As many species are merged and treated in similar agencies, but carrying different names and administrative structures, the financial resources for squirrels in North America within those agencies can usually not get well tracked and accounted for; associated budgets are probably very low either way. The way how those species are truly managed, and priorities are set, remains rather disputable, and is far from transparent, nor does it follow good management practices. It is a beast of the past to reckon with.

And so, one may slice it or dice it in many ways, but the bottom line remains that squirrels (or rodents, or small mammals) are not reflected in the title of their assigned governance agencies, nor are proper, modern, and fully devoted structures and titles chosen that give those species full justice. Arguably, squirrel ministries, universities, or institutes are non-existing nor can one get a degree in 'squirrels' or 'squirrelology'. The real expert decisions on conservation are usually done by legal experts, and with financial budget limits on a daily basis, but not taking into account squirrel research or realities. The legal apparatus is essentially totally divorced from squirrels in the field!

One may easily take issue with the notion that squirrels are 'game', e.g. similar to elk, deer, or moose. This is just an old English and royal term from many centuries ago (Taber and Payne 2003). We also would not treat squirrels as a natural resource, as there is virtually no component of the North American society that truly lives off squirrels (e.g. a squirrel pelt in Alaska fetches 30 cents and thus is not worth the effort, or bullet for that matter). And who really has sustainable principles and metrics for let's say, Flying Squirrel harvest?

Many other problems can be found when digging deeper. For instance, the underlying administrative and governance structures between divisions, departments, and commissions are not obvious either. But if words have a meaning, it cannot all mean the same for the governance of the same squirrels? It is highly confusing and highlights the family's marginalization.

And then there is a recent link and connection between parks and tourism. Many members of society might well welcome such a link. But in reality, it hardly means a protected area or a National Park (which is usually federal and has its own agency, governance, and legacy), nor are tourists and more humans often a good thing for wild squirrels.

So where does that leave us in the future?

In order to sustainably conserve our planet's species, we believe that humanity must go back to its roots, synergize with nature, respect its natural boundaries, and minimize our human impact. Consumption is to be reduced in a sustainable way. Only by doing this, the rapid decline of species richness across the globe can be slowed down and the global biodiversity can remain relatively intact, without suffering to an irreversible extent from human impacts and their overall '*laissez-faire*' approach to living respectful on this earth (Ceballos et al. 2015; Wyner and DeSalle 2020).

References

Abreu-Jr EFD, Pavan SE, Tsuchiya MT, Wilson DE, Percequillo AR, Maldonado JE (2020) Spatiotemporal diversification of tree squirrels: is the South American invasion and speciation really that recent and fast? Front Ecol Evol 8:230

Acharya KP, Chand R, Huettmann F, Ghimire TR (2022) Rabies elimination: is it feasible without considering wildlife? J Trop Med 2022:1–6

Agan SW, Treves A, Willey L (2020) Wild red wolf Canis rufus poaching risk. bioRxiv 18:202

Agan SW, Treves A, Willey LL (2021) Estimating poaching risk for the critically endangered wild red wolf (Canis rufus). PLoS One 16(5):e0244261

Alaska Department of Fish & Game (2008) Red squirrel. https://www.adfg.alaska.gov/static/education/wns/red_squirrel.pdf. Accessed 14 July 2022

Ali S, Hussain T, Zhang G, Nurunnabi M, Li B (2018) The implementation of sustainable development goals in "BRICS" countries. Sustainability 10(7):2513

Allendorf FW, Hard JJ (2009) Human-induced evolution caused by unnatural selection through harvest of wild animals. Proc Natl Acad Sci 106(Supplement 1):9987–9994

Anekwe L (2019) We're pushing 28,000 species closer to extinction. NewScientist. https://www.newscientist.com/article/2210437-were-pushing-28000-species-closer-to-extinction/. Accessed 14 July 2022

Angst D, Chinsamy A, Steel L, Hume JP (2017) Bone histology sheds new light on the ecology of the dodo (Raphus cucullatus, Aves, Columbiformes). Sci Rep 7(1):1–10

Atwood A (2016) Glaucomys volans. Animal Diversity Web. https://animaldiversity.org/accounts/Glaucomys_volans/. Accessed 14 July 2022

Banks PB, Hochuli DF (2017) Extinction, de-extinction and conservation: a dangerous mix of ideas. Aust Zool 38(3):390–394

Bartels L (2013) Human rights and sustainable development obligations in EU free trade agreements. Legal Issue Econ Integr 40(4):297–313

Bednekoff PA (2011) Nature's Ghosts: Confronting Extinction from the Age of Jefferson to the Age of Ecology. Great Plains Research, 21(1), 107

Börzel TA, Risse T (2004) One size fits all! EU policies for the promotion of human rights, democracy and the rule of law. In Workshop on democracy promotion (4, 5). Stanford: Stanford University

Brown JH (2014) Why are there so many species in the tropics? J Biogeogr 41(1):8–22

Buckley M, Lama D (2021) This fragile planet: his holiness the Dalai Lama on environment. Sumeru Press Inc., Bangkok. ISBN 10-1896559735

Burgin CJ, Wilson DE, Mittermeier RA, Rylands AB, Lacher TE, Sechrest W (eds) (2020) Illustrated checklist of the mammals of the world. Lynx

Burney DA, Flannery TF (2005) Fifty millennia of catastrophic extinctions after human contact. Trends Ecol Evol 20(7):395–401

Butchart SH, Walpole M, Collen B, Van Strien A, Scharlemann JP, Almond RE et al (2010) Global biodiversity: indicators of recent declines. Science 328(5982):1164–1168

Butchart SH, Lowe S, Martin RW, Symes A, Westrip JR, Wheatley H (2018) Which bird species have gone extinct? A novel quantitative classification approach. Biol Conserv 227:9–18

Buzas MA, Collins LS, Culver SJ (2002) Latitudinal difference in biodiversity caused by higher tropical rate of increase. Proc Natl Acad Sci 99(12):7841–7843

Cabecinhas R, Brasil JA (2019) Social representations of history in the global South: remembering the colonial past from the margins. Editorial de Belgrano, Buenos Aires

Cardoza JE, Langlois SA (2002) The eastern cougar: a management failure? Wildl Soc Bull:265–273
Ceballos G, Ehrlich PR, Barnosky AD, García A, Pringle RM, Palmer TM (2015) Accelerated modern human–induced species losses: entering the sixth mass extinction. Sci Adv 1(5):e1400253
Clausen R, York R (2008) Global biodiversity decline of marine and freshwater fish: a cross-national analysis of economic, demographic, and ecological influences. Soc Sci Res 37(4):1310–1320
Connell R, Dados N (2014) Where in the world does neoliberalism come from? Theory Soc 43(2):117–138
Conteh FM, Maconachie R (2021) Artisanal mining, mechanization and human (in) security in Sierra Leone. Extr Indus Soc 100983
Coscieme L, Mortensen LF, Anderson S, Ward J, Donohue I, Sutton PC (2020) Going beyond gross domestic product as an indicator to bring coherence to the sustainable development goals. J Clean Prod 248:119232
Crundwell F, Moats M, Ramachandran V, Robinson T, Davenport WG (2011) Extractive metallurgy of nickel, cobalt and platinum group metals. Elsevier
Cuker BE (2020) Passenger pigeon and waterfowl: flights to extinction and not. In: Diet for a sustainable ecosystem. Springer, Cham, pp 155–174
Czech B (2017) The steady state economy. In: Routledge handbook of ecological economics: nature and society. Routledge, pp 467–476
Daly HE (1993) Steady-state economics: a new paradigm. New Lit Hist 24(4):811–816
Diamond J (2011) Guns, germs and steel: the fate of human societies. Random House
Dietz R, O'Neill D (2013) Enough is enough: building a sustainable economy in a world of finite resources. Routledge
Fleischer RC, McIntosh CE (2001) Molecular systematics and biogeography of the Hawaiian avifauna. Stud Avian Biol
Flessa KW, Calderon-Aguilera L, Cintra-Buenrostro CE, Dettman DL, Dietl GP, Goodwin DH et al (2019) Vaquita face extinction from bycatch. Comment on Manjarrez-Bringas, N. et al., Lessons for sustainable development: Marine mammal conservation policies and its social and economic effects. Sustainability 2018, 10, 2185. Sustainability 11(7):2161
Fulton GR (2017) The bramble cay melomys: the first mammalian extinction due to human-induced climate change. Pac Conserv Biol 23(1):1–3
Gintis H (2000) Beyond Homo economicus: evidence from experimental economics. Ecol Econ 35(3):311–322
Gippoliti S, Cotterill FP, Groves CP, Zinner D (2018) Poor taxonomy and genetic rescue are possible co-agents of silent extinction and biogeographic homogenization among ungulate mammals. Biogeographia – J Integr Biogeogr 33
Goheen JR, Swihart RK (2005) Resource selection and predation of North American red squirrels in deciduous forest fragments. J Mammal 86(1):22–28
Goodman A (2018) Elegy for the eastern cougar: forgotton souls of appalachia. Doctoral dissertation, Appalachian State University
Goossens B, Ambu LN (2012) Sabah Wildlife Department and 10 years of research: towards a better conservation of Sabah's
Graf R, Fischer C (2021) Atlas der Säugetiere: Schweiz und Liechtenstein. Haupt, Schweizerische Gesellschaft für Wildtierbiologie
Hansen LP, Nixon CM (1985) Effects of adults on the demography of fox squirrels (Sciurus niger). Can J Zool 63(4):861–867
Haughland DL, Larsen KW (2004) Ecology of North American red squirrels across contrasting habitats: relating natal dispersal to habitat. J Mammal 85(2):225–236
He F, Langhans SD, Zarfl C, Wanke R, Tockner K, Jähnig SC (2021) Combined effects of life-history traits and human impact on extinction risk of freshwater megafauna. Conserv Biol 35(2):643–653
Huettmann F (2017) Climate Change Effects on Terrestrial Mammals: A Review of Global Impacts of Ecological Niche Decay in Selected Regions of High Mammal Importance. Encyclopedia of the Anthropocene, Volume 2, 2018, Pages 123–130
Huettmann F (2018) Climate change effects on terrestrial mammals: a review of global impacts of ecological niche decay in selected regions of high mammal importance. Encyclopedia of the Anthropocene 2:123–130
Huettmann F, Shrestha M, Puri G, Regmi GR (2020) Hydrodams again: a closer look at expensive gharials, lost dolphins, dissected fish migration, and destructive energy for non-sustainable mining and societies abroad fueling a non-democratic governance for the designed 'homo electricus'. In: Hindu Kush-Himalaya watersheds downhill: landscape ecology and conservation perspectives. Springer, Cham, pp 649–661
Hurly TA (1987) Male-biased adult sex ratios in a red squirrel population. Can J Zool 65(5):1284–1286
Jacques CN, Zweep JS, Jenkins SE, Klaver RW (2017) Home range use and survival of southern flying squirrels in fragmented forest landscapes. J Mammal 98(5):1479–1488
Johnson LK (1992) The Buddhist perception of nature: implications for forest conservation in Thailand. Trumpeter 9(1)
Kenward RE, Hodder KH, Rose RJ, Walls CA, Parish T, Holm JL et al (1998) Comparative demography of red squirrels (Sciurus vulgaris) and grey squirrels (Sciurus carolinensis) in deciduous and conifer woodland. J Zool 244(1):7–21
Kirch PV (1996) Late Holocene human-induced modifications to a central Polynesian island ecosystem. Proc Natl Acad Sci 93(11):5296–5300
Kirsch S (2014) Mining capitalism. University of California Press
Kline PD (1964) Iowa squirrels: hunting statistics, sex and age ratios, and the influence of mast and agriculture. Proc Iowa Acad Sci 71(1):216–227
Kloke-Lesch A (2018) Why is the EU failing to champion the SDGs? Horizons: J Int Relat Sustain Dev 12:144–159
Koch PL, Barnosky AD (2006) Late quaternary extinctions: state of the debate. Annual Review of Ecology, Evolution, and Systematics, p 37
Koenig-Archibugi M (2011) Is global democracy possible? Eur J Int Rel 17(3):519–542
Koju N, Chalise MK, Kyes RC (2021) Pikas and people: human-Pika (Ochotona) interaction in the Himalaya of Nepal. Biodivers J Biol Divers 22(11)
Kolipakam V, Singh S, Pant B, Qureshi Q, Jhala YV (2019) Genetic structure of tigers (Panthera tigris tigris) in India and its implications for conservation. Global Ecol Conser 20:e00710
Koprowski JL, Nandini R (2008) Global hotspots and knowledge gaps for tree and flying squirrels. Current Science, pp 851–856
Koskimäki J, Huitu O, Kotiaho JS, Lampila S, Mäkelä A, Sulkava R, Mönkkönen M (2014) Are habitat loss, predation risk and climate related to the drastic decline in a Siberian flying squirrel population? A 15-year study. Popul Ecol 56(2):341–348
Krishna V, Bansal N, Morjaria J, Kaul S (2022) COVID-19-associated pulmonary mucormycosis. Journal of Fungi, 8(7), 711
Lair H (1985) Length of gestation in the red squirrel, Tamiasciurus hudsonicus. J Mammal 66(4):809–810
Lanz B, Dietz S, Swanson T (2018) The expansion of modern agriculture and global biodiversity decline: an integrated assessment. Ecol Econ 144:260–277
Laurance WF, Useche DC, Rendeiro J, Kalka M, Bradshaw CJ, Sloan SP et al (2012) Averting biodiversity collapse in tropical forest protected areas. Nature 489(7415):290–294

Laves KS, Loeb SC (2006) Differential estimates of southern flying squirrel (Glaucomys volans) population structure based on capture method. Am Midl Nat 155(1):237–243

Lees N (2021) The Brandt Line after forty years: The more North–South relations change, the more they stay the same? Rev Int Stud 47(1):85–106

Lennard CJ (1997) The causes of avian extinction and rarity. Master's thesis, University of Cape Town

Levitt SD, List JA (2008) Homo economicus evolves. Science 319(5865):909–910

López AJ (2007) Introduction: the (post) global south, vol 1. The Global South, pp 1–11

Maconachie R (2009) Diamonds, governance and 'local'development in post-conflict Sierra Leone: lessons for artisanal and small-scale mining in sub-Saharan Africa? Res Policy 34(1–2):71–79

Magris L, Gurnell J (2002) Population ecology of the red squirrel (Sciurus vulgaris) in a fragmented woodland ecosystem on the Island of Jersey, Channel Islands. J Zool 256(1):99–112

Mahoney SP, Geist V (eds) (2019) The North American model of wildlife conservation. Johns Hopkins University Press

Majumder N (2021) Paradigm war: a distinct indigenous vision. Acad Lett 2

Malamuth E, Mulheisen M (2011) Glaucomys sabrinus. Animal Diversity Web. http://www.biokids.umich.edu/accounts/Glaucomys_sabrinus/. Accessed 14 July 2022

Malcolm JR, Markham A (2000) Global warming and terrestrial biodiversity decline. World Wide Fund for Nature, Gland

Malhi Y, Doughty CE, Galetti M, Smith FA, Svenning JC, Terborgh JW (2016) Megafauna and ecosystem function from the Pleistocene to the Anthropocene. Proceedings of the National Academy of Sciences, 113(4), 838–846

Malm O (1998) Gold mining as a source of mercury exposure in the Brazilian Amazon. Environ Res 77(2):73–78

Marchetti R (2008) Global democracy: for and against: ethical theory, institutional design and social struggles. Routledge

Mari V, Martini S, Romeo C, Molinari A, Martinoli A, Tosi G, Wauters LA (2008) Record litter size in the Eurasian red squirrel (Sciurus vulgaris). Hystrix: Ital J Mamm 19:61–65

Matschiner M (2021) Museum specimens tell the history of rhinoceroses. Cell 184(19):4841–4842

Mercer JM, Roth VL (2003) The effects of Cenozoic global change on squirrel phylogeny. Science 299(5612):1568–1572

Michael K (2007) Homo electricus and the continued speciation of humans. In: Encyclopedia of information ethics and security. IGI Global, pp 312–318

Michael K (2014) Beyond human: lifelogging and life extension. IEEE Technol Soc Mag 33(2):4–6

Milberg P, Tyrberg T (1993) Naïve birds and noble savages – a review of man-caused prehistoric extinctions of island birds. Ecography 16(3):229–250

Mitlin D, Satterthwaite D (2012) Urban poverty in the global south: scale and nature. Routledge

Moher D, Olkin I (1995) Meta-analysis of randomized controlled trials: a concern for standards. JAMA 274(24):1962–1964

Moore JC (1957) The natural history of the fox squirrel, Sciurus niger shermani. Bull AMNH 113:article 1

Munck T (2000) Enlightenment. The Oxford Handbook of the Ancien Régime

Munguia-Vega A, Esquer-Garrigos Y, Rojas-Bracho L, Vazquez-Juarez R, Castro-Prieto A, Flores-Ramirez S (2007) Genetic drift vs. natural selection in a long-term small isolated population: Major histocompatibility complex class II variation in the Gulf of California endemic porpoise (Phocoena sinus). Mol Ecol 16(19):4051–4065

Neutrino Energy Group (2018) The Era of the 'Homo Electricus' and Neutrino Energy. www.prnewswire.com/news-releases/the-era-of-the-homo-electricus-and-neutrino-energy-300753068.html. Accessed 2 Sep 2022

New Hampshire PBS (NA) Northern Flying Squirrel – Glaucomys sabrinus. https://nhpbs.org/natureworks/northernflyingsquirrel.htm#:~:text=Breeding%20season%20runs%20from%20March,period%20of%20about%2040%20days. Accessed 14 July 2022

Nijman V (2022) Harvest quotas, free markets and the sustainable trade in pythons. Nat Conserv 48:99–121

Nixon CM, McClain MW (1975) Breeding seasons and fecundity of female gray squirrels in Ohio. J Wildl Manag 39:426–438

Nogues-Bravos D, Rodriguez J, Hortal J, Batra P, Araujo MB (2008) Climate change, humans, and the extinction of the woolly mammoth. PLoS Biol 6(4):685–693

Nriagu JO (1994) Mercury pollution from the past mining of gold and silver in the Americas. Sci Total Environ 149(3):167–181

Olson SL (1976) Oligocene fossils bearing on the origins of the Todidae and the Momotidae (Aves: Coraciiformes). Collected Papers in Avian Paleontology Honoring the 90th Birthday of Alexander Wetmore

Organ JF, Geist V, Mahoney SP, Williams S, Krausman PR, Batcheller GR et al (2012) The North American model of wildlife conservation. Wildl Soc Tech Rev 12(04)

Ortiz-Alcaraz A, Arnaud G, Aguirre-Muñoz A, Galina-Tessaro P, Méndez-Sánchez F, Ortega-Rubio A (2017) Diet and home-range of the feral cat, Felis catus (Carnivora: Felidae) on Socorro Island, Revillagigedo Archipelago, Mexico. Acta Zoológica Mexicana 33(3):482–489

Outram D (2019) The enlightenment. Cambridge University Press

Parr M, Juniper T (2010) Parrots: a guide to parrots of the world. Bloomsbury Publishing

Patterson JE, Patterson SJ (2010) Multiple annual litters in Glaucomys sabrinus (northern flying squirrel). Northeast Nat 17:167–169

Pearson DE (2000) Evidence of autumn breeding in red squirrels, Tamiasciurus hudsonicus, in western Montana. Can Field-Nat 114(4):703–704

Pest Strategies (2022) How long do squirrels live? (It surprisingly varies…). https://www.peststrategies.com/pest-guides/squirrels-guides/lifespan/. Accessed 14 July 2022

Petersmann EU (2021) The UN sustainable development agenda and rule of law: how to limit global governance failures and geopolitical rivalries? EUI Department of Law Research Paper, (2021/10)

Pogge T, Sengupta M (2015) The sustainable development goals (SDGs) as drafted: Nice idea, poor execution. Wash Int'l LJ 24:571

Porter R (1990) The enlightenment. Macmillan International Higher Education

Purkayastha J (2018) Urban biodiversity: an insight into the terrestrial vertebrate diversity of Guwahati, India. J Threat Taxa 10(10):12299–12316

Rich R (2001) Bringing democracy into international law. J Democr 12(3):20–34

Ripple WJ, Miller SD, Schoen JW, Rabinowitch SP (2019) Large carnivores under assault in Alaska. PLoS Biol 17(1):e3000090

Rosshart SP, Herz J, Vassallo BG, Hunter A, Wall MK, Badger JH et al (2019) Laboratory mice born to wild mice have natural microbiota and model human immune responses. Science 365(6452):eaaw4361

Saad G (ed) (2011) Evolutionary psychology in the business sciences, vol 197. Springer Science & Business Media

Saad G (2014) The evolutionary instincts of Homo consumericus. The Interdisciplinary Science of Consumption, vol 59. MIT Press, Cambridge, MA
Safran RJ, Nosil P (2012) Speciation: the origin of new species. Nat Educ Knowl 3(10):17
Salsberg CA (2000) Resurrecting the woolly mammoth: science, law, ethics, politics, and religion. Stan Tech L Rev, 1
Saunders DA (1988) Adirondack mammals. State University of New York, College of Environmental Science and Forestry. 216pp
Seinfeld J (1999) Sciurus vulgaris. Animal Diversity Web. https://animaldiversity.org/accounts/Sciurus_vulgaris/. Accessed 14 July 2022
Shar S, Lkhagvasuren D, Bertolino S, Henttonen H, Kryštufek B, Meinig H (2016) Sciurus vulgaris (errata version published in 2017). The IUCN Red List of Threatened Species 2016: e.T20025A115155900. https://doi.org/10.2305/IUCN.UK.2016-3.RLTS.T20025A22245887.en. Accessed on 19 Feb 2022
Shorten M, Elton C (1951) Some aspects of the biology of the grey squirrel (Sciurus carolinensis) in Great Britain. Proc Zool Soc Lond 121(2):427–459. Oxford: Blackwell Publishing Ltd
Silvy NJ (ed) (2020) The wildlife techniques manual: Volume 1: Research. Volume 2: Management, vol 1. JHU Press
Smith CC (1970) The coevolution of pine squirrels (Tamiasciurus) and conifers. Ecol Monogr 40(3):349–371
Smith WP (2007) Ecology of Glaucomys sabrinus: habitat, demography, and community relations. J Mammal 88(4):862–881
Smith MJ, Betts MG, Diamond AW, MacLean DA, Erdle T, Courtenay SC (2011) Effects of habitat amount and configuration on the survival and movement of the northern flying squirrel (Glaucomys Sabrinus) in a forest mosaic. Doctoral dissertation, University of New Brunswick, Faculty of Forestry and Environmental Management
Stapanian MA, Smith CC (1978) A model for seed scatterhoarding: coevolution of fox squirrels and black walnuts. Ecology 59(5):884–896
Steele MA (2008) Evolutionary interactions between tree squirrels and trees: a review and synthesis. Current Science, pp 871–876
Steiner M, Huettmann F (2021) Justification for a taxonomic conservation update of the rodent genus Tamiasciurus: addressing marginalization and mis-prioritization of research efforts and conservation laissez-faire for a sustainability outlook. Eu Zool J 88(1):86–116
Stork NE (1997) Measuring global biodiversity and its decline. In: Biodiversity II: Understanding and protecting our biological resources, vol 41. JH Press, Washington DC, pp 41–68
Sumanarathna AR, Katupotha J, Abeywardhana K, Madurapperuma B (2017) Extinction of quaternary mammalian habitats of Megafauna in Sabaragamuwa Basin, Sri Lanka. J Eco Astron 1:16–31
Taber RD, Payne NF (2003) Wildlife, conservation, and human welfare: a United States and Canadian perspective. Krieger Publishing Company, Malabar
Takahashi B, Duan R, Joshi A, Van Witsen A, Freedman E (2017) Framing extinction: societal attitudes toward the passenger pigeon in editorials and opinion pieces. In: Biodiversity, conservation, and environmental management in the Great Lakes Basin. Routledge, pp 63–74
Taliaferro J (2019) Grinnell: America's environmental pioneer and his restless drive to save the West. Liveright Publishing
Thaler RH (2000) From homo economicus to homo sapiens. J Econ Perspect 14(1):133–141
Thomas JE (2018) Evolution & extinction of the Great Auk: a palaeogenomic approach. Bangor University (United Kingdom)
Thomson R (1992) The wildlife game. Nyala Wildlife Publications Trust
Thorington RW Jr, Koprowski JL, Steele MA, Whatton JF (2012) Squirrels of the world. JHU Press
Turkia T, Selonen V, Brommer JE (2016) Large-scale spatial synchrony in red squirrel (Sciurus vulgaris) sex ratios. J Mammal 97(3):744–752
Turvey ST, Crees JJ (2019) Extinction in the Anthropocene. Curr Biol 29(19):R982–R986
Twining JP, Montgomery WI, Tosh DG (2020) The dynamics of pine marten predation on red and grey squirrels. Mamm Biol 100(3):285–293
University of Wisconsin-Stevens Point (NA) Sciurus niger – Eastern Fox Squirrel. https://www.uwsp.edu/biology/VertebrateCollection/Pages/Vertebrates/Mammals%20of%20Wisconsin/Sciurus%20niger/Sciurus%20niger.aspx#:~:text=The%20gestation%20period%20for%20Fox,as%207%20being%20extremely%20rare. Accessed 14 July 2022
Van de Kaa G, De Vries HJ, van Heck E, van den Ende J (2007) The emergence of standards: a meta-analysis. In: 2007 40th Annual Hawaii International Conference on System Sciences (HICSS'07). IEEE, pp 173a–173a
Van Horne B (2008) Conservation of ground squirrels. In: Rodent societies. University of Chicago Press, pp 463–471
Van Wieren SE (2012) Reintroductions: learning from successes and failures. Restoration ecology: the new frontier. Wiley, Hoboken, pp 87–100
Waller NL, Gynther IC, Freeman AB, Lavery TH, Leung LKP (2017) The Bramble Cay melomys Melomys rubicola (Rodentia: Muridae): a first mammalian extinction caused by human-induced climate change? Wildl Res 44(1):9–21
Wang JY, Riehl KN, Yang SC, Araújo-Wang C (2017) Unsustainable human-induced injuries to the critically endangered Taiwanese humpback dolphins (Sousa chinensis taiwanensis). Mar Pollut Bull 116(1–2):167–174
Wardle DA, Bardgett RD (2004) Human-induced changes in large herbivorous mammal density: the consequences for decomposers. Front Ecol Environ 2(3):145–153
Wikipedia (2022) Niche differentiation. Wikipedia. https://en.wikipedia.org/wiki/Niche_differentiation. Accessed 14 July 2022
Willems W (2014) Beyond normative dewesternization: examining media culture from the vantage point of the Global South. Global South 8(1):7–23
Wishart AE, Williams CT, McAdam AG, Boutin S, Dantzer B, Humphries MM, Lane JE (2018) Is biasing offspring sex ratio adaptive? A test of Fisher's principle across multiple generations of a wild mammal in a fluctuating environment. Proc R Soc B 285(1891):20181251
Wodak J (2020) (Human-inflected) evolution in an age of (human-induced) extinction: synthetic biology meets the anthropocene. Humanities 9(4):126
World Life Expectancy (NA) Eurasian Red Squirrel. https://www.worldlifeexpectancy.com/mammal-life-expectancy-eurasian-red-squirrel. Accessed 14 July 2022
Wyner Y, DeSalle R (2020) Distinguishing extinction and natural selection in the anthropocene: preventing the panda paradox through practical education measures: we must rethink evolution teaching to prevent misuse of natural selection to biologically justify today's human caused mass extinction crisis. Bioessays 42(2):1900206

Part IV
First Conclusions and the Way Forward

Chapter 14
A Conservation Management SWOT Analysis for Over 230 Squirrels of the World Using 132 GIS Layers Confirming the PESTLE Assessment

Abstract There is a lot of theory on how to manage and administer a precious resource. However, despite being in the Information Age, few of these management tools are applied to sustainability assessments, conservation, and species management yet. Here we run a SWOT and PESTLE analysis for the best publicly-available 132 GIS habitat layers – conservation content and potential – for over 230 squirrels on a global scale. While somewhat indirect and using proxies, our findings show that those tools provide progress and new information; we find that SWOT is the more simplistic analysis and PESTLE is better suited for more holistic, environmental questions. However, when seen from the perspective of ecological economics, none of the two provide substantially new information and approaches on how to tackle finite resources in the Anthropocene for conservation and sustainability, e.g. applied to squirrels, leaving the traditional problems widely unresolved.

Keywords SWOT analysis · PESTLE analysis · Squirrel conservation management · Management theory

14.1 Introduction

There are approximately 300 squirrel species in the world (Chap. 1 in Steiner and Huettmann 2023); they are naturally distributed globally (except Antarctica and Australia & Oceania) (see Black 1972; Koprowski and Nandini 2008; Palmer et al. 2007; Rajaratnam and Redman 2001; Chap. 3 and 12 – Steiner and Huettmann 2023), and their individual numbers easily reach in the billions (Watts 2020). However, many squirrel populations are stressed, many are endangered, wilderness habitats got transitioned and lost, and even more, they are not assessed well for effective conservation management (= >app. 95% of the species – see Chap. 1 in Steiner and Huettmann 2023). Many relevant conservation management issues can be named that are either not done, not developed, or not developed well enough, or done to satisfaction and wide agreement showing a justified 'success' in the public eye. In the majority of cases, conservation management for squirrels takes a shot in the dark (Table 14.1). Globally speaking, squirrel research is widely not done, often one-sided or underfunded, and such species are not on most agendas, certainly not in the tropics or in rich nations where actually most squirrels live and species diversity is rather high.

Mankind lives already in the age of 'enlightenment' using science and reasoning (e.g. Carlson 2012, Danis et al. 2013), with gigantic research budgets, e.g. CERN in the EU with c. 1 billion EUR (http://home.cern) per year, or the National Science Foundation NSF U.S. with c. $8.5 billion (National Science Foundation Centers – NSF https://www.nsf.gov › about › budget › pdf). Spending 2% of the national GDP on research is not unheard of. And while large economic models exist to 'manage' and to apply management theory to topics of concern, this actually is not applied much to nature (Daley and Farley 2011), or squirrels yet (Steiner and Huettmann 2021). It also widely fails on global sustainability questions, e.g. climate change (Stern 2008). This is a holistic problem because if humans continue to capitalize on nature and its resources, many of the currently highly funded projects will hardly be needed for humanity if their surrounding environment is naturally inhabitable for humans.

And although we are living in the Information Age, data on squirrels are even less solid to use and obtain, e.g. online and with a proper documentation to understand their use and constraints (Huettmann 2009, 2015). Often those data are simply incorrect, confusing (Steiner and Huettmann 2021 for taxonomy), misleading, or completely missing, as an IUCN Red List status review easily shows 35 squirrel species as data deficient (10% of the world's squirrels). In the meantime, proxies or relevant datasets do exist and can be used for habitat and similar assessments to get at those conservation management

Supplementary Information The online version contains supplementary material available at https://doi.org/10.1007/978-3-031-23547-4_14.

M. Steiner, F. Huettmann, *Sustainable Squirrel Conservation*, https://doi.org/10.1007/978-3-031-23547-4_14

Table 14.1 Known shortcomings for squirrel management, considering advanced management theory

Conservation management topic	Details	Citation	Comment
Population assessments	Populations are to be known for valid management, e.g. for sustainable harvest	Silvy (2020)	Neither population sizes nor harvest rates are known for most squirrel species in the world (95%+)
Species and individual density measures	Species number or individuals per area is needed for valid impact assessments etc.	Buckland et al. (2015)	None of those metrics are known for most squirrels (and their populations) in the world.
Midden density	While the midden density is secondary it remains a central item in research for several squirrel species.	Robold and Huettmann (2021)	While other metrics are missing, midden densities are often better known. However, the true link between middens and actual squirrel occurrence remains unclear.
Demographics	Linked with habitat, demographic metrics are vital for any species management	Silvy (2020)	While those metrics are widely unknown for most squirrel species in the world, they are relatively easy to obtain and the bread-and-butter of many wildlife management agencies and their research.
Laws specific to squirrels	Squirrel management needs laws to operate under	Koprowski and Nandini (2008)	Squirrels usually fall under blanket laws for wildlife.
Laws specific to squirrel habitat management	Squirrel management needs laws to operate under.	Koprowski and Nandini (2008)	Habitat laws for wildlife are relatively rare.
Local, national, and international alignment	Locations and nations do not agree with each other on how to manage species, even if it is the same taxon and similar habitat and ecological niche.	Adamowicz (2016)	Mis-alignments are usually related to laissez-faire and create massive management obstacles to proceeding well. It is a professional fault usually.
Taxonomic disagreement	Large taxonomic disagreements for the same species etc.	Steiner and Huettmann (2021)	Without an agreed taxonomy, species management is not able to proceed (unless a habitat path is taken and prioritized)

questions. Over 100 of those data are found online and readily available though (Sriram and Huettmann unpublished). An even larger set – the biggest available in the world – was compiled by MS consisting of 132 GIS layers available for good use and wider assessment (see Chaps. 3 to 7, where it was applied in good measure, and made freely available for mankind).

Further, management and leadership theory has developed steadily, and many new tools are used and applied in those textbooks. However, the wildlife management textbook (Silvy 2020), now in eighth edition, has not implemented those concepts yet and, less so applies them. The same can be observed for many mandated agencies and NGOs for governance and policy (see squirrel survey in Chaps. 11 and 13 – Steiner and Huettmann 2023).

Here we look at a basic but powerful and promoted textbook management tool 'SWOT', to assess the status of squirrels of the world and the environmental data available for them. SWOT looks at four aspects, Strengths, Weaknesses, Opportunities, and Threats (Piercy and Giles 1989). By now it is an established basic but widely used and understood method to get a baseline for a *status quo*, a decision, and a situation; usually applied to business and competitive assessments. It is essentially a rapid-assessment parsimonious tool and summary template based on a two-by-two matrix and is well-known to have weaknesses (e.g. Omer 2018). While a well-done SWOT analysis can be somewhat holistic, in reality, it is often pretty coarse and leaves shortcomings. However, it can be informative, helps to find and assess gaps, and adds new insights. Usually, it is the first strategic step of assessing and understanding businesses, competitors, strategies, etc.

Therefore, for advancement, other more inclusive and finer-detailed approaches have been proposed in Management theory to be more holistic and meaningful. And so we then go one step further beyond SWOT and look at a PESTLE (Political, Economic, Social, Technological, Legal, and Environmental) analysis for 230+ species and their data. In the absence of good management data in modern times, we base our species assessment on the best-available GIS habitat data (environmental predictors used for the global SDMs). This is also done to assess whether the conclusions and findings match the SWOT analysis findings across those management tools for these species.

Overall, the SWOT and PESTLE approach is mostly part of the social sciences and equals more a meta-analysis (Moher and Olkin 1995; Van de Kaa et al. 2007), as well as a trend assessment rather than traditional quantitative and 'precise' science following a hypothesis testing path (the latter is widely promoted in the discipline, such as by Silvy 2020 for instance for the professional society of wildlife biologists). Still, SWOT and PESTLE are standard in any planning and for business and administration, and they are applied worldwide in business, economics, management, etc.

To start the process, here we pursue this approach overall because most squirrels are not in the mainstream research agenda nor managed so well; a pro-active approach – as mandated by the U.N. (1992 Pre-cautionary Principle) and Andorno (2004) – is widely missing. In addition, the discipline of wildlife management widely falls short in promoting and using the latest management theory (Silvy 2020), while the Anthropocene has broadly moved into a global environmental crisis (Hamilton et al. 2015; Houston 2013). We hope by using widely used business analysis tools to get new insights and a wider buy-in and become more proactive.

14.2 Methods

14.2.1 SWOT Analysis

For the SWOT analysis, we used the 132 GIS habitat layers compiled by MS (see Appendix 14.1). Those layers were classified as multi-dimensional habitat assessments and impacts for squirrels according to the four SWOT aspects of Strength, Weakness, Opportunities, and Threats (Madsen 2016; Piercy and Giles 1989).

In the SWOT analysis, we asked the question of whether the actual existence and content of those GIS habitat layers meet the four SWOT questions for squirrel conservation management. For instance, "Is the current 'landcover' – or road information – in the Anthropocene a threat for squirrels?", etc.

14.2.2 PESTLE Analysis

Like in the SWOT analysis, we used the 132 GIS habitat layers and classified their status and meaning as assessments for squirrels according to the six PESTLE aspects: Political, Economic, Social, Technological, Legal, and Environmental (Perera 2017).

In the PESTLE analysis, we asked the question of whether the content of those GIS habitat layers meets the six PESTLE questions for squirrel conservation management. For instance, "Is the current 'landcover' – or road information – in the Anthropocene political for squirrels?", etc.

In order to achieve a generic conclusion, for SWOT and PESTLE, we added up the majority of the metrics for a yes vs no matrix table. This matrix can be found in Appendix 14.2 for a detailed overview.

14.3 Results

14.3.1 SWOT Analysis

As shown in Appendix 14.2, we find that most typical factors for species conservation are also emphasized in the SWOT analysis. It is rather similar. While factors related to weaknesses and opportunities add a new insight and management options, little new really comes to light with few options left when acting under the assumption that resources are finite, as emphasized for a long time in Ecological Economics (Darley and Farley 2011; Farley 2014; for Steady-State Economics (SSE) see Czech 2017; Czech and Daly 2004; Huettmann and Czech 2007).

This is fully reflected in the summary columns of each of the four metrics in SWOT. For strengths, the majority of the 132 GIS habitat layers reflect that. For weaknesses, most of the 132 GIS habitat layers show no major weaknesses for conservation. Opportunities are present only for the minority of the 132 GIS habitat layers in relation to squirrel conservation management. The majority of the 132 GIS habitat layers – their content and how used – present threats.

14.3.2 PESTLE Analysis

From the PESTLE analysis, Appendices 13.3a and 13.3b allow us a more nuanced view compared to the SWOT analysis. While the specific environmental question in PESTLE itself adds the environment specifically to the agenda, the other factors are not really new, except perhaps the political and social ones. Still, from a more holistic perspective PESTLE provides a more rounded assessment and forces it onto the result.

The summary columns of the PESTLE analysis show us this pattern even stronger: The majority of habitat layers show a political aspect, just like the economic column, but not the social metric. The majority of environmental habitat layers do not show technological, or legal aspects, but certainly, the environmental one does.

14.4 Discussion

Conservation management is part of the wider management theory. The latter has evolved significantly, specifically for business and administration, with an economic focus. It is somewhat divorced from sustainability and developed into a mindset of its own. However, for ecology and conservation, there are still few of such concepts available or implemented yet. Tools like SWOT and PESTLE are well established outside of wildlife and conservation management but not within such sectors (e.g. Silvy 2020). Here we assess and show an application for SWOT and PESTLE and what those can bring to the table, using squirrels of the world and their available data and content.

By using business and management analysis tools, it allowed us to use 'state-of-the-art' concepts in management. Here we looked at the best-available GIS habitat dataset and content for squirrels and their ecological niche aspects in a different light. We focus on Hutchinson's n-dimensional niche (Hutchinson 1957). The 132 GIS habitat layers we used are the best publicly-available online data sources for habitat descriptions on a 2.5 km pixel scale. Those have not yet been truly used for assessments of the world's squirrels and their conservation status; nor are SWOT and PESTLE used for species conservation management (compare with Silvy 2020).

It becomes clear though that those 'modern' management tools are mostly designed for human business aspects, not for wildlife, sustainability or ecology and conservation, e.g. the economics or social views, especially the technology metrics in PESTLE. Still, we were able to obtain new insights overall, which can be described in the following:

We found that the SWOT analysis was not able to describe most of the habitat issues for the squirrels of the world in more depth. But we found that the PESTLE analysis helped to put more aspects on the issue of squirrel habitats. Namely, the specific focus on political and environmental metrics makes it a more holistic tool and better suited for ecological and species management questions in the real world.

However, neither SWOT nor PESTLE showed a good insight into squirrel habitats nor provided a solid outlook or differed much in the acknowledged meta-analysis result for squirrel conservation management (see this book for overall outcome and message). That is specifically the case when using Ecological Economics as the platform to work from. New options for squirrel conservation remain few, regardless of the data and assessment tools used.

With these two widely used management tools, we tried to investigate and potentially establish the very basis for 'modern' squirrel management theory. From there, it would be the task of follow-up studies and research to apply more ecologically adequate management tools and applications to the global squirrels, their habitats, conservation, and generally the ecological sector. Possible follow-up research can include Analytical Tools for Environmental Design and Management in a Systems Perspective (Wrisberg et al. 2002), Environmental Management Tools (see Balkau 2005; Sheppard 2018), and similar more advanced tools. Such tools are generally more inclusive in regards to habitat and ecology questions, and good underlying data, but they are significantly less known and developed compared to SWOT and PESTLE analyses giving them less buy-in with the public and managers. What would be helpful is a good and updated training on the actual latest and most effective management tools for wildlife.

References

Adamowicz WL (2016) Economic analysis and species at risk: lessons learned and future challenges. Can J Agric Econ/Revue canadienne d'agroeconomie 64(1):21–32
Andorno R (2004) The precautionary principle: a new legal standard for a technological age. De Gruyter Publisher, New York
Andreadis KM, Schumann GJP, Pavelsky T (2013) A simple global river bankfull width and depth database. Water Resour Res 49(10):7164–7168
Balkau F (2005) International frameworks for environmental solutions. In: Environmental solutions. Academic Press, pp. 401–434
Black CC (1972) Holarctic evolution and dispersal of squirrels (Rodentia: Sciuridae). In: Evolutionary biology. Springer, New York, pp. 305–322
Buckland ST, Rexstad EA, Marques TA, Oedekoven CS (2015) Distance sampling: methods and applications, vol 431. Springer, New York
Carlson DJ (2012) IPY 2007–2008: where threads of the double helix and sputnik intertwine. In: Huettmann F (ed) Protection of the three poles. Springer, Tokyo. https://doi.org/10.1007/978-4-431-54006-9_2
Czech B (2017) The Steady state economy. In: Routledge handbook of ecological economics. Routledge, Abingdon, pp. 467–476
Czech B, Daly HE (2004) In my opinion: the steady state economy—what it is, entails, and connotes. Wildl Soc Bull 32(2):598–605
Daly HE, Farley J (2011) Ecological economics: principles and applications. Island press

Danis B, Van de Putte A, Renaudier S, Griffiths H (2013) Connecting biodiversity data during the IPY: the path towards e-polar science. In: Verde C, di Prisco G (eds) Adaptation and evolution in marine environments, From Pole to Pole, vol 2. Springer, Berlin/Heidelberg. https://doi.org/10.1007/978-3-642-27349-0_2

Farley J (2014) Steady state economics. In: Degrowth. Routledge, pp. 77–80

Fick SE, Hijmans RJ (2017) WorldClim 2: new 1km spatial resolution climate surfaces for global land areas. Int J Climatol 37(12):4302–4315

Hamilton C, Gemenne F, Bonneuil C (eds) (2015) The Anthropocene and the global environmental crisis: rethinking modernity in a new epoch. Routledge, London

Houston D (2013) Crisis is where we live: environmental justice for the Anthropocene. Globalizations 10(3):439–450

Huettmann F (2009) 4 the global need for, and appreciation of, high-quality metadata in biodiversity. In: Data mining for global trends in mountain biodiversity. CRC Press, Boca Raton, p 25

Huettmann F (2015) On the relevance and moral impediment of digital data management, data sharing, and public open access and open source code in (tropical) research: the Rio convention revisited towards mega science and best professional research practices. In: Central american biodiversity. Springer, New York, pp 391–417

Huettmann F, Czech B (2007) The steady state economy for global shorebird and habitat conservation. Endanger Species Res 3(3):89–92

Hutchinson G (1957) Concluding remarks. Cold Spring Harb Symp Quant Biol 22:415–427

Jenkins CN, Pimm SL, Joppa LN (2013) Global patterns of terrestrial vertebrate diversity and conservation. Proc Natl Acad Sci 110(28):E2602–E2610

Jones P, Wint W (2015) Data set produced by Waen Associates for Environmental Research Group Oxford, Limited, funded by the International Research Consortium on Dengue Risk Assessment, Management and Surveillance (IDAMS), European Union's Seventh Framework Programme for research, technological development and demonstration under grant agreement no 281803

Koprowski JL, Nandini R (2008) Global hotspots and knowledge gaps for tree and flying squirrels. Curr Sci 95:851–856

Madsen DØ (2016) SWOT analysis: a management fashion perspective. Int J Bus Res 16(1):39–56

Moher D, Olkin I (1995) Meta-analysis of randomized controlled trials: a concern for standards. JAMA 274(24):1962–1964

Omer SK (2018) Swot analysis; the tool of organizations stability (kfc) as a case study. J Pro Manag New Technol 6(4):27–34

Palmer GH, Koprowski J, Pernas T (2007) Tree squirrels as invasive species: conservation and management implications

Perera R (2017) The PESTLE analysis. Nerdynaut

Piercy N, Giles W (1989) Making SWOT analysis work. In: Marketing Intelligence & Planning

Robold RB, Huettmann F (2021) High-resolution prediction of American red squirrel in interior Alaska: a role model for conservation using open access data, machine learning. GIS and LIDAR PeerJ 9:e11830

Sheppard C (ed) (2018) World seas: an environmental evaluation: volume iii: ecological issues and environmental impacts. Academic, London

Silvy NJ (ed) (2020) The wildlife techniques manual: volume 1: research. volume 2: management (vol. 1). JHU Press

Sriram S, Huettmann F (unpublished) A global model of predicted peregrine falcon (Falco peregrinus) distribution with open source GIS code and 104 open access layers for use by the global public. Earth System Science Data Discussions:1–39. https://doi.org/10.5194/essd-2016-65

Steiner M, Huettmann F (2021) Justification for a taxonomic conservation update of the rodent genus Tamiasciurus: addressing marginalization and mis-prioritization of research efforts and conservation laissez-faire for a sustainability outlook. Eur Zool J 88(1):86–116

Stern N (2008) The economics of climate change. Am Econ Rev 98:1–37

UNEP-WCMC, IUCN (2020) Protected Planet: the World Database on Protected Areas (WDPA) [Online], September 2020, Cambridge, UK: UNEP-WCMC and IUCN. Available at: www.protectedplanet.net

Van de Kaa G, De Vries HJ, van Heck E, van den Ende J (2007) The emergence of standards: a meta-analysis. In: 2007 40th Annual Hawaii International Conference on System Sciences (HICSS'07), pp 173a. IEEE

Watts A (2020) How many squirrels are there in the world? Quora. https://www.quora.com/How-many-squirrels-are-there-in-the-world. Accessed 30 Aug 2022

Wrisberg N, de Haes HAU, Triebswetter U, Eder P, Clift R (eds) (2002) Analytical tools for environmental design and management in a systems perspective: the combined use of analytical tools, vol 10. Springer

Rajaratnam SM, Redman JR (2001) Circadian locomotor activity rhythms of the diurnal Indian palm squirrel in constant light. Chronobiol Int 18(1):47–60

Chapter 15
First Conclusions, Success Stories, and A Good Calls-to-Action for the Conservation of the World's Squirrels: From 'Squirrelology' and 'Ministry of Squirrels' to a 'Global Squirrel Agenda'!

Being pragmatic does usually not solve the initial problem and it creates additional ones
Source unknown

Abstract For the progression of science and society, it is essential to use the synthesized information and to present good 'calls to action' for the future. With science-based 'calls to action' we believe to push for – and hopefully to achieve – that urgently needed progress in science and society, for squirrels and beyond. Therefore, here we synthesize the main topics addressed in this book, which cover the *status quo*, shortcomings, conservation issues, governmental failures on sustainability, and the future. Additionally, regions of high risk have been identified, and some relevant issues of governmental actions and policy are elaborated. For constructive progress and improvements, here we will provide a small series of success stories and details with a narrative, where conservation activities have been executed successfully and can lead as an example. Projects like the first steps of the recovery and conservation of prairie dogs are described in this chapter (after culling more than 95% of the prairie dogs in the USA, e.g. using Strychnine and human pursuit). It will be analyzed what has been done in a good way and as a foundation to work from, what went well, and what can still be improved in future conservation programs. In order to present the global situation in a more complete fashion, we invited several squirrel and small mammal experts from different continents across the world to provide us with their unified opinion on the future fate of squirrels and on the conservation progress towards a sustainable 2100. Additionally, we discuss the serious cases of concern where immediate conservation actions are necessary to be taken in order to prevent the further extinction of several squirrel species. This consists of a set of good "calls-to-action" which include clear instructions and examples of how to improve the conservation of threatened species. Examples of threatened squirrel species that are included in this discussion consist of - but are not limited to - Delamare fox squirrel (*Sciurus niger cinereus*), Mt. Graham red squirrel (*Tamiasciurus fremonti grahamensis*), Eurasian Red squirrel (*Sciurus vulgaris*) (threatened in some areas), Giant tree squirrel (*Ratufa indica*) (in Sarawak), and Squirrels in Israel and Jordan. These good "calls to action" also include suggestions on how to manage the threatened habitats in the future to preserve and conserve the species more sustainably and holistically. However, here we are not only discussing the situation of the threatened squirrel species, but in order to achieve a sustainable balance of biodiversity, our 'calls to action' also include suggestions for invasive species such as the Eastern grey squirrel (*Sciurus carolinensis*) and Pallas squirrel (*Callosciurus erythraeus*). Last but not least, a concise way forward is presented to conclude this book chapter, and this book. Our studies show that the well-being of squirrels and the fate of humanity are linked.

Keywords Successful conservation · Calls to action · Squirrel conservation · Synthesis · Meta-analysis · Ministry of Squirrels

Supplementary Information The online version contains supplementary material available at https://doi.org/10.1007/978-3-031-23547-4_15.

M. Steiner, F. Huettmann, *Sustainable Squirrel Conservation*, https://doi.org/10.1007/978-3-031-23547-4_15

15.1 Introduction

15.1.1 Everybody Has Probably Seen a Squirrel, No?

Squirrels are ubiquitous species interacting with mankind. Except for Antarctica, one can encounter them virtually anywhere (Bertolino and Lurz 2013; Thorington et al. 2012). Squirrels are detected across most biomes (Koprowski and Nandini 2008; Menéndez et al. 2021), also in cities and in your average urban downtown park e.g. from Novosibirsk to Vancouver, New York, and London (Agafonov and Yerdakov 2013; Bertram and Moltu 1986; Bonnington et al. 2014; Ten Hwang and Larivière 2006). One can find them likely also on the local high school and college campuses (Robold and Huettmann 2021; and see for instance at UC Berkeley – Campus Movie Fest 2022). Squirrels are part of our human life and are well-liked; they can make your day. But then they are also readily poached, killed, overrun by cats, stripped for the fur, and eaten - globally.

Squirrels should be a prime learning topic for lower-academic educational field projects and biology graduation degrees. However, beyond the surface, reality shows us that most students never really looked at them closer, never meaningfully engaged in the squirrel world and ecology, nor do adults and the professional life really and truly deal with squirrels (in contrast, see what role squirrels play in indigenous people's lives and cultures for millennia; see Fonyuy 2020). Who lives and breathes squirrels? If wildlife professionals would truly have engaged here more, we would see more squirrel-related knowledge, awareness, publications, more funding, less marginalization, and an emphasis on social behaviors and individualism in squirrel ecology with underlying shared data for improved conservation (similarly to primates, let's say; a discipline that is way ahead of 'squirrelology'). It is clear that the usual western society does not only ignore and marginalizes squirrels, but it also actually works against them (see Table 15.1), as evidenced in the widely uncontrolled and not monitored use of squirrels for meat, as target practice, and as pests. Squirrels are a fair and easy game for the taker.

So we looked at much of the existing knowledge about the world's squirrels and their conservation, e.g. in Thorington et al. (2012) and similar, as well as the IUCN Red List (2022). We did this with the good intention for future generations to enjoy those species, to use the squirrel data and findings for progress, and to observe them in nature, assuming squirrels are to be well looked after as a trusted public resource by a mandated governance.

But when we did this, we got very disappointed, finding little digital science data, just a few meaningful maps in geographic information systems (GIS) - or Google Earth for that matter - for relevant conservation, and how humans relate to the environment, including squirrels. Like with legal policies or lawsuits favoring the sound conservation management of squirrels, 'the cloud' as the latest research tool, was widely empty of science-based squirrel data, policies, and research

Table 15.1 Science-based Conservation Management topics known to be missing for the world's squirrels

Conservation management topic	Details	Reference	Comment
Taxonomic data	www.ITIS.gov	Balakirev and Rozhnov (2019), Burgin et al. (2020), de Abreu-Jr et al. (2020), Koprowski and Nandini (2008), Steiner and Huettmann (2021) and Steiner and Huettmann (2023) (Chap. 1).	While taxonomy is not well resolved, and hardly valuable, an alternative approach using pixels, habitats, and landscapes – as a conservation container for species - is to be pursued instead
Population sizes	www.iucnredlist.org	Flyger (1959) and Greene and McCleery (2017)	Even if available for very few species, it is widely outdated
Species density	www.iucnredlist.org	Baltag et al. (2014) and Parker and Nilon (2008)	This is needed for most environmental impact assessments
Laws specific to squirrels	See responses to letter survey in Steiner and Huettmann (2023) – Chap. 11; Harvey (2021)	Cohen (2014) and Gurnell (1996)	The vast majority of online records concern the US, UK and a couple European countries. All other global countries are widely absent, thus far.
Laws specific to squirrels and their habitat management	not available	Gurnell (1996), Patton (1977), Pierce (2012) and Robbins (2007)	A concept and a vision on international, national, and local squirrel laws does not exist
Squirrel-specific web portals	www.iucnredlist.org, www.gbif.org, www.squirrel-net.org		Those existing websites cover only narrow information about squirrels, leaving habitat and man-made climate issues out
Squirrel degrees in schools and education pathways	Squirrels offer themselves as easy and excellent study animals	Hamilton (1987), Shamhart (2010) and Young (2013)	A potential is widely unused

projects (One may search there online with Microsoft AZURE, ORACLE Cloud Services, or any Supercomputing Center, etc. for squirrel projects widely absent)! While this species group is widely abundant in nature with relatively high diversity, it is widely hunted and shows very high mortality rates caused – not exclusively – by road kills (driving a car is what many people participate in daily; Desmond 2013; Smith-Patten and Patten 2008).

And so it is not surprising that we find in our assessment that relatively little research has been conducted on squirrels; virtually no ecology, the family is widely underfunded, and data is outdated and rarely shared (Bourtis 2021). Additionally, as discussed in Chap. 2 (Steiner and Huettmann 2023), squirrels have been recorded in their ancestral form for already approx. 40 million years in modern Central USA (Emry and Korth 2007; Janis 1993; Pečnerová 2011). This is believed to be the place where their evolution started; a cradle (Mercer and Roth 2003; Pečnerová and Martínková 2012). Not respecting and trying to protect these species seems to indicate a strong detachment from nature, from squirrels, from a meaningful science, and a sustainable management in modern times (Steiner and Huettmann 2021, and see groups like Facebook 2022a, b). Despite the huge popularity of squirrels, modern times, and governance widely failed squirrels. The underlying industrial business model and employment of an 8-17 h, 5 days a week – even if unionized – destroys squirrels (e.g. Taber and Payne 2003; see also Czech 2000). This detached approach to nature, and to squirrels, seems also to be reflected in the management practices and efforts for these approx. 300 cosmopolitan species of the world. As a study guide for the future, an overview of topics that are known to be poorly researched and connected with squirrels has been presented in Table 15.1; it can be seen as a good research agenda to work from in a constructive fashion.

In such an environment, even if one wants to, it will be difficult to make good, sound and sustainable decisions for squirrels. As a matter of fact, squirrels are mostly not on the public agenda at all, far away from a 'Ministry of Squirrels' (but compare that with a 'Ministry of Fish and Game ', a concept which dominates the wildlife and biodiversity governance for many rich nations). Squirrels are 'cute', but they are also quite 'dominant/big' and rather overlooked predators in the wider ecosystem (Callahan 1993). Within the squirrel family, body lengths of up to 1.3 m can be observed and with body weights up to 8 kg (Choudhury 2002; Jackson 2012; Kryštufek and Vohralík 2013; Payne et al. 1985; Thorington et al. 2012). Already the amounts of predating mushrooms, or songbirds, and many other smaller species by squirrels remain widely unknown (Schmidt and Whelan 1999; Thompson 2007), but one may believe it is on the level of a significant ecosystem engineer; on a landscape-scale. This is only one of the many topics which still remains unknown but without a solid plan and outlook on research effort investments (see Table 15.1).

Our work in this chapter, and book, attempts to tackle some of these issues and draws a baseline of the performed research, with the intention to present and address a few of the many significant gaps in modern squirrel science, – politics and – management. Throughout this book, the world's squirrel species have been assessed, done from different angles, in order to address a wide spectrum of issues in many globally important regions. To summarize the essence and key messages of each chapter, here in this concluding chapter, a synthesis is presented (see Table 15.2).

The aim of this work is to provide the reader with a broad spectrum of state-of-the-art conservation assessment tools and approaches which can also be used as a certain template for virtually any other species, genus, or family on the planet. Additionally, it serves as the first-ever holistic conservation assessment of the world's squirrels. Throughout this text, it might seem that only negative aspects of squirrel conservation and management are being presented. This is partly true, as reality shows no other than what has been presented throughout these last 14 chapters; so are the facts, and the state of the science affairs in 2023. However, to also present some of the few existing positive moments in squirrel research, conservation, and decision-making, in the following part, some 'success stories' are presented.

Additionally, by combining the gathered knowledge and findings from the last 14 chapters and the success stories reported in the literature, here a final possible future outlook and suggested way forward is painted on a canvas. We all can get there. These final comments include input from both authors and additional expert profiles from all over the world who are considered to have deep knowledge on squirrel conservation, management, decision-making, legal protection, as well as indigenous practice, and culture. It is what keeps the world and mankind alive.

15.2 Methods

To conclude this book, in this final chapter, we synthesize the essence and main findings of each chapter. This has been done using a meta-analysis approach.

Additionally, we compiled from various sources available to us 'success stories' of squirrel research, conservation, and decision-making by performing a literature review on published records about this topic. The literature resources mainly consist of a short list of academic books about squirrels. Additionally, scientific papers have been retrieved from Google Scholar, Scopus, Google, and the world wide web (www). We also added narratives that we were aware of.

Lastly, in order to move forward sustainably from this point, a suggested path forward is being presented, including expert perspectives. Beyond ourselves, this has been achieved by gathering three outside expert opinions from different societies, countries, and continents on the suggested way forward for squirrel conservation and their fate on this planet. Every expert has answered/ commented on a series of questions we posed to them. These questions include topics such as the biodiversity trend of the global squirrels for the future, the expected shift of distribution ranges, the influence of climate change and urbanization on squirrels, assumed future trends for zoonotic disease transmissions, etc. The full survey, including the seven questions we asked, can be found in Appendix 15.1. The detailed information and answers to the questions we posed to the three profiles from this short survey can be found in Appendix 15.2.

15.3 Results

The synthesis of the first 14 book chapters of Steiner and Huettmann (2023) can be found in Table 15.2.

Table 15.2 Meta-Analysis of the first 14 book chapters and the messages of Steiner and Huettmann (2023)

Chapter #	Topic	Take-home message
1	Introduction & Taxonomic overview	The global squirrel taxonomy is widely disagreed upon, and taxonomic lists vary greatly in numbers. The suggested way forward includes a proper large-scale assessment, to determine a mutually agreed-upon taxonomic list that can from then onwards be updated when needed by skilled experts in this field.
2	Evolution & Conceptual evolutionary tracking models	The squirrel's evolution is complex but follows the commonly known principles of speciation over millions of years. Further in-depth analysis and models can show great insights into future studies. Here the authors present a base model which all future in-depth studies can use as a reference.
3	Species distribution Models for all global squirrels with available data (233) using 132 environmental predictors	Squirrels are predicted to be able to survive in almost all global habitats, except Antarctica. The global SDMs created here are the largest constructed dataset to this day for squirrels, small mammals, and likely all animals. This study sets the ideal base for all further in-depth studies addressing species distribution within the squirrel family. Additionally, the methods used here can be replicated for virtually any other faunal or floral family on this planet, using this study as a template.
4	Meta-analysis of Big Data Mining of Global Predicted Squirrel Distribution Models	On a global scale, the environmental variables "World soil characteristics", "Climate Classes", "Proximity to Protected Areas", "World Threatened Mammal Density", "Altitude", "Slope", "Proximity to major cities", "Human Influence Index", and many climate variables seem to have the most influence on the distribution of the world's squirrels.
5	Global squirrel distribution and predictions in cities (urban and suburban areas)	Human distribution and abundance are continuously increasing with the modern expansion of urban and suburban areas, infrastructure development, etc. This in turn increases the road network development and the accessibility of natural prestige landscapes. These aforementioned factors all influence the ongoing habitat fragmentation and destruction with major influences on the global squirrel population. Long term, the authors predict that human influences bring several squirrel species near cities in great danger and reduce overall population sizes due to the effects of human development. Also, likely more squirrels will live in urban and suburban areas in the future, partly because they can/ must adapt and survive there, and partly because their natural habitat has been eradicated by human development.
6	Squirrels in the tropics: A specific review of their fate, stress, declines, and extinctions	The authors predict that the squirrel distribution in the tropics drastically decreases based on a 'business-as-usual' climate warming scenario, even more drastically for a scenario with global warming of 3+°C. To conserve these prestige habitats (tropics) and the species within, the authors strongly suggest investing increased efforts and funding in tropical research and conservation. Additionally, resolving modern issues in the tropics (poverty, poor politics, war, corruption, etc.) would significantly improve the outlook of the conservation situation in the tropics
7	Squirrels on islands: The effect of a 'laissez-faire' approach from governments	Islands are amongst the most common habitats for squirrels. The authors predicted that island environments pose a significant threat to squirrels, due to their limited size and isolation which does not allow species to significantly move/ track their niche in response to the ongoing changing (warming) climate and its effects on e.g. sea level, frequency & magnitude of floods, storms, hurricanes, monsoons, green-house gas composition, etc. Thus, island-endemic species (in this case specifically squirrels), are under a major threat to their surrounding natural environment and are predicted to be among the most negatively impacted species in response to human-accelerated climate change. Additionally, the authors identified that squirrels on islands are not resolved or managed well for conservation. In conclusion, squirrels on islands should be treated with extra care, while deserving a spot high up on the global conservation agenda.

(continued)

Table 15.2 (continued)

Chapter #	Topic	Take-home message
8	Squirrels in old-growth Forest landscapes	Old-growth forests are globally vanishing, and with them also the habitat-endemic species, this includes multiple-dozens squirrel species. Therefore, to avoid this, the authors suggest permanently protecting and conserving these habitats and species on a global scale. This is only possible using science-based approaches, as applied by the authors in this study. It shall also function as a role model for any further, more in-depth studies analyzing old-growth habitats.
9	Strategic conservation Planning of squirrels in old-growth forests	The authors discussed how the old-growth biodiversity-valuable forests can be better protected and what the influences of squirrel conservation and management are on such endemic biodiversity-rich forests of global relevance. Because the study area (Tanana Valley state Forest – Alaska, US) includes 20% indigenous land ownership, the authors show that approx. 30% of the 'best' solution planning units (PUs) for a possible protected area fall on indigenous land, raising cultural and holistic awareness for the required protection efforts of squirrels, old-growth forest reserves, and indigenous land. An attempted 50% protection scheme seems to be impossible to implement, considering indigenous land, old-growth forest, and timber harvesting.
10	Squirrel economics. A global and national cross-scale assessment of GDP vs conservation status	Globally, it can be observed that countries with a higher GDP and species richness do not have necessarily more budget assigned to squirrel conservation. The authors also observe that the countries with the highest species richness, especially of endangered species, have relatively high GDPs. This leads to the conclusion that high GDPs and human development efforts can be associated with a high percentage of endangered species.
11	Global hunting situation of the world's squirrels	Squirrels are widely hunted, mostly without data, lack of strict regulations and very poor – impossible- law enforcement. Approximately more than 85% of the squirrel species have not been reported in the responses to the authors' letter survey (with an overall 40% response rate). This not only indicates marginalization, but also '*laisse-faire*' conservation approaches by governmental entities and agencies in charge of species conservation, law enforcement, and conservation. For a sustainable future, the authors suggest properly including all squirrels within the occurring mandate area in the corresponding hunting/ trapping/ fur-bearing regulations, especially the species that shall be protected from hunting/ trapping/ fur-bearing practices. This shall be true for all global countries where squirrels are known to be occurring.
12	Global squirrel hotspots and climate change. Distribution forecast for the year 2100 of all global squirrels	Here, the authors produced a global species distribution forecast analysis for the year 2100, using climate predictors of three different climate scenarios. These scenarios were: Slight cool-down (0.5–1°C), warming - business as usual (1.5–2°C), and high warming (3+°C). The vast majority of the global squirrel species reacted to the warming scenario with a predicted dispersal trend poleward, and upward (altitude). Also, an overall decrease in predicted dispersal has been observed. While considering the concepts of polar amplification, and the "escalator of extinction"this trend is definitely not promising, especially while considering the earth has finite resources. Major changes are predicted to occur in the ecological niches of the world's squirrels.
13	Squirrel's marginalization	Here, the authors analyzed the sustainability of global squirrel management, which resulted in a relatively negative outlook. Only a minuscule percentage of conservation entities are even aware of the 17 UN-based SDG goals, and even fewer entities, if any, base the squirrel management on these 17 SDGs. The authors also analyzed the assigned budget for squirrel conservation in most global countries in which squirrels naturally occur. Also, here again, only a small percentage of the respondents mentioned that their entity has a budget assigned to the conservation management of the locally occurring squirrel species. Of the ones that responded with an assigned budget, three mentioned numbers were around $30,000, whereas all others were significantly below that. In conclusion, the authors suggest educating conservation entities better when it comes to sustainable conservation management, and additionally, ask these entities to assign more budget to squirrel conservation management. Ideally, also publicly share the conservation budget balance sheets to support open-access data sharing approaches.
14	SWOT and PESTLE analysis of the world's squirrels and the 132 environmental predictors used for modeling the species' distributions	The authors applied two widely known economic analyses (SWOT & PESTLE), however, when seen from the perspective of ecological economics, none of the two provide substantially new information and approaches on how to tackle finite resources in the Anthropocene. No real surprise there. For better insights into the world's squirrels from an ecological economics perspective, the authors suggest more detailed ecological economic analyses which are less globally known and developed.

The literature review of so-called 'success stories' in squirrel management, and generally squirrel science, yielded solid and documented insights into which countries put effort into squirrel conservation, who pays attention to their citizens and scientists, and which countries place squirrels on their science agenda, and which methods have been successfully used to conserve squirrels/improve their life quality. The detailed results of the literature review, describing the conservation actions taken to achieve conservation success, and details about the specific projects can be found in Textbox 15.1.

Textbox 15.1: Success Stories in Squirrel Conservation and Management Are Far and Few: Simply Ask the Squirrels Instead!

Most of the squirrel species of the world are highly marginalized. This has been shown several times throughout multiple chapters in Steiner and Huettmann (2023). Especially when it comes to specific conservation actions, squirrels are mostly left out. That is certainly true globally, where over 95% of the squirrels are essentially ignored. An example of this can be seen in the fact with the International Union for Conservation of Nature (IUCN) that has a specific subdivision devoted to just one vulnerable single species: Polar bear (*Ursus maritimus*) (IUCN SSC Polar Bear Specialist Group – IUCN (2022)). Whereas for 300 species of squirrels, let alone individual species (many of which are considered Endangered and Critically Endangered), there are no such specialist groups in place whatsoever; certainly no funding for them. The most-specific group for squirrels – as good as it gets - is the IUCN Small Mammal specialist group (IUCN SSC Small Mammal Specialist Group – https://small-mammals.org/), being in charge of a pool of over 2800 species (likely an incomplete set) with a world-wide focus. It seems obvious that a single IUCN group, in charge of almost 3000 species has a harder time providing time, results, and significant tools to successfully conserve all of its assigned species, compared to another species specialist group in charge of only one species. One can easily see here that it is widely imbalanced and cannot perform equally, or well. Impacts then are easily seen in their habitats, their conservation status, and their position on the scientific agenda.

However, despite this poor setup for squirrel conservation, some 'success stories' have been reported which are discussed here in this textbox. But most so-called squirrel conservation "success stories" have just been reported in Europe, more specifically, in the United Kingdom and Italy. Such examples have been presented by Bertolino and Genovesi (2003), Parrott et al. (2009), Signorile et al. (2014), Wauters and Martinoli (2021). Bertolino and Genovesi (2003) describe "*conservation success*" mostly by eradicating the invasive species Eastern grey squirrel (*Sciurus carolinensis*) in Italy to avoid that the Eurasian red squirrel (*Sciurus vulgaris*) being pushed out of its endemic habitat. The same has been described by Signorile et al. (2014), as well as by Wauters and Martinoli (2021). On the other hand, Parrott et al. (2009) review the conservation activity of *S. vulgaris* in northern England. But the primary focus seems to be on the eradication of *S. carolinensis* in order to 'save' the endemic species *S. vulgaris*. And that is essentially 'it' for the acclaimed 'success' in squirrel management and conservation.

Apart from these described conservation measures that promote locally endemic species by eradicating their invasive competitor in two of the many EU nations, there have been a few other conservation measures recorded that have shown positive results throughout the last years. An example of this is the recovery of the Black-tailed prairie dog (*Cynomys ludovicianus*) population in the Western USA. This "conservation success" has been thoroughly explained in a video created by Defenders of Wildlife (YouTube 2022). This Prairie dog population has been severely diminished (by more than 95% !) in terms of population numbers and habitat extent in the last few decades by enforcing eradication programs, lethal control, sylvatic plague, and the conversion of grasslands into croplands and rangelands. Whereas, this species is considered a keystone species, influencing over 100 other species in American planes (Bergstrom et al. 2014), but yet they are still considered Least Concerned by the IUCN Red List, even after their population crash (which seems to have no relevance on the status *per se*?) (Cassola 2016). After recognizing this fatal population crash, this conservation program seems to be in good hands with the Defenders of Wildlife – an NGO – (https://defenders.org/), on a successful road again. However, starting from a 95% loss is not a good position to be in, nor a recovery success when considering the massive habitat loss and climate change going on unabated.

Apart from these described conservation initiatives, there are a few more conservation efforts described in the literature. For example, Selonen and Mäkeläinen (2017) describe the ecology and conservation initiatives for the Siberian flying squirrel (*Pteromys volans*) in Finland. Another example is the Vancouver Island marmot (*Marmota vancouverensis*), which is one of the very few squirrel species that is considered Critically Endangered, that has received some media attention in the last few years and multiple facilities support in-situ and ex-situ conservation initiatives (Aaltonen et al. 2009; Casimir et al. 2007; Jackson et al. 2016; Werner 2006).

Another reintroduction program of the Tarbagan marmot (*Marmota sibirica*) in Mongolia has recorded success in the last few years after initially failing to communicate conservation plans to the local community (www.steppewildlife.org/conservation). Some efforts have also been observed in Nepal with the Ukali project (www.ukali.org/), including the Himalayan Marmot (*Marmota himalayana*), and also some other closely related species e.g. Pikas (*Ochotona spp*).

An additional successful reintroduction program has been recorded in northern Italy with the Alpine marmot (*Marmota marmota*) (Dolomiti Bellunesi Parco Nazionale 2022).

Apart from these few reported "conservation success stories", there are unfortunately very few conservation initiatives reported, especially with positive results. To move forward from this, we highly urge to pay more attention to all of the 300+ squirrel species and their conservation on this planet. The synergy matters for the vast amount of squirrel species, specifically in the tropical regions.

Finally, squirrels know their protection best. If conservation would simply be put in the squirrels' hands, most likely the current conservation efficiency would be significantly increased, and the same would most likely be suggested by a "*world squirrel court*". Squirrels rule.

As a summary of the so-called 'success stories', it can be stated that the majority of those are simply just focused on the eradication of previously imported/ invasive species. Virtually none of them deal with wider and fundamental changes in doing business or tackling finite resources and habitats, or human culture. Additionally, it is notable that virtually all of those referred to 'success stories' are recorded in Europe, with the prime countries of the United Kingdom, followed by Italy (Barkham 2021; Bertolino and Genovesi 2003; BES Press Office 2022; Cumbria Wildlife Trust 2018; Farming Independent 2006; National Trust for Scotland 2022; Parrott et al. 2009; Signorile et al. 2014; Stokstad 2016; Trees for Life 2022; Wauters and Martinoli 2021). These 'success stories' described in the literature are mainly focused on the omnipresent and popular species of the Eurasian Red Squirrel (*Sciurus vulgaris*), following the trend in the remaining research sector where just a few species are largely studied and "preferred", whereas research and conservation are widely absent for the remaining approx. 300 species (98%) (Koprowski and Nandini 2008; Ramos-Lara and Koprowski 2014). It is an utterly biased picture and not a successful one whatsoever for the squirrels of the world.

Arguably, those promoted 'success stories' are only successful because they focus on the local and incomplete, parsimonious perspective; they are divorced from wider linkages, from a relevant ecology (Naess 2009), from the world of squirrels, and from what is possible and what really matters for progress: how humans and squirrels co-exist on earth in future generations. Thus far, we find virtually no global success in squirrels.

The summary table of the three profiles of expert answers regarding the future fate of the squirrels can be found in Table 15.3.

The detailed comments and answers for each profile we received can be found in Appendix 15.2.

In short, it can be observed from all evidence compiled, that the expert outlook on conservation success and species numbers until 2100 is rather negative, with the squirrels considered moderately-severely threatened.

It seems that all experts across the continents agree on the trend of habitat fragmentation (decreasing wilderness patches), and land change that forces squirrels to inhabit primarily urban areas in the future. The overall squirrel abundance is communally agreed to be decreasing (!), whereas the squirrel species diversity is assumed to decrease less severely. Confirming our earlier assessment, the overall outlook on the fate of squirrels seems to be relatively negative, with an identified lack of sufficient and meaningful conservation efforts, likely leading to several species extinctions.

15.4 Discussion

In this study, we synthesized the main findings on squirrel conservation and so-called 'success stories'. We presented a good 'call-to-action' with take-home messages for Chaps. 1 to 14 by Steiner and Huettmann (2023). Additionally, we presented science-based conservation management topics known to be missing for the world's squirrels. There, we identified that basic life-history traits are widely unknown for the majority of squirrel species. This includes the overall solid taxonomy, phylogeny, population sizes, species densities, laws specific to squirrels, laws specific to squirrel habitat management, squirrel funding, squirrel governance structures, squirrel-specific web portals, and squirrel degrees in schools and education pathways for a solid 'career' in squirrels to serve mankind and well-being. Another topic that has barely ever been discussed for squirrels focuses on impact assessments. This has been described in Textbox 15.2, where all its aspects and difficulties have been described.

Table 15.3 Summary of the comments/answers of the profiles regarding the future fate of the squirrels and a suggested way forward

Experts	Positive outlook on conservation success and species number until 2100	How threatened are squirrels considered now and in 2100?		Preferred habitat for squirrels in 2100?		Will the squirrel diversity change by 2100?		Overall positive outlook on the fate of squirrels?	Are the current conservation projects benefiting squirrels long-term?	Are those conservation projects sufficient to avoid extinctions in the upcoming 80 years?
		Severely threatened	*Mildly threatened*	*More urban-oriented*	*More untouched nature-oriented*	*Decrease*	*Remain stable*			
1	No	No	Yes	Yes	No	Yes	Yes	Yes	No	No
2	Yes	Yes	–	Yes	No	Yes	No	No	No	No
3	Yes, for abundant species, no for species with narrow niches	No	Yes	–	–	Yes	Yes	Yes	Yes, in the expert's local jurisdiction	No

Textbox 15.2: Impact Statements *ad absurdum*: Squirrel Abundance Estimates Are Unlikely a Good and Sustainable Answer to a Global Onslaught

Falk Huettmann, Moriz Steiner

In order to assess and mitigate impacts, to do no harm, in the western world impact, assessments are to be done. Here it is assumed those are done in a proactive fashion, assessing and avoiding risks before any damage is done. That is a common procedure – best-professional practice – in industrial development projects, done on a national level by most advanced countries and is to provide a public and balanced approach to "*sustainable development*" (Note: This globally-done approach – also suggested by the U.N. and its entities - fully accepts destruction and a bad impact, but then it tries to minimize it after the act; the weaker partner tends to pay the prize (Taber and Payne 2003; see also Czech 2000)). An entire profession of impact and environmental consultants is working on this topic and tries to achieve such work under a legally approved and allowing framework (that is, a framework that fully accepts negative impacts). Such assessments usually involve quantifications and thus, statisticians are employed to obtain those numbers in a reliable fashion with confidence. While the 'school of statistics' used there widely relies on 'frequency statistics' with p-values and finding 'significant' impacts, the wider and more holistic approaches are widely missing. Landscapes, the world – the Anthropocene and its atmosphere – look accordingly then. Industrialization comes with a legally approved destruction.

For squirrels, one is even further away from the concept of 'do no harm'. As a matter of fact, there are very few impact assessments known by the authors that truly have squirrels on the agenda at all and even fewer ones where squirrels and impacts are quantified. The quantification matters for legal compensation procedures and quantifying the monetary value to the damage, as part of a legal and ethical requirement in the western world (the entire business scheme of (life) insurances for instance is built on it).

How to quantify squirrels and their biology remains a topic of discussion though; it is rarely attempted and less so achieved. As a matter of fact, in most wildlife studies 'quantification' relates to populations, territories, abundance, and life history parameters, namely trends, densities, fecundity, life expectancy, survival, cohorts etc. Whereas the wider spatial aspects – ecology – remain unaccounted for. However, when it comes to the quantification of squirrels, e.g. for red squirrels, one can find studies about midden densities. It is one of the easiest metrics to obtain and thus done: counting middens per km^2, let's say. It is an apparently obvious feature to obtain; a low-hanging fruit. For red squirrels, it can be approx. 435 middens per km^2 (Robold and Huettmann 2021). From that, one could then estimate how much a development project area would affect squirrels, and compensate accordingly for squirrels lost.

Unfortunately, this method has serious flaws. There are just a few studies done on midden densities, and they are difficult to generalize to other (forest) areas. Middens are not equally distributed; they can be clustered in a landscape. Midden detection on a landscape scale also remains difficult and one has to use 'detection surveys' or models (see Robold and Huettmann 2021 for an example). Those detections can be as low as 20% and thus rely on large extrapolations difficult to prove. And even if midden densities are known, those do not indicate how many squirrels are in the area. Associated squirrel territory estimates remain vague and have a high variance. The frequent assumption is that a midden means a 'nest' and thus a pair of squirrels. One may perhaps add some non-breeders and a floating population. But already that essential correction factor is not known, nor how viable the midden sites are over time (some can get over 50 years old, others depend on the annual pine cone cycles or predation). Squirrel turn-over is not described much. Squirrel patches – such as in old-growth stands – can blink in or out. One will even have severe difficulties already in defining 'what a midden is' because they can overlap and be a network of middens. Middens are certainly an indicator of 'squirrel presence', but midden counts are kind of pointless to obtain for true and easy-to-conclude abundance estimates of midden and squirrels with confidence. Midden maps – a squirrel atlas – for large landscapes remain missing. Middens as such remain easy squirrel features to detect, but without getting closer to the squirrels themselves, it has little meaning and precision.

So how to get closer to the squirrels? Modeling provides an answer, albeit still a rough one. To get at the numbers, Robold and Huettmann (2021) reported for interior urban Alaska 157–629 squirrels per km^2, with other authors from the 1960s reporting approx. 163 squirrels per km^2. Those are the plain numbers to go with, and when considering forestry, urbanization, etc. destroying the habitat for these species in that area to be compensated, the outlook remains hardly bright.

Another issue to address when looking at approximating population metrics for squirrels using this midden-proxy analysis as the main method is that only a very small number of squirrel species create middens. According to literature (see Steiner and Huettmann 2021 and references within), only two squirrel species create middens (North American red squirrel (*Tamiasciurus hudsonicus*), and the Douglas squirrel (*Tamiasciurus douglasii*)). So the question remains, what about the remaining approx. 305 species (99%+)? And how to go about impact assessments for squirrels?

With that, one is still not much closer to the core question: how harmful is development, or habitat loss for the world's squirrels, and how to compensate, if ever? One thousand dollars for compensation has no meaning for squirrels, nor a million dollars, once the habitat is widely gone. Squirrels and the financial world do not mix well, nor do squirrels and economic growth, or the currently practiced sustainable development. And there is little wiggle room in that. This is especially true because for the vast majority of squirrel species population numbers, and trends are unknown, with ongoing (often unregulated) hunting and development pressure.

The situation can be seen and considered to be in a certain state of chaos, with currently no relevant acknowledgments and subsequent initiatives set up to escape this state of chaos. Squirrels and many other species (as part of our global ecosystem) are severely disregarded, all while other topics such as the financial market well-being and GDP decimals are discussed daily in thousands of different TV channels, newspapers, and journals even though, those topics are just not part of multiple global ecosystems, spiritual believes, tribal, and indigenous communities for millennia, etc. (like squirrels are).

So what will the future for squirrels then look like? Well, judging by the recent and ongoing global habitat conversion and habitat loss, and unless cultural change occurs, it looks rather grim, with no good end in sight. The true impact is not really quantitative or financial, but cultural.

We additionally performed a survey where we asked squirrel experts from three different continents about their opinion on the future fate of the squirrels. Overall, we observed that the experts had a relatively positive outlook on the survival of widely abundant species until 2100. However, species with low abundance and small distribution ranges are believed to be in great danger. Amongst the experts, it is widely assumed that the global squirrel biodiversity and abundance will decrease in the upcoming 80 years and that in the absence of relevant governance, the squirrels must adopt their habitat preferences more and more to urban environments. These experts also agree with each other – and with our own generic findings - that the current conservation initiatives are not sufficient to ensure the survival of all squirrel species till 2100.

Lastly, here we would like to present a prioritization setup of conservation management action items for the future well-being of the world's squirrels, and beyond. A squirrel agenda! There, we primarily included the following topics:

- addressing the over-consumption and waste of resources for sustainability,
- addressing how humans better interact with the living world and how they position themselves,
- sustainably within it,
- mitigating man-made climate change, e.g. CO2 and GHG emissions,
- finding long-term sustainability (harmony with Mother Earth and the Universe), and focusing on
- conservation in the habitats of/under high risk.

After reviewing the core findings of this study and the synthesis of Chaps. 1 to 14 in Steiner and Huettmann (2023), we fully acknowledge that these previous 14 chapters are incomplete in regards to the world that squirrels live in, experience, and have to survive in. However, it indeed represents a first painting on a canvas that has been blank for decades, even centuries, and already from that incomplete perspective, it starts to paint us a pretty bleak picture (see Table 15.2).

So what has happened in the last four decades, when many parts of the world experienced unprecedented wealth? The GDP figures for many of the so-called first 'World Nations' show us no other: nature was sacrificed for growth as a global agenda. We also had major Green parties and the Environmental and Biodiversity Movements on the rise, ongoing for easily over 40 years (where much of the major destructions happened in parallel). What have they really achieved, for squirrels or beyond?

Looking at Table 15.2 from a meta-analysis perspective, indicating the overall take-home message, several major profiles become clear for the *status quo*:

1. Squirrels of this world are poorly managed for their conservation, a convenient *laissez-faire* concept prevails for squirrels (in the majority of the world's countries and the world's squirrel species). All of this occurs while resources are not well spent and distributed to account for the value and role squirrels play for our own lives and the survival of future generations, for the Earth.

2. An additional crucial topic to mention is the extremely large differences between the 'cosmopolitan'/ 'western' squirrel species efforts and all the others (c. 98%). It becomes clear to everybody by simply looking at governmental spending (Chap. 11), Hunting regulations (Chap. 11 - Steiner and Huettmann 2023), and the lack of research equality in literature (Koprowski and Nandini 2008; Mendes et al. 2019; Ramos-Lara and Koprowski 2014), proper well-balanced efforts (de Abreu-Jr et al. 2020; Steiner and Huettmann 2021 and references within), and conservation initiatives (Robold and Huettmann 2021; Weigl 2007). This evidence can further be seen in the high ratio of Data Deficient species on the website of the IUCN Red List (www.iucnredlist.org), low funding priorities (Merrick et al. 2021; Parrott et al. 2009), and most crucially, the utter lack of proper care for squirrels within society (Benson 2013; Crowley et al. 2018; Flynn 2002; and see much more on the internet and Chap. 11 of this book).

Now, the global resources and administrative budgets have been rather big, but almost next to none – a tiny amount – got spent on squirrels, or squirrel habitats (Steiner and Huettmann 2021, see budget assignments and survey responses in Chap. 11 of this book). All relevant metrics show us no other.

15.5 What Is Needed for Good Squirrel Conservation Management? Calling for the 'Ministry of Squirrels'

Arguably, we are far away for students to have a 'Degree on Squirrels', a Ph.D. in Squirrel'ology, or have an IVY League School for an Education that takes squirrels and their needs seriously; far from it. While this sounds like a somewhat funny idea to some people, the reality is that the western world – and its teaching – has lost most contact with the living world, nature biology, and Mother Earth. It is totally divorced from nature, nature deficit disorder rules! But this stands in contrast with the initial teachings of Aldo Leopold, John Muir, Arne Naess, or Francis of Assisi, let's say. A relevant spiritual connection, with wilderness and its components, got lost in globalization (but must not be; Lama and Buckley 2020). Already just a quick glimpse at your favorite textbook on the economy, ecology, and wildlife conservation, shows us no other. Shiny presentations of many incomplete and unsustainable pseudo-concepts and values drive the agenda!

While the indigenous world has a link with squirrels (Chetia et al. 2021; Fonyuy 2020), much of the western world, or the recent Asian world (e.g. China, Singapore, Japan), does not have such a link (Coates 2015; Steiner and Huettmann 2021). Squirrels simply got done and away with; it seems like squirrels are of no relevance; a faulty approach though. In upper-management circles – the well-trained leaders of the global enterprise - squirrels do essentially not receive the slightest blink of attention on the agenda; as they virtually create no big money (only for small-scale bush meat businesses in Africa and Asia, and -mostly- illegal fur trades) (see some web pages e.g. with Lake Texoma 2022; Mitchell 2019, etc.).

Here we argue that the lack of squirrels – and similar species – on the agenda make for a poor life and indicate the failure of such management schemes and such lives.

We are not arguing to be pragmatic and parsimonious about the current squirrel failure (being pragmatic does not solve the initial problem and creates new ones). Clearly, not many people are interested, or willing, to devote time to squirrels, or their conservation. However, to outline some future action points, Table 15.4 has been created, as a minimum list for the future betterment of squirrel science.

From the three profiles and their statements on the possible future fate of squirrels and a suggested constructive way forward, it can be concluded that the 'modern' outlook for squirrels and beyond is rather negative with an overall decline in squirrel abundance and a conservation effort doomed to fail (e.g. see issues similar to the ones described by Kachamakova et al. 2022).

Table 15.4 Prioritization of Conservation Management action Items for the World's Squirrels

Conservation topic for Squirrels of the World	Suggestion	Example of the topic already achieved	Outcome	Suggested reading
Addressing waste and over-consumption of resources	Education on the negative influences of "business-as-usual", trophic theory of money, etc.	Steady state economics, ecological economy	Relatively sustainable future (by mainly looking at the environmental and economical aspects of sustainability).	Czech (2000), Daly (2014) and Daly and Farley (2011)
Addressing how humans interact with the living world in a good and sustainable way	Resolve detachment from nature, educate on the influences of unsustainable exploitation of nature, and how it backfires on the Earth's flora and fauna, including humans	Pika-human interaction (Chap. 13), sustainable hunting practices in indigenous cultures (see Inuit, Chap. 13 for Sierra Leone, etc.)	Humans respect nature and its taboos, leaving space for financial inefficiency, and allowing nature to recover. Greater appreciation and overall "caring" for our earth's nature.	Bruchac (2014), Freeman et al. (1998), Koju et al. (2021), Murray (2003), Native American Squirrel Mythology (2022) and Robinson and Bennett (2000)
Mitigate man-made climate change, e.g. CO_2 emissions	Decrease eradication and deforestation of old-growth habitats & forests	Law enforcement with REDD(+) strategy	Increased carbon storage, decreased greenhouse gasses, less severe climate warming, and less severe climate-change-related catastrophes.	Andoh and Lee (2018), Köhl et al. (2009), Poffenberger (2009) and Tacconi et al. (2019)
Finding long-term sustainability (harmony with mother earth and the universe)	Taking a step back from the daily busy routine, personal education on examples provided in the next column, find inner harmony. Once inner harmony has been achieved, such attitude and respect can be shared with fellow humans and our surrounding nature. Generally, moving away from materialism and capitalism, towards a respectful anti-materialism.	China government policy "live in harmony", basic principles of Buddhism, Hinduism, Francis of Assisi (Christianity), Liberation theology (Catholicism), some Shamanism	Societal Well-being, intact natural ecosystems	DuBois (2009), Han (2008), Mahatthanadull (2020), Rajagopalachari et al. (1970), Sorrell (1988) and Stenberg (2006)
Habitats of/ under high risk/ concern	Focusing conservation on urban environments, old-growth habitats, the tropics, and islands	Full conservation of island fauna, development of conservation strategies using GIS for tropical environments, old-growth forest evaluations, and planning of conservation units (see Chap. 9).	Risk decrease of the mentioned habitats and their inhabiting species. Enrichment and conservation of the global biodiversity for future generations.	Gangoso et al. (2006), Ramos-Lara and Koprowski (2014), Rhodes and Wilson (1995), Savitsky and Lacher Jr. (1998), Steiner and Huettmann 2023 – Chaps. 3–8 and Wirth et al. (2009)

15.6 Final Conclusions

Therefore, as a unified suggested way forward, the authors promote moving into a new culture where humans accept the limits of finite resources, in a mildly fluctuating stable economy, with a good spiritual base, and try to live in harmony with nature and with themselves; squirrels included. While these are just a few words, it is actually a rather big step forward for mankind, much bigger and more relevant than flying to the moon (which is kind of irrelevant for conservation and mankind, and for squirrels, whatever got stated to the contrary).

"That's one small step for man, one giant leap for mankind."...

Neil Armstrong's words when he became the first person to set foot on the Moon

"We abuse land because we regard it as a commodity belonging to us. When we see land as a community to which we belong, we may begin to use it with love and respect."

Aldo Leopold (Words of Wisdom 2015)

Moriz Steiner and Falk Huettmann, Fairbanks Alaska 7th September 2022

References

Aaltonen K, Bryant AA, Hostetler JA, Oli MK (2009) Reintroducing endangered Vancouver Island marmots: survival and cause-specific mortality rates of captive-born versus wild-born individuals. Biol Conserv 142(10):2181–2190

Agafonov GM, Yerdakov LN (2013) Cycles of abundance dynamics of squirrel (Sciurus vulgaris, L.) in Transbaikalia, Russia. In: 2nd International Symposium on hunting "Modern aspects of sustainable management of game populations" p 153

Andoh J, Lee Y (2018) National REDD+ strategy for climate change mitigation: a review and comparison of developing countries. Sustainability 10(12):4781

Balakirev AE, Rozhnov VV (2019) Taxonomic revision of beautiful squirrels (Callosciurus, Rodentia: Sciuridae) from the Callosciurus erythraeus/finlaysonii c omplex and their distribution in eastern Indochina. Raffles Bulletin of Zoology 67:459–489

Baltag EŞ, Zaharia G, Fasolă L, Ion C (2014) European ground squirrel (Mammalia: Rodentia) population from eastern Romania: density, distribution and threats. Eur Sci J 10(10):94–101

Barkham P (2021) Scottish forests could save red squirrel from extinction. The Guardian. https://www.theguardian.com/environment/2021/aug/04/scottish-forests-could-save-red-squirrel-from-extinction. Accessed 19 July 2022

Benson E (2013) The urbanization of the eastern gray squirrel in the United States. J Am Hist 100(3):691–710

Bergstrom BJ, Arias LC, Davidson AD, Ferguson AW, Randa LA, Sheffield SR (2014) License to kill: reforming federal wildlife control to restore biodiversity and ecosystem function. Conserv Lett 7(2):131–142

Bertolino S, Genovesi P (2003) Spread and attempted eradication of the grey squirrel (Sciurus carolinensis) in Italy, and consequences for the red squirrel (Sciurus vulgaris) in Eurasia. Biol Conserv 109(3):351–358

Bertolino S, Lurz PW (2013) Callosciurus squirrels: worldwide introductions, ecological impacts and recommendations to prevent the establishment of new invasive populations. Mammal Rev 43(1):22–33

Bertram BC, Moltu DP (1986) Reintroducing red squirrels into Regent's park. Mammal Rev 16(2):81–88

BES Press Office (2022) New research shows that current national red squirrel conservation strategies likely to undermine species survival in future. https://www.britishecologicalsociety.org/new-research-shows-that-current-national-red-squirrel-conservation-strategies-likely-to-undermine-species-survival-in-future/. Accessed 19 July 2022

Bonnington C, Gaston KJ, Evans KL (2014) Squirrels in suburbia: influence of urbanisation on the occurrence and distribution of a common exotic mammal. Urban Ecosyst 17(2):533–546

Bourtis E (2021) Underfunded but passionate, Native American conservationists call for more support. Mongabay. https://news.mongabay.com/2021/11/underfunded-but-passionate-native-american-conservationists-call-for-more-support/. Accessed 19 July 2022

Bruchac M (2014) Indigenous knowledge and traditional knowledge. Pennsylvania University, Cham

Burgin CJ, Wilson DE, Mittermeier RA, Rylands AB, Lacher TE, Sechrest W (eds) (2020) Illustrated checklist of the mammals of the world. Lynx, Barcelona

Callahan JR (1993) Squirrels as predators. In: The Great Basin Naturalist. pp. 137–144

Campus Movie Fest (2022) Campus Pets: The Squirrels of UC Berkeley. https://campusmoviefest.com/movies/50048-campus-pets-the-squirrels-of-uc-berkeley. Accessed 19 July 2022

Casimir DL, Moehrenschlager A, Barclay RM (2007) Factors influencing reproduction in captive Vancouver island marmots: implications for captive breeding and reintroduction programs. J Mammal 88(6):1412–1419

Cassola F (2016) Cynomys ludovicianus (errata version published in 2017). The IUCN Red List of Threatened Species 2016: e.T6091A115080297. https://doi.org/10.2305/IUCN.UK.2016-3.RLTS.T6091A22261137.en. Accessed on 03 May 2022

Chetia H, Chatakonda MK, Koprowski JL (2021) Squirrels and Tribes–A Cultural Analysis From Indian Eastern Himalayas

Choudhury A (2002) Petaurista nobilis singhei-first record in India and a note on its taxonomy. Journal-Bombay Natural History Society 99(1):30–34

Coates P (2015) A tale of two squirrels: a British case study of the sociocultural dimensions of debates over invasive species. Invasive species in a globalized world: ecological, social, and legal perspectives on policy. University of Chicago Press, Chicago, pp 44–71

Cohen T (2014) Grey squirrel law scrapped: Ministers finally revoke 1937 legislation that made failure to report presence of rodent on your land a criminal offence. Mail Online. p 19

Crowley SL, Hinchliffe S, McDonald RA (2018) Killing squirrels: exploring motivations and practices of lethal wildlife management. Environment and Planning E: Nature and Space 1(1–2):120–143

Cumbria Wildlife Trust (2018) Red squirrel conservationists welcome success stats. https://www.cumbriawildlifetrust.org.uk/news/red-squirrel-conservationists-welcome-success-stats. Accessed 19 July 2022

Czech B (2000) Shoveling fuel for a runaway train: errant economists, shameful spenders, and a plan to stop them all. Univ of California Press.

Daly HE (2014) From uneconomic growth to a steady-state economy. Edward Elgar Publishing, Cheltenham

Daly HE, Farley J (2011) Ecological economics: principles and applications. Island press

de Abreu-Jr EF, Pavan SE, Tsuchiya MT, Wilson DE, Percequillo AR, Maldonado JE (2020) Museomics of tree squirrels: a dense taxon sampling of mitogenomes reveals hidden diversity, phenotypic convergence, and the need of a taxonomic overhaul. BMC Evol Biol 20(1):1–25

Desmond J (2013) Requiem for roadkill: Death and denial on America's roads. In: Environmental anthropology. Routledge, Abingdon, pp. 60–72

Dolomiti Bellunesi Parco Nazionale (2022) Reintroduction of the Marmot (Marmota marmota) in Dolomiti Bellunesi National Park. www.dolomitipark.it/en/park-authority/conservation-projects/reintroduction-of-the-marmot-marmota-marmota-in-dolomiti-bellunesi-national-park/. Accessed 19 July 2022

DuBois TA (2009) An introduction to shamanism. Cambridge University Press, Cambridge

Emry RJ, Korth WW (2007) A new genus of squirrel (Rodentia, Sciuridae) from the mid-Cenozoic of North America. J Vertebr Paleontol 27(3):693–698

Facebook (2022a) Extreme Squirrel Hunting. https://www.facebook.com/extremesquirrelhunting/. Accessed 19 July 2022

Facebook (2022b) Northeast Squirrel Hunters. https://www.facebook.com/Northeast-Squirrel-Hunters-304180919720570/. Accessed 19 July 2022

Farming Independent (2006) Italians take serious steps to deal with grey squirrel. https://www.independent.ie/business/farming/italians-take-serious-steps-to-deal-with-grey-squirrel-26369781.html. Accessed 19 July 2022

Flyger VF (1959) A comparison of methods for estimating squirrel populations. J Wildl Manag 23(2):220–223

Flynn C (2002) Hunting and illegal violence against humans and other animals: exploring the relationship. Society & Animals 10(2):137–154

Fonyuy BE (2020) Building Intercultural Dialogue through Indigenous Justice Systems: A Study of the Ambivalences Governing the Concept and Practice of Justice in Ambanasom's Son of the Native Soil. In: International Journal of Humanitatis Theoreticus. vol. 3. iss. 2

Freeman MM, Bogoslovskaya L, Caulfield RA, Krupnik II, Egede I, Stevenson MG (1998) Inuit, whaling, and sustainability, vol No. 1. Rowman, Altamira

Gangoso L, Donázar JA, Scholz S, Palacios CJ, Hiraldo F (2006) Contradiction in conservation of island ecosystems: plants, introduced herbivores and avian scavengers in the Canary Islands. Biodivers Conserv 15(7):2231–2248

Greene DU, McCleery RA (2017) Reevaluating fox squirrel (Sciurus niger) population declines in the southeastern United States. J Mammal 98(2):502–512

Gurnell J (1996) The grey squirrel in Britain: problems for management and lessons for Europe. European mammals. In: Mathias ML, Santos-Reis M, Amori G, Libois R, Mitchell-Jones A, Saint-Girons MC (eds) Museu Bocage. Lisboa, Portugal, pp 67–81

Hamilton C (1987) Fox Squirrel Interaction with Electrical Transformers. (M.S. Thesis, University of Nebraska-Lincoln). Under the supervision of Professor Ron J. Johnson. pp 76

Han AG (2008) Building a harmonious society and achieving individual harmony. J Chin Polit Sci 13(2):143–164

Harvey F (2021) Red squirrels and pine martens could lose protection in UK review, say experts. The Guardian. https://www.theguardian.com/environment/2021/jul/02/red-squirrels-and-pine-martens-could-lose-protection-in-uk-review-say-experts. Accessed 19 July 2022

IUCN (2022) A word from our co-chairs. https://www.iucn-pbsg.org/#:~:text=The%20Polar%20Bear%20Specialist%20Group%20(PBSG)%20is%20composed%20of%20skilled,and%20Natural%20Resources%20(IUCN). Accessed 19 July 2022

IUCN Red List (2022) Family Sciuridae. https://www.iucnredlist.org/search?taxonomies=101297&searchType=species. Accessed 19 July 2022

Jackson SM (2012) Gliding mammals of the world. Csiro Publishing, Collingwood

Jackson CL, Schuster R, Arcese P (2016) Release date influences first-year site fidelity and survival in captive-bred Vancouver Island marmots. Ecosphere 7(5):e01314

Janis CM (1993) Tertiary mammal evolution in the context of changing climates, vegetation, and tectonic events. Annu Rev Ecol Syst 24(1):467–500

Kachamakova M, Koshev Y, Rammou DL, Spasov S (2022) Rise and fall: results of a multidisciplinary study and 5-year long monitoring of conservation translocation of the European ground squirrel. Biodiversity Data Journal 10:e83321

Köhl M, Baldauf T, Plugge D, Krug J (2009) Reduced emissions from deforestation and forest degradation (REDD): a climate change mitigation strategy on a critical track. Carbon Balance Manag 4(1):1–10

Koju N, Chalise MK, Kyes RC (2021) Pikas and people: human-Pika (Ochotona) interaction in the Himalaya of Nepal. Biodiversitas Journal of Biological Diversity 22(11):5085–5091

Koprowski JL, Nandini R (2008) Global hotspots and knowledge gaps for tree and flying squirrels, vol 95. Current Science, pp 851–856

Kryštufek B, Vohralík V (2013) Taxonomic revision of the Palaearctic rodents (Rodentia). Part 2. Sciuridae: Urocitellus, Marmota and Sciurotamias. Lynx, series nova, p 44

Lama and Buckley (2020) xxx I had sent you the reference

Mahatthanadull S (2020) Teaching document: concepts and principles of Buddhism. Mahachulalongkornrajavidyalaya University Press

Mendes CP, Koprowski JL, Galetti M (2019) Neosquirrel: a data set of ecological knowledge on Neotropical squirrels. Mammal Rev 49(3):210–225

Menéndez I, Gómez Cano AR, Cantalapiedra JL, Peláez-Campomanes P, Álvarez-Sierra MÁ, Hernandez Fernandez M (2021) A multi-layered approach to the diversification of squirrels. Mammal Rev 51(1):66–81

Mercer JM, Roth VL (2003) The effects of Cenozoic global change on squirrel phylogeny. Science 299(5612):1568–1572

Merrick MJ, Morandini M, Greer VL, Koprowski JL (2021) Endemic population response to increasingly severe fire: a Cascade of endangerment for the Mt. Graham Red Squirrel BioScience 71(2):161–173

Mitchell T (2019) Twelve things about Squirrels that will blow your mind. Todd Mitchell Books. http://toddmitchellbooks.com/twelve-things-about-squirrels-that-will-blow-your-mind/. Accessed 19 July 2022

Murray M (2003) Overkill and sustainable use. Science 299(5614):1851–1853

Naess A (2009) The ecology of wisdom: Writings by Arne Naess. Catapult

National Trust for Scotland (2022) Red squirrel success story at Inverewe. https://www.nts.org.uk/stories/red-squirrel-success-story-at-inverewe. Accessed 19 July 2022

Native American Squirrel Mythology (2022) Native American Legends About Squirrels. www.native-languages.org/legends-squirrel.htm. Accessed 19 July 2022

Parker TS, Nilon CH (2008) Gray squirrel density, habitat suitability, and behavior in urban parks. Urban Ecosyst 11(3):243–255

Parrott D, Quy R, Van D, Lurz P, Rushton S, Gurnell J, et al. Reynolds J (2009) Review of red squirrel conservation activity in northern England. Natural England Commissioned Report

Patton DR (1977) Managing Southwestern ponderosa pine for the abort squirrel. J For 75(5):264–267

Payne J, Francis CM, Phillipps K (1985) Field guide to the mammals of Borneo. Sabah Society, Kota Kinabalu

Pečnerová P (2011) Phylogeny of squirrels (Sciurus, Rodentia) based on supertrees reconstruction (Doctoral dissertation, Masarykova univerzita, Přírodovědecká fakulta)

Pečnerová P, Martínková N (2012) Evolutionary history of tree squirrels (Rodentia, Sciurini) based on multilocus phylogeny reconstruction. Zool Scr 41(3):211–219

Pierce RA (2012) Tree squirrels: managing habitat and controlling damage (2012). Natural resources

Poffenberger M (2009) Cambodia's forests and climate change: Mitigating drivers of deforestation. In: Natural Resources Forum. Oxford: Blackwell Publishing Ltd., vol. 33, No. 4, pp 285–296

Rajagopalachari C, Diwakar RR, Munshi KM (1970) Hinduism: doctrine and way of life. Bharatiya Vidya Bhavan

Ramos-Lara N, Koprowski JL (2014) Deforestation and knowledge gaps threaten conservation of less charismatic species: status of the arboreal squirrels of Mexico. Mammalia 78(4):417–427

Rhodes TC, Wilson PN (1995) Sky islands, squirrels, and scopes: the political economy of an environmental conflict, vol 71. Land Economics, pp 106–121

Robbins K (2007) Missing the link: the importance of keeping ecosystems intact and what the endangered species act suggests we do about it. Environmental Law 37:573–594

Robinson J, Bennett EL (eds) (2000) Hunting for sustainability in tropical forests. Columbia University Press, New York

Robold RB, Huettmann F (2021) High-resolution prediction of American red squirrel in interior Alaska: a role model for conservation using open access data, machine learning. GIS and LIDAR PeerJ 9:e11830

Savitsky BG, Lacher TE Jr (eds) (1998) GIS methodologies for developing conservation strategies: tropical forest recovery and willdlife management in costa Rica. Columbia University Press, New York

Schmidt KA, Whelan CJ (1999) Nest predation on woodland songbirds: when is nest predation density dependent? Oikos:65–74

Selonen V, Mäkeläinen SLM (2017) Ecology and protection of a flagship species, the Siberian flying squirrel. Hystrix It J Mamm 28(2):134–146

Shamhart J (2010) Isotopic analysis of arctic ground squirrel tissues and potential food sources. University of New Hampshire

Signorile AL, Paoloni D, Reuman DC (2014) Grey squirrels in Central Italy: a new threat for endemic red squirrel subspecies. Biol Invasions 16(11):2339–2350

Smith-Patten BD, Patten MA (2008) Diversity, seasonality, and context of mammalian roadkills in the southern Great Plains. Environ Manag 41(6):844–852

Sorrell RD (1988) St. Francis of Assisi and nature: tradition and innovation in Western Christian attitudes toward the environment. Oxford University Press, Cary

Steiner M, Huettmann F (2021) Justification for a taxonomic conservation update of the rodent genus Tamiasciurus: addressing marginalization and mis-prioritization of research efforts and conservation laissez-faire for a sustainability outlook. The European Zoological Journal 88(1):86–116

Stenberg SJ (2006) Liberation theology and liberatory pedagogies: renewing the dialogue. Coll Engl 68(3):271–290

Stokstad E (2016) The bloody battle to save the red squirrel. Science. https://www.science.org/content/article/bloody-battle-save-red-squirrel. Accessed 19 July 2022

Taber RD, Payne NF (2003) Wildlife, conservation, and human welfare: a United States and Canadian perspective. Krieger Publishing Company, Malabar, Florida

Tacconi L, Rodrigues RJ, Maryudi A (2019) Law enforcement and deforestation: lessons for Indonesia from Brazil. Forest Policy Econ 108:101943

Ten Hwang Y, Larivière S (2006) A test of interspecific effects of introduced eastern Grey squirrels, Sciurus carolinensis, on Douglas's squirrels, Tamiasciurus douglasii, in Vancouver. British Columbia The Canadian Field-Naturalist 120(1):10–14

Lake Texoma (2022) Attention squirrel hunters. https://www.laketexoma.com/news%2D%2Dhunting%2D%2DAttention-squirrel-hunters/1954. Accessed 19 July 2022

Thompson FR (2007) Factors affecting nest predation on forest songbirds in North America. Ibis 149:98–109

Thorington RW Jr, Koprowski JL, Steele MA, Whatton JF (2012) Squirrels of the world. JHU Press

Trees for Life (2022) Red squirrel reintroduction success with breeding and natural expansion. https://treesforlife.org.uk/red-squirrel-reintroduction-success-with-breeding-and-natural-expansion/. Accessed 19 July 2022

Wauters L, Martinoli A (2021) A golden cage for the European red squirrel in Italy? Proposal for a targeted control of the grey squirrel. Biodiversity 22(1–2):87–90

Weigl PD (2007) The northern flying squirrel (Glaucomys sabrinus): a conservation challenge. J Mammal 88(4):897–907

Werner JR (2006) The endangered Vancouver Island marmot: Allee effects and reintroduction success (Doctoral dissertation, University of British Columbia)

Wirth C, Gleixner G, Heimann M (2009) Old-growth forests: function, fate and value–an overview. Old-growth forests. pp. 3–10

Words of Wisdom (2015) Aldo Leopold: The Earth Doesn't Belong to Us, We Belong to It. Big Think. https://bigthink.com/words-of-wisdom/aldo-leopold-the-earth-doesnt-belong-to-us-we-belong-to-it/. Accessed 5 Sep 2022

Young A (2013) Foraging Behavior of Eastern Gray Squirrels on the University of Maine Campus

YouTube (2022) Webinar – prairie dog restoration. https://www.youtube.com/watch?v=8rSjgTARDOA. Accessed 19 July 2022

Index

M. Steiner, F. Huettmann, *Sustainable Squirrel Conservation*, https://doi.org/10.1007/978-3-031-23547-4